geographical perspectives in the soviet union

a selection of readings

Edited and translated by

GEORGE J. DEMKO AND ROLAND J. FUCHS

 OHIO STATE UNIVERSITY PRESS:COLUMBUS

Library of Congress Cataloging in Publication Data

Demko, George J   1933-   comp.
    Geographical Perspectives in the Soviet Union.

                        Includes bibliographical references.
1.   Russia—Economic conditions—1965-   Ad-
dresses, essays, lectures. 2. Geography, Economic—
    Addresses, essays, lectures. I. Fuchs, Roland J.,
        joint comp.   II. Title
    HC336.23.D39   309.1'47'085   74-9853

ISBN 0-8142-0196-2

Manufactured in the United States of America

geographical perspectives
in the soviet union

To Yulian G. Saushkin, head of the
Department of Economic Geography of
the USSR, Moscow State University,
advisor and friend,

and

the late Nikolay N. Baranskiy (1881–1963),
first occupant of the Chair of
Economic Geography, Moscow State
University and founder of Soviet
economic geography

Contents

Preface

In the Soviet Union the system of geographical sciences is commonly divided on a first order basis into physical geography and economic geography. As a result, the field of Soviet economic geography is a somewhat broader discipline than its Western counterpart in that it encompasses population geography and elements of cultural and political geography, which are generally accorded recognition as separate subdisciplines in Western geography.

Despite its breadth of topics, Soviet economic geography retains a high degree of cohesion based on its attempt to employ a Marxist-Leninist philosophical framework and in its practical orientation to *socialist-construction,* that is problems of national planning and economic development. Its self-image is that of a *constructive* and *transformative* discipline that, since the founding of the Soviet regime, has helped provide "the scientific basis for the new planned territorial organization of productive forces of the socialist society and their rational location throughout the country."[1] Unlike Western geography, with its traditional view of science as a search for understanding without the necessity of immediate practical significance, Soviet economic geography is frankly acknowledged as "an applied field which justifies itself according to its service to the state, particularly in terms of contributing to a rational distribution of productive forces and in the development of natural resources."[2]

Because of its actual and potential contribution to planning and development, economic geography has advanced its relative importance in Soviet geography. Initially less favored in terms of the number of personnel, facilities, and resources, it has benefited from a shift in the "balance between physical and economic geography in favor of the latter." Today economic geography is the most widely represented geographic discipline in institutions of higher learning of the USSR with programs offered in more than two hundred universities, colleges, and faculties.[3]

Soviet economic geographers, including those employed in educational institutions, specialized research facilities, and planning agencies, undoubtedly represent the largest concentration outside of the English-speaking nations. They have contributed an already large and rapidly growing body of literature to the field of economic geography most of which in the past has remained inaccessible to the non-Russian reader.[4] This situation was not particularly serious in its implications for Western geography at an earlier date when the greater part of the Soviet literature was characterized by a high degree of ideological content, a descriptive approach, and verbosity. Its continuation would be unfortunate at a time when the Soviet literature has improved in content and approach and many geographers of the West share with their Soviet colleagues an interest in the application of their sciences to practical problems of economic development and planning. An additional convergence of interest is afforded in the growing application in Soviet research of mathematical and statistical methods that transcends ideology and promises to "further increase interactions among professional geographers at the international level."[5]

The present book of readings is offered by the editors as an attempt to make available to the English language reader a selection of representative readings on the various topics that make up Soviet economic geography. It is offered in the hope of increasing communication and dialogue between the economic geographers of the Soviet Union and those of the English-speaking and other nations. The book is expected to be of particular interest to professional economic geographers, planners, economists as well as to Soviet area specialists and students of Soviet geography. For these various readers it should provide an introduction to the distinctive philosophy and approaches employed by Soviet geographers.

The book contains sections devoted to philosophy and methodology, economic regionalization, resource management, agricultural geography, industrial geography, transportation geography, population geography, urban geography, and historical economic geography; each section contains an introductory statement by the editors followed by several translated readings. The introductory statements, which provide brief summaries of the development and current state of each subfield, are intended to serve as background for the reader to help place in context the selected translations. Footnote citations in the introductory statements provide reference to critical reviews and other relevant studies particularly those available in the English language.

The readings consist of translations of selected articles and excerpts from longer works originally published in Russian by Soviet geographers. The original materials are scattered in a variety of journals and publications, many rarely found in libraries outside of the USSR and include some now fugitive items. The majority of these items represent original translations, heretofore unpublished; some previously translated items are included to illustrate the scope of Soviet work or because adequate alternative selections were not available. Although most articles have been published without abridgement, some have been abridged or otherwise modified as noted. In an attempt to render the material more readable, great liberties have been taken in the translation process. Although attention has been paid to retaining the true meaning and even nuances of the original version, a loose rather than literal translation was employed wherever useful and appropriate. The transliteration system used is that of the United States Board on Geographic Names as modified and employed by *Soviet Geography: Review and Translation.*

The items selected, of course, form only a small proportion of the available literature. Selection of some items has been based on their methodological or substantive value; wherever possible stress has been placed on articles of more general theoretical significance. However it is recognized that other articles, unknown or unavailable to the editors, may have served as well and the policy of favoring previously nontranslated work has resulted in the exclusion of writings by other well known Soviet economic geographers which, had they not already been translated in *Soviet Geography: Review and Translation*, would have been included.

The problem of selection was made more difficult because of the current transitional state of Soviet economic geography, which is being transformed by the adoption of mathematical and statistical techniques. The content of the book reflects this transitional state; the selections include some older classical works, traditional conceptual works, more recent analytical works, and some items selected largely for their substantive content. To supplement the readings included here the reader is referred to the journal *Soviet Geography: Review and Translation*, edited by Theodore Shabad and published by the American Geographical Society, which has made available selected translations of Soviet writings in all fields of geography since its initiation in 1960. Since any work on a rapidly evolving field is to some extent dated as soon as it is completed the reader should consult *Soviet Geography* for continuing developments in the field.

As may be expected in the course of preparing such a volume a great number of people have been involved. Most have been of invaluable service, others have been a distressing hindrance. Dismissing the latter, the editors would like to acknowledge the help of many friends and colleagues. Particular thanks are due to Professor Yu. G. Saushkin for initial advice and continued support as well as to Professor Chauncy D. Harris, and Theodore Shabad for their help and encouragement. We are grateful also to Professors David Hooson, W. A. D. Jackson, Jerome Fellman, Richard Lonsdale, Robert Taaffe, Allan Rodgers, Jeremy Anderson, Edward Espenshade, Frank Leversedge, Paul Lydolph, John Morrison, Anthony French, Robert Jensen, and Robert Lewis for their suggestions and comments.

In addition, we wish to acknowledge the assistance of the Office of Science Information, National Science Foundation, which provided support for the costs of editing and translating associated with the preparation of the manuscript. We would also like to express our deep gratitude to Ms. Sally Millett of the Ohio State University Press. Her encouragement and long hours of editorial work, her attention to detail and nagging were essential to the completion of the project and immeasurably improved the final product. Special thanks are also due the secretarial and cartographic staffs of the Department of Geography, University of Hawaii, for their kindness in preparing tables and maps. Finally, we wish to thank our wives, Jeanette and Gay, and the children who silently and uncomplainingly contributed so much.

In conclusion the editors wish to note that this effort has been overly long in execution with the tedium involved far exceeding anything envisaged when the project was conceived. But should the book succeed in bringing to the reader a better understanding of Soviet geography and make a contribution to increased communication among Soviet and Western geographers we will consider ourselves amply rewarded.

George J. Demko
Roland J. Fuchs

Columbus, Ohio

1.   Yu. G. Saushkin, "A History of Soviet Economic Geography," *Soviet Geography: Review and Translation* 7, no. 8 (October 1966): 2.

2.   R. G. Jensen and G. J. Karaska, "The Mathematical Thrust in Soviet Economic Geography; Its Nature and Significance," Discussion paper, no. 2, Regional Science Research Project, Department of Geography, Syracuse University, p. 17.

3.   O. A. Konstantinov, "Economic Geography in the USSR on the 50th Anniversary of Soviet Power," *Soviet Geography: Review and Translation* 9, no. 5 (May 1968): 423–24.

4.   For a review of methodological contributions to regional science by Soviet economic geographers and economists, which includes an extensive bibliography, the reader is referred to R. S. Mathieson, "The Soviet Contribution to Regional Science: A Review Article," *Journal of Regional Science* 9, no. 1 (1969).

5.   Jensen and Karaska, "The Mathematical Thrust in Soviet Economic Geography," p. 20.

SECTION I

History, Philosophy and Methodology

Introductory Note

A distinctive characteristic of economic geography in the USSR, notes Konstantinov in a recent review article, is that it does not owe its development solely to leaders of institutions of research and higher education. "A major contribution has also been made and continues to be made by high Party and government circles, as well as by the planning and design agencies that do the practical work on the location of productive forces of the USSR."[1] Soviet economic geography, not only in growth of personnel and academic significance, but also in the development of its unique philosophy and methodology, has been clearly affected by the singular social, economic, and political context in which it has arisen.[2]

In prerevolutionary Russia economic geography had not yet established itself as a distinct discipline and was taught only in commercial schools and in a few faculties of economics. After the revolution of 1917, its practical significance for the development of the national economy was recognized, and it rapidly achieved a significant position in school curricula. Particularly critical in this respect was the decree, signed by Lenin in 1921, concerning the inclusion of economic geography in the required program of higher educational institutions that is considered as "the foundation for the spread of the teaching of the discipline in the higher schools."[3] Further impetus was received from the May 1934 decree of the Party and government "Concerning the Teaching of Geography in the Primary and Middle Schools of the USSR."

Both the pedagogical and research orientation of Soviet economic geography were molded soon after the October Revolution by the methodology of the state plan for electrification (the GOELRO Plan) of 1920-21 and the Gosplan's (State Planning Commission's) project of economic regionalization (1921).[4] The practical orientation to questions of the rational distribution of economic activity and problems of economic regionalization was further reinforced by the implementation of long-range national economic planning in 1928. The discipline has continued to remain highly responsive to resolutions of Party congresses and of the Party's Central Committee, for example, the resolutions of the Twenty-third Congress "concerning improvement of the basic proportions of economic growth, improvement of economic location, and of the integrated development of economic regions."[5]

Soviet economic geographers view their field as fundamentally different from prerevolutionary Russian economic geography because it proceeds from Marxist-Leninist philosophy and is based on the tenets of dialectical and historical materialism.[6] Under conditions of such fundamental philosophical unity it is claimed that within Soviet economic geography mutually hostile trends or fundamentally different schools cannot exist. Disagreements are considered possible only on specific questions.[7] In the past, however, there have actually been serious cleavages on fundamental issues.

Among the heatedly discussed theoretical questions of Soviet economic geography has been that of the role of the natural environment in problems of economic geography.[8] In the early 1930s a severe struggle took place between the group (later labeled as *leftists* or *nihilists*) that denied any relation between environment and the structure of society and viewed economic geography as essentially an economic science rooted in political economy and their opponents lead by Baranskiy who, though opposed to any determinism, favored a positive consideration of the role of the environment. The Party decree of 1934, which established basic guidelines for geography teaching, represented a victory for the Baranskiy viewpoint. Apparently there was later a temporary resurgence of the nihilist view in the latter years of the Stalin regime.[9]

Echoes of the earlier dispute are found in the more recent controversy generated by the writings of V. A. Anuchin, which has received wide coverage in American journals.[10] In his book, *Teoreticheskiye problemy geografii* [Theoretical Problems of Geography] published in 1960, Anuchin sees the study of the geographical environment (as opposed to the natural environment) as the basic task of the geographic sciences. Study of the geographical environment, because it requires a knowledge of both natural and social laws, is to provide a unifying basis for the fragmented, specialized discipline of geography. Anuchin's views received strong support from such prominent economic geographers as Baranskiy and Saushkin but were attacked by others including Vol'skiy (in an article included in this section) on the grounds that physical geography, which deals with physical laws, and economic geography, which as a social science deals with social laws, cannot be unified without violating basic tenets of Marxism. In more recent years the argument, though still not ended, appears to have become somewhat muted.

A possible resolution to the controversy, for practical purposes if not in theory, may have been offered by an Anuchin opponent, I. P. Gerasimov, director of the Institute of Geography, Academy of Sciences of the USSR, and a major spokesman on Soviet geography, who has suggested that the future development of geography be outlined "not in terms of its traditional subdivisions, but in terms of its major research problems." These he views as:

1. The development of theory and the elaboration of scientific programs for a planned transformation of nature needed for effective resource use, the formation of new and the reconstruction of existing territorial-production complexes.

2. The elaboration of a theory and creation of territorial models of social production for rational development and location of the economy and its major regional subdivisions.

3. Study of patterns of settlement and of the development of populated places under various geographic conditions and the elaboration of scientific programs of local-area planning envisaging the creation of favorable conditions for the life of people.[11]

Through an emphasis on these major problems and problem approaches the separate branches of geography would be integrated, though not unified.

Of recent methodological developments in Soviet economic geography, the most noteworthy and potentially most significant has been the growing adoption of mathematical and statistical methods.[12] Although the number of researchers employing such methods is still small there is every likelihood of a rapid growth since the methodology has such clear application to the problems of optimal spatial organization of the Soviet economy that are basic to the field. Although the adoption of mathematical methods has not been entirely without controversy, it has been very slight in relation to that generated by the Anuchin proposal of a unified geography or to that generated earlier in the United States in a similar period of mathematization and quantification of the field. One requirement that the introduction of mathematical methods will most certainly force upon the discipline, Mayergoyz has noted, is a clarification and formalization of terminology, concepts, and theories, which at present are often far from precise.[13]

The articles included in this section relate to the themes discussed above. In "A Brief Outline of the Development of Economic Geography" Baranskiy examines the historical development of the field from its embryonic beginnings in the literature of voyages through the development of Soviet economic geography. His major theme is that "no discipline develops in a vacuum" and that "economic geography, as all sciences, has been created under the influence of the practical demands of life." In each previous stage of its evolution— voyage literature, commercial geography, the sectoral-statistical approach, regional geography, location theory—including the Soviet period, economic geography is viewed as a practical field that has been responsive to *social commands.*

In the second selection, "Consideration of the Natural Environment in Economic Geography," Baranskiy attempts to develop the Marxist-Leninist position on the relation of natural environment and society. He rejects geographic fatalism (determinism) as philosophically unacceptable and even harmful under Soviet conditions "since it generates disbelief in the strength of socialist construction." Also he rejects geographic nihilism, which he considers theoretically incorrect and injurious "because the ignoring of natural conditions leads in practice to failures in construction." Baranskiy sees the "correct Marxist-Leninist position" in this question as equally distinct from both approaches; it is essentially an engineering approach in which one studies the natural environment in order to overcome it.

In the third article, "Scientific Problems of Geography," Kolosovskiy discusses questions of the borders, content, and method of geography. He is particularly concerned with the question of dualism versus unity in the field. In his view the dualism of physical and economic geography was characteristic of prerevolutionary Russian geography and today represents an unacceptable bourgeois survival because Marxism-Leninism emphasizes the unity of a materialist world view. He is aware that many geographers accept the concept of unity but practice duality. He suggests that unity can be achieved through "the dialectical elaboration of boundaries" and the "erosion of watersheds" between subfields if geography takes as its object of study the "regional material-technical basis of production," which is subject to the effect of natural laws and is at the same time a part of "the productive forces of social labor."

Vol'skiy in his article "Some Problems of Theory and Practice in Economic Geography" takes a counter position and attacks the concept of a unified geography on the grounds that it "stems from the absolutely incorrect assumption that society is part of nature." Although economic geography cannot operate "without close contact and interaction with physical geography," it is a "social science concerned with the study of laws of location, association and interaction of productive forces." The discussions associated with the concept of a unified geography, Vol'skiy charges, have served to divert the attention of practitioners from practical tasks and have resulted in justified criticisms by Party officials, including former Premier Khrushchev, of *serious shortcomings* in the field.

The final article by Vasilevskiy is a general treatment of the nature and application of mathematical methods to problems in Soviet economic geography. A close reading of the article indicates that the quantitative revolution in Soviet economic geography may take a considerably different tack than it did earlier in American geography.[14] Soviet economic geography will apparently stress mathematical methods in association with optimal solutions to spatial planning problems rather than the statistical, quantitative techniques that have characterized American geography. The expected emphasis on normative approaches is, of course, consistent with the Soviet intent to develop economic geography as a *transformative* science riented to planning and development needs. The mathematical th st thus is "a development of major significance" that should "greatly enhance the role of Soviet economic geography in its contribution to national planning and development."[15]

1.   O. A. Konstantinov, "Economic Geography in the USSR on the 50th Anniversary of Soviet Power," *Soviet Geography: Review and Translation* 9, no. 5 (May 1968): 418.

2.   For a detailed account of the development of Soviet economic geography see Yu. G. Saushkin, "A History of Soviet Economic Geography," *Soviet Geography: Review and Translation* 7, no. 8 (October 1966): 1–103.

3.   O. A. Konstantinov, "Economic Geography" in *Soviet Geography: Accomplishments and Tasks*, trans. Lawrence Ecker, ed. C. D. Harris, American Geographical Society Occasional Publication, no. 1 (New York, 1962), pp. 35–36.

4.   Yu. G. Saushkin, "Economic Geography in the USSR," *Economic Geography*, 38, no. 1 (January 1962): 29 ff.

5.   L. P. Altman, et al., "Economic Geography at Leningrad University," *Soviet Geography: Review and Translation*, 9, no. 1 (January 1968): 1.

6.   N. P. Nikitin, "A History of Economic Geography in Pre-Revolutionary Russia," *Soviet Geography: Review and Translation* 7, no. 9 (November 1966): 3–37. See also J. E. Chappell, Jr., "Marxism and Geography," *Problems of Communism* 14, no. 6 (November–December 1965): 12–22 and David J. M. Hooson, "The Development of Geography in Pre-Soviet Russia," *Annals of the Association of American Geographers* 58, (1968): 250–72.

7.   For detailed commentary on the question of Marxism and environmentalism, see I. M. Matley, "The Marxist Approach to the Geographical Environment," *Annals of the Association of American Geographers* 56 (1966): 97–111; J. E. Chappell, Jr., "Marxism and Environmentalism," *Annals of the Association of American Geographers* 57 (1967): 203–6; and I. M. Matley, "Comment on Note by Chappell," *Annals of the Association of American Geographers* 57 (1967): 206–7.

8.   Yu. G. Saushkin, "A History of Soviet Economic Geography," pp. 28–49.

9.   Ibid., pp. 43–44.

10.   See for example: V. A. Anuchin, "The Problem of Synthesis in Geographic Science," *Soviet Geography: Review and Translation* 5, no. 4 (April 1964); D. J. M. Hooson, "Methodological Clashes in Moscow," *Annals of the Association of American Geographers* 52 (1962): 469–75, and V. A. Anuchin, "On the Criticism of the Unity of Geography," *Soviet Geography: Review and Translation* 3, no. 7 (September 1962).

11.   I. P. Gerasimov, "The Past and the Future of Geography," *Soviet Geography: Review and Translation* 7, no. 7 (September 1966): 12–13.

12.   For detailed commentaries see R. G. Jensen and G. J. Karaska, "The Mathematical Thrust in Soviet Economic Geography: Its Nature and Significance," Discussion Paper no. 2, Regional Science Research Project, Department of Geography, Syracuse University; and R. S. Mathieson, "Quantitative Geography in the Soviet Union," *The Australian Geographer* 11, no. 3 (1970): 299–305.

13.   I. M. Mayergoyz, "Foreword," *Soviet Geography: Review and Translation* 7, no. 4 (April 1966): 35–37.

14.   The reader is referred to the perceptive interpretation of Jensen and Karaska, "The Mathematical Thrust in Soviet Economic Geography."

15.   Ibid., p. 20.

N. N. BARANSKIY

A Brief Outline of the

Development of Economic

Geography

Economic geography[1] is the study of the economic singularity of countries and regions, of spatial differences in economy on the earth's sphere, that is, differences from place to place, and also of spatial combinations within an economy. If these differences did not exist there would be no economic geography; it could not exist, for there would be no special object for investigation. In order to study these spatial differences, they must first be stated, then described as exactly as possible, and finally, the causes of these differences must be investigated. Ultimately, having investigated all these causes one must develop the mechanisms of these changes in the economy from place to place. In this way one can begin to formulate the problems of economic geography.

It must be emphasized that the differences from place to place in an economy are not due solely to differences in natural conditions. Let us assume that all differences in natural conditions have for some reason or another ended, that all climates have been blended, the seas have been divided evenly with land areas, and the mountains have been erased; even in such a case, contrasts in economy from place to place appear in each given country and at each given moment. These con-

Translated and abridged from *Ekonomicheskaya geografiya. Ekonomicheskaya kartogra-fiya* [Economic Geography and Economic Cartography] (Moscow, 1960), pp. 7–35.

trasts depend not only on the differences in natural conditions but to a great degree they also depend on the development of productive forces, on the past investment of labor on any part of the earth's surface, and on a number of other socio-historical features.[2]

THE NECESSITY OF A HISTORICAL APPROACH

In order to more closely approach and more clearly visualize the problems of economic geography, it is necessary to consider this question historically. Science is not the concern of only one person and a small segment of time; science is a collective experience, the experience of many generations. Only a historical approach to science can show the paths of its development, detect the impasses that must be avoided and, on the other hand, acquaint us with past achievements and thereby prevent the labor of discovering America for the second time. This is why a historical approach to this discipline is needed.

Besides this, we know very well that no discipline develops in a vacuum, that ideas do not fall from the sky, that every system of scientific views and opinions develops in a definite social environment, and that "the demands of life" are the main factors in the development of any science and any ideology.

The stronger the association of a given science with practice, the stronger is the influence of the demands of life. These demands of life in a class society are in the end nothing more than the changing interests of the ruling classes; in this one must see, if not the only, certainly the main stimulus to the development of some sciences. We know that even the emergence itself of the sciences was the result of a social need for them. So geometry appeared, as its name implies, from the word geodesy, measurement of the earth, as the result of a completely practical necessity to divide irregular pieces of land into plots. It is true, geometry has gone far beyond land measurment, but the first social command that gave birth to geometry was geodesy.

Economic geography, as all sciences, was created under the influence of the practical demands of life, and in its development has experienced incessant impulses from social life, definite commands, and has had to react to those practical needs which occurred in a specific historical era. Thus, "a social command" must guide the Marxist in approaching the development of any kinds of sciences and ideas.[3]

With this kind of approach we turn to our subject.[4] What forms did economic-geographic thought take on in the process of its development? What stages did economic geography pass through as a science studying differences from place to place in the economy? First we will give a general outline of the development of economic-geographic learning on the world scale, and then note the special features of the development of this learning in our country.

ELEMENTS OF ECONOMIC GEOGRAPHY
IN THE LITERATURE OF VOYAGES

First of all we note the embryonic period of the development of economic geography, when as such it did not yet exist independently but only as part of a single complex physical geography and ethnography, in the form of travel journals. Such descriptions of voyages were not only the original written sources of geographic information (in particular, information from the field of economic geography) but also information referring to other disciplines that later evolved and separated from each other. Not only the science of geography (including physical, economic, and cultural geography) stems from such sources, but other sciences as well, for example, ethnography.[5]

A description of a voyage is a very complete form of narrative. A person comes to a strange country and everything there is of interest to him. He is surprised and describes everything that he sees: he does not select the facts by categories as do specialists who are interested in only nature (even its individual elements), or only the economy (or only its individual branches), or only the life and customs of the population; the traveler describes absolutely everything that strikes his imagination—mountains, vegetation, climatic phenomena—and if he encounters something surprising in the economy or in the life and customs of the people, he then includes that in his description also. Writing this up along his route he interweaves everything to such an extent that no one can discern where physical geography ends and where economic geography or ethnography begins.

It was quite natural that the descriptive form of voyages was the initial form for geographic and ethnographic sciences. Why? We have already agreed that the essence of geography lies in the study of differences from place to place. But when a person lives in one place he gets accustomed to his surrounding world, to the surrounding nature and forms of economic activity and does not see anything surprising, especially interesting, or worthy of description. But when that person finds himself in other places where there are other natural conditions, other people, and another economy, everything amazes him, and he begins to notice and describe it. Thus it is natural that literature devoted to voyages became the first form of geography.

This form of literature (descriptions of voyages) is as old as the voyages themselves. Literary monuments of this kind have remained from the ancient Greeks. In the last analysis, the *Odyssey* is nothing but a poetic description of voyages. Descriptions of voyages have remained also from the Egyptians, Carthagenians, and the Arabs. The largest number of voyage reports were made during the era of geographic discoveries, beginning at the end of the fifteenth century with the voyages of Vasco de Gama and Columbus and in the early part of the sixteenth century with the first voyage around the world by Magellan.

Since in the voyage reports the economic-geographic information was not separated specifically but was mixed with other aspects, it is rather difficult to get a full and systematic notion of the economic peculiarity of each country from the description of the traveler who took one route and was not especially concerned with the economy of the country. On the other hand, however, what one has seen with one's own eyes can, of course, be described in a livelier, more buoyant, and more interesting manner, and these voyage descriptions combine economics, nature, and ethnography, all of which make them much more engrossing.

The value of a voyage description depends primarily upon the traveler himself, on his ability to observe, the breadth of his outlook, on the stock of general perceptions that guide him in processing his observations.[6] The very content of the voyage depends on the predominant interests of the traveler. Interest in trade, imperialistic expansion, land acquisition, and so forth had an influence not only on the organization of the voyages and expeditions but also on the direction of the voyages. Many voyages were in essence nothing but a preliminary survey for seizures. The aims of the surveys—military, commercial, and others—were often cloaked under scientific geographic investigations.

FINANCIAL STATISTICS

Economic geography attained for the first time a more developed and independent form in the so-called financial statistics that developed in Germany at the beginning of the eighteenth century as the result of special needs of the "police state" that had risen on the ruins of feudalism.[7] Under the feudal system the resources of the country could be made known without a special science, without forming a special branch of learning, and without creating any university courses. In the feudal era the land of the whole country was divided into comparatively small fiefs. Moreover, each feudal lord was not simply a landowner but also the governor of his land; he had the right to collect taxes, to recruit, and to judge those who lived on his territory. All of this belongs, in our opinion, to the area not of private rights but of the rights of the state. Yet the lord's sovereign, the feudal king or duke, did not interfere in his internal affairs; the so-called right of immunity existed at that time, which was one of the privileges of the feudal lord.

Immunity as an institution of state law meant the freedom of the feudal lord from interference by the king; it meant also that the king's officials had no right to interfere in the internal affairs of the fiefs.

Under such a system each feudal lord could obtain directly from his lower lord information concerning how much could be taken from a given land in assessments and recruits and what resources the land had. Thus, the king could find out from each of the feudal lords under his power how much each could give in soldiers, money, and other resources. Hence within a feudalistic "coarse grained" system of government all this information could be obtained informally in a volume sufficient for practical purposes.

With the collapse of the feudal system, the governmental power began to reach the population directly as the government demanded taxes to draft recruits and passed judgment through its officials. As soon as a sizable territory with a large population came within the framework of a single national state so that a government economy could rapidly develop, everything became much more complex. A governmental apparatus had to be created for the administration of this huge state; in order to carry out the functions that had been those of the feudal lords and their administrators, new special cadres of educated officials were needed.[8]

And in Germany a new science was created, *financial statistics.* Achenwall, the father of financial statistics, called this science "a descriptive government science of individual countries." His work, *Abrisse der neuesten Staatswissenschaft*, which initiated financial statistics, appeared in 1739. The aim of financial statistics was to provide the government with information and to train young officials by giving them a university education.

What were the structure and content of financial statistics? Attention was given to the territory of the state, its borders, and its administrative divisions; then came finances, budget, and individual branches of the economy—agriculture, industry, transport, communication, and so forth; then military forces, the fleet, relations with other countries; in one word, the governmental machinery considered as a whole. Thus, the subject of financial statistics was the state,[9] but not so much from the legal viewpoint as from the viewpoint of the resources of the state.

This was a descriptive governmental science. Indeed, in order to understand any state one must know all that has been enumerated, that is, territory, borders, population, form of governmental system, all branches of the economy, military forces, diplomacy; all this is essential. Each educated official must clearly understand what relations his government has with other governments, what financial resources it has, and related information. All of this belongs to the field of education in government.

There were very few geographic elements in financial statistics: mainly directions for the distribution of individual branches of the economy, and sometimes descriptions of separate administrative units were included. Financial statistics, created in Germany for the organization and strengthening of the police state, spread from Germany to other countries.[10]

Financial statistics developed its own methodology. These were the times when science was permeated with the scholasticism of the Middle Ages and when methodological rules were brief formulas in Latin. Financial statistics had a formula, "Vires unitae agunt," translated literally: "United forces lead." As a whole this formula seems to recommend: "Describe first the material resources of the country, then describe the governmental organization on this territory, and then show all this in action." Clearly, the methodological formulas were not all useless. In fact, however, it did not go farther than endless lists, and, numerical information.[11]

What was the fate of financial statistics? It had to utilize a large amount of every kind of numerical data. The methods of statistical investigation worked out on this basis were then applied to a number of natural sciences, to the study of genetics by botanists, zoologists, and others. As a result it turned out that financial statistics gave its "soul" so to speak, its method of investigation of mass phenomena, to the discipline we call statistics.[12] And the perishable "body" that was left without a soul, that is, the collection of all kinds of reference information about one country or another, has become an essential element of what is called the sectoral-statistical approach in economic geography,[13] and in this form it drags on its existence to this day.

COMMERCIAL GEOGRAPHY

Another impetus for the development of economic geography was the development of trade relations that followed the era of great discoveries. In the course of the sixteenth, seventeenth and eighteenth centuries there occurred a tremendous expansion of seaways and a corresponding development of transoceanic trade by the leading Western European countries. Earlier the range of European sea trade was limited to the North by the North and Baltic seas, to the South by the Mediterranean Sea.[14] The farthest expeditions within these limits were those between the Mediterranean and North Seas. Beginning with Vasco de Gama and Columbus, little by little, routes were discovered around Africa to India, then from Europe across the Atlantic Ocean to America and, finally, to the islands of the Pacific Ocean. Thus, the range of trade grew greatly, crossed national boundaries and took on an international character. This basic fact could not but affect the demands life placed on economic-geographic knowledge.

Actually, in small trade conducted on a local scale there is no need for any special science for the training of agents. Imagine a trade establishment under prerevolutionary conditions in a village that has a trade consisting of two kinds of operations: on the one hand, buying up local agricultural products and on the other hand, selling city goods such as cloth, ready-made clothes, utensils, and agricultural equipment. What knowledge must a man have who performs trade operations of this type and at this scale? He must know the volume of supply and demand within the limits of his trading sphere, that is, where and when he must buy up the surplus of local agricultural production and the dimension, composition, and season of demand for city goods. His whole scope may be limited to one or two districts. No special science would be needed; since he could get all the necessary information personally. It is quite different when the market widens and even extends beyond national borders. Crossing state borders we find different trade conditions. There are new measures of goods, new monetary systems, special trade rights, new legal procedures, and custom duties. The situation is completely different: quite different natural conditions, other trade customs, approaches, and so forth. To study all the diversity of trade conditions, the whole assortment of goods, and, finally, to study where to buy cheaper and where to sell at a higher price within the framework of the whole world is already a science that demands special studies, a special literature, and special schools.

Just as the interests of the police state gave rise to the demand for educated officials and brought financial statistics to life, so the interest of a trade extending to international dimensions gave rise to the large-scale demand for educated merchants and business agents and resulted in the creation of commercial geography.[15] Here again, there is clearly a transition from quantity to quality.

What then must be the scope of geographic knowledge demanded of an educated merchant as distinguished from the scope of knowledge demanded of an educated official? An educated official could largely limit himself to information within the boundaries of his country, but the interests of an important merchant extended over the whole world, and included various countries with completely different natural conditions. Therefore, interest in natural conditions increased.

In order to systematize this entire mass of diverse trade knowledge, it is necessary to understand why in one country one can buy such products as sugar cane, coffee, or cocoa, and in another, fur or lumber; it is indeed necessary to have some geographic knowledge. This cannot be done through mechanical cramming of price-lists, reference books, or dictionaries. This knowledge must contain a great deal of geography, including physical geography.

On the other hand, those sections of financial statistics that dealt with governmental organization, finances, army, navy, and diplomatic relations, information that the official needed, were of less interest to the merchant. Thus, the structure of this kind of course changed considerably and was no longer financial statistics but commercial geography; many chapters became superfluous, but the number of countries studied rose, attention to natural conditions, geographic elements, and foreign trade increased.[16] A characteristic feature of textbooks on commercial geography, especially English ones, was the special attention given to a detailed analysis of foreign trade (by countries and commodities). Thus, in addition to financial statistics (and often together with it), commercial geography was created.[17]

This was the time of growing capitalism, when it was still opposed to autocratic monarchical power, when the ideology of the bourgeoisie was in its youth, and its ideologists still believed that the new bourgeois economic order was bringing harmony and universal happiness to mankind. It was assumed therefore that the less there was of official pressure and regulation, the sooner this harmony and general happiness would be established on the earth. This was the time of the dominance of the theories of "laissez faire, laissez passer," a time when one was inclined to look upon the government as a night watchman, and the young bourgeoisie wanted to limit its functions to the preservation of security while preventing its interference in the economy. Hence one can understand the somewhat disdainful attitude of the leading bourgeoisie and its ideologists toward government and its functions, toward the officials and the knowledge that they needed.

If one says that a merchant is most interested in where to buy cheaply and sell dearly, one is basically correct. This is his main interest and to a considerable degree it predetermines the content of courses in commercial geography. In commercial geography there is surprisingly little or no discussion of the general characteristics of the economy. There are chapters discussing the general character of agriculture and of industry, but the discussion is for the most part limited to strictly tangible branches such as wheat or rye production. Why? It is, quite simply, that agriculture is not traded, but wheat and rye are; and these practical mercantile interests influence to a considerable degree the content of courses in commercial geography. The same preponderance of mercantile interests compelled the inclusion of reference information useful to the merchant, especially that concerning the science of staple commodities, in the courses in commercial geography. Commercial geography in its more practical forms is quite close to a descriptive science of staple commodities in their geographic aspects. And often scientists concerned with staple commodities undertook instruction in commercial geography.

In order to conduct trade, one needs more than a mere knowledge of where to buy cheaply and sell dearly; even for commercial capitalism it was of interest to know the conditions of police security in the country where one traded and that is why merchants were interested in law and order in foreign countries.[18]

Another demand on geographic knowledge, in addition to that of an economic geographical order, is the demand for a knowledge of military defense. In this respect the information on the size of the army and military equipment is of prime importance, but this information does not exhaust the question. In order to judge the military strength of a country, it is necessary to know everything that enters into the idea of war potential; and this is a very broad idea, including not only war industries but the whole economy of the country and its concrete spatial manifestation. More than that, it is also important to know national policy, both domestic (including relationship of classes and parties) and foreign (including not only formal declarations but also interests). Thus it is clear how broad and diverse the demands are, from the military-defense point of view, on geographic knowledge, including economic-geographic.[19]

After the period of trade capitalism came that of industrial and financial capitalism when the interest in foreign countries was no longer limited to just trade but included foreign capital investments, which in time acquired greater and greater significance.[20] However, if one is interested in foreign countries not as a merchant but as a representative of financial capital looking for investment possibilities, then his interest in these countries is broader. All the natural resources of a country, of industrial as well as agricultural significance, are of interest: the level of development of productive forces of the country; the levels of consumption of its population; also its labor skills and much else are of interest, and not only in general and on the average but in its spatial distribution and combination in the various parts of the country.

One will also want to know whether a revolution could occur that can cause him to lose everything he has invested. This is an interest of a very general order. It is natural, therefore, that present works of this kind include, together with all kinds of practical information that is frequently quite general, detailed and more or less specific descriptions of individual countries. Thus the interests of a modern merchant or financial man with respect to foreign countries can be rather broad and extend far beyond the framework of a single reference book about where and what to buy more cheaply and where and what to sell more dearly.

THE SECTORAL-STATISTICAL APPROACH

One must also bear in mind that these various "orders" to financial statistics and commercial geography are not divided by some kind of barrier. They have influenced and penetrated one another and created something with a common nature. It is possible to train simultaneously an official, a merchant, and a military man. Elements of economic geography initiated by various demands have somehow united and become part of general education.

The majority of courses of economic geography that have evolved from financial statistics or commercial geography have retained traces of one approach. In these courses of a so-called traditional or sectoral-statistical approach, prime attention characteristically is given not to separate regions of a country but to individual sectors of the economy. The object of study is not the territory itself but the economy in this territory, not as a whole but by individual sectors.

Economic geography in works of this sectoral-statistical approach is treated as the science of the state of individual sectors of the economy. The economy here is thought of as split into separate sectors, and each chapter is a description of the state of a given sector at a given moment in a given country, but the question of the distribution of this sector and the geography of production is only one of many and diverse interests guiding the author. Technology, history, and economic policy of the government in relation to the economic sector are all treated in the form of endless numerical references and enumeration, and only within all this is there also found some treatment of distribution. However, a minimum of space is given to questions of distribution, and even then the treatment of these questions is for the most part limited to simple references such as where a given industry is located and in what quantity, without explaining the reasons for the distribution.

It is in such consolidated and somewhat modified form—sectoral-statistical economic geography—that these two approaches (financial statistics and commercial geography) exist to this day. Both as financial statistics and as commercial geography, economic geography remained pure description without any inherent regularities, without a real scientific foundation. Economic geography in both these forms is better considered as simply a kind of somewhat dull literature rather than a science.[21] If in these surveys there were some valuable generalizations with respect to science, it was not due to the method but in spite of the method; it was due to the personal talent of the author.

The scientific thought of economic geography did not progress within economic geography itself but in disciplines allied to it, or simply in an amateurish way when scientists of other fields for some reason were temporarily interested in questions of economic geography and contributed some valuable thoughts as was the case with Lomonosov and later Mendeleyev.

ECONOMIC-GEOGRAPHIC DESCRIPTIONS IN THE REGIONAL
GEOGRAPHY [STUDY OF COUNTRIES] LITERATURE

Besides financial statistics and commercial geography, general geography, due to its complex content, also had an influence on the development of economic geography. General geography (including general earth science and especially regional geography) had become one of the general-education subjects in public schools during the eighteenth century and even earlier in a number of countries.

With the development of transport, international trade, and general international relations, the need for geographic knowledge began to grow and spread. A certain minimum became necessary not only for persons of narrowly defined professions such as state and trade representatives but for all educated people regardless of profession. This minimum included information on mathematical and physical geography, as well as economic geography. From this complex the school subject evolved under the general name of geography. The main content of this subject consisted of a description of individual countries, including information on the natural environment, population, and economy of each country. Two or three generations later, when geography was firmly established in the curriculum of the secondary school and the number of secondary schools had increased, the time had come to train schoolteachers of geography, and in the course of the nineteenth century geography penetrated higher educational institutions, including the universities.

The immense amount of factual material collected since the beginning of the era of great geographic discoveries about the surface of the earth was so diverse that it demanded generalization and thereafter attracted the interest of brilliant minds who tried to find definite regularities in the distribution of natural phenomena and human society as a whole, and in their combinations within each separate country and in each separate region. With the development of secondary and higher education and with the increased number of educated people, the public demand for geographic literature grew. University courses and compilations on regional geography appeared more and more substantial in volume and content; they no longer had as their aim the training of narrow specialists but rather a general education in geography. The development of such a general, university-level geography could not but exert an influence also on the development of the then incipient forms of economic geography.

Instruction in economic geography often had to be turned over not to statisticians, students of commodities, or practitioners (agents or merchants) but to geographers—that is, regional specialists. The same occurred with the composition of textbooks. And if, on the one hand, geography in the form of regional specialization influenced financial statistics and commercial geography, then the content of courses in financial statistics and commercial geography conversely penetrated into regional geography. Since no special theoretical knowledge was needed for an economic-geographic description in the sectoral-statistical style, regional geographers could take on this job with little concern. And who was not intelligent enough to find references in statistical collections? For their part, the regional geographers contributed their knowledge of physical geography, knowledge of maps, and sometimes certain skills in general geographic thinking.

Some of the regional specialists who undertook to describe economic activities would adopt, without much concern, the traditions and the structure of the sectoral-statistical school, and thereby ceased for a time to be geographers. Others, on the contrary, tried to describe the economy by taking account of general geographic practice. that is, by locating precisely the objects and phenomena described, by using maps widely, by attempting to tie in the features of the economy with those of natural conditions and, finally, by attempting to rearrange this description from a sectoral framework to a regional one. At first such regional geographers who switched to economic geography from physical geography used a network of natural, physical-geographic regionalizations. But after the rapid development of transportation in the era of capitalism and the growth of a geographic division of labor within large national states, there came into being sharply specialized economic regions, and the geographers began a transition from natural regions to economic regions in the economic-geographic section of their descriptions. A new and very positive contribution made by these geographers to economic geography consisted in the strengthening of the geographic components in it.

Many of the major regional geographers who were more interested in the geography of economic activity (or in a wider sense, the geography of man) and better trained in this activity made a sizable contribution to the development of economic geography with their general methodological statements as well as with concrete examples of the economic-geographic characteristics of countries and regions. Of the Russian geographers one can mention K. I. Arsen'yev, P. P. Semenov, V. P. Mechnikov and A. I. Voyeykov; of the French, Reclus, Vidal de la Blache and his students Demangeon and Brunhes, as well as Blanchard and Baulig; of the Americans, Smith, Ekblaw, Bowman; of the English, Mackinder, Stamp, Beaver, Herbertson, and others. As examples of great works on regional geography (and the most interesting with respect to economic geography) one can cite *Rossiya* in Russian literature, edited by Semenov-Tyan-Shanskiy, and in French literature, the multivolume *Geographie Universelle*. The French *Geographie Universelle* has many valuable economic-geographic studies of whole countries and individual regions. Of special note is the work of Henri Baulig on North America and of Demangeon on the British Isles.

STANDORT [LOCATION] THEORIES

Relatively recently economic geography received one more impetus from the practical demands of life, from capitalist monopolies and trusts acting within the framework of an entire country and sometimes extending beyond its borders. Monopolistic groups developed primarily in the United States and Germany in the last decades of the previous century.

The trust, as the highest form of capitalistic union, in contrast to the cartel and the syndicate, assumes a *complete* merger of enterprises. A new stockholding company is created, and each participant sells to it his enterprises for a specific package of stock, and from the moment of founding of the trust the owners of the enterprises that are combined in the trust are no longer owners. A new elective board is created, handling all business, and the former owners become shareholders, with the right to receive dividends and to influence the composition of the board during elections. On this basis, one of the world's largest trusts, the Steel Trust, was created in America in 1901. This was such a powerful organization that at one time its budget exceeded that of France. The main task of a trust is to maximize profits; the means for doing this are to raise prices, lower wages, and decrease production costs. One of the first tasks is the reduction of production to the level of demand and the concentration of production in enterprises with the least production costs.

The experts of such an organization calculate the productive capacity of all plants and come up with, say, 80 million tons. Then they calculate the purchasing power of the market, adjust for growth, and arrive at a total of 40 million tons. Obviously, in the process of capitalistic competition production had been inflated and did not conform to the market. The main problem, as the trust sees it, is overproduction, and the main task is to decrease production, to close superfluous establishments, especially those which are less well equipped and those with higher costs. After these plants have been closed, the productive power is still, let us say, 60 million tons, and the market demands only 40 million tons, which means the productive power must be reduced by another 20 million tons. What other criterion must be taken into account? Another criterion is that of the convenience of location; an enterprise is closed not only if it is more poorly equipped but also if it is inconveniently located. Thus, with the appearance and development of capitalist trusts there arises a very interesting question for the economic geographer: What must be the rational distribution of a given sector in order to maximize profit, that is, for a maximum reduction of production costs and transportation expenditures?

If we compare this demand of life with the preceding, we see that there is a great qualitative difference. In the demand for educated officials and agents, the demand was mainly for strictly factual knowledge, purely empirical reference information. However, when we are concerned with the demand for captains of industry, managers of such tremendous unions, we are dealing with a demand not for reference material, but for an entire theory. It concerns the creation of a theory of the most rational distribution of industry under the given capitalist conditions.

The so-called *standort* theories served as an answer to this question posed by monopolistic capitalism.[22] Best known is the theory of Weber in his main work, *Theory of the Location of Industries*. Weber created a whole school in Germany. On the basis of Weber's theory his students compiled a number of monographs on individual branches of German industry. Then there is a theoretical work by Englander, the works of Schroeder, Schilling, and others, and a number of articles in the German journal *Wirtschaft und Statistik*. All these location theories tend to anwer the demands for the most advantageous distribution of industry under capitalist conditions, demands that arose as the result of the creation of industrial trusts. Although they exerted a certain influence by strengthening the theoretical elements in economic geography, they did not create a separate school. The location theories were used in economic-geographic works to explain the existing distribution of industry; examples of such use can be found in *Geographie Universelle*.

Now let us turn to our own country and briefly give the essential characteristics in the development of economic geography before the revolution and then the main directions of development of our economic geography during the Soviet period.

In speaking of characteristic features in the development of economic geography in Tsarist Russia, one must first of all note the role of an exceptional intellectual giant of thought, M. V. Lomonosov.[23] A scientist-encyclopedist in the full sense of the word, Lomonosov, besides conducting investigations in the fields of nature and technology, also worked in philosophy, history, and philology. He also was very much interested in economic questions and initiated the economic-geographic study of Russia.

Taking account of the needs of economic development in Russia, Lomonosov emphasized the acute necessity of developing industry, especially metallurgy. In a work titled "First Foundations of Metallurgy, or Ore Industries,"[24] he wrote: "Warfare, trade, navigation and other necessary government institutions require metals." Lomonosov saw clearly that "the well-being, glory and flourishing of the country" depend on the development of trade, "on the mutual exchange of internal surpluses with distant peoples, by means of trade." Hence arose the necessity of a complete study of the country. Directing the Geography Department of the Academy of Sciences, Lomonosov considered it necessary to compile an atlas with "a political and economic description of the whole Empire, including Siberia." For this purpose Lomonosov compiled a remarkable questionnaire consisting of thirty questions—a whole program for the economic-geographic study of Russia. Lomonosov had planned to compile an economic lexicon of Russian products and evidence of their internal and external communications with attached maps. This would have been a detailed trade geography of Russia. In 1764 Lomonosov offered in the Academy of Science an opinion on the utilization of a census for a geography of Russia. Lomonosov was the initiator of expeditions for the geographic study of our country, moreover, he understood this in a comprehensive sense. He gave great attention to the study of the North.

From these brief references, the number of which could be greatly increased, it is clear how broadly Lomonosov understood the tasks of the new science that he had begun to create and for which he used the term *economic geography* for the first time in the 1760s, a term that was based on a correct understanding of problems in the development of Russia. One cannot possibly say that Lomonosov coined the expression *economic geography* by chance. This was an exact definition of the new science that was fully developed in his mind and was filled with rich and varied content.

How broadly Lomonosov understood geography as a whole can be seen from his "Word of praise . . . to Ylizaveta Petrovna" [daughter of Peter the Great] where he wrote: "What is useful to mankind for the mutual sharing of their surpluses, what is more necessary to travelers in foreign lands, than to know the places, the course of rivers, the distances between cities, the size, abundance and proximity of various lands, the customs and habits and governments of various peoples? Geography, which at one glance reveals the vastness of the universe, shows this clearly."[25] This remarkable characterization of geography, "which opens the whole vastness of the universe at a single glance," is also not a random expression by Lomonosov.

In his larger works of great geographic significance, such as *On the Earth's Layers, On Electric Atmospheric Phenomena* and *On Northern Navigation to the East Along the Siberian Ocean*, Lomonosov invariably emphasized the interconnection of phenomena, the necessity to find and explain these relationships, the necessity of a historical approach to phenomena, and the exceptional importance in geography of maps, including economic maps. It must be noted, however, that Lomonosov in his view was so far ahead of his time that his brilliant ideas remained without further development for a long time.[26]

Nevertheless, it is to Lomonosov, and no one else in all of world science, that credit belongs both in terms of a broad view of geography, which "reveals the vastness of the universe at a single glance," and in the use of the term economic geography itself, and to the formulation of its main tasks.

The prerevolutionary history of our economic geography was also characterized by the exclusive attention paid to questions of regionalization of the country, including economic regionalization. Regionalization of the country was of great interest in prerevolutionary Russia, greater than in almost any country (with the exception perhaps of only the United States). The reason for this, one must assume, lay in the vastness of the country itself and the diversity of its parts. Under such conditions it is impossible to know the country without distinguishing the various regions in it; regionalization becomes not only a necessary premise for this knowledge but one of the most important methods. It is not by chance that the greatest number of works on the regionalization of Russia began to appear in the nineteenth century, when, with the development of capitalism, the geographic division of labor rapidly developed, and also the differentiation and specialization of regions. About fifty attempts to regionalize prerevolutionary Russia can be cited. Some of them, it is true, have a narrowly specialized character and give a regionalization based on only one component of the natural environment (climate or soil),[27] but for the most part these attempts give a regionalization either by totality of all features (nature, population, and economy) or by one predominantly based on the character of the economy.

The best known are the attempts of K. A. Arsen'yev (1818, 1848), P. P. Semenov-Tyan-Shanskiy (1871–80),[28] D. I. Mendeleyev (1893, and then from census materials in 1897),[29] and A. N. Chelintsev (1910, 1916). Of great methodological interest was the article by N. P. Ogarev on questions of economic regionalization, published in *Moskovskiye Vedomosti* in 1847. Also of great methodological significance is the classical work by V. I. Lenin, *The Development of Capitalism in Russia*, on questions of economic regionalization. The work showed the possibility and necessity of studying the economy of a capitalist country by regions and at the same time it gave a classical example of Marxist methodology by turning major attention to the singularities of the historical development of the country and its sociopolitical order.

As a whole, the prerevolutionary works on regionalization made a great contribution to the geographic knowledge of our country and in addition helped to inject the principle of regions into the thinking of our geographers.[30]

Our most outstanding geographers, beginning with V. V. Dokuchayev and ending with L. S. Berg, quite definitely stated that the task of geography is not to study individual components of the geographical environment only in their distribution but to study the interconnections of these components in definite spatial combinations.

As early as 1912, A. A. Borzov wrote in his article "Geography" (in the *Practical School Encyclopedia*):

Geography is, consequently, a science of the earth's surface and deals not with its individual phenomena but with the totality, taking it as it is found in the various parts of the earth. In other words, geography does not deal separately with ores, plants, animals, or man and does not even study the prevalence of these individual phenomena —all this is the affair of other sciences. Geography examines the forms and landscapes into which all these phenomena are organized in the various places of the earth's surface, how soil, river and water basins as a whole, climate, plants, animals and man are interrelated in each locality, in each landscape and in what forms their influence on each other is expressed; it reveals the influence of an individual landscape on another and tries to describe the kinds of landscapes of which the surface of the earth is composed and to explain the life of each landscape individually as well as the law of their distribution and mutual influence.

During the Soviet period, economic regionalization has received a new meaning and a new significance. Formerly it was the major method of comprehensive geographic perception of territory, but under Soviet conditions, it became the necessary premise of planned development for the whole of Soviet economic planning. In the State Plan for Electrification of Russia (GOELRO), established according to the conception and plan of Lenin, who called it *the second program of the party*, regionalization of the country was planned in connection with the plan for building electric power stations. Based on the transmission of electric current over a distance of not more than 300 to 400 kilometers, regions were planned in such a manner that within each region there would be a correspondence between intake of electrical energy (from various sources) and output (for various needs).

This first draft of Soviet economic regionalization quite clearly reflected the characteristic features distinguishing it from the regionalization experiments of prerevolutionary time, namely: (1) a forward-looking nature, a concern with transformation as well as understanding; (2) the conception of a region not as some "statistical homogeneity" but as an economic unit complete in itself (in this case according to the balance of electric energy). In 1919 the work on economic regionalization was given over for completion to the section of regionalization in the Gosplan (State Planning Committee), which in 1921 drew up a projected economic regionalization scheme in which the USSR was divided into 21 regions. The report of the Subcommission of the All-Russian Central Executive Committee, under the chairmanship of M. I. Kalinin, which examined this project, stated:

> A distinctive and, so far as possible, economically complete territory of the country must be set apart as a region and would represent one of the links in the chain of the national economy because of a combination of natural characteristics, cultural traditions, and the specialized training of its populace for productive activity. This principle of economic completeness makes it possible to construct a project of economic development for the region that best utilizes all possibilities with minimum expenditures on the basis of a well-selected complex of local resources, capital values, and the general state plan of the national economy. Other important results are achieved by insuring that regions specialize to a certain degree in those branches that can be most fully developed there and that the exchange between regions is strictly limited to the necessary quantity of expediently directed goods.

Thus, in the Gosplan project of 1921, a region was understood to be a productive territorial entity, which was so far as possible complete in itself (but not closed), with productive connections within it developed to a maximum extent, and with a specialization on an All-Union scale.

In contrast to the regions projected in the experiments of prerevolutionary regionalization schemes, a Gosplan region was thought of not as an area on a map and not as a totality of statistical indices but as an administrative-political unit with an organ of local power actively engaged in a struggle for the fulfillment of its share of assignments in the All-Union plan, actively transforming its economy according to the planned specialization assigned to it by the center and linking all other branches of its economy with this major specialization. A region in the Gosplan conception must not only be homogeneous, as demanded in prerevolutionary regionalization systems, but, as a general rule, very heterogeneous; for the most complete industrial combine can be obtained most easily from a combination of heterogeneous parts that complement each other.

A region as an industrial territorial complex with a specialization on an All-Union scale is not only a necessity but also a basic principle of regional development. Each of our regions must develop in such a manner that its economy can become more comprehensive, due to more complete utilization of its resources, and as self-sufficient as possible.

The Gosplan definition of a region contains not only the law of its development but also methodological directions on how this region must be studied and how to draw up and interpret its character.[31] In its time, the 1921 Gosplan project of economic regionalization, which provided a complete and clear methodology linking theory with practice, caused a great deal of methodological and factual literature to be published. The economic-geographic literature connected with GOELRO (State Commission on Electrification of Russia) and the Gosplan project of 1921, together with related works, created a Soviet regional approach that replaced the sectoral-statistical approach. The new approach was accepted by the Party and government in its resolution of 16 May 1934, and in the resolution of the Presidium of the Committee on Courses in Higher Institutions of Learning and Higher Technical Schools of 14 July 1934, "On the Teaching of Geography in Higher Schools of Learning and Higher Technical Schools." In accordance with these resolutions, the entire content of instruction in economic geography in secondary and higher schools of learning was reorganized.

In the process of the work on economic regionalization all knowledge of our territory had to be reexamined when questions were to be solved concerning the network of districts, the borders, and the choice of a center for each district.

Questions on economic regionalization were also posed at the Twelfth Party Congress and at the fifteenth through twenty-first congresses; a fact that clearly shows the great significance attached to these questions by the supreme organs of our country. Questions on the rational distribution of industry and agriculture were ranked first in the political report of the Central Committee of the All-Union Communist Party at the Sixteenth Party Congress.

It must be said that the economic geographers, for example N. N. Kolosovskiy, made a great contribution to economic regionalization, rational distribution of the economy, and other practical questions of economic development.

Detailed large-scale investigations of the land, especially of a comprehensive economic-geographic nature, that were conducted as a result of the practical requests emanating from socialist construction of the economy had a great methodological significance for the development of Soviet economic geography. The economic geographers who formerly used secondary materials in the form of statistical information by large administrative units (by oblast), could now in these expeditionary investigations collect in the field the necessary material with the required detail. Large-scale field work was thereby initiated in economic geography making it possible to discover the regularities not evident in small-scale studies of the literature and statistical tables. Field investigations bring the geographer face to face with nature, with the map, with the appearance of correlations, with conditions at the microscale, and thereby help to strengthen the geographic elements of economic geography.

Socialist economies, having separated land from the realm of private property, also create a demand for economic maps, including large-scale maps that under our conditions have the same function as the plan had under capitalistic conditions. Field preparation of large-scale economic maps, which no one had previously even thought about, has received an impetus from the works on the detailed study not only of agriculture but also of drainage and irrigation, and of improvement and development of paved roads. The development of economic cartography, especially on a large scale, connected with the field collection of data contributes even more to strengthening the geographic elements in economic geography and is of great methodological significance. A large-scale economic map plotted in the field fixes the distribution of natural conditions, population, and economy fully, exactly, and to a degree of detail unavailable in textual description. Such an economic map thereby compels the economic geographer to transfer his main emphasis from describing a distribution to explaining it.

Formerly economic geographers could succeed by empirically describing where something could be found, thereby considering their work finished. Now, however, with the appearance of such economic maps it is impossible to limit distribution to a bare description. Stimulating economic geography to explain the distributions on a large-scale economic map by the fullness and detail of portrayal simultaneously helps to point out the pertinent material. In addition, in the process of drawing up economic maps, methodological questions and needs in economic geography arise, and sometimes their solution is hastened. One such problem involves the typology of industrial types of agriculture.

All of the above-mentioned factors—economic regionalization under conditions of a planned socialist economy, the rational distribution of industry, the correct regional specialization of agriculture and consolidation of subregions, and the large-scale economic cartography connected with it—greatly stimulates the development of economic geography and raises and expands the demand for economic-geographic knowledge. The principal distinguishing features of Soviet economic geography, with regard to the way it has been shaped by the enumerated factors, can be formulated as follows: First its activity and close connection with the practice of socialist construction of the economy; the influence of socialist requirements and the dissatisfaction with bare statements of what is where; the desire for an exhaustive explanation on a Marxist-Leninist basis; the establishment of regularities making it possible to establish standards; and the scientific determination of further development plans for each region, large and small. Second, the characteristic feature of Soviet economic geography is its comprehensiveness and, at the same time, its concrete and detailed nature, which is achieved by field investigations and the mapping of the territory. Third, and inseparable from the first two, is the ever-growing technical training of our economic geographers, which is necessary in order to understand more fully the geography of production and to plan the future more correctly.

Under Soviet conditions, both people and occupations grow in conformity with the tasks put upon them by our great era. Therefore, in spite of all kinds of survivals from the past in our discipline, there is no doubt that Soviet economic geography will in coming years be equal to the demands put upon it by our socialist development.

1. This article does not intend at all to discuss the whole process of development of economic geography with all factors that have influenced it and all shades of economic-geographic thinking. This could only be done in a special investigation. Here the attempt is to outline schematically the major approaches and the major stimuli that caused them.

2. It must be remembered that natural conditions have an influence on the economy not by themselves and not directly, but are refracted through a given structure of productive forces and industrial relations, so that one cannot dispense with this structure.

3. This point of view does not deny the presence in every science of its own internal logic of development that compels it to put certain problems in a certain sequence. The internal logic of development compels the specialists in a given field, directing its interests in a definite direction according to the received command, to attend to specific problems and expend energy on the solution of precisely this and not some other problems. An overly narrow concern with the practical hinders the development of science, and this is reflected in the application of science to practice. One can cite many examples to show that theorists have given to practice incomparably more than the most practical practitioners. In order to be convinced one need only to compare Farady and Maxwell on the one hand and Edison on the other.

4. In a larger work specially devoted to economic geography one should, of course, account in more detail for the influence originating from the internal logic of development, but in a brief outline this is not possible.

5. It is interesting to note that in earlier times the various branches of economic activity were looked upon as one of the ethnographic features of some peoples. Hence to this time the expression "occupations of the population" has been retained instead of "branches of the economy." This may signify that every nation is engaged in one or another branch of economy by, so to speak, innate inclination as a beaver builds his dams. For a certain state of development this point of view had its bases. Working skills transferred from generation to generation can for a long time retain their significance so to a certain extent they must be taken into consideration. Ethnography for a long time was a part of our Geographic Society on a basis of equal rights with geography.

6. A classic example are Humboldt's voyages.

7. Earlier roots in this direction can be found in the works of Konring (Germany, seventeenth century) and still earlier in Botero's works (Italy, sixteenth century).

8. Here one has to keep in mind that the replacement of feudalism by capitalism was by far not as simple historically as the replacement of one theater set by another. Very often feudalism retained its position. The centralizing state power continued with one hand to support (and even strengthen) serfdom and with the other to create large enterprises of military industry at government expense, and thereby prepared the beginning of capitalism. As an example, the reign of Peter I can be cited.

9. Precisely a state, not a country. From the point of view of financial statistics the economic concession of one state to the territory of another state should be referred to the description of the state that had obtained the concession, not the one that had given it.

10. In particular, its influence also affected Tsarist Russia. One must keep in mind that, beginning with Peter I up to World War I, Russia was a country where German scientists had a kind of "seasonal work." Our Academy of Sciencies during its first years of existence was full of Germans and with them came financial statistics. Among the German scientists who worked in Russia in this specialty was Reichel who gave the first course in statistics at Moscow University, then Hachman, Miller, Schletzer, Heim and others. Statistics, in the definition of Reichel, is "the knowledge of the state of governments and republics . . . an exact and certain explanation of the present state of the most significant possessions in Europe." Concerning the practical significance of statistics the same Reichel said that it is "necessary and useful for ambassadors, ministers, citizens and scientists . . . to discover its benefit and have its advantages."

11. This kind of literature was described by Marx as "a mixture of the most diverse information, the purifying fire that each candidate for the German bureaucracy who is full of hopes must withstand."

12. Statistics in the contemporary view is not a material science, it is a method for the investigation of mass phenomena of whatever nature, natural or social.

13. The sectoral-statistical approach inherited its content not only from financial statistics but also from commercial geography.

14. Europe, in particular the Mediterranean countries, has had connections with Asian countries since early times (not only from the eighteenth century but much earlier). These connections, quite distant and difficult in every respect, were limited to a few, very valuable goods (silk, spices, and so forth) and were of no great significance.

15. Among the predecessors of economic geography, besides financial statistics and commercial geography, one can consider also political arithmetic, the best representative of which was the Englishman William Petty (seventeenth century). Marx pointed to him as one of the forefathers of political economy.

16. In the various countries of Western Europe foreign trade began to develop at different times. This is the order: Venice, Holland, England; in approximately the same order appeared works with information useful for trade having some reference to commercial geography. In the Italy of the Renaissance era, the closer acquaintance with the geographic works of antiquity, Greek and Roman, had an influence on the development of geography.

17. Rudiments of commercial geography can be found before the creation of financial statistics in Germany in connection with the development of sea trade by the Venetians, Dutch, and especially the English. Of interest in this respect are the instructions of an important Englishman, Hakluyt, to English sailors who set out to find a northeast passage.

18. There were many works on commercial geography in Western Europe, especially in such countries as Venice, Holland, England. In Russia the greatest work of this kind was: *Istoricheskoye opisaniye Rossiyskoy Kommertsii* [An Historical Description of Russian Commerce], by M. D. Chulkov (1781–88).

19. See the chapter on "Military Geography" in *American Geography: Inventory and Prospect*, ed. P. E. James and C. F. Jones (Syracuse, N. Y., 1954).

20. In the course of recent decades in the United States many books were published on South America as a whole and on its individual states. The authors give to Americans most detailed descriptions of Paraguay, Uruguay, Chile, Peru, and other countries. Evidently in the United States there is a considerable demand for South American material. This demand is dictated by the presence in the United States of large amounts of capital available for investment abroad and above all in the countries of South America. Compared with prewar times United States investments in the countries of South America have increased more than ten times and have exceeded in the majority of states of South America such an "old firm" as England.

21. With the transition from financial statistics to commercial geography only the object of enthusiasm changed for the compilers of surveys; if for financial statistics such an object was the native country with its military and financial might, later its place was taken by native trade and industry with its trade.

22. From the German word "standort," which literally means "place of standing," or "location." [Hereafter the term *location* is used instead of standort.]

23.   As predecessors of Lomonosov in the field of science of interest to us, one should mention I. K. Kirillov and V. N. Tatishchev. Kirillov composed the first atlas of Russia and its description by guberniyas [provinces]. Tatishchev, a great statesman with a well-rounded education worked in the field of history and geography: he compiled a very detailed questionnaire for gathering information on questions of geography, economy, ethnography, and the *Leksikon rossiyskiy istoricheskiy, geograficheskiy i grazhdanskiy* [A Russian Historical Geographical, Political and Civil Lexicon] (up to the letter K).

Kirillov and Tatishchev clearly realized the diverse practical significance of economic-geographic knowledge and made a major contribution. In connection with this one should also mention P. I. Rychkov who in his *Toporgrafiya Orenburgskaya* [Topography of Orenburg] gave the first sample of regional description with an emphasis on economy using locally collected materials.

24.   M. V. Lomonosov, *Sochineniya* [Works] (Moscow, 1934) vol. 7.

25.   Ibid. (1898), 4:267.

26.   What is now combined under the name of economic geography developed in the educational literature not according to instructions of Lomonosov but in the manner of directions already familiar to us: financial statistics (for example, the textbook of Heim), commercial geography (textbooks by Subbotin, Morev, Sobolyev), and the sectoral-statistical approach (Den). Only at the turn of the nineteenth and twentieth centuries did there appear in our educational literature good geographical textbooks using the approach of a science of countries with a rich economic-geographical content.

27.   Here belong, for example, the geobotanical regionalization of Trautfetter (1850) and the climatic regionalization of Kemets (1860).

28.   The network of regions by P. P. Semenov-Tyan-Shanskiy was long-lived. It was used in the works of the Central Statistical Administration until the end of the 1920s.

29.   The experiments in regionalization by D. I. Mendeleyev are distinguished by the fact that he paid great attention to industry while the great majority of other authors concentrated their attention on agriculture.

30.   Not without reason was it stated in many works devoted to the history of our geography that the history of geography of our country is to a considerable extent the history of its regionalization.

31.   For more detail of the economic regionalization of 1921, see my book *Ekonomicheskaya geografiya SSSR: Obzor po oblastyam Gosplana* [Economic Geography of the USSR: A Survey by Districts of Gosplan] (Leningrad, 1927).

N. N. BARANSKIY

Consideration of the Natural Environment in Economic Geography: Formulation of the Problem

When the role and significance of the natural environment is spoken of and argued about, it concerns in most cases all those influences of the natural environment on human society as well as those of human society on the natural environment. However, two clearly different issues can and, in my opinion, must be distinguished. One issue is the influence of the natural environment on the development of human society, that is, on the change of social structures, the transition from one into another, and so forth. Marxism, although acknowledging the natural environment as one of the necessary and constant conditions for the material life of society that influences its development, does not consider this influence to be a determining one.

The other issue is the influence of differences in the natural environment upon spatial variations in the productive trend of the economy within the context of a given social structure, the character of which as a whole also determines the character of utilization of the natural environment.

The first question in its general form is under the purview of philosophy, and, in its specific formulation, under the purview of

Translated from *Ekonomicheskaya geografiya. Ekonomicheskaya kartografiya* [Economic Geography and Economic Cartography] (Moscow, 1960) pp. 36–54.

history, which investigates the processes of social development and the change of social structures. In such a formulation, economic geography has no direct relation to this question, not because it ignores the differences in social conditions from place to place, but because in accounting for them it considers them as given facts without investigating the question of their origin; this question is left for the historians.[1]

The second question, the investigation of the influence of spatial differences in the natural environment on the spatial variations in the productive direction of the economy is, on the contrary, a fundamental task of economic geography, with respect to which an economic geographer can say: If not we, who else?

It is not difficult to see that though the questions formulated above are in some measure related, they are at the same time somewhat different. It is one thing to consider what combination of natural conditions could accelerate the development of capitalism or, in general, the formation of a stable human society, but it is something completely different to consider, for example, what natural conditions, within the borders of Soviet Central Asia, are more suitable for cotton growing, which for wheat, and which for livestock-raising. One can ascribe a very modest role to the influence of natural conditions on the development of social structure and at the same time consider it indisputable that, for example, low-lying land in Central Asia accessible to irrigation is most suitable for cotton growing, whereas high mountain pasture land in the summer and unirrigated lowlands the rest of the year are most suitable for livestock-raising. Other conditions being equal, the differences in natural conditions can very often prove to be decisive for the explanation of variations in the favored products of agriculture from place to place.

It would be incorrect to think that the influence of natural conditions on the productivity of the economy is so great only in regions with such sharp contrasts in nature as Central Asia. New England in the northeastern United States in this respect is completely different from our Central Asia; however, in the selection of plots for the growing of vegetables in New England even the smallest differences in the level of the land, soils, insolation, microclimate, and other conditions are very carefully taken into account. In a number of other regions of the United States with highly intensive agriculture, for example California, one also observes a careful consideration of natural conditions in the choice of agricultural products.

The greatest specialist in the geography of agriculture in the United States, O. E. Baker, analyzed the situation as follows: "The influence of natural conditions on agricultural development has increased instead of diminished with the progress of science and technology. The development of commercial agriculture and the resulting sharp competition between various regions make the production of any crop sensitive to the slightest natural asset or liability of a given place and thereby cause shifts in the distribution of crops or in the use of the land with a rapidity not known in former centuries."

There are so many differences in human life, economy, socio-political structure, and culture, from country to country and often within the borders of the same country, and these differences are often so great that they could not but interest everyone who encounters them. In discussing the reasons for these differences, the variations in natural conditions are for the most part taken into account in one way or another. The founders of Marxism-Leninism and our present Party and government leaders continually emphasize the necessity of considering local, natural, and economic conditions.[2] In order to be able to take into account local conditions, however, one must first of all know them well, and this is the affair of geography and geographers. If the question is formulated in this manner then it is quite clear that the influence of local conditions, including natural conditions, is not denied at all in our country, but on the contrary is admitted. If this influence were denied then why speak of the necessity of taking it into account?

GEOGRAPHIC FATALISM AND GEOGRAPHIC NIHILISM

Concerning the issue of the influence of the natural environment, there exist two diametrically opposed systems of errors; one, generally speaking, involves the overestimation of the role of nature and originated in antiquity; and the other, quite new, again generally speaking, involves the underestimation of the role of nature.

The first position, although not represented by anyone in our modern literature,[3] is still very much alive in educational practice. Very often teachers of economic geography, especially those with a biased physical-geographic training, give explanations such as the following: "In such and such a place fishing is developed because there are many fish; in another there are more forests and there the lumber industry is developed; in a third there are chernozem soils and therefore much grain is produced."[4]

In its time, this leftist school [the second system of errors above] was represented by a textbook for Leningrad aspirants [graduate students] under the editorship of Minuskin and Poleyess, by textbooks of Setyukov and Danilov, and by articles in 1931 and 1932 in the journals *Na fronte kommunisticheskogo prosveshcheniya* [On the Front of Communist Enlightenment] and *Obshchestvovedeniye, istoriya i geografiya* [Social Science, History, and Geography]. The most classic formulation of the leftist position concerning nature was given in an article by Gurevich in the journal *Na fronte kommunisticheskogo prosveshcheniya* (nos. 4–5, 1931), which states:

"Nature in general, without reference to a definite social structure, does not exist."

Both points of view are incorrect in theory and harmful in practice. Geographic fatalism, first of all, is theoretically incorrect because it completely ignores human society, which has its own principles, denying thereby also all sociohistorical sciences. A physical geographer, with his store of knowledge, attempting to judge the phenomena of an economic-geographic character commits a great error in ignoring the social sciences, as does a person who approaches the phenomena of the organic world of plants and animals with only chemistry and physics.

> Materialism of the eighteenth century was predominantly a mechanical materialism because of all natural sciences only mechanics and only the mechanics of solid bodies (terrestial and celestial) or in short, the mechanics of gravity, had by that time reached a certain stage of completion. Chemistry was in its infancy; the phlogiston theory was still accepted. Biology was in diapers; the plant and animal organism had been little investigated; its functions were explained by purely mechanical causes. In the eyes of the materialists of the eighteenth century man was a machine, as were animals in the eyes of Descartes. The exclusive application of the criterion borrowed from mechanics to chemical and organic phenomena, that is, to such phenomena where mechanical laws, although continuing to act, recede to the background before other, higher laws, is the first specific, unavoidable feature of the restrictiveness of classical French materialism.[5]

A similar error is committed by the representatives of geographic materialism who attempt a judgement on phenomena of social, in particular of economic, geography, with only the criteria borrowed from physical geography.[6] "In order to produce, people enter into definite ties and relations and only by means of these social ties and relations does their relation to nature exist and does production occur."[7]

In practice, geographic fatalism is harmful, because, by attaching absolute, decisive importance to natural conditions, it creates a state of mind in which the fate of each people is once and for all predestined by the natural conditions of its country. Geographic fatalism thereby inhibits the activity of the people in the struggle with nature, in overcoming difficulties caused by natural conditions. Such a state of mind is especially harmful in our country's conditions, since it generates disbelief in the strength of socialist construction. The whole falsity of the position of geographic fatalism can be seen vividly in the example of our country that has developed its productive forces so rapidly and sharply.

Geographic nihilism is theoretically incorrect, because denying any significance whatsoever to natural conditions and removing human society from the material environment of its existence and development inevitably leads to Menshevik idealism. In practice it is harmful because the ignoring of natural conditions leads to underestimation of these conditions, to underestimation of local peculiarities of geological structure, relief, climate, soil, and other factors, and to resulting failures and disruptions in construction, of which numerous examples can be cited.

The correct Marxist-Leninist position in this question is equally distant from each bias. One cannot ignore natural conditions either in theory or in practice, but one cannot attach absolute significance to them. They must be studied, and studied very carefully; but in taking into account their influence, it is necessary to keep in mind the decisive regularities of social development and to understand that at each of the various stages the same combination of natural conditions can have a completely different significance and exert a different influence. Before mankind mastered the technology of maritime navigation, the sea separated peoples, but since mankind has mastered the technology, the sea has united rather than separated. Formerly waterfalls were only a hindrance to navigation, whereas now waterfalls are a large and very cheap source of electric energy.

The Bolshevik approach to nature is above all an engineering approach: it does not bow before nature as do the Buddhists, but studies it with the aim of utilizing it for the good of mankind. Here is a mountain that must be conquered. He who is accustomed to bowing before nature advises us to bypass it. He who is used to ignoring nature will go into it without looking; for him it is best if it is hidden by clouds. A reasonable person, however, first investigates and studies the mountain and then, on the basis of this study, draws up a plan and overcomes the difficulties. The Soviet conquest of the Arctic may serve as a classic example of a truly Bolshevik attitude toward overcoming difficulties that nature puts before us.

In spite of their apparent polar opposition, the systems of distortions elaborated above are much closer to each other than each is to the correct Marxist-Leninist position. On the theoretical front, the representatives of geographic fatalism, for the most part academically inclined physical geographers, are in full agreement with the leftists from economic geography in that they favor a complete break between economic and physical geography. The symbiosis of an "inhuman" physical geography with an "unnatural" economic geography, based on the premise of not hindering one another,[8] is a touching illustration of mutual forgiveness of errors.[9] As a result there is a complete break between physical and economic geography that is extremely harmful both theoretically and practically.

The leftist system of distortions is especially dangerous, because by representing something "new in the literature," it appears under the mask of Marxism and hides behind a thick smokescreen of vociferous verbiage. These leftist theories and tenets were designated as anti-Soviet and effectively discredited by the resolution of the Central Committee of the All-Union Communist Party and the Council of People's Commissars of 16 May 1934, in the editorial in *Pravda* of the same date, by the resolution of 14 July 1934 of the Presidium of the Committee on Technical Education, and in the editorial in *Pravda* of 10 September 1937.

There is nothing surprising in the position taken by the party and government in the documents enumerated above. For proof it is well to cite a number of statements from the works of Marx and Engels.[10] For convenience I divide these into two categories: (1) statements examining the question in general form without application to any individual case; (2) statements referring to specific situations. I believe that for us geographers, the second is no less important than the first, because we are concerned precisely with the application of these general statements to concrete cases.

Finally, in order to resist the leftist distortions, it is necessary for young people to know, even in extract form, the words of Marx and Engels. In "German Ideology" we read:

> The first prerequisite of any human history is, of course, the existence of live human personalities. Therefore, the first concrete fact to be established is the physical organization of these personalities and their relationship, specified by the organization, to the rest of nature. Of course we cannot be concerned here with either the physical properties of the people themselves or with the natural conditions, geological, oro-hydrographical, climatic and other relations, which the people find. Any historical description must proceed from these natural bases and their modifications in the course of history due to the activity of the people.[11]

What is Marx saying here? He is telling historians that they must proceed from natural relations, and the geographers, one would think, must do that all the more. What kind of economic geographers are those who fulfill this legacy of Marx to a lesser degree than do the historians, or, to be more exact, do not fulfill it at all? They are the real internal enemies of geography! In *Das Kapital*, Marx stated:

> In reality, the realm of freedom begins only where work which has been dictated by need and external expediency ceases; consequently,

by the nature of things, it lies beyond the sphere of proper material production. As a savage, in order to satisfy his needs, in order to maintain and reproduce his life, must struggle with nature, so a civilized man must struggle in all social forms and by all possible methods of production. With his development this realm of natural necessity enlarges, because his needs enlarge; but at the same time the productive forces which meet these demands also broaden. Freedom here can mean only that socialized man, associated producers, rationally regulate their relation with nature, put it under their common control instead of having it dominate them like a blind force; they accomplish it with minimum effort and under conditions most worthy of their human nature. But nevertheless it remains a realm of necessity.[12]

Thus, as a final result a complete rejection of nature seems impossible to Marx, the materialist, even under conditions of socialism. In another quotation from the works of Marx and Engels, Engels writes to Bebel:

Between individual countries, districts and even localities, there will always exist a certain inequality in living conditions which can be reduced to a minimum but can never be completely removed. Inhabitants of the Alps will always have living conditions different from those of the inhabitants of the plains.[13]

We cite one more of the more important quotations taken from a letter of Engels to Starkenburg. The letter by Starkenburg seems to be unpublished. From Engels's answer one can guess that Starkenburg had written approximately thus: "Life is very complicated, history is also; is it possible that you reduce everything to economics alone?" In his answer to Starkenburg, Engels says:

Under economic relations, which we consider the determining basis for the history of society, we mean the method by which the people of a given society produce the means for subsistence and exchange products (insofar as there exists a division of labor). Here then enters the whole *technology* of production and transport. This technology, according to our view, determines the method of exchange and then distribution of products, and, consequently, after the disintegration of the tribal structure, also determines the division into classes, relationships of domination and subordination and, consequently, the state, politics, law, etc. The idea of economic relations includes also the geographic *basis* on which these relations develop and the remains of former stages of economic development, which in fact have come over from the past and which continue to be retained often only by tradition or due to inertia, and also, of course, the external environment which surrounds this social form.[14]

Finally, in *Das Kapital* is this remarkable statement:

> One and the same economic basis—one and the same from the stand-
> point of essential conditions—because of infinitely different em-
> pirical circumstances, natural conditions, race relations, historical
> influences acting from without, etc.—can reveal infinite variations
> and gradations that can be understood only by means of analyzing
> these empirically given circumstances.[15]

This thought by Marx expresses in essence what is important and
basic in the analysis of questions concerning society, namely the
laws of social development characteristic for a given social structure; but in this general context there can arise a mass of gradations and variations that must also be studied in order to understand
a given society at a definite place and at a definite time. Who must
study this? Certainly the geographer, among others, but the geographer above all. If the geographer will repeat only the general and
leave out the specifics of a given country, a given place, then why
is he needed? Truth is concrete and the geographer calling himself
a Marxist must provide this concreteness and specific character.
"The main statement of dialectics: there is no abstract truth, truth
is always concrete."[16]

With this we will conclude the quotations referring to the general
formulation of the question and turn to those that examine particular
issues of this problem.

First, I shall dwell on the statement in which Marx tries to determine what totality of natural conditions and what specific character
of natural environment can assist specifically in social development.
This is how the question is stated.

> Too extravagant, nature leads man, like a baby, by the apron strings.
> It does not make his own development a natural necessity. The tem-
> perate zone, not the tropical climate with its mighty vegetation, was
> the birthplace of capitalism. Not the absolute fertility of the soil but
> its differentiation and the diversity of its natural product, form the
> natural basis for the social distribution of labor; due to the change
> of natural conditions under which man has to conduct his economy,
> this diversity helps to multiply his own needs, capabilities, means,
> and methods of labor.[17]

The leftists categorically prohibited the establishment of any kind
of correlation between nature and social structure. Marx says directly: "The temperate zone, not the tropical climate with its
mighty vegetation, was the birthplace of capitalism." This is the
first thing that must be noted. Marx continues:

The necessity to control socially any natural force in the interests of economy, the necessity of utilizing it or subordinating it with the help of large-scale equipment erected by the hand of man, plays a decisive role in the history of industry. Serving as examples are the regulation of water in Egypt, Lombardy, Holland, India, Persia, and so forth, where irrigation by man-made canals not only supplies the soil with water necessary for plants but at the same time brings, with the silt, mineral fertilizer from the mountains. The secret of the economic prosperity of Spain and Sicily during the supremacy of the Arabs lay in artificial irrigation.[18]

In the works of Marx and Engels, we find the following concerning the comparison of the Old World and America:

But the Eastern mainland, the so-called Old World, possessed almost all animals capable of domestication and all types of cereal grains useful for cultivation, except one; the Western mainland, America, however, had of all the mammals capable of domestication only one, the llama, and then only in one part of the South; and of all cereal grains only one, but the best: corn. From this time on, as a result of this difference in natural conditions, the population of each hemisphere goes its own special way and the border posts marking the individual stages of development become different for each of these cases.[19]

This quotation is clear without any comments.

I shall cite one more very interesting excerpt from "Letters from Engels to Marx" of 6 June 1853 in which singularities of countries of the East are discussed:

The absence of private ownership of land is indeed the key to the understanding of the whole East. Here lies the root of both the political and religious history. But how is one to explain that in the East they have not come as far as private property, even feudal property? It seems to me that the answer rests mainly in the climate in connection with the character of the soil, especially with the huge deserts which, beginning with the Sahara, stretch through Arabia, Persia, India and up to the highest Asiatic plateaus, Tataria. Agriculture is based mainly on artificial irrigation and irrigation is already the affair of the community, district or central power. Governments in the East have always had only three departments: finance (robbery of their own population), war (domestic and foreign robbery) and the department of social work (the concern for reproduction). . . . Free competition has become a laughingstock. The artificial fertilization of the soil which stops immediately as soon as the water conduit breaks explains the curious phenomenon that whole areas of land, covered at one time with luxurious vegetation (Palmira, Petra, the ruins of Yemen and a number of places in Egypt, Persia Hindustan), are now barren deserts.[20]

Here again we see a correlation between the purely social feature, the question of the presence or absence of feudalism and the feature of natural conditions, climate, and soils. For an explanation of the singularities of social and historical development, Engels considers it possible to use differences in natural conditions. And this is not by chance.

Marx and Engels very carefully, and sometimes in great detail, studied natural conditions. Read, for example, the work by Engels "The Po and the Rhine," and you will be amazed at the minuteness of his examination of the mountain passes across the Alps. And how Engels describes small landscapes of France in his account of his walking tour from Paris! These are brilliant geographical vignettes, and our nihilists of geography at one time especially attacked geographic vignettes, thinking that they contain some kind of special perniciousness.

In "The History of the Revolution and Counterrevolution in Germany," Engels points out the singularities of the sociohistorical development in Germany as compared with France and England. Germany in 1848 was, of course, a backward country compared with France and England. Engels writes in this respect: "There were many reasons for this backwardness in German industry. It is sufficient to name two: the unfavorable geographic position of the country far from the Atlantic Ocean which had become the great route of world trade and the constant wars into which Germany was drawn and which were fought on its territory from the sixteenth century to the present time."[21]

As one can see, for the solution of a concrete question concerning the singularities in the development of Germany, Engels took into account the singularities of its geographic situation. It is necessary to call attention to the fact that Engels gives a formulation that makes it impossible to interpret his answer in the sense of geographic fatalism.

The works of Marx and Engels from which the above excerpts were taken encompass a large time interval beginning with the early works of Engels, continuing through their joint activity for more than three decades, and ending with the last years of Engels after the death of Marx. Thus, it is incorrect to say that the point of view developed in these excerpts concerning the role of geographic environment is characteristic only for one period of their activity. The works from which these excerpts were taken are diverse in character and content; there are journal articles, letters, and substantial works, including *Das Kapital*; among them are political, historical, economic, and philosophic works.

However, the main position on the issue of interest to us remained unchanged. And what is especially important, they not only expressed their position in general form but also applied it to concrete cases with respect to the most diverse countries. Among these countries we find Germany, Spain, Lombardy, Holland, Central Asia, Egypt, India, Persia, America, and our Russia. Thus, one cannot speak here of chance or exception, and we are fully convinced that on the question of the role of the geographic environment, the views of the leftists have decidedly nothing in common with the position of Marx and Engels. And the fact that the leadership of our party so sharply criticized the position of the leftists in the 1930s is not at all surprising to the Marxist.

THE INFLUENCE OF NATURAL CONDITIONS IN THE PRODUCTIVE DIRECTION OF THE ECONOMY

What must be kept in mind without fail when speaking of the influence of the natural environment on the productive direction of the economy (and all the more on the social structure) is the fact that nature only creates conditions and possibilities, and nothing more than that. The realization of these possibilities is determined wholly by sociohistorical features that may vary greatly.

Natural environment is not a cause in the strictly logical sense of this word (a totality of circumstances inevitably evoking a certain event),[22] but only an assisting (or impeding) feature; therefore, one does not have to speak about causal dependence but only about correlation, that is, correspondence. The economic geographer juxtaposes spatial differences, that is, variations from place to place, in the productive direction of the economy with spatial differences in the natural environment, and thus identifies definite regularities.

It is necessary to keep in mind that, in addition to differences in natural conditions, the differences in productive directions in the economy are always influenced by other factors, as for example the differences in the economic-geographic situation of the territory under investigation with respect to the market, the historically accumulated productive skills of the population, and much, much more.

Let us clarify our thought with a concrete example. In order to explain the productive specialization of the northern European part of the USSR in lumbering, reference to the presence of huge forests there is clearly inadequate. Huge forests also exist in Yakutia, but as yet a lumber export specialization has not been planned. What other features must be taken into account? First of all, of course, one must consider the proximity of the northern European part of the USSR to Western European countries, to the large consumers of lumber—Germany, France, England; next the small capacity of the domestic market of northern European part due to the sparse population and limited industrial development; and then the direction of the river flow northward to the sea and the poor transport connections with the domestic market to the South. Finally one must take into account our need to have foreign currency for economic development, which was very urgent during the First and Second Five-Year plans. Now if one adds to the abundance of forests all these additional features, then the lumber-export specialization of the European North becomes understandable in all respects. And consequently one can understand not only why the European North was assigned such a specialization at the beginning of the First Five-Year Plan, but also why in the Third Five-Year Plan when, due to the expanded economic development, the domestic need for lumber materials rose, a considerable part of the lumber from the European North was shipped to the domestic market, thereby introducing an essential correction into the specialization of the European North.

It must also be realized that both concepts—the concept of natural conditions and the concept of productive tendencies of the economy—are quite complex. Therefore it is expedient to begin the investigation on a narrow basis, that is, from the study of the relation of any individual element in the natural environment to some individual narrow sector of the economy, going from one element to the other and from one sector to the other, in order first to get experience in this approach and second, to gather sufficient factual data. But this is good only for the initial stage of investigation. One cannot stop there, because in reality we always have to deal with some combination of these elements.

In analyzing natural influences on the economy one cannot take only one of these features but must take the whole complex of natural factors important in this respect and examine them in combination. At the same time one must keep in mind that a specific combination of natural conditions even under the same social conditions and the same technological level does not predetermine the productive direction of the economy but leaves a rather wide margin for choice depending on a number of additional features of various kinds. For example, the narrow strip of easily irrigated land with fertile gray desert soils that stretches in southern Kazakhstan along the ranges of the Tyan'-Shan' has possibilities for the development of various crops, but the question of precisely which of these crops to choose for this strip remains open. In most places in this strip, sugar beets, tobacco, vegetables, sunflowers, and a number of other crops including cotton can be grown with almost equal success. If we examine how this question was being solved in practice, we see that during recent years changes in specialization occurred many times; moreover in assigning a specialization, local and regional considerations were not as much of a guiding factor as were general and national considerations; specialization was assigned according to the greatest need of the All-Union economy at a given time.

However, as the result of such changes, more or less justified under conditions of a new region, experience was accumulated in the course of a number of years that made it possible to assign a more precise specialization based on a number of features formerly not taken into account. Included in such features are, for example, the areal differences in the water reserves for irrigation and in temperatures (beets need more water and cotton more heat). However, in addition to features of the natural order, other features must be taken into account: proximity to an urban market (important for vegetables), the presence or absence of manpower reserves, skilled labor, and traditions of the population, the possibilities of matching production with other branches of the economy (for example, the sugar-beet crop with intensive livestock-breeding).

In all this complex totality of various factors and features, the natural factors always play some role, not as the only ones but in conjunction with a number of others although of a different order: sociohistorical or transport-market factors. In all investigations of the question of the influence of natural conditions on the production tendencies of the economy, it is necessary to keep in mind the technology of production, which in turn is closely related to the social structure. Before the revolution, under conditions of individual peasant households with a low capacity and primitive technology, there were great difficulties in utilizing the soils of the nonchernozem zone for wheat. However, under the conditions of socialist agriculture with its advanced technology, wheat is sown quite successfully on nonchernozem soils and gives increased and stable yields.

In the area of mining industry, too, we can observe a sharp reappraisal of values in connection with the perfection of technology. For example, the ores considered unprofitable to exploit under conditions of primitive technology become quite profitable under a perfected technology. Sulphur deposits in Texas and Louisiana lying at a depth of 100 and more meters had a very low evaluation before the invention of a special method of mining as compared to the deposits in Sicily; but with the invention of this method their value has risen sharply.[23]

The evolution of new forms of economic utilization of the same natural environment can be seen even more clearly in the example of new, rapidly growing regions. In recent decades it is easiest to find them in our country (in the Kuzbass, Maritime, and Ural regions) and in the people's democracies.

From the preceding remarks it is already clear that research must be pervaded by history. First, it is necessary to point out here that natural conditions can influence some aspect of the economy only to the extent that they become known to man as a result of his practical experience or as a result of scientific investigation. Deposits of useful minerals, no matter how large and valuable, cannot have the slightest influence on any economy as long as they remain unknown. Geological survey is, strictly speaking, the first stage of the mining industry. Therefore, in ascertaining the natural conditions of a country or region, there is already a historical feature present, since useful minerals can exert their effect on the economy only if they have been surveyed.

In nature nothing at all may have changed, yet each generation of people enters a somewhat new geographic environment. An economic evaluation of natural conditions cannot be absolute for all times but must be historically generated.

The calculation of product cost is a necessary condition for scientific investigation concerning the relations between the natural environment and productive directions of the economy. Economics begins only with the appearance on the scene of a measure of value—the ruble. The proposition that there is a greater or smaller advantage or disadvantage in some natural condition for a production branch of agriculture, as well as for a region, for the development of an industry, or the construction of an industrial enterprise, remains a general phrase until it is translated into the exact language of the ruble. In solving concrete questions of this kind clarity can be achieved only by the calculation of comparative production costs. Such calculations are necessary and are in fact performed not only in capitalist countries but also under conditions of socialism in the USSR. "The lowering of cost is the central problem of industry and all other tasks must be subordinated to the solution of this problem," were directives of the Fifteenth Congress of the All-Russian Communist Party as contained in its resolution.

The difference, and an essential one, lies in the character of this calculation and in its aims. Under capitalist conditions, the question is whether it is profitable for the individual capitalist enterprise for the near future, perhaps even for the next year; under these conditions arithmetic definitely solves the question. However, under our conditions calculation is necessary not only for a specific enterprise for the near future but also for the more distant future and the impact on neighboring industries and regions.

One can say that the question of the influence of the geographic environment on human society and, in particular, on the differences in the productive directions of the economy is inseparable from the question of the reciprocal influence of human society on the geographic environment. The first connection between these two questions is that if the geographic environment had no influence on human society then there would be no need for human society to try to modify this environment, to adapt it to its needs, or in one way or another to exert an influence on it.

One should also consider that historically the influence of human society on the geographic environment is just as old as the influence of geographic environment on human society. When we examine the question of geographic environment, we can say as a general rule that we have to deal with a geographic environment to some degree modified by the influences of human society. Sections of the feather-grass steppes in the southern part of the Ukraine can at present be found only in special preserves, and they have been maintained only because of conscious intervention on the part of human society. The primitive landscape (Urlandschaft) is no longer a reality but a product of complex mental work on the part of specialists on restoration.

In the process of human history the dimensions of this reciprocal influence of human society on geographic environment grow at an increasing speed parallel to the growth of human societies and their technology. Let us cite several examples.

In the process of soil formation man becomes one of the most powerful factors; the portion of arable land in many regions and countries reaches 60–70 percent of the total area; the crop layer in places of ancient culture, for example in China, India, and Egypt, is more than ten meters; the volume of land turned over annually in the process of plowing is 3,000 cubic kilometers, though all rivers annually carry into the sea only two to three cubic kilometers of land; just as large are the figures expressing the work performed by humanity in the process of extracting mineral wealth. In the words of academician Fersman, "Man, as a new 'geological factor' acting in the arena of history, is changing the face of the earth." Finally man is changing the very contours of the land. The construction of canals, the Suez (1869) and Panama (1914), is a rather significant modification of the earth's surface.

Concerning this question, we read in "Dialectic of Nature" by Engels: "Very little has remained of the 'nature' of Germany as it was in the era when Germans moved there. The surface of the land, the climate, vegetation, animal world, even the people themselves

have changed infinitely . . . while the changes without human assistance which have occurred during this time in the nature of Germany are negligibly small."[24]

With the transition of all humanity to communism and the cessa-
tion of wars, the effect of humanity on nature will increase even
more; one can foresee changes not only of a quantitative but also of
a qualitative order; undoubtedly the effect of humanity on nature
will assume entirely new forms. The main point to keep in mind is
the direction of the effect. In a class society the effect of man on
nature was very often destructive, predatory and, generally speaking, negative. In a communist society this effect will be creative,
beneficial, and, generally speaking, positive. From this it is clear
that intervention by man in nature has gone very far and in the future
will go considerably farther. The Marxist economic geographer must
never overlook in his reasonings, especially in determining prospects for the future, the growing power of organized human society
in its effects on surrounding nature.

Above all, fully realizing the growing power (with respect to both
quantity and quality) of this effect of man on nature it is necessary
to acknowledge: (1) that this effect is far from unlimited (one cannot
speak of any escape from nature nor of any miracles); (2) that with
the development of human society not only its power expands but also
its needs; and (3) that with the development and complication of technology and the power of man over nature his ties with nature not only
will not diminish but on the contrary will be increased and more
complicated, for, an increase of power of man over nature, in the
scientific understanding of this process, means not freedom of man
from nature but only a wider, fuller, and more expedient utilization
of nature. As long as man did not use either coal or oil, his economy
could not depend on the deposits of these minerals. However, now
that these deposits are being used, their distribution has begun to
influence strongly the distribution of industry and transport and
hence also the distribution of human economy as a whole.

1. The only thing that can justify the interference of geographers in this question is
the ignoring of it by historians. Political geography, if one were created, could deal with it
in a Marxist framework. In the correspondence of Marx and Engels, one finds many remarks on the differences between countries in sociopolitical respects. The systematization
of these differences, their explanation and the development of regularities would be the
tasks of such a Marxist political geography that, of course, would have nothing in common
with bourgeois geopolitics.

2. See, for example, the reports on questions of the progress of agriculture at the January plenum of the Central Committee of the CPSU in 1955.

3. One cannot help noting that in the foreign bourgeois geographical literature there have
recently also appeared quite a few statements against geographic determinism. Most indica-

tive in this respect are the statements made in the first chapters of the large composite work *American Geography*.

Dickinson writes even more strongly: "Geographical determinism is as dead as the dodo."

4.   Expressions similar to those cited above in the process of teaching in school very often do not reflect even an atom of ill intention on the part of teachers and students of geographic materialism but are only the result of carelessness and haste. In the same manner such phrases in school textbooks of geography are for the most part the result of a pursuit of brevity at any cost.

5.   Karl Marx and Friedrich Engels, *Sochineniya* [Works], 14:647–48.

6.   My textbook reads: "We must take into account the geographic factor, not isolated from everything in the world, not metaphysically as many 'nature philosophers' did, but dialectically in association with the state of technology, in association with productive relationships, in association with the whole socio-historical setting of a given country at a given stage of its development."

7.   Marx and Engels, *Sochineniya* (1929), 5:429.

8.   Otherwise economic geographers would have to study physical geography and in addition, geology, geomorphology, and climatology; the physical geographers, economic geography and also economics and history; and both of them, technology; they would also have to harness themselves into applied work with all the resulting consequences, but as it is, one can live quietly.

9.   By the way, it is possible that some "brain" will find an illustration to the law concerning the unity of opposites. Having turned their backs on each other, neither physical nor economic geography alone can give a complete idea of the appearance of a country or region, not to speak of the solution of any concrete practical question, because a certain degree of interpenetration and contact are necessary.

10.   This is not only useful but necessary because one cannot fight with those who distort Marxism in any other way than by comparing the distortions with the original. The reader will do well if, not satisfied with the cited quotation, he turns to the original work itself.

11.   Marx and Engels, *Sochineniya* (1933), 4:10–11.

12.   Karl Marx, *Das Kapital* (1955), 3:833.

13.   Karl Marx and Friedrich Engels, *Izbrannyye proizvedeniya* [Selected Works] (1955), 2:32

14.   Ibid., pp. 483–84.

15.   Marx, *Das Kapital*, 3:804.

16.   V. I. Lenin, *Sochineniya* (1946), 7:380.

17.   Marx, *Das Kapital*, 1:517

18.   Ibid.

19.   Marx and Engels, *Sochineniya* (1937), 16:11.

20.   Ibid., 10:721–22

21.   Ibid.

22.   This would be an absurd formulation of the question, because from an arid climate one cannot, of course, obtain a Merino sheep any more than one can obtain a collective agricultural economy from the Verkhoyansk mountain range.

23.   It happens, however, that with the development of technology some resources formerly used prove to be unprofitable, for example, small beds of iron ore (lake ores and others) and copper-bearing sandstones in the Urals, which were used in the eighteenth century.

24.   Friedrich Engels, *Dialektika Prirody* [Dialectics of Nature] (1953), p. 193.

N. N. KOLOSOVSKIY

Scientific Problems of Geography

NEW TIMES MEAN NEW DIRECTIONS IN SCIENCE

The Nineteenth Congress of the Communist Party passed a resolution concerning the necessity of developing Soviet science to the level that would enable it to take first place in the world. The efforts of Soviet scientists must be directed toward a more rapid solution of the problems of utilization of the immense natural wealth of our country. The congress proposed a strengthening of the creative cooperation of science with industry. These directives also refer in full measure to geography, which is called on to study territorially the productive forces and natural environment of our homeland, as a whole as well as by republics and regions, in connection with the socialist method of production and its technology.

In the Soviet era the development of geography has reflected the passion of the revolutionary struggle of the old against the new, of socialism against bourgeois ideology. The transformation of the world and not its description represents one of our struggles with bourgeois survivals in geography. The unity, in principle and practice, of geographic knowledge about nature, social life, technical creativity of people, and the physical-material-technical basis of

Translated from *Voprosy geografii* [Problems of Geography], no. 37 (1959), pp. 129–50.

production is the second line of the continuing struggle for the new geography of communism. With respect to nature and economy, the study of huge areas of the Asian parts of the Soviet Union and the Far North, which has filled gaping blanks in our knowledge of our homeland as it existed before the revolution, falls within the sphere of practical geographical work in the broadest sense, as it has developed during Soviet time. The colossal amount of work put into the geodetic survey and mapping of previously inaccessible and immense spaces in the Central Asian deserts, Siberia, the Urals, the North, and elsewhere must also be mentioned. One must also remember the study of resources of the earth's surface and interior, which solved the problems of industrialization and electrification of the country, and the achievement of economic independence of our homeland. The scientific and practical achievements in the field of economic regionalization of the country based on a planned economy must also be mentioned as well as achievements in the field of comprehensive development of regions of intense activity such as the Urals, Donetsk, Kuznetsk, Eastern Siberia, and others. Before the revolution geography was unaware of such problems and did not know how to approach them.

A natural result of the great practical work of Soviet scientists and practical workers in fields new to geography with new demands was the large amount of accumulated, but as yet unsolved, theoretical problems in geographical science and the presence of certain discrepancies in the formulation of the problems concerning practical affairs in academic institutions, in higher schools of learning, and in research institutes. It is noteworthy that in recent years the interest in general theoretical questions of geography has greatly increased. A few of these problems as the author sees them are as follows: (1) the problem of boundaries, content and method of geography, and its relation to adjacent fields of knowledge, natural science, the social disciplines, and technology; and (2) the influence on geography, the oldest of sciences, of new ideas and decisions in adjacent fields. Of course, he does not pretend to be infallible even in the statement of problems at this period of great new conquests in geography.

THE PRINCIPLE OF UNITY IN GEOGRAPHY

The principle of unity in geography is a legacy from the distant past of Russian science, one that had originated by M. V. Lomonosov's time in the eighteenth century. In the nineteenth century it was fully developed and resolved differently at various times. There is no full agreement on its resolution even at the present time. Some Soviet geographers are firmly convinced that there exists a single geographical science that is supposed to describe the contemporary life of man and of nature on the earth and to give an integrated view by separate countries and regions, distinguishing, of course, countries with various means of production and social structure (socialism, capitalism). Other authors hesitate and sometimes are inclined to maintain that there can be no unity of geography since the laws of nature and human society differ in substance. In their opinion there must be two sciences: one, physical geography, concerned with nature, and the other, economic geography, concerned with the life of human society. Consequently, they argue, there must be two geographical pictures of contemporary life and not one. Such views can hardly be expected to bear up under Marxist theory. It will be of interest to recall some facts from the history of geography in Russia.

Before the revolution such a dual view of geography was accepted as almost the only possible form of geographic knowledge by Russian bourgeois science. At that time physical geography was studied in universities by the natural history faculties. Economic geography, however, was considered one of the economic sciences in economic textbooks and secondary schools of learning. The first was most often based on the philosophy of the so-called natural-historical materialism according to which, in the final analysis, nature and natural conditions determine the development of human society. The second reflected in one form or another the various Neo-Kantian doctrines, contrasting in principle the natural sciences with the social sciences. As is known, this break between natural and social sciences received final formulation in the so-called Baden school of Neo-Kantianism (Wildenband, Rickert).

The direct result of this idealistic dualism in Russian prerevolutionary geography was the notion that it was impossible to understand the very essence of the external world—that of nature—and objectively existing territorial combinations of productive forces developing according to definite laws did not exist; hence the denial of the objectivity of economic regionalization and the acceptance of it only as a method of investigation chosen by the author on subjective grounds. In a word, the incompatibility of the two branches of geography within a single science, as established by the bourgeois Russian geographers on the basis of their idealistic philosophical position, is completely unacceptable to Marxism.

The founders of Marxism-Leninism emphasized often and very strongly the principle of the unity of the materialistic world view. "We know," wrote Marx and Engels, "only one single science, the science of history. Examining history from two sides one can divide it into the history of nature and the history of people. However, both these sides are inseparably bound so far as there exist people; the history of nature and history of people condition each other."[1] As is known, the vehement objections of V. I. Lenin and his definition of Marxist teaching as "materialistic monism," were opposed to any manifestation of dualism in the Marxist world view.[2]

The same integrity and universality of scope of the world of nature and the world of people was emphasized by Stalin: "Marxism in the science of the laws of development of nature and society."[3] The denial in principle of the possibility of presenting an integral view of the world in geographical works and the acceptance instead of the necessity of being satisfied with two views of the world, not interconnected in one system, must be taken as a direct defense of philosophical dualism that is out of place in Soviet science.

The unification of the study of physical and economic geography in the geographic faculties of universities and pedagogical institutes, introduced during Soviet time, must contribute to the establishment of a correct direction in the development of geography. However, it must be emphasized that a real unity of geographical science cannot be obtained by itself. A declaratory acceptance of such a unity of organizational measures is not enough, and the present state of the science leaves much to be desired in this respect. In fact we have economic geography and physical geography apparently developing in parallel ways.

Moreover, there is not complete unity of views on the direction of theoretical investigations in the area of common questions in geography. Some think that one must proceed, so to speak from the bottom, from particular questions of the individual geographical disciplines, processing the vast factual material from Marxist positions and approaching general conclusions with great caution. Others try to solve the problem of geography abstractly, "from the top," and by ambush. A number of unsuccessful works of this type have given rise to a very cautious attitude toward theoretical synthesis in geography, in spite of the fact that the necessity of such generalizations is realized and arouses great interest.

First of all we clearly hear the suspicious voice of the practical, the majority of geographers, engaged in teaching or in applied geographical work. The majority of them accept the advantage of the approach to geography as a single system of knowledge. At the same time, however, a number of reservations are made, concerning the factual state of the science. They say, "We are for unity but via complexity [composite approach—kompleksnost]." And complexity is understood in various ways. Some understand it only as a formal collection of diverse geographical data. For others complexity must lead to an integrated picture of the world, that is, revealing an organic connection between the components. In the opinion of some, a general theory of geography is needed to achieve this integration.

Others think it is sufficient for physical and economic geography to each develop separately on Marxist theoretical principles. But physical geography by its own means must take into account the influence of society on nature, and economic geography in its own way must take into account the influence of the natural environment on the development of society. Under such conditions, they think, a well-educated geographer, equipped with general Marxist theory, can give an integrated comprehensive picture of the world.

The applied workers usually do not conceal the contradictions of their position and its vulnerability to criticism, but they justify their views by saying that it is exceedingly difficult for geographers to independently develop a bridge because of the diverse principles of development in the natural sciences and the social sciences. This, they say, is a problem clearly beyond geographical science.

With all this contradiction of argument, as we see, the applied workers have serious reservations deserving attention. At the same time it must not be forgotten that the empirical way of generalizing data from separate geographical disciplines, developed by the applied group, too often leads to an eclecticism, to a kind of impressionism in geographical works, especially if the authors, without serious training, undertake broad and versatile themes of contemporary Soviet development. The defects of such literary attempts have been noted by our critics more than once. Their cause is the same: an insufficient specific knowledge of the natural environment and economy of a region and its interrelations, and a lack of capacity for scientific synthesis.

And so, we must once more note that acceptance of the unity of geography in principle alone is insufficient to bring it about; the actual present state of the science, I repeat, is far from such a unity although much has already been done in the sense of ideological preparation. It is important to come to an agreement on how it must be approached in order not to make grave errors.

In the natural sciences there was a period, well described by Engels in his "Dialectics of Nature," when physics and chemistry were considered sciences differing in principle from each other. The former was the study of continuously changing forces whereas the latter was the study of particles of matter, atoms and molecules changing with an increase or decrease by entire units, that is, discretely, in an intermittent manner. At that time it seemed we had to deal with completely noncomparable laws, but then the phenomenon of the decomposition of chemical solutions by electric current was discovered. In analysis, chemical and physical laws had to be used simultaneously. However, even this seemed to be insufficient. The whole mechanism of interaction of both laws had to be established, and these laws augmented by a new physical-chemical law of the molecular structure of electricity itself. So for the first time electrons were discovered, and since then physics and chemistry have become a single system of physical and chemical principles laying the basis of modern natural science. At the interface of the two sciences a special scientific discipline developed—physical chemistry. This did not come about immediately, nor without a heated difference of opinions.

During the short time since the publication of the last edition of Engels's[4] work the process of "breakdown of barriers" between the various scientific disciplines, which had already begun in the nineteenth century, proceeded at an ever increasing tempo. The dialectical singularity of all human knowledge becomes ever more evident. The road that all sciences follow to their final unity and always with positive results is that of the dialectical elaboration of the boundaries of two adjacent sciences and of the contradictions in their approach to a given theme. This is the process that Engels explained using the example of physics and chemistry. It proved to be the correct process for other sciences. The question is: Why should it be wrong for the two branches of geography? For the solution of the problem one must evidently find an object of geographical study that would be subject to the effect of natural laws and at the same time would be a part of the productive forces of social labor, subject to the effect of the laws of social development. In that case such an object of study could be examined simultaneously in both physical and economic geography, and one can attempt by means of comparison to establish mutual relations and ties. Such an object of study can, in fact, be selected and is a familiar aspect of our lives. It is the regional material-technical basis of production on a contemporary Soviet scale and technological level. Dams on a river, a canal, a regional electric system, mines, or a railroad network are all subject to laws of nature. It is the same nature but in a form changed by man. The engineer who supervises construction or exploitation and who designs these structures is guided by laws of nature. By means of physical units of measure such as meters, kilograms, kilowatts, and calories, he measures the natural forces that give "free service" (Marx). These structures and enterprises can be studied geographically (with only a few adaptations) by methods familiar to physical geographers. However, simultaneously these same elements of the material-technical base are an inherent part of the productive forces of social labor and consequently can be studied in economic geography, from the economic point of view. Their development, thus, is also subjected to the effect of social laws. In this social sphere are measures of quantity in the form of the amount of working time per unit of product, that is, man-hours, or in terms of value, we can use, rubles. The relationship between the two systems of study and the calculation of quantities, physical and economic, is determined by calculating the ratio of the two, for example:

$$\frac{\text{quantity of expended kilowatt-hours}}{\text{quantity of expended man-hours}} = e$$

where e is the index of energy consumption per unit of labor, a known measure of industrialization of social production introduced into our science and practice at the time of Lenin's plan of electrification and the five-year plans.

Logically this is the same method that was used in the study of interaction between physical and chemical units in the process of electrolysis, pointed out by Engels. That relationship was calculated in the following manner:

$$\frac{\text{quantity of expended energy}}{\text{quantity of decomposed molecules}} = e$$

where e is the entire quantity of electrons.

Thus it would seem that the material-technical basis of production on a contemporary scale can be looked upon as an intermediate link of the geographical study of the ties between the local natural conditions and resources and the economics of the region. In general outline, this is one of the possible directions to search for a solution of the problem concerning interrelationships between natural and social phenomena in geography. As we see here, we once more return to the ideas that once were expressed in the GOELRO plan [State Commission on Electrification of Russia]. Indeed, in the work of Academician G. M. Krzhizhanovskiy we find scientific generalizations and explanations of the strong ties between the technical, economic, and social aspects of the regional electrification of our country according to the GOELRO plan.

The socialist method of production presumes the organization of energy production in social regional forms using the most modern technology and electrification. The general electrification of industry, agriculture, and transport on this basis leads to consistently large growth in the productivity of social labor and to an increase in the population's capacity to work. Krzhizhanovskiy explained that in science the very term *energy* signifies the capacity to produce work, in other words, efficiency; in the regional organization of energy utilization, this means the efficiency of the population of the region. Thus, the term *energy production* has not only a technical but also an economic value. The GOELRO plan provides for economic regionalization of the country on an energy basis, also taking into account natural conditions and resources; precisely these ideas were developed by Soviet economic geography serving as a starting point of reconstruction in a socialist direction. Up to now physical geography has considered that the ideas of the GOELRO, as economic ideas, have no relation to its problems. On the basis of what has been written above one can assume that the geography of the material-technical base of regional electrification and production in the sense of the GOELRO plan can serve as an intermediate link for the establishment of ties between physical and economic geography. Physical geographers should consider these ideas.

SOME GENERAL AND STABLE PARTS OF GEOGRAPHY

Geography differs from other sciences in a number of special characteristics; however, the determination of these characteristics is not an easy task, all the more so since, in the course of time, changes occur in the content and form of the science itself. Nevertheless, there exist some common features of geography that have been retained for centuries and that make it possible to determine accurately whether a given work of an early author or that of a contemporary, belongs to the field of geographic knowledge or to some other science. These characteristics are: the unity of a territory and territorial combination in the study of phenomena and facts; all phenomena described and analyzed must be handled simultaneously and contemporaneously with the author; the integrity, unity, and complexity of the whole geographical picture; and a clear tie to the history of the development of nature and society in the territory under study.

The Territorial Principle

Phenomena and facts studied in geography are always related to a definite territory. This territory may be small or, on the contrary, vast. It can be a region, part of a region, a country, or the earth's sphere as a whole. However, in every case the facts and phenomena under study must belong (or be capable of being intellectually related) to a definite territory. Otherwise the very essence of geography is lost.

It must be noted that territory is understood physically, as the surface of earth, not abstractly in the form of geometric space. Territory is always represented in an actual material form, that is, valleys, lowlands, mountains, and so forth.

As it was in ancient times when human science was yet in its infancy, so it was in the Middle Ages and during the period of capitalism. During the capitalist period, under the influence of the need to extend and systematize the continually growing amount of scientific material, a number of disciplines separated from geography as, for example, geology, botany, zoology, and others. At one time, geography itself seemed to break up into a number of separate sciences— the geographical complex having unclear relationships between its individual components. However, whereas this did not occur on the whole front of the science, two of its divisions, physical and economic geography, separated considerably from each other. The territorial principle was retained in both branches, although the natural region received its meaning via comparison with the regions of social life (economic, political, cultural-national).

Contemporary Phenomena

Geographical science has two means for depicting the results of its investigations: verbal language, that is, the written or spoken word; and graphic language, that is, maps. These two languages must express one and the same thing—a picture of contemporary life interconnected simultaneously with existing events, phenomena, and facts. The past is of interest to the geographer only as far as it is needed for the explanation of the present; the future, in so far as it is needed to conjecture where the development of the present [trends] will lead. Only in special cases does the geographer engage in paleogeography, the geography of the past. Often paleogeographical surveys are done not by geographers but by historians or geologists.

However, it must be stipulated that the clear differentiation of the history of human society and the history of the geological structure of the earth from geography, in the proper sense of the word, became established only in the era of bourgeois science. There are works by ancient writers in which history and geography are closely related. In the best current historical works the geographic element is always present and, conversely, in good geographical works it is unthinkable to do without significant excursions into the field of the history of nature and society in a given place. As the physical geographer cannot do without the history of the earth in terms of geology, geochemical regions and centers, and paleography of the organic world, so too, economic geographers cannot explain phenomena and facts without showing their historical dependence. No sharp rules of separation between history and geography can be pointed out here; this separation is accomplished by the scientific perception of the author. However, there is no need at all to fuse history and geography.

The Historical Approach

In connection with what has been said, a question arises for the Marxist geographer: Can the Marxist historical method of investigation be utilized in geography? The answer, of course, must be affirmative. The metaphysical division of time and space has been rejected by Marxist science. In spite of the territorial principle in geography and the limitation of time to the narrow boundaries of the recent past and future, geography examines all contemporary phenomena in their development, both as regards dialectical relations and the struggle of contradictions, that is, by using methods of dialectical and historical materialism. The principles of development and the ready-made conclusions of the adjacent historical sciences concerning nature and society are widely used. From our point of view one may call geography "the evolving history of today, examined territorially [spatially]." With its transformational tendency Soviet geography also dares to examine the future geographical picture of the world.

Unity and Complexity

After the period of metaphysical division of human knowledge during the last century and the appearance of new sciences, a reverse tendency could be noticed: that of uniting particular sciences into one general system on the basis of the materialistic conception of the world. The separate branches of the natural sciences—mechanics, hydraulics, thermodynamics, instruction in electricity, magnetism, light radiation—have gradually overcome their boundaries and have become one system of physics, far removed from the recent mechanical conceptions and, what is more, united with the principles of chemistry.

Parallel to this, the separate branches of the human sciences, such as history, political economics, sociology, philology, law, and others, by great efforts of the founders of Marxism-Leninism, have become one system of science with the common basic objective of discovering laws of social development similar to the laws of natural science, which are determined by the material conditions of society, by the state of productive forces, and by productive relations. Moreover, as Marx and Engels have established, the development of the natural environment and human society mutually condition one another.

With similar premises of general growth, the geographic sciences cannot remain in the state of ideological parceling that was characteristic of Russian bourgeois geography during the second half of the nineteenth and the beginning of the twentieth centuries. It cannot help but strive toward a development by which a dialectic integrity of the whole picture of life of nature and people would be achieved, similar to ancient classical geography, but on the incomparably higher level of present scientific achievements and based on the vast accumulations of factual material. At the present stage of its development Soviet geography can achieve this by establishing organic interconnections between the individual geographical disciplines, and if required, by developing new disciplines, or bridges, at the intersection of two existing disciplines. This refers alike to the group of physical-geographical disciplines, economic-geographical disciplines, and to the ties between them. A possible unity of the whole under such conditions must be achieved by complexity, that is, by a combination of data from the various disciplines for the characteristics of the whole. However, complexity is not a goal in itself; the goal is the unity of geography, and that is not the same. Complexity is a way to achieve unity, but a complexity that does not lead to this unity is not needed in geography. And at the same time in the establishment of interconnections and complexity between the natural and social components, within these groups as well as between them, lies one of the characteristic features of geography. Geography studies not any single form or process but their interconnections as a whole on a given territory.

ON THE GEOGRAPHY OF THE NATURAL ENVIRONMENT

In recent years Soviet geographers have come to the conclusion that it is necessary to define physical geography more exactly as the science of the natural (geographic) environment of human society with regard to its existence and development, thereby making use of the same term employed in political literature, in particular, in the works of Marx and in the *History of the All-Union Communist Party: Short Course*. In the opinion of the originators of this term, it most correctly characterizes the basic subject of study for the totality of individual physical-geographical disciplines, including general physical geography, geomorphology, land hydrology, oceanography, climatology, biogeography, and other fields.[5] Indeed, this term has a number of advantages over the others, such as "the outer sphere of the earth" or the "geographic envelope." In all definitions of this kind, we cannot avoid a certain hint of mechanical views on the nature of the earth, and there is an artificial division of the surface of the earth from its interior and of matter from the energy resources of the earth.

It is important also to note that the new definition emphasizes the ultimate connection between physical geography and economic geography. Indeed, the natural environment is one of the necessary conditions for the existence and development of human society. This is the totality of resources and conditions, matter and energy of the earth, that is important for life and the productive activity of human society. Into it enter not only the resources of the atmosphere and the surface and interior of the earth but also the energy of solar radiation, magnetic and electrical fields of the earth, and the geochemical and atomic energy resources of the earth.

It is known that such prominent soviet scientists as V. I. Vernadskiy and A. Ye. Fersman have often appealed to geographers, through their public speeches and published works, to accept a more modern and wider conception of the content of geography. It seems to me that such a conception is the only correct one from the point of view of modern science. Moreover, such a viewpoint more fully reflects the current practical needs for the development of communism in our country. Indeed, this development is already occurring through an unusually rapid growth in the application of scientific knowledge to effect changes in nature. The material-technological means become so great that one can compare them with elemental forces of nature, such as the movement of air and water masses, and eruptions of volcanoes. We are already drawing up projects to reconstruct the natural conditions of entire regions by utilizing modern powerful technological and biological means of influence. If we think of atomic energy, the prospects for its use are indeed enormous. In the renovation of the means of production, regional energy production directed by man determines everything. Such is the revolution prepared by modern science and technology in the relationships between man and the natural environment. Yet the science of the natural environment—physical geography—continues to be the captive of definitions of its content, with limitations and methods that were established in the nineteenth century and that were once progressive but cannot be considered so today. In some cases the unconditional opposition of physical and economic geography, being completely incomparable in the essence of their laws, continues up to the present time. In principle, as shown above, such a contrast is based on philosophical dualism, is theoretically incorrect, and is harmful for practical purposes.

Some geographers are inclined to limit the problems of physical geography to so-called landscape science. With all due respect to this field of knowledge, it must be recognized that this is only one of the branches of geography. Landscape science arose in the nineteenth century on the basis of the very important achievements of the theory of zonal phenomena of life on the earth and the very close relationship between geographical latitude (and surface elevation), and climate, soils, and vegetation. We already find the ideological sources of this theory in Russian science in the second half of the eighteenth century. We can trace its development in the course of the first half of the nineteenth century and the influence on it of the physiocratic ideas, Hegelianism, Russian philosphical thought of the nineteenth century, and the general achievements of science. On the basis of the theory of zonality, there developed in the second half of the nineteenth century such closely connected Russian sciences as soil science and landscape science. The names of famous Russian scientists, such as P. P. Semenov-Tyan-Shanskiy, V. V. Dokuchayev, A. I. Voyeykov, and D. N. Anuchin, are associated with these achievements. A further use of these theories by Soviet geography, agronomic science, and biology enriched Soviet science and led to a number of achievements in Soviet agriculture and forestry. However, it must not be forgotten that the Soviet scientists had to make substantial corrections in the theoretical constructs of the past in view of the clear traces of idealistic philosophical views, especially of so-called natural-historical materialism. The extreme expressions of these teachings in the forms of so-called geographical materialism led such great geographers as V. V. Dokuchayev and A. I. Skvortsov to assertions about the decisive significance of natural conditions for social development. In connection with this, the doctrine about landscapes as formulated in the past was revised in Soviet physical geography.

Using the laws of nature for its own purposes, human society completely changes landscapes in the course of time. The landscapes of the Chinese lowland, the Eastern European Plain, Mesopotamia, Egypt, and France have very little in common with their historical appearance. The jungles and marshes of Mesopotamia and the delta of the Nile, the gigantic marshes of Belorussia, the primeval forests, and soils of Gaul will never be restored to their former appearance. Nature could not have independently created the vast fields of Soviet wheat and other crops that do not exist in a wild state. Nature could not have created the modern industrial landscapes of the Ruhr or Donetsk basins or modern large cities. All this is the handiwork of man. If man's works are capable of crumbling in the course of centuries, the works of nature, too, are changeable and not eternal. The experience of human history shows that people never cease developing a country or region once it has been undertaken, and in place of destroyed cities and installations something new always arises that is also vital and human, not something born blindly, automatically, or naturally. The history of nature and the history of human society condition each other. All that has been said refers even more emphatically to the most recent Soviet communist era, when new energy resources and technology have acquired a previously unknown power and are subject to planned designs of social development. In studying the physical geographical conditions and resources of regions of the Soviet Union, the geographer of today cannot ignore the laws and demands of social development.

For the study of nature in parts of the earth not touched by man—and there are many in the immense expanses of the Soviet Union—the physical geographer can still use the zonal theory and methods of study, well worked out in the past, which are of "canonical" character. For the study of nature in regions completely developed for agriculture the geographer must know well the history of development of the region, its present character, the basis of modern Soviet agrotechnology, agricultural energetics, and the economic assignments for the future of the region. Only then can he get an idea of how and what to study in the nature of the region. The "canon" of the complex is different and in each case specific, depending on the assumed material-technological basis of reconstruction.

For the study of nature in regions of industrial development, the physical geographer must know not only the resources of the surface but also the subsurface resources exploited in the region. Biogeographical zonality in such a region can recede to the background for the geographer as compared with the geography of the geochemical zones related to deposits of useful minerals. In a number of cases the network of water currents, soil cover, vegetation, and fauna can be completely destroyed by industry, roads, and cities, the relief changed, whole mountains flattened, and therefore the landscape zones can be studied only paleologically.

Thus, for an industrial region the natural environment is usually presented from entirely different aspects. Its study demands a selection of other components of the natural environment and from other points of view. The biological-geographical key evidently does not fit here, and the usual zonal theory is not the only means of general conceptualization. From this derive the important creative tasks of reconstructing physical geography in such a manner that it corresponds to the present demands of life.

Let us explain. The integrity of the picture of nature for us is not something received by way of direct perception as it was for the ancient geographers. For us it is a unity, analytically dismembered into component parts—"the unity of diversity." This diversity is tied into a single whole by internal relationships that are governed by law and yet alive and changing. Consequently, the reflection in science of this unity of nature must be achieved by intricate and minute logical work that makes use of Marxist dialectical materialism. In order to know how to discriminate in this multiplicity and to avoid obtaining, instead of organic unity and integrity, a simple piling up and collection of facts, phenomena, and independent principles, valid only for individual components, it is necessary to use Lenin's instruction concerning the significance of the "leading links of interconnections," which, if grasped, enable one to pull out the entire chain.

The phenomena of nature and the effect of the material-technological framework on them are to be studied not only qualitatively but also quantitatively by applying physical units of measure both to nature and technology and, in necessary cases, comparing expenditures of energy (natural and technical) with expenditures of human labor.

In correspondence with the present state of science, the area of physical geography must be expanded to include the water and energy components of the earth, which comprise the natural environment for modern methods of science and technology. In conformity with this goal, the contemporary, one-sided, terrestrial canons of investigation must be reworked.

SOME QUESTIONS OF ECONOMIC GEOGRAPHY

With respect to the task and content of Soviet economic geography various opinions exist at the present time. This impels me to explain briefly their historical formation.

The Influence of the GOELRO Plan and the Work of Gosplan in the Twenties

In the very first years of Soviet power great shifts occurred in the formulation of economic geography. The directive of Lenin to study the productive forces of Russia on a wide scale and to recruit the country's scientific forces brought a response from Moscow University, specifically from professor of geography, D. N. Anuchin, who said that henceforth geography should become a science of productive forces. However, the ideological centers of work at that time were not the academic establishments but the commissions engaged in working out the program of Gosplan [State Planning Commission] of the USSR and the congresses on the study of productive forces.

In the fate of every science the essential role is played by its initial, fundamental ideas and the demands of life. Only later do these initial thoughts and ideas usually take on the appearance of fully developed academic theories. So it was in this case. Bourgeois Russian economic geography, removed from physical geography by the ideological barriers of Neo-Kantian dualistic theories, could not cope with the new demands of life raised by the socialist revolution. In this dramatic conflict of contradictions between life and science, Soviet economic geography was born. Its ideological foundation was the GOELRO plan and the works of the Gosplan in the twenties on economic regionalization, which were based on a complete, monistic conception of the world and Marxist philosophy.

In these works the problem of the relations between the productive forces of labor, the material-technological base of production (on the basis of its electrification), and the natural environment of the regions was being solved on the plane of philosophical and practical unity and also with specific examples. Evidently, this indirectly prepared the way toward a solution of the problem of the relations between physical and economic geography, which was correct from the Marxist view of the necessity of their interconnection and unity principle.

The problem of territoriality, a second basic question of geography, was being solved by GOELRO and Gosplan with the help of a socialist theory of economic regionalization based on new energy bases.

This was a radical solution of an age-old problem of Russian geographical science, a solution that was impossible in the past within the framework of capitalist science because of theoretical difficulties in the area of the relationships between the natural and social elements in geography. In the scope and scientific depth of its formulation the theory of Gosplan had great advantages over bourgeois theories. For the first time in history economic regionalization acquired not only a descriptive but also a transformational significance.

The problem of the historical approach in geography (a third geographical problem) was resolved by acknowledging the variability of the structure and boundaries of regions and their relations with other regions and by introducing into the method of study of regionalization the past as well as the future of the region (prospects of the economy's development). The problem of complexity was understood as integrity and confinement of the development of regions and relationships to nature and technology.

The problem of the geographical division of labor under the conditions of the USSR with its vast spaces and gigantic distances was being solved by GOELRO by comparing the economic needs of the Soviet state for long-distance hauling given the dimensions of the territory and the new means of technology (electric railroads, superhighways), which will radically solve the problem of interregional connections and of the new geographical division of labor in the USSR.

Thus the GOELRO plan and the works of Gosplan created for geographical science a reliable system of brilliant solutions and fundamental ideas on productive forces. The totality of these new ideas completely changed the dimensions of geographical science familiar to bourgeois Russian geography; they related it to life and to revolutionary, transformational activity.

The University Approach

The ideas of the GOELRO plan influenced geographical science in the various scientific establishments in various ways and at different times. On the whole, the greatest influence was on the instruction of geography at the pedagogical institutes and universities, especially at Moscow University, where the ideas of the GOELRO plan, due to the efforts of Professor N. N. Baranskiy in 1925–29, were proclaimed the basis of development for Soviet economic geography. These ideas found protection here even in moments of violent attacks by individual economists in the thirties and in the postwar period. The symbiosis of the scientific legacy of past Russian geographical science and leading Soviet theories was most successful here; the regional method and the concrete interpretation of geographical reality, the use of maps and field observations— all this was in proper accordance with directives by the Party and government on the teaching of geography in schools of higher learning.

However, there remain deficiences that must be corrected in the near future. They are the result of ignoring the need for a further elaboration of theoretical questions in geography. During the recent years, for example, university geographers have continued the well-known methodological separation of physical and economic geography and the breakup of the science into separate specialties, especially in physical geography. The general future line of development of geography seems to have been lost. All this, intensifying the break between geographical disciplines, inhibits the development of Soviet geography in the proper direction in accordance with the demands of life. A narrow professional specialization of knowledge is not a university approach to science.

The Economic Direction in Geography

In the twenties, the economic geographers and economists of the bourgeois school were hostile toward the Gosplan approaches to the interpretation of geographical questions in the GOELRO plan and the works on economic regionalization. In the report of Gosplan at the third session of the All-Russian Central Executive Committee special sections are devoted to the criticism of bourgeois approaches to regionalization that are responses to this hostile attitude. The bourgeois economists pointed to the inadmissibility of introducing into science a subjective evaluation of the prospects for economic development, as well as degrading science with technicism (in the spirit of the GOELRO plan), and attempting to bring technical methods, that is, methods of energy production, into purely economic questions of economic regionalization. Such a criticism was natural for scientists of bourgeois thought. But since the thirties there has been unexpected criticism of the Gosplan school in geography by economists who are seemingly in the Marxist camp, and sometimes with a repetition of arguments from bourgeois economists. On the one hand, an inertia of thought was apparent: before the revolution economic geography was being elaborated and taught by economists as an economic discipline. It is incomprehensible they thought, that it should be given over to the natural scientists at the university. On the other hand and more significantly, the criticism was theoretically based, as they implied, on Marxist political economy.

Economic geography, in their opinion, was an economic science (which is true, but with substantial additions) based on laws of social development that are studied precisely by political economy (also true but incomplete). This is, consequently, the "applied part" of political economy, its supplement to the study of the geography of social phenomena and, moreover, its inherent part (a hasty conclusion). The incorrect logical formulation of these authors begins with the very first statement on the character of economic geography. Economic geography is indeed an economic science, but one so closely connected with natural science (natural environment) that its theoretical basis must be not only political economy but also natural science (physical geography). Therefore, it is impossible to construct Marxist geography without basing it on Marxist philosophy as a whole, on dialectical and historical materialism.

An incorrect initial assumption about economic geography as an addition to political economy leads to a number of other erroneous assumptions that for more than twenty years now have been repeated by some economists involved in geography. Indeed, technicism, for which geographers are blamed, is inadmissible in political economy, the borders and content of which are clearly defined by classical Marxism. From the point of view of geographers, a regard for technical problems is obligatory even in the concrete form of factories, installations, and enterprises. This is not accepted by the economists.

Geographical materialism, for which the geographers are also blamed, is inadmissible in political economy and geography. However, whereas in political economy digressions into the laws of nature are not at all obligatory, in physical and economic geography the determination of the mutual influence of nature and society for a given place is obligatory from the geographical point of view. This connection can be established strictly in a Marxist way, without eclecticism.

In the heat of the polemics one heard opinions that the GOELRO plan was out of date and had long since been more than fulfilled, that the theory of regionalization of Gosplan was a quasi-bourgeois theory, since in the era of communism productive forces will be distributed evenly and the boundaries of regionalization will disappear together with the survivals of capitalism. The answers to these questions, the formulation of which is caused by a shallow and even superficial study of the Marxist-Leninist theory, were given by geographers at the time, but traces of these accusations have remained. Some comrades are inclined to think that something is wrong in university economic geography from the Marxist point of view, even if only in a vague way. There is, indeed, something wrong but on a different level. The noisy criticisms of the economists in the press and at scientific conferences more than once evoked among individual workers in geography a trend toward overly literary stylization of geographical works, in order to gloss over deficiencies by the authors on theoretical questions; often they gave rise to biased descriptive works in evident contradiction to the wide scale of reconstruction in our country. Thus, in a number of cases, a hasty criticism was not useful, but harmful. By itself the economic trend in geography proved exceedingly fruitless, both practically and scientifically.

GEOGRAPHICAL FEATURES OF PRODUCTIVE FORCES

Economic geography is called on to study spatially, by countries and regions, the productive forces in relation to the natural environment, the means of production, and the related technology. Now that the sufficiently long discussion of geographic problems in recent years has introduced us to our subject, the tasks of economic geography can be formulated in terms of these three functions.

Some geographers and economists even now think that it is preferable not to say "productive forces" but "distribution of productive forces," using the corresponding term [the latter] employed by Engels. We think this should not be done since distribution proves to be an idea not so simple, precise, and clear as it seems at first glance. It demands a great deal of explanatory work before one can introduce it into a definition understandable to all. It is simply inconvenient in the definition, and then it evokes attempts at a limited formal interpretation of geography, excluding from it, for example, questions on the structure of productive forces that also have a geographical facet, although they do not treat questions of distribution. Let us stop to examine in greater detail the question of productive forces in geography.

Marx gives a developed definition of productive forces in his works "Wages, Price and Profits" and in the first volume of *Das Kapital.*

> If one leaves aside the differences of natural characteristics and the acquired productive habits of various peoples, then the productive forces of labor must depend mainly: (1) on *natural* conditions effecting labor, such as fertility of soil, wealth of mineral resources, and others; (2) on the progressive improvement of the social forces of labor, which is the result of large-scale production, the concentration of capital, the combination of labor, the division of labor, machines, perfected methods of production, the use of chemical and other natural forces, the shortening of time and space by means of communication and transport and all other inventions by means of which science compels the forces of nature to serve labor and due to which the social or co-operative character of labor develops.[6]

> The productive force of labor is determined by complex circumstances, among others, the average degree of skill of the workers, the level of development of science and the degree of its technological application, the social combination of the productive process, the dimensions and effectiveness of means of production and, finally, natural conditions.[7]

Economic geography is supposed to study in detail these geographic facets of productive forces (with regard for productive relations) as defined by Marx with all the specific singularities of geographical science discussed earlier, that is, spatially accounting for influences of the natural environment, the means of production and its technology, in a composite way, on the basis of historical and dialectical materialism; predominantly contemporary phenomena and not the history of change of social structures are hereby introduced into the study.

In summing up the discussion on problems of political economy, Stalin gave this definition: "The subject of political economy is production and the economic relations of people; (1) the forms of ownership of the means of production; (2) the resulting position of various social groups in production and mutual relations or, as Marx says, 'the mutual exchange of activities'; (3) the resulting forms of product distribution."[8] Prime attention is given to the laws of "social production and distribution of material wealth at various stages of development of human society."[9] Thus, in political economy we have a science studying the social productive relations in their interaction with the productive forces on the historical plane of a succession of social forms. In correspondence with all its aims, in studying the concrete reality, facts, and phenomena of life in order to establish common laws, political economy must be abstracted from all local historical and geographical characteristics. Economic geography acts conversely, obtaining in ready-made form from political economy the basic economic laws of development; geographic science concentrates on revealing and explaining all local historical, geographical, and national-cultural influences and peculiarities. We think that the demarcation of the two sciences must be found along this line, remembering that it is conditional upon the principle of unity of Marxist science.

Let us go further. Geography is a concrete science. The productive forces of social labor, understood concretely and geographically, refer above all to their association with a definite territory (and to its natural environment) delimited precisely on the map and understood broadly enough to ensure the social character of production and exchange, their complexity, their developed internal and external economic ties and relations, and also their sufficient correspondence to local natural conditions and resources. Finally, area characteristics of cultural and political boundaries and connections must be taken into account. Geographization of the abstract political-economic idea of productive forces prompts obligatory reference to a specific territory with a specific natural environment with boundaries established while taking account of the location of the material-technical basis of production, the settlement of people, and the limits of expansion of national culture.

In this manner we have arrived at a conceptualization of territorial forms of productive forces of the earth, of countries and states, and of the economic regions of these countries and their parts. For the geographer the development over time of the productive forces of social labor means changes in the system of economic regionalization. Sometimes these changes will be changes in the external boundaries of the regions, and sometimes they will be changes in the structure of production and in the geography of its connections. Economic regionalization as studied by Soviet geography is at present a vast and independent segment of geography of the greatest significance for the construction of a communist society. The present state of this question [economic regionalization] merits a separate article devoted to it.

1.   Karl Marx, and Friedrich Engels, *Sochineniya* [Works] (1933),4:8.
2.   V. I. Lenin, *Materializm i empiriokrititsizm* [Materialism and Empiriocriticsm] (Gospolitizdat, 1948), p. 74.

3. I. V. Stalin, *Marksizm i voprosy yazykoznaniya* [Marxism and Problems of Linguistics] (Gospolitizdat, 1951), p. 54.

4. In 1894.

5. This proposal was made by the geographers of Moscow University and the Institute of Geography of the Academy of Sciences of the USSR as the result of a lengthy discussion concerning the principles of geography. In *The History of the All-Union Communist Party: Short Course*, "nature surrounding society" and "geographical environment" were used as interchangeable concepts. In the context of this article, where both branches of geography are touched upon—physical and economic—the author prefers to speak about nature, or natural-geographic environment, since there also exists a social-geographical environment or social milieu. Consequently, the term geographical environment is not ambiguous from this point of view.

6. Karl Marx, *Zarabotnaya plata, tsena i pribyl'* [Wages, Price, and Profit] (Gospolitizdat, 1940), pp. 32–33.

7. Karl Marx, *Das Kapital* (1953), 1:46.

8. I. V. Stalin, *Ekonomicheskiye problemy sotsializma v SSSR* [Economic Problems of Socialism in the USSR](Gospolitizdat, 1952) p. 73.

V. V. VOL'SKIY

Some Problems of Theory and Practice

In Economic Geography

The construction of communism reserves an honorable and responsible place for Soviet science, which is becoming increasingly a direct productive force. The geographical sciences, too, are called upon to make a substantial contribution to the creation of the material and technical basis of communism. The geographical sciences can and must take part in the solution of two closely related problems within the overall national objective: on the one hand, in uncovering and assessing natural resources and possible ways of making integrated utilization of such resources, and in the scientific justification of ways of achieving an integrated transformation of nature so that it may be better utilized; and on the other hand, in the process of utilization of natural resources and territory for the production of material goods, the geographical sciences must justify an application of productive forces that is the most rational and economically most effective from the point of view of the aims and needs of society. The first range of problems is the concern of physical geography, and the second of economic geography.

Reprinted with modifications from *Soviet Geography: Review and Translation* 4, no. 8 (October 1963):14–25, by permission of the publisher. The article originally appeared in *Vestnik Moskovskogo Universiteta*, seriya geografiya [Herald of Moscow University, Geography series], no. 4(1963), pp. 14–24.

The practical tasks of economic geography are to raise the economic effectiveness of the national economy through an optimal utilization of a territory and its resources, and of the entire geographic environment that arose within a given territory as a result of the development of nature and its transformation by man. The analytical work of economic geography consists of the study of the economy of a given territory and of its laws of formation as man has used and transformed the geographic environment in various stages of development of society. On the basis of these analytical data and of recommendations from physical geography regarding the most rational utilization of the geographic environment from the point of view of the laws of nature, economic geography should in its synthesis seek to justify economically a distribution, combination, and interaction of newly created productive forces that would be most advantageous from the point of view of society.

Since the more penetrating the analysis, the more justified and easier the synthesis, the ability of economic geography to carry out such a synthesis determines its transforming power as a science. It must be acknowledged, however, that the practical application of economic geography in the Soviet Union still falls far short of the scope of economic construction now under way. That is why the Communist party devoted serious attention to the unsatisfactory state of affairs in this field. Speaking at the conference of officials concerned with the industry and transportation of the RSFSR, Khrushchev said on 24 April 1963; "We know how important the correct distribution of productive forces is for the development of the nation's economy. However, there are serious shortcomings in this field that prevent us from making full use of the advantages of the planned socialist economic system. . . . In the next few years we will be building thousands of new industrial plants. The location of these plants must be economically well founded. But not all is well with the economic geography of industrial location." Khrushchev stressed particularly the immediate need for working out long-range integrated plans for the development of major economic regions.

The same point was made by L. F. Il'ichev a secretary of the Party's Central Committee, when he said: "In general we must emphasize the absolute necessity of sharply increasing the regional, territorial approach in economic research."[1]

A good example of what could be achieved by economic-geographic research and its introduction into practical economic development was given in *Ekonomicheskaya Gazeta* of 18 May 1963 by P. Borozdin, chief specialist of the Teploenergoproyekt Institute [Institute for the Designing of Thermal Power Stations]. In his view the lack of economic evaluation of geographic conditions for the construction of thermal power plants is causing great harm to the national economy. Borozdin estimated that "a rational site selection for a power station from the geographical point of view would lower construction costs by 15 to 20 percent. That is the tremendous reserve we hold in our hands!"

A large share of the responsibility for the situation in economic geography and for its place among the builders of communism rests on the Geography Faculty of Moscow University, the largest geographic institution of the Soviet Union. Lately the forces of that institution have been diverted from practical tasks by discussion of the abstract theory of a unified geography, which is being insistently advocated by V. A. Anuchin.[2] The relation of that theory to the immediate tasks of science and practice may be judged from the author's own attempt at justification: "The practical purpose of scientific development must, of course, not be oversimplified. Specific scientific research may not have any immediate practical significance. The process of cognition of the objective world may take place outside of the sphere of immediate practical demands and may not improve life even for many generations to come."[3] The appearance of this type of statement in Soviet geography has been given a far clearer and unambiguous interpretation in the West.

David Hooson, in his survey of Soviet geography, wrote about the supposed rapprochement of viewpoints between American and Soviet geography: "Soviet geography is more severely practical than American geography and hedged in by the requirements of party-set tasks. In recent years, however, even this condition has been considerably modified by the introduction of more purely cultural facets of the subject, such as the historical geography of towns, and above all, by the time taken up in methodological discussion. Soviet geographers have moved closer to the normal Western view of an academic subject as a more or less 'pure' investigation of a field of knowledge rather than as an instrument of policy."[4] Since the party has firmly vowed to build communism in the Soviet Union within the time span of one generation, Anuchin and his followers risk being rather late with their contribution to the cause of building communism if that is how they understand and expect to realize the ties between science and practice.

The principal theoretical problem in geography, according to Anuchin, is the interrelationship between the geographical sciences, especially economic and physical [geography], and clarification of the question whether these sciences have the right to exist independently or should be auxiliary branches of and purveyors of data to a nonexistent unified geography. Anuchin tries to demonstrate the need for creating such a geographical superscience, a unified or "monistic" geography (according to the terminology of Anuchin, who uses this philosophical term incorrectly to designate the unity of geography as a science), which would be concerned only with synthesis.

The idea of a unified geography is not new. It has long been the official doctrine of bourgeois geography. But, in Anuchin's view, the great shortcoming of bourgeois unified geography is that its so-called monism "cannot be justified theoretically" because American geographers have not mastered materialist dialectics and therefore use monism only intuitively, haphazardly, and declaratively.[5] As for Anuchin, he seeks to justify the monism of geography theoretically and to create a Marxist concept of a unified geography. (David Hooson, in a review of Anuchin's book, writes in this connection: "Indeed it is difficult to pin down anything exclusively Marxist in the arguments." The reviewer holds that although Anuchin's ideas "are just one more belated attempt to 'catch up' with the United States," they are entirely suitable for use in American geography.)[6]

Anuchin's theory stems from the absolutely incorrect assumption that society is part of nature, part of the landscape envelope of the earth ("although individual persons are not such parts," Anuchin says), part of the geographic environment. From this assumption follows the only possible conclusion that the development of a part of society follows not only its own internal laws but also the laws of the whole of nature and the geographic environment. Changes in one part of the entirety affect all the other parts: "Changes in nature affect the conditions of life of society, and consequently, indirectly affect society itself."[7] The unity of the geographic environment "from relief to human society inclusive" assumes the existence of general laws of development, which, in Anuchin's view, should be the subject of study of a unified geography.

Regarding economic geography as a branch of a unified geography, Anuchin asserts that society, which he falsely considers to be an association of biological individuals, follows the laws of physics, chemistry, and biology.[8] It is therefore quite natural that Anuchin sees the lack of validity of bourgeois geographical determinism only in its mechanistics[9] and believes he injects that geographical determinism with a dialectical content by stating that "the natural environment as a whole in a number of cases plays a *decisive role, determining* the possibilities of concrete interrelationships between society and nature,"[10] that "interaction between society and nature depends on the geographic environment,"[11] that "the geographic environment may consequently determine the economic specialization of individual countries and regions."[12] The only legitimate practical conclusion deriving from these statements is that the object of study of economic geography is the geographical environment, whose characteristics would then determine the course of economic development of a given region. The aims and tasks of society, social relationships, and the mode of production would supposedly be of no significance: "The geographical division of labor in its pure form does not depend on the mode of production. It arose at the dawn of human history and will exist as long as human society."[13] "Reindeer raising and fishing will retain their importance in the life of the people of the tundra even under communism *because these activities correspond in the highest degree to the geographical conditions of the region.*"[14]

But if geographical conditions predetermine once and for all the interregional or international division of labor, then what is supposed to be the practical role of economic geography? If we were to accept Anuchin's "law of a cause-and-effect relationship in the development of individual elements making up the geographical environment," then the social economy and its specialization could change only with changes in nature, in the geographic environment. Evidently economic geography is to be concerned with the development of the geographic environment and with the possibilities that it offers: "The dependence of society on the environment *increases* rather than weakens as productive forces grow. The geographic environment has the ability to expand and this quality has long been used in the practice of economic activity."[15]

Anuchin is quite aware of the practical significance of his concept: "General geographic investigations have for the most part only *indirect* significance in practical activity, and therefore the need for such investigations is not always realized."[16] Nevertheless, he holds that geography can make progress only on the basis of his monistic approach.

The interpenetration of sciences and the realization of interdisciplinary research have become increasingly important and are assuming increasing scope at the present time. But when we talk about an interdisciplinary approach, it is not to demonstrate its usefulness, which is well known to all, but to show how such an approach can be implemented in practice and to work out its methodology. In such an approach, separate sciences remain separate and the task of the researcher becomes more concrete and complex. Therefore, Anuchin does not call for interdisciplinary work in geography but for a unity of geography and proclaims as his slogan N. N. Kolosovskiy's statement, "An *interdisciplinary* approach is not an aim in itself: the aim is the unity of geography, and that is not the same thing."[17]

Anuchin's concept has found support among some economic geographers of our Geography Faculty. Yu. G. Saushkin, in discussing Kolosovskiy's ideas, holds, for example, that the solution of practical problems requires "in the first place the firm theoretical foundations of the unity of the geographical sciences" so that one may have "not two isolated pictures—of the world of nature and of the world of social production—but one integrated picture of the world of developing nature in the broadest sense of the word."[18]

But there is also another view, namely that Anuchin's concept does not promote the urgently needed strengthening of ties between the geographical sciences and the needs of the national economy, that it falsely drives practical activities into a passive adaptation to geographical conditions, and that it limits the sphere of production to those possibilities that are supposedly granted to man by the geographical environment in the course of its self-development. Nor does the concept promote the development of a theory of geography because it is founded on a false interpretation of the meaning of society and incorrectly views the interaction between society and nature. In addition, Anuchin tries to blot out the entire Soviet theory of economic geography, calling it dualistic and idealistic, and scornfully labeling its followers "locationists" and "followers of a split geography." He tries to make it appear as if the independent existence of economic and physical geography as sciences makes impossible their close cooperation in integrated geographic research. Anuchin holds that before him there was no correct conception of geography in the Soviet Union, that it existed only abroad, but that "the monistic concept of geography lacks serious theoretical foundation abroad and Soviet geographers are now only beginning to approach it from a 'new' philosophical viewpoint."[19] Therefore Anuchin sees his task in trying not only to introduce the monistic concept into Soviet geography but to provide for the first time in its history a materialistic approach to that science. In the foreword to his book, he says quite openly: "The content of this book was dictated by the author's desire to provide a monistic, materialistic approach to geographic science."[20]

Anuchin's pretensions to priority in introducing materialism into Soviet geography cannot but cause surprise. Soviet geographers have long been relying on the firm base of Marxist materialism, and they have always approached both a unified geography and the theory of geography in general from the viewpoint of these time-tested and certainly not new philosophical principles. Anuchin's new philosophical principles merely confuse the entire question of the theory of geography. Soviet geography has clearly established views on the problem of interaction between nature and society, on economic and physical geography, on the subject of economic geography, and on the character of the laws with which it is concerned. These views have emerged in the course of practical research and have provided a firm basis for the further development of the science and its practical applications. Some of the basic principles of the theory of Soviet economic geography and its relations with physical geography should also be recalled in the present paper.

The common aspect of the process of production of material goods for any period in history, for any social-economic structure is the interaction between society and nature, an interaction in which, as Karl Marx stated "the subject is mankind and the object is nature, and they are always the same."[21] However, the concrete content of that interaction and its character change substantially from period to period depending on changes in productive forces and in the productive relationships among people. Study of the specific content of the interaction between society and the surrounding environment in all its diversity and complexity is the principal object of the system of geographic sciences. That system is made up basically of physical geography and its related disciplines and of economic geography and its branches.

Physical geography belongs to the natural sciences. It studies nature with a view to transforming it from a thing in itself into a thing for us, that is, with a view to transforming it into a geographic environment, into a nature that can be utilized by man. Physical geography seeks new ways of intensifying man's supremacy over nature and provides a scientific foundation for the multipurpose transformation and utilization of nature in various natural regions and zones. Man has long been utilizing and transforming nature for his own purposes, but only scientific development makes it possible to estimate the long-range consequences of man's interference in the development of nature when immediately sought results are achieved.

In terms of its research goals, physical geography as a science is thus intimately linked with the practical aspects of economic development of society. But such close ties with economics and the necessity for taking account of economic developments and needs do not prevent physical geography from being a purely natural science. The laws of nature that are the object of study of physical geography are the same in all countries irrespective of their social system. And no matter how much society may transform nature, it will always remain nature, and it can be utilized and transformed further only on the basis of knowledge of its own laws.

Economic geography is a social science. The character of economic utilization of the geographic environment and of natural resources depends on the aim of such utilization and on the aim of production. These aims, in turn, vary with types of society and with social relationships. But if the aims of capitalist and socialist production vary, so do the methods of the one and the other. These differences appear mainly in the character of production relationships among people, and through them in the relationship of people toward the utilization of nature. Marx wrote: "In order to produce, people enter into certain ties and relationships, and only through these social ties and relationships are people related to nature, and does production take place."[22]

In any society the relationship between people and nature is expressed by means of productive forces. The productive forces of society and their level of development are a result of the striving of people to control the forces and materials of nature and are a measure of the degree of dominance of people over nature. In the hands of society, these forces are a tool with which society takes from nature all that is needed for existence and transforms these raw materials into products ready for consumption. If economic geography wants to trace the economic utilization of natural resources up to the consumption stage, it must therefore study in effect the utilization of productive forces in the process of producing material goods.

A clear distinction must be made between such concepts as *geographic environment* and *economic utilization of the geographic environment*, between *productive forces* and *utilization of productive forces*. Both in a capitalist and in a socialist country, the principles of organization, the capacity of machines, and the technical qualifications of workers may be identical. But fundamental differences arise when we speak of the utilization of machines themselves. Powder is powder, no matter whether it is used to wound a person or whether it is used to heal the wound of that person."[23] That is why anyone who maintains that economic geography should be concerned with the study of the geographic environment, or with some of its "social elements,"[24] or simply with productive forces, inevitably deprives economic geography of its social-class content. Economic geography is expected to be concerned with study of the utilization of productive forces in the process of utilization of the geographic environment and the production of material goods. And all economic questions of the application of productive forces in their relation to the geographic environment can be reduced in practice to the question of location of productive forces, their association (formation of territorial production complexes), and their interaction (intraregional, interregional, and international economic exchange and the geographical division of labor). In a socialist (planned) economy all three problems are intimately related: the site of an enterprise must not be selected without preliminary planning of how the new plant will be associated and will interact with other plants. In a capitalist economy, the location of individual plants is determined only from the point of view of the owner's interests; the formation of production complexes and the geographical division of labor arise in the haphazard fashion inherent in capitalism.

The substance of economic geography and its basic content can be defined as follows: economic geography is a social science concerned with the study of laws of location, association, and interaction of productive forces in the process of society's utilization of the geographic environment, that is, in the process of production of material goods, at various stages of development of human society.

Economic geography cannot exist outside of the sphere of study of the process of production, without firmly relying on knowledge of the economic laws of the development of production, or without a clear differentiation of the aims of social production in each given type of society. The practical task of economic geography is to justify economically the optimal location of the economy and its integrated development and to provide a link between concrete economics and the land, or a given territory. This means that economic geography cannot operate without knowledge of a territory and its natural resources, without close contact and interaction with physical geography. In a socialist economy, where scientific progress is expected to serve practical achievements, the interaction between economic and physical geography is far more essential than it is under capitalism. For a capitalist, in his economic activity, it is sufficient to know enough about nature to produce the desired immediate effect from its utilization, namely profit. Under socialism, the joint task of all geographic disciplines is to insure a system of economic management in which the contemporary utilization and transformation of nature in the process of production would have desirable natural and social consequences for future generations. And there can be no more solid and vital justification of the unity of the entire system of geographic sciences than this expectation of our social system.

All economic sciences regard the process of production primarily from the point of view of interrelationships between productive forces and production relationships. The specifics of economic geography lie in the fact that it links the process of production with a third element, the geographic environment. That environment, on the one hand, serves as the source of the objects of labor and raw materials used directly in the production process, and, on the other hand, provides the material conditions that do not enter directly into the production process but are essential to it, namely the territory in which production is located and operates. What are the relationships between these three elements that are the concern of economic geography? What is the character of these relationships? What role is assigned to the geographic environment?

Let us start with the last question (which is the one to which Anuchin seeks to find a new answer, differing from the one generally accepted in Soviet geography). Can the character of the geographic environment really determine the development of social production within a given territory? Yes, it can, but only in a negative sense: you cannot catch fish where there are no fish. But where there are fish, you may not fish at all or you may fish in one of several ways—wisely, rationally, or wastefully. Cuba, which is surrounded by seas abounding in fish, previously had no fishing industry, and the fishing industries of Japan and the Soviet Union, though located in the same seas of the Far East, differ sharply in terms of fish resources. The utilization or lack of utilization of any given natural conditions and the choice of ways of utilizing them depend entirely on society, on the state of social production, and on its aims.

Still, if society merely adapted itself to the natural environment, then the absence or presence of any given natural prerequisites might have a decisive and permanent effect on the economy and its specialization. The point is that in practice the aims of production always go far beyond what nature has to offer. Therefore contradictions always exist between society and nature, and they are solved in the process of creating new productive forces. If this is in accord with the economic aims of society, then it will seek and ultimately find ways of overcoming natural obstacles to realize its aims; it will itself create the conditions that were lacking in nature. There was a time when fish could not be caught where they were not provided by nature; now man has learned to breed fish, including the use of artificial reservoirs. In the seventeenth and eighteenth centuries, the level of transportation promoted rapid development of trade and capitalism only in the maritime countries of Europe; the natural conditions of Austria isolated it from the sea and from the rapidly developing countries and slowed the development of capitalism in it. The invention of the locomotive, the automobile, and the airplane put an end to the geographical causes of isolation. Nowadays, there can only be social-economic causes for the isolation of countries and regions, as, for example, colonialism and its consequences.[25]

The influence of the geographic enviroment, of course, leaves its imprint on the historical process of development of society, but it can never be decisive in that process and, moreover, depends on the development of society itself and on its productive forces. Those geographical conditions that may have a favorable or adverse effect on the development of society at one stage of development of productive forces may lose their significance at another stage. In view of the contemporary level of productive forces, it would be laughable to assert that the mountainous character of Peru or Honduras is to blame for the lag in capitalist development in those countries.[26]

The task of economic geography is not at all to establish the absence of a given set of conditions in the geographic environment and on that basis to draw conclusions regarding differences in the economies of, say, Honduras and Arkhangel'sk Oblast [of the northwestern section of the RSFSR]. The banal fact that bananas are raised in one of these areas and cannot yet be raised in the other was known long before the appearance of modern science. The geographical sciences are expected to provide a correct evaluation of the natural prerequisites in a given territory and to suggest ways for transforming nature with a view to achieving new possibilities of benefit to society; they are expected to justify economically a utilization of the natural potential that would be the most rational, the most comprehensive, and at the same time economically the most effective.

To carry out the tasks it now faces, economic geography must uncover those social laws that guide the location, association, and interaction of productive forces within a given territory. For example, Soviet economic geography bases itself in its research on the existence of an objective law that was uncovered by the classic writers of Marxism-Leninism. Its basic principles are as follows: In any given territory, in a given geographic environment, the character of utilization of productive forces, their location, association, and their interaction in the process of production of material goods always depend decisively on the form of ownership of the tools and means of production prevailing in that territory. And, inversely, in any geographic environment, a certain type of exploitation of natural resources corresponds to each given type of productive relationship and to each given form of ownership of the tools and means of production.[27]

Marx had the following to say about the role of ownership in the production process: "Any production represents the mastering by an individual of the objects of nature within and by means of a given social form. In that sense it would be a tautology to say that ownership (mastering) is a condition for production. . . . There can be no production, and thus no society, if there is no form of ownership."[28]

The forms of ownership determine the interests of the owner in the production process, whether it is a small peasant or a people's state, and determine the aims of production and its methods. Economic geography does not concern itself especially with study of the internal laws of development of production relationships, but takes them ready-made from political economy and, on the basis of the spatial distribution and changes of various types of production relationships, seeks to establish types of utilization of productive forces in each given territory.

The effect of these principles can be illustrated in the case of agriculture, a branch of the economy that, more than any other, is affected by natural influences. Nevertheless, the character of utilization of the land is always to a decisive degree dependent on the question of who owns the land and in whose interests it is used. In one and the same country, in one and the same region with identical natural conditions, one may find different forms of land ownership. And the utilization of a given piece of land by a small peasant will certainly differ from its utilization by a big landholder, and the latter will differ from utilization of the land by a capitalist. At the same time, in all parts of the country, and even in various countries, the use of the land by small peasants will be identical in its essential features and will belong to one and the same type; it will always be a consumer type of crop-raising designed to feed a family in which, year in and year out, one and the same food crop is planted on all or most of the land, whether it is rice after rice, or potatoes after potatoes, or manioc after manioc; on such landholdings, tillage practices will always be rudimentary, and the input of labor extremely high.

Nature undoubtedly plays a part in the choice of the specific crop or animal raised on a given landholding, but it never plays a decisive part: wherever crops can be raised, it is always possible to raise a wide range of crops, and that range increases with improved tillage methods, or the availability of hybrid varieties. Under these conditions, only American bourgeois geography could go so far as to say, for example, that the monoculture of sugar cane in Cuba was to be explained by the fact that the natural conditions of the island were most suitable for that crop. Revolutionary Cuba, after having carried through a land reform, quickly refuted that thesis: it has already demonstrated that it can grow excellent crops of rice, beans, tomatoes, cotton, henequen, and others. The share of sugar cane in the total crop area dropped from more than 58 percent in 1958 to 42 percent in 1963, and is supposed to be reduced to one-third under the 1965 plan.

Each social-economic formation, each type of economy follows distinct laws that determine the choice of crops within the limits permitted by nature at the given level of development of productive forces. For precapitalist forms of economy these laws would be the influence of local and ethnic traditions; for a capitalistic commercial economy, the influence of the market; for a socialist state economy, the influence of planning based on the most rational utilization of land resources for the production of all needed farm products at minimum labor input per unit of output.

Economic geography is a social science. Its genuine philosophical and methodological bases could be formulated only with the appearance of Marxism. The work of Marx, Engels, and Lenin laid a solid foundation for all social sciences, including economic geography, and continues to serve as a true directional beacon for Soviet economic geographers. For that reason alone, the materialistic outlook on geography in the Soviet Union requires no further justification. It does not mean that all the theoretical problems of economic geography and its integration with other geographic disciplines have been solved. Science cannot halt in its development because life itself constantly poses increasingly complex and responsible tasks. It is in the study of life, in close connection with practical aspects, that economic geography must gain its strength. Only under such conditions can it make a worthy contribution to the building of communism.

1. L. F. Il'ichev, *Obshchestvennyye nauki i kommunizm* [The Social Sciences and Communism](Moscow, 1963), p. 70.

2. V. A. Anuchin, *Teoreticheskiye problemy geografii* [Theoretical Problems of Geography](Moscow, 1960).

3. Ibid., p. 163.

4. D. Hooson, *Annals of the Association of American Geographers* 49 (March 1959):81.

5. V. A. Anuchin, *Teoreticheskiye problemy geografii*, pp. 108, 112.

6. *Annals of the Association of American Geographers* 52 (December 1962):469–75.

7. V. A. Anuchin, *Teoreticheskiye problemy geografii*, p. 132. If we were to carry this thought further, it would follow that there is no need for any revolutions to change society, all one has to do is change the geographic environment.

8. Exactly the same view of society and its laws was once advanced by A. Bogdanov in his *Osnovnyye elementy istoricheskogo vzglyada na prirodu* [Basic Elements of a Historical View of Nature] (St. Petersburg, 1899), p. 153, and was exhaustively criticized by V. I. Lenin in his "Materialism and Empiriocriticism" (*Sochineniya* [Works], 14:313–15). Society is an association of carriers of social relationships—workers, peasants, students, and others—and not of biological individuals. Of course, society cannot exist without living people, but neither the social origin nor the social position of man depends in any way on his qualities as an organism or as a biological individual. The existence of society and social life do not depend on individual persons, or on physiological processes in their organisms, or on their consciousness. On the other hand, the social system and the social position of man have a tremendous influence on his health and his biological life. Society follows its own law of development; to apply the laws of nature to it means not to understand the essence of society.

9. V. A. Anuchin, *Teoreticheskiye problemy geografii*, p. 150.

10. Ibid., p. 154 (Anuchin's italics).

11. Ibid, p. 183.

12. Ibid, p. 147.

13. Ibid, p. 183.

14. Ibid., p. 146 (Vol'skiy's italics).

15. Ibid., p. 128 (Anuchin's italics).

16. Ibid., p. 88 (Anuchin's italics).

17. Ibid., p. 136.

18. *Vestnik Moskovskogo Universiteta,* seriya geografiya [Herald of Moscow University, Geography series], no. 6 (1961), p. 8.

19. V. A. Anuchin, *Teoreticheskiye problemy geografii*, p. 112.

20. Ibid., p. 5.

21. Karl Marx and Friedrich Engels, *Sochineniya* [Works] vol. 12, pt. 1, p. 175.

22. Ibid., 2:81

23. Ibid., p. 24.

24. V. A. Anuchin, *Teoreticheskiye problemy geografii*, pp. 158–59.

25. Anuchin is of another opinion on that score; "Usually a lag in the development of individual countries is related to the effect of geographical conditions that isolate the given country from the rest of the world. It is precisely the effect of the geographical factor that helps to explain why there are still tribes in parts of the world that live at a stage of low or medium degree of barbarity or even wildness" (ibid., p. 149).

26. But this is precisely what Anuchin contends: "Individual countries, usually mountainous in character, may be slow in advancing from feudalism to capitalism when the geographic environment retards their progress" (ibid.).

27. They were already formulated in Soviet economic geography, notably by R. M. Kabo in *Voprosy Geografii* [Problems of Geography], no. 5 (1947), p. 16.

28. Marx and Engels, *Sochineniya* [Works], vol. 12, pt. 1, p. 177.

L. I. VASILEVSKIY

Some Possible Approaches to the Study Of The Territorial Structure of the Economy Using Mathematical Methods

Present-day Soviet economic geography, which we can already call classical (and whose methodology we tend to counterpose to the new mathematical methods) originated relatively recently in a struggle against the old commercial geography and against the statistical approach of V. E. Den (1867–1933). Since then, thanks to the work of N. N. Baranskiy and his followers, Soviet economic geography has grown and developed, achieving major theoretical and practical results and attaining maturity.

There is no foundation for the view found among scholars in other disciplines, who are unfamiliar with economic geography (particularly, among mathematicians, physicists, and some representatives of the engineering disciplines), that traditional economic geography, in addition to its descriptive and qualitative aspects, also makes wide use of classifications of phenomena and processes, frequently involving quantitative gradations; it also seeks to reveal and study structural characteristics, again using quantitative criteria. Such divisions of economic geography as the theory of economic regionalization, economic cartography, urban and population

Reprinted, with modifications, from *Soviet Geography: Review and Translation* 7, no. 4 (April 1966): 38–49, by permission of the publisher. The article originally appeared in *Kolichestvennyyee metody issledovaniya v ekonomicheskoy geografii* [Quantitative Methods of Research in Economic Geography] (Moscow, 1964), pp. 8-29.

geography, the geography of industry, and transportation have advanced far beyond the stage of the descriptive disciplines and have reached the level of the nomothetic disciplines, which study causal relationships and the laws of development of phenomena. Nevertheless, economic geography as a whole, from the point of view of its research method, is still closer to the humanities than to the exact sciences.

Economic geography in the West, including so-called regional science, has also progressed. It is quite clear that bourgeois economic geography, by not relying on the methodology of Marxism-Leninism, is unable to provide broad theoretical generalizations and to solve fundamental problems. But Western economic geography and regional science have made substantial progress in the study of particular phenomena and processes, largely through the use of new quantitative research methods. These results could not have been obtained either by the statistical-descriptive methods of the old commercial geography or by the humanistic descriptive-typological methods of regional geography.

Soviet economic geography undoubtedly stands above Western bourgeois economic geography in its initial methodological positions; we have long overcome the pitfalls of the old commercial geography, whereas Western economic geography still suffers from the survivals of that approach. Nevertheless, we are lagging behind in individual areas, particularly in the application of the new mathematical methods in economic geography, which is still in its infancy in the Soviet Union, although there are some studies on individual problems that may even be superior to similar work abroad.

Quite naturally, many economic geographers have lately shown a desire to enrich economic geography with new, more effective, and more exact research methods, bringing it closer to the exact sciences. This desire has been strengthened in part by the experience of neighboring sciences, such as biology and linguistics, which have been applying exact methods, and especially by geography's nearest neighbor, economics, which adopted this approach earlier and has now moved ahead of geography in that respect. They have laid the basis of a mathematized economic science (mathematical economics or econometrics, as it is often called) and have used the new methodology in a number of concrete problems in various fields of research (for example, in the problem of interindustry balances).

On the other hand, the logical course of the development of geography inevitably leads to a search for mathematical research methods and, in general, to an enrichment of the methodological arsenal of economic geography. Starting in the 1920s, and then in the 1930s and after World War II, a number of attempts were made in that direction. For various reasons they did not bear fruit: partly because of dogmatic prohibitions in the days of Stalin's personality cult, partly because of the immaturity of the attempts themselves. Nevertheless, work continued along these lines, largely on the initiative of a few scholars. These early attempts included the centrography school, which, despite the naivete and immaturity of its theoretical basis, did make a certain contribution to the study of the territorial structure of the economy and population and showed that it was possible to represent at least some aspects of the distribution of many economic phenomena by means of simple physical-geographical models.

Now, the situation has apparently changed radically. The greatest obstacle has become the inadequate training of economic geographers, and perhaps also the natural inertia of old traditions and habits. Familiarity with the experience and achievements of past work in this area could help overcome that inertia. Therefore, in addition to working out new applications for mathematical methods in economic geography, it would be beneficial to collect, reassess critically, and generalize the past efforts of Soviet scholars in that area in the past. In view of the generally unfavorable attitude toward new methods (especially mathematical) in economics and economic geography before the Twentieth Congress of the Soviet Communist Party, many scholars were unable to publish their work or even bring it to the attention of the scientific community. Being ignorant of what scholars in neighboring disciplines were doing and unfamiliar with work under way abroad, these scholars in many cases "discovered" what was already known and needlessly duplicated parallel research; nevertheless, their work apparently contained a number of original aspects. Only careful study will show which part of their work is of limited historical interest and which part is capable of enriching geography and deserves to be continued and developed.

What is it that does not satisfy us, or does not fully satisfy us, in the present methodology of economic geography? We must, first of all, distinguish two closely related, but not identical, concepts: quantitative research methods and mathematical research methods.

We know that quantitative indicators were also widely used in classical geography to distinguish classificatory gradations and to characterize types of phenomena and processes. For example, on the basis of the relationships among the principal branches of the economy in the total volume of production or in the total employment and by using quantitative statistical indicators, we can divide countries and regions into these categories: industrial, industrial-agrarian, agrarian-industrial, and agrarian. Similarly, quantitative gradations have been used to distinguish types of populated places and the like. But all this does not constitute an application of mathematical methods. Cause-and-effect relationships in the usual economic-geographic study are established largely in *qualitative* terms and are not raised to the level of *quantitative* laws; therefore, they have not been expressed, even approximately, by mathematical functions, which can be represented analytically by formulas and graphically by nomograms. The relative importance and influence of various factors has also been evaluated qualitatively, whereas the quantitative aspect has been expressed in vague terms (major, important, or, on the other hand, secondary, or associated, causes or factors). The use of such imprecise terminology still distinguishes economic geography from scientific work not only in physics, astronomy, and chemistry but also in industrial engineering-economic research in which the economic and engineering sciences find a meeting ground in deriving quantitative (even if approximate or empirically derived) laws and relationships that can be expressed by formulas and nomograms.

Mathematics is now penetrating many disciplines, even such fields as biology and linguistics, which previously seldom carried their research on cause and effect to the stage of establishing quantitative relationships, patterns, or scientific laws that could be expressed by mathematical formulas (similar to the laws of physics, astronomy, or chemistry). There is no reason to regard economic geography as less prepared at present for the introduction of mathematical methods than is industrial economics. There is no doubt that in economic geography, too, the study of scientific laws and cause-and-effect relationships can be carried to the stage of establishing quantitative relationships that can be expressed by graphs, empirical formulas, and, in some cases, by analytical formulas. Quantitative laws in economic geography could be applied not only in explaining the existing territorial structures of the economy, analyzing developmental trends, and making relevant forecasts but also in finding optimal solutions for economic location, regionalization, and other problems of the territorial organization of the economy.

When we speak of mathematical methods and, more broadly, of quantitative methods or even new methods in general, we mean mathematics in the widest sense, with all its new ramifications, including mathematical logic and even information theory.

Economic geography, theoretical economics, and applied industrial economics are closely interrelated and interwoven. It may seem best, therefore, to apply to economic geography all that has been done in the field of industrial economics, for example, the use of linear programming to solve planning problems of the location of new enterprises or of the new production capacity of existing enterprises. Such a mechanical transplanting into economic geography of methods used in industrial economics may be of some use. These methods, however, are unsuitable for research in fundamental problems of economic geography, in basic questions of the territorial structure and the territorial organization of the economy, which are specifically geographic in character and distinguish economic geography from pure economics. Therefore, such a transplantation would not solve the problem.

Economic geography will evidently have to find its own ways of applying mathematics, information theory, and the like, not only by borrowing and critically reworking, but by adding to the experience of mathematization of other disciplines (and not only of economics).

In the study of capitalist countries, we are concerned with the existing territorial structure of the economy, its processes of formation, and tendencies of future development. In the study of the economic geography of socialist countries, we are, in addition, concerned with problems of a constructive character. We are looking for optimal solutions in the area of the organization of an economic territory for purposes of short-term or long-term planning, the location of individual branches of production, the specialization of economic regions for planning purposes, and also for industrial-design solutions. To a certain extent, similar problems may be faced by Soviet economic geographers working on the developing nations, where such studies may be of practical value to agencies giving economic and technical assistance to these nations.

The present article does not pretend to be a comprehensive survey of what has been done in the Soviet Union and abroad in the area of application of mathematical methods to problems in economic geography. It is not intended to be an informational paper. Its aim is to offer for discussion some ideas and thoughts on possible directions of application of the new methods in economic geography, generalizing the experience in this field, and some results of completed studies and plans for future work.

If we want to use mathematical research methods in economic geography, we must first find the "transmission belts" from the mathematical apparatus to economic-geographic data. Any application of mathematics to the specific material in any field requires preliminary formalization of that material. Mathematical research methods are, after all, a kind of mechanization of reasoning.

The raw material of economic geography cannot be processed directly by the powerful, but strictly formalized, mathematical apparatus. Therefore, the raw factual and numerical data must first be processed by the usual methods of economic-geographic research, mainly selection and generalization. In contrast to the raw material, not only the ultimate findings of classical economic-geographic research, but also the intermediate results obtained at various stages of the selection, generalization, and abstracting process evidently lend themselves well to the formalization needed for the application of mathematical methods.

The following objects of economic-geographic research lend themselves particularly well to the application of the new methods.

1. *Laws of geographic location* (both static and developmental) of various territorial-economic parameters (especially, dasymetric [from the word dasymeter, used in early physics experiments] parameters, that is, indicators of crowding and density) and their relationships. Among the principal dasymetric parameters are: density of population, manpower resources, and labor inputs; the density of national wealth and national income (or just the consumed part of that income); the density of aggregate production and of the output of individual branches of the economy (in value, weight, or other physical indicators).

In addition to dasymetric parameters, there are other important parameters of the territorial structure of the economy that change from place to place; they can be studied by means of the same mathematical methods used for the study of density indicators. They are:

a. Local structural parameters (for example, indicators of population composition; the industrial distribution of output or employment in industrial centers; the composition of crops [by sown area or by harvest] and of livestock by minor civil division or of industrial enterprises in an industrial center);

b. Local parameters of geographic concentration, from the simplest—the percentage share or rank order of each place in terms of an absolute indicator (for example, the output of a certain branch of industry)—to the complex indicators of geographic concentration used by Florence, Isard, and others;[1]

c. Local dynamic indicators (for example, rates of growth of population, industrial output, and income in various places);

d. Local parameters of intensity and potential. These include primarily level indicators (of prices, costs, labor productivity, per capita income, and population mobility). Some can be modeled by a potential field in the physical sense, which makes it possible to use the mathematical apparatus of the relevant branches of physics.

A distinctive group of indicators are the parameters of potential in the economic-geographic meaning of that term; the size of such an indicator at every point expresses the aggregate influence of a particular economic phenomenon distributed over the entire territory, in relation to the distance to the particular point. For example, the demographic potential (population potential) introduced by Stewart of the jth point of a territory is expressed by the formula

$$V_j = \frac{\sum_i P_i}{D_{ij}}$$

[This equation should read,

$$V_j = \sum_{i=1}^{n} \frac{P_i}{D_{ij}}$$

where $j \neq i$. Eds.], where P_i is the population of the ith point and D_{ij} is the distance from the ith point to the jth point.[2]

2. *Territorial-economic relationships and interactions* of various types: international, interregional, and interpoint. They include freight flows, commodity flows in value terms, product flows in terms of various physical measures[3] (for example, energy flows in tons of conventional fuel, megacalories, or kilowatt hours), information flows, population migrations, passenger flows, and financial, and credit relationships.

The formulation of a mathematical theory of transport-economic relationships and other territorial-economic relationships would provide a new approach to the study of the geographical division of labor (international, interregional, and intraregional). At the present time, we have found ways (and that but recently and in preliminary form) to make only a quantitative estimate of the intensity of the territorial division of labor. In addition to pure statistical relationships that depend on the system of regionalization employed (the share of output produced for export, and the share of imported goods in the aggregate consumption of a region or place), we can now use indicators that are unaffected by the regional system and its boundaries. They include the geographical transfer distance [l_r (weight) and l_r (value)], that is, the average distance between points of production and consumption (calculated per ton of output and per unit of value, respectively) and its relationship to the linear dimensions of the study area $\delta = l_r / k \sqrt{S}$, where S is the effective area of the freight-exchange territory, k is a coefficient that depends on the configuration of the territory, and l_r is defined by the formulas:

$$l_Q = l_\tau \text{ (weight)} = \frac{\Sigma ql(1-i)}{\lambda \Sigma Q}$$

$$l_c = l_\tau \text{(value)} \quad \frac{\Sigma cql(1-i_c)}{\lambda_c \Sigma CQ}$$

where q is the weight of the freight, c is the value of one ton of the freight, l is the hauling distance, Q is the weight of the output [in tons], C is the value of one ton of output, i and i_c are the percentage of nonrational hauls in the aggregate freight transfer, expressed in ton-kilometers or in ruble-kilometers, respectively; λ and λ_c are the mean excess of the transfer distance compared with the orthodromic distance calculated per ton or per ruble of freight moved, respectively.

However, the coeffcient of intensity of the territorial division of labor thus calculated is a simple scalar; it reflects only one aspect of this complex and many-faceted phenomenon. After all, for a given mean intensity of the territorial division of labor, it may be significant to know precisely which regions and places are closely connected with one another by transport-economic reglationships, resulting from the division of labor; to what extent these relationships are symmetrical, and what the commodity composition of the connecting freight flows is. In other words, what is needed is not only the mean intensity of the territorial division of labor but the direction and the composition of the flows, that is, a complete description of the complex anisotropic fabric of transport-economic relationships, its spatial and commodity structure. We do not yet know how to express (or study) that structure mathematically, and we are forced to resort to verbal descriptions, enumerations, and bulky tables of numerical data (matrices of relationships between regions and between points for each type of product). Even sadder is the fact that we are still unable to determine the economic effect of the geographical division of labor except in extremely rough and unsatisfactory form.

3. *Local and territorial complexes* (places, microregions, economic regions of various orders, countries and groups of countries.

4. *Categories of economic-geographic situation, the distance factor* (in units of distance or travel time) and other major categories in economic geography.

5. *Classificatory, genetic, and structural systems* of varying content. Such systems can be investigated, in particular, by the methods of one of the new mathematical sciences, such as graph theory.

6. *Research methods and techniques of traditional economic geography*, for example, the methodology of economic regionalization.

7. *Models of territorial-economic complexes and "economic spaces"* as a whole.

One of the major areas of economic geography in which the new mathematical methods can be applied is economic regionalization. The major and most difficult task here is to work out an algorithm for establishing the boundaries of economic regions by applying a system of objective criteria to a sufficiently rich body of diversified data, that is, boundaries that would reveal to a maximum extent objectively existing territorial-economic complexes.

Of great usefulness would be the creation of a mathematical theory of selection and generalization. Selection and generalization still constitute the basic method for the economic geographer who is working with concrete factual data, not only in mapping but also in writing textual descriptions and characterizations. As N. N. Baranskiy once convincingly demonstrated, identical principles underlie cartographic generalization and the generalization of textual description.[4] Selection and generalization represent an essential preliminary stage in the processing of any geographic material and in its preparation for analysis, which is intended to establish causal relationships and patterns. However, despite the great contributions made by a number of Soviet geographers to the theory of selection and generalization, the application of these theories remains to a large extent an "art." The creation of a logical-mathematical theory of selection and generalization would make it possible to formalize this processing of geographic data to such an extent that it may be expressed by an algorithm suitable for use in computers.

The new methods of research in the territorial structure of the economy fall into the following basic approaches.

1. *Statistical-analytical methods.* These are used to establish the existence of interrelationships or cause-and-effect relationships between phenomena by establishing a correlation between statistical series.

A correlation of time series makes it possible to establish a direct causal relationship between the phenomena under study, or at least the existence of common factors producing similar changes. But far more important in the analysis of economic-geographic data (and also far more difficult) is the establishment of a correlation of spatial series. Among the statistical-analytical methods is factor analysis, which can be considered transitional to the following group of methods.

2. *Methods of functional analysis of empirical relationships.* On the basis of empirical statistical series, these methods make it possible not only to establish the existence of correlations but to find mathematical functions that approximate the interconnections and causal relationships among the phenomena and processes under study (these functions may be expressed analytically or graphically).

3. *Economic-analytical methods.* In these methods, a mathematical form of functional relationship is given to patterns and cause-and-effect relationships that have already been established in qualitative terms on the basis of economic and geographic reasoning and the logical analysis of the essence of the phenomena under study. On the basis of the character of these phenomena and an understanding of the mechanism of their operation, one can then derive quantitative laws that may be expressed in analytical terms. These deductively established quantitative laws are then simply tested with statistical data.

4. *Methods of finding invariants, principles of conservation, and monotonic sequences.* An example of an invariant indicator is the above-mentioned transport-geography parameter, the mean radius of transport-economic links, that is, the mean distance between places of production and places of consumption. This indicator (which has been worked out by the author and calculated for the USSR and several capitalist countries) also characterizes indirectly the level of the geographic division of labor independent of the network of economic regionalization employed or the boundaries of the economic regions.

5. *Methods of physical models* of economic-geographic phenomena and processes. A particular case of physical model-building is geometrical model-building. An example of the use of this method is the above-mentioned modeling of geographic differences in price levels or cost levels by means of a potential field, which is used in some studies in mathematical economics.

In addition to the physical concept of potential, attempts have been made to apply related physical concepts to economic-geographic research. For example, Stewart,[5] in addition to the demographic potential

$$V_j = \frac{\sum\limits_{i} P_i}{D_{ij}}$$

[Again, this should be

$$V_j = \frac{\sum\limits_{i=1} P_i}{D_{ij},}$$

where $i \neq j$. Eds.] introduced the concept of demographic force

$$F = G \ \frac{P_i P_j}{D_{ij}^2}$$

and of demographic energy (the energy of demographic interaction)

$$E = G \ \frac{P_i P_j}{D_{ij}}$$

[In these formulations G is a constant often expressed as K; P_i and P_j are population points i and j respectively; and D_{ij} is the distance between points i and j. Eds.]

The method of physical modeling has its strong and weak sides. The use of physical models of economic and economic-geographic phenomena and processes does have the advantage of using the rich mathematical apparatus of physics for work in economic geography. The main shortcoming of this method lies in the fact that the model reflects reality only within certain limits. Beyond those limits the analogy is incomplete, and the use of models may yield incorrect conclusions in the economic-geographic problem under investigation. This shortcoming is true to an even greater extent in the geometric modeling of geographic phenomena. A good example of this method can be found in centrography, an approach that has been undeservedly forgotten in the Soviet Union in the last few decades.

The idea of the centrographic method originated with Mendeleyev; it was further elaborated by Svyatlovskiy and other Russians and achieved noteworthy results in the 1920s and 1930s. Subsequently, work was halted and almost completely forgotten. At the same time the centrographic method became widespread in the United States and other Western countries, where it was used by economic geographers, demographers, and statisticians, and applied in economic and demographic cartography.[6] Despite the primitive character of the centrographic method in its original form, it may deserve further study involving application of the mathematical apparatus.

6. *Application of the theory of graphs*, and especially of a new mathematical apparatus that is still in the process of being developed, the study of graphoids.

A graph is, in descriptive terms, a collection of objects that are represented in a figure by points (vertices of the graph), some of which are connected by line segments that may be nondirected (edges) or directed (arcs). No significance is attached to the position of the vertices or to the form of the line segments; it is only essential to know which vertices are connected with one another.

Strictly speaking, a graph G represents the mapping of a certain set "into itself" (as mathematicians put it):

$$G = (X, \Gamma),$$

where X is a set of elements x_1, x_2, x_3, . . . , x_n, and Γ is a function mapping these elements into the set, that is, the association (not necessarily single-valued) between each element x; and a certain well-defined subset of the same set X. The subset may include one or more elements, or, in a particular case, no element at all (the so-called empty subset). Further,

$$\Gamma X = \overset{n}{\underset{i=1}{\cup}} \Gamma x_i$$

where the symbol $\cup$ stands for the union of all subsets Γx_i, from $i = 1$ to n.[7] The theory of graphs has already found a number of applications. In particular, it has been proven that with its help one can solve problems of optimization (including the classical transport problem) more simply and better than is done by the use of linear programming of the simplex method.[8]

A special, and very simple, type of graph (the so-called tree) is represented by all classification schemes, genealogical trees, and systems of administration and control. This allows application of graph theory to logical schemes and classifications in a wide range of disciplines.

Any transport network (considered, of course, apart from the concrete qualitative and quantitative characteristics of the various nodes and line sections of the transport routes) can also be represented as a graph. The theory of graphs can be used to solve problems on a schematic map of a road network. Other map diagrams are also susceptible to study by means of graph theory. This opens tempting prospects for the practical application of the theory of graphs in economic-geographic research through the use of maps.

Unfortunately the number of theorems in graph theory (that is, studied properties and relationships) is still quite limited and, what is most important, the methods used in the study of graphs and in the solution of various applied problems vary in quality and do not yet constitute a single rigorous theory.

On the basis (and within the framework) of graph theory, a distinctive method has begun to take shape: the study and application of marked graphs or graphoids. Graphoids (the term is not widely accepted, but it is convenient) are graphs in which the elements (vertices and/or arcs) are characterized, in addition to their basic property (their connections in the system of the graph), by a parameter, that is, a quantitative and/or qualitative value. All theorems of the theory of graphs also apply to graphoids; moreover, graphoids, in view of their extraordinary flexibility, are very suitable for practical application in a wide range of disciplines. The prospects for the use of graphoids are especially promising in economic geography since they are easily applied to the economic geographer's principal tool, the economic map. It may be hoped that the application of graphoids will become an effective and fruitful method for solving both research problems and applied problems (planning and design) in economic geography.

7. *Cartometric methods.* These methods correspond best of all to the specific character of economic-geographic research (in contrast to industrial economics) and may play a key role in the mathematization of economic geography.

Cartometry is already being widely used in physical geography. The possibilities of its application in economic geography are apparently even greater. However, thus far only negligible use has been made of this method. Economic maps are being used primarily for illustrative, reference, and teaching purposes. Maps still play a subordinate role in economic-geographic research: they serve largely as heuristic aids, helping to form intuitive conclusions. Economic maps are rarely used for cartometric purposes and, in any case, are still far from becoming an essential tool for the study of basic problems.

The use of economic maps as precise research tools using cartometric methods requires the creation of new types of maps, including geonomograms (map nomograms) and a general enrichment of economic cartography with more precise and flexible methods of mapping the territorial structure of the economy.

Among these new types of economic maps (similar to a map nomogram) are the maps using varied value [variavalent] projections, worked out by the present author. In these maps, the mapped economic indicator is introduced in the actual map projection (that is, into the mathematical method used to map a territory). For example, in the equidemographic [ekvidemicheskoy] projection, the area of any section of a mapped territory is proportional to its population; the population of any administrative, physical-geographic, or economic region within any arbitrarily assigned set of boundaries can be easily determined on the map with a planimeter, and all other mapped economic indicators can thus be regarded as related to population rather than territory.

Inadequate use is also being made for research purposes of other well-known mapping methods: the isopleth (isoline) method, transplanted to economic cartography from physical-geographic cartography, the dot-map method, and so forth.[9]

8. *The application of self-programming (self-testing) computers* in economic geography. This is a new, independent, and highly promising area of the science of the future. There is no longer any doubt that such devices can be created in principle. However, tremendous technical problems remain to be overcome. The use of self-programming computers for research purposes may mark a new stage in the development of many sciences.

The time within which self-teaching computers suitable for research purposes may be created depends on the rate of advance of technology. It is up to the economic geographers to be prepared to have their problems ready in a form suitable for study by self-programming computers, when these become available for research use.

1. P. S. Florence, "Measures of Industrial Distribution," in *Industrial Location and Natural Resources* (Washington, 1943); H. H. McCarty, *The Measurement of Association in Industrial Geography* (Iowa City, 1956); W. Isard, *Methods of Regional Analysis* (New York, 1960).

2. O. D. Duncan, R. P. Cuzzort, and B. Duncan, *Statistical Geography: Problems in Analyzing Areal Data* (Glencoe, Illinois, 1961), pp. 82–83.

3. Problems of the general mathematical theory of flows and relationships using physical models are discussed in W. Bunge, *Theoretical Geography* (Lund, Sweden, 1962).

4. N. N. Baranskiy, "Generalization in Cartography and in Geographic Textual Description," in his book *Ekonomicheskaya geografiya. Ekonomicheskaya kartografiya* [Economic Geography and Economic Cartography] (Moscow, 1958), pp. 219–317.

5. J. Q. Stewart, and W. Warntz, "Physics of Population Distribution," *Journal of Regional Science* 1 (1958): 99–123; see also Isard, *Methods of Regional Analysis*.

6. Duncan et al., *Statistical Geography*, pp. 81–82.

7. Claude Berge, *Theory of Graphs and Its Applications*, Russian edition (Moscow, 1962).

8. Ibid. Appendixes 1 and 2, pages 250–60 and 261–69, and also additions to the Russian edition, pp. 291–92.

9. Several possibilities of using cartometric methods in economic geography are discussed in Bunge, *Theoretical Geography*, especially in the chapter on "metacartography."

SECTION II

Economic Regionalization

Introductory Note

One of the most insistent and important themes in Soviet economic geography is that of economic regionalization. Even the most cursory perusal of the geographical literature in the Soviet Union would indicate that regionalization can be construed as the basis or core of Soviet economic geography.[1] In fact, one of the taxonomic divisions most frequently cited by Soviet geographers is that of general and regional geography in which the former corresponds closely with the systematic subdivisions in the Western sense, and the latter centers on questions of regionalization and the study of separate areal units and economic regions. However, Konstantinov states in his essay on economic geography that although "both divisions are of equal significance, . . . actually, in this country chief attention is paid to the regional division."[2]

Such a pivotal role for regionalization and the region in Soviet geography is not suprising given the size of the country, its economic, cultural, and physical diversity, and the centralized political and economic system. A regional system in this context is considered essential in the location and management of industry and agriculture and particularly as a tool in the planning of the national economy and the economy of separate areas of the country. The history of regionalization schemes can be traced to the earliest years of the Soviet regime when the GOELRO (State Electrification) Plan distinguished eight economic regions for planning purposes. After GOELRO the responsibility for regionalization rested with the most important planning body in the country, Gosplan, which delimited a twenty-one-region scheme in 1921 and has been revising the regional system since that time.

Compared to Western geography, the Soviets not only place greater emphasis on regionalization and the region but also explain and define these concepts differently. The region is considered an actual (the term concrete is used most often in Soviet terminology) existing territorial unit and the process of regionalization discovers the objectively existing system of regions in the country. This concept of the region is at variance with the common Western notion that regions are mental constructs. In the Soviet context the core of the *economic region* is the territorial-production complex, a concept first put forward by N. N. Kolosovskiy.[3] A territorial-production complex is defined as an association of mutually complimentary and economically related activities within a given area. Further, the process of regionalization discovers the existing or potential territorial-production complexes, which because of their characteristics, interrelationships, and quantitative parameters lead to the delimitation of the existing network of economic regions.

The concern with economic regionalization in Soviet geography has led to the development of a set of methodological principles to be adhered to in such research. These principles include:

    1. the objective character of the existence of regions as living realities (including the existence of a hierarchial system of regions);
    2. the unity of the regional and sectoral structure of the economy (an economic region must contain a number of economic sectors and the location of any sector is nothing but a distribution of enterprises by regions);
    3. the existence of a close relationship between the regional concept and the socialist mode of production;
    4. the importance of the long-range aspects and the constructive character of economic regionalization;
    5. the importance of economic geographic *processes* in regionalization;
    6. the close relationship between economic regionalization and the administrative-territorial regionalization of the country.[4]

In summary, the concept of region and the method of regionalization is an essential part of Soviet geography, which is closely related to almost all geographic research in the Soviet Union. Much of the work by specialists in transportation, urban, population, industrial resources and agricultural geography is directed toward the ultimate goal of discovering the correct system of economic regions in the country. Thus an understanding of the method and principles of regionalization is essential in order to understand and follow current trends in Soviet economic geography.

The five articles in this section have been selected so as to present a developmental view of the work in Soviet economic regionalization. The section begins with Kolosovskiy's classic article on territorial-production complexes and ends with a rather sophisticated formal analysis of the regional concept by Rodoman.

In the initial article Kolosovskiy establishes the precise meaning of the term territorial-production complex and demonstrates its applicability in delimiting economic regions. He distinguishes types of production processes that utilize related types of raw materials and power which, in turn, allows generalizations to be made about production processes. Eight distinct cycles are identified, ranging from a pyrometallurgical cycle of ferrous metals to the hydro-reclamational industrial agrarian cycle. The remainder of the study centers on a comparison of regionalization methods using first the (then) traditional system utilized by Gosplan and then the production-cycle method.

The article by Saushkin and Kalashnikova, "Basic Economic Regions of the USSR," serves a dual purpose in that it applies the territorial-production-complex principle and provides a very useful description of economic regions of the USSR. It should also be noted that the article was in part inspired by the creation of the economic-administrative regionalization scheme (sovnarkhozy) produced as a result of the 1957 economic reforms in the USSR. (The sovnarkhozy were abolished in 1965 when a modified version of the sectoral-ministerial system was established.)[5] The authors note that the administration of industry and construction under the 1957 reforms was related to administrative-political units that were created as a *convenience* for administration and transport rather than on the basis of production complexes. They then delimit a system of 29 basic economic regions and groups (the latter being combinations of union republics and labelled groups in order to preserve their autonomy) and proceed to examine and justify each member of the system.

The third article, by Varlamov, "Geographic Features of Territorial Ties Between the Industrial-Territorial Complexes of the Western and Eastern Regions of the USSR," is a study of spatial interaction that demonstrates the close relationship of transport geography and economic regionalization. In his study of interregional and interzonal ties between the European and Asiatic sections of the USSR the author argues the greater value of the regional rather than the sectoral approach. The regional approach allows an examination not only of freight structures in terms of surpluses or deficiencies but also the connectivity of the regions.

The article by Budtolayev, Novikov, and Saushkin, "On Methods of Drawing Up a Territorial (Spatial) Model of the National Economy of the USSR," is concerned with a spatial model of the economy in the form of a chain of connections between production and consumption that would greatly improve a sectoral or branch model. In essence, two descriptive models are put forth. The first is a structural model that illustrates the economic links beginning with nature and the exploitation of resources to the final process of consumption. The second is a parallel, spatial model in which the country is divided into six economic districts. The economic indices of these districts are examined in great detail, including such factors as the *economic distance* between regions. As the authors point out, the major limitation of the spatial model is the lack of refined data to analyze the interregional chain reactions.

The last selection by Rodoman, "Mathematical Aspects of the Formalization of Regional Geographic Characteristics," represents the most recent trend in Soviet geography, the application of mathematical concepts and methods. The author attempts to demonstrate the importance of applying such mathematical concepts in the characterization of areas and the process of regionalization. The utilization of set theory, operations research notions, and graph theory are discussed and examples are given.

1. See for example, Yu. G. Saushkin, "A History of Soviet Economic Geography," *Soviet Geography: Review and Translation* 7, no. 8 (October 1966):3–104; and V. V. Pokshishevskiy, "Economic Regionalization of the USSR," *Soviet Geography: Review and Translation* 7, no. 5 (May 1966):4–32.

2. *Soviet Geography: Accomplishments and Tasks*, trans. Lawrence Ecker, and ed. Chauncy D. Harris, American Geographical Society Occasional Paper, no. 1 (New York, 1962), p. 32.

3. N. N. Kolosovskiy, "On the Question of Economic Regionalization," *Problemy Ekonomiki* [Economic Problems], no. 1 (1941); idem, "The Territorial-Production Complex in Soviet Economic Geography," *Voprosy Geografii* [Problems of Geography], no. 6 (1947), included as the first selection in this section; idem, *Osnovy eknomicheskogo rayonirovaniya* [Fundamentals of Economic Regionalization] (Moscow, 1958).

4. V. V. Pokshishevskiy, "Economic Regionalization of the USSR," p. 19.

5. For a discussion of the economic-administrative regional scheme, see Z. Mieczkowski, "The Economic Administrative Regions in the USSR," *Tijdschrift Voor Econ. en Soc. Geografie*, July–August 1967, pp. 209–19.

N. N. KOLOSOVSKIY

The Territorial-Production

Combination (Complex) in Soviet

Economic Geography

INTRODUCTION

At the beginning of the great work of reconstruction, that is, during the First Five-Year Plan, a new term, the production complex, reflecting a new phenomenon in our construction, entered into Soviet economic-geographic literature.

Still earlier, in the Gosplan [State Planning Committee] work on the economic regionalization of Soviet Russia, the notion of the region as a production combine became current. The latter term, in the author's intention, was to designate a tightly knit (technical-economic) combination of the various branches of the economy in regions primarily focused around electric power. However, this term did not prove very satisfactory for designating territorial-production combinations arising in Soviet reality, and did not gain a foothold, partly, it appears, because of its insufficient flexibility in capturing the various nuances of economic relations, which by nature are quite different from a combine. But in a restricted sense, the term com-

Reprinted, with extensive modifications and alterations, from the translation by Lawrence Ecker appearing in the *Journal of Regional Science* 3, no. 1(1961):1–25, by permission of the publisher.

bine has nevertheless become current, though not for every combination of lines of production.

It must be considered that these circumstances were the cause of the appearance of the term production complex. It is still used to designate groupings of industries, both those related by a common territory and those not related in this way. One speaks of a complex of ferrous-metallurgy industries, regardless of the location of the individual branches of this complex; of a complex of all the industries in the vast territory of the Ural economic region; or of the Novo-Tagil complex or the Lys'va-Chusovoy complex, located on more restricted territories.

From the very beginning, the exact meaning of the term was not established, and indeed it was hardly possible to do so because of the novelty of the planning and construction of interrelated groups of various industries, and because of the novelty of the economic and technical problems that arose. Hence, it is natural that different authors came to use the term to designate different concepts. By territorial-production complexes they came to understand, on the one hand, groups of enterprises organically related to one another, and on the other, groupings that had almost no visible production ties with each other except a common territory or the exploitation of the same natural conditions and common services, for example, the possibilities of having a common supply of water or electricity or a common labor supply source. They came to speak of a complex of branches of the natural economy or a complex of productive forces along with the combined use of raw materials and power. Sometimes it is difficult to determine from the context what a given author means by such terminology. In a word, just as sometimes it is hard to distinguish the inscriptions on a coin worn smooth from long circulation, even if they existed earlier, so too the terms *complex* and *integration* do not now possess the proper degree of precision.

At the same time, the knowledge and experience in socialist construction and operation of enterprises have grown considerably, making it necessary to distinguish with sufficient precision all the diverse forms of cause-and-effect relationships and interdependence for essentially different territorial-production groupings. From this arises the need, especially in the scientific literature, for a more precise use of words.

THE COMBINATION OF PLANTS AND ENTERPRISES

In my article, "On the Question of Economic Regionalization,"[1] an attempt was made to define the cognate concepts of the socialist combine and the complex.

What is meant by the production combine is a [tightly knit][2] production combination or union of technological or energy processes in which two or more kinds of useful products [or results of labor] are obtained from one and the same material. In comparison with specialized production processes, such a combination yields a definite savings in raw materials, energy, labor and transportation, and a reduction of production losses. The simplest examples of combination of production are the simultaneous production of steam and hot water for heating processes and of electricity in the central thermoelectric systems, and the Orkla method used to recover copper and elemental sulfur at the same time.

This definition contains a number of inaccuracies, and it remains incomplete as it leaves out important questions of the combination of processes not only in one plant but also in several plants and enterprises located in the immediate vicinity of one another in the same economic region. To develop the above, let us add that the chief feature of a combination (association) is the appearance of an additional (or the result of labor) product simultaneously in material and in value form.

The combining of specialized plants and enterprises, both those located very near to one another and those farther apart, is usually associated with the installation of intricate transportation networks for the transportation of raw materials and products, with the joint use of the so-called waste of specialized enterprises, or with the joint use of sources of energy. Combination is an important means of combating the dissipation of resources that takes place in the production processes of even the newest capitalist enterprises. In many cases we are for the time being unfortunately obliged to use these production processes in our Soviet plants as well, since the process of breaking out of the framework of capitalist production experience on a massive scale has only just begun. Here it should be noted that construction during the past three five-year plans has not been accompanied by the widespread efficiency of combining plants, even those located on the same or adjacent grounds, unless these plants belonged to the same official department. What is more, plants listed as combined in the plans and projects have sometimes had to be split in the process of construction.

There are two causes here: one in substance and one in form. In substance, let us note that in combining several different processes at separate plants it is necessary to reckon with both quantitative and spatial problems. If the distance of reciprocal transmission is small and the plants have a small scale of production, all the possibilities for combination are usually present, and the undertaking becomes technically practicable and economically advantageous. But if we have to deal with an aggregate of modern plants, on a very large-scale, each of which occupies an area of 50 to 150 hectares or more (the usual size of the site of our metallurgical plant is 170 hectares), the distances between the plants and shops to be combined sometimes prove to be many kilometers, with the industrial areas of these combines extending for 10 to 15 kilometers. Meanwhile, the amount of waste, by-products, steam, hot water, gases, and other substances to be transported prove to be enormous. The organization of the transportation of these masses, sometimes passing or crisscrossing each other, proves to be an uncommonly complex technical task, sometimes doubtful as to its economic results, and sometimes downright unprofitable and unjustified by the advantages of technological combination.

An effective means of facilitating the combination of plants is their continuous-operation arrangement, such as was adopted, for example, in planning the coal-chemical combinations at Cheremkhovo, under the following setup: coal-concentrating station, electric station operating on concentration wastes, coking plant, and the organic synthesis from gases. This sort of arrangement in space, facilitating the successive transportation of the products and wastes of the various stages of production, can be recommended in a number of instances.

A second example of a rational combination is the modern metallurgical plant, which consists essentially of several shops capable of existing separately and actually doing so at several places: coal concentration, coking, central power system, blast-furnace plant, open-hearth mill, rolling mill. The material and financial savings from combination are based on energy. They are brought about by a unified power system with the use of all the thermal wastes (coke dust, fumes, furnace and coke gases) and by economizing on heat (for heating metal) in various forms.

However, it is still necessary to supplement the definition of a production combine given earlier and include in it the case of the rational combination of several plants located near each other and considering the limitations above. The category of spatial forms of combination also includes various regional power associations along with the corresponding networks: the power system association on a regional scale, the gas supply association, and others that are accompanied by fuel savings and by a rise in the overall efficiency of the system as against the isolated management of individual enterprises.

The Gosplan term *regional combine* thus may have energy and transport significance. With respect to the other elements of the production processes of the regions, it has a very restrictive interpretation for the present stage of development of the national economy. At the same time, for the long run of socialist construction, let us say for the stage of carrying out the main economic task, the term may be retained in the broader sense, but again under certain technical and organizational conditions.

For a considerably broader range of territorial groupings, the term *complex* and sometimes *territorial grouping* and *hub* may be retained. In the article quoted earlier, we find the following definition of a complex: "The name production complex is given to an economic (interrelated) combination of enterprises at an industrial site in an entire region, whereby a definite economic effect is achieved through a successful (planned) choice of enterprises in accord with the natural and economic conditions of the region, and with its transportation and economic-geographic situation."[3]

This definition clarifies the term complex by the Russian term interrelated (interdependent) combination in contrast to the simple coexistence of industries on a specific territory; the latter is better rendered by the word grouping. It goes without saying that this implies a historical approach. Thus, the territorial-production combination (complex) may, in certain cases, include both the economic formations of past historical epochs, especially those reconstructed by Soviet rule if they combine well with the economy of the region, and new formations that though they have not yet had time to develop all the "satellites," "neighbors," or "levels" of the production processes, as well as all the technical-economic ties with the existing economy of the given region, clearly have certain elements of the necessary ties.

In contrast to these, the spatial concentration of various industries, which have either remained from the past or have arisen in the Soviet epoch and are without mutual ties or connections with the resources or economic-geographic situation of their region (even in the future), cannot from this point of view be called a combination (complex). This is a mere territorial group of industries. Such cases of the appearance of simple groupings may occur in a planned socialist economy; they arise because of the natural or economic peculiarities of a given industrial location, or because of market conditions, or deliberately, as at the beginning of future industrial formations that are not yet completed.

The development of the new economic regionalization of the USSR proposed by the government to the Academy of Sciences has precisely the objective of establishing the scientific grounds for rational combination of the productive forces of the economic regions and the Union republics in order to bring industry closer to the sources of raw materials and power, and achieve the highest productivity of labor throughout the USSR as a whole, and also to achieve full economic equality of the nationalities as a basis for the development of the economy in the national republics. Thus, all the diverse forms of territorial-production ties must be involved.

The concept of the territorial-production combination, provided it undergoes a sufficiently profound study, will prove extraordinarily fruitful for economic-geographic investigation. It does not and cannot serve as a substitute for the theory of economic regions, nor does it abolish the regions themselves; but it does permit one to introduce organically into economic geography and regionalization the notions of industrially organized forces and forms of social labor, which act directly and consistently upon the process of the formation of regions and may modify both the internal structure of the region and its external boundaries and interregional relationships.

Whatever taxonomic units may be established by Soviet science for internal economic regionalization, it will be impossible to dispense with the notion of "nuclei," "foci," and "centers" that form regions. It is impossible to reduce the process of regionalization to the mere establishment of the boundaries of regions by formal methods or to the study of the geographical division of labor between regions established almost arbitrarily (so-called economic-geographic areas, chosen for study under one circumstance or another).

The territorial-production combination, that is, social human labor organized in definite technical forms, with its power and machine equipment and its application to a definite combination of natural resources, is precisely the foundation of the geographic region-forming process that is being discussed a great deal in the geography of our time.

The concept of territorial-production groupings and combinations (complexes) may be used in studying the process of formation of the body of the major economic regions of Gosplan and of the smaller taxonomic units of internal regionalization, all the way to an analysis of concrete groupings of production around one small urban settlement. The difference will consist solely in the degree of generalization of the production processes basic to the economy of the given combination (complex), that is, in methodological adjustment to the scale of the task.

Generalization is admissible and justified when and if the study process can be generalized with its aid. It is also useful in cases where it is necessary to determine the processes not only qualitatively but also quantitatively whereas the processes interesting us are too complicated for direct mathematization. These considerations hold in this case, since the ties between the combinations (complexes) are indeed extremely diverse.

As may be seen from the foregoing, the combine may be regarded as a particular case of the complex, and the latter in turn as a particular case of the grouping. Production ties and general economic ties for regional complexes (combinations) may include the use of transportation, power, raw materials, semifinished products, building materials, parts of machines and articles, and manpower; the use of common sources and installations of water supply, housing, food, and cultural and scientific forces and means; in a word, they are very multifarious. A special place among them is occupied by vertical production ties, from raw materials up to the finished article; iron ore, pig iron, steel, rolled metal, castings, machinery, hardware; or coal, coke, coal-tar chemistry, pigments, plastics.

Ties also arise horizontally, between the branches of the vertical series (for example, coking, coking gases, open-hearth production; nonferrous metallurgy, sulfur gases from roasting, sulphuric acid chemistry, and fertilizer chemistry). Farther removed from the finished article (finished machine or complicated chemical products) there are created production branches for making auxiliary parts, materials, chemicals, reagents, equipment, instruments, and so forth; so-called neighbors and satellites—plants arising in the process of cooperative production, owing to the necessity of having close by a whole complex of parts and equipment for making any article of the vertical series.

CYCLES OF PRODUCTION

The multiformity of production ties of the regional economy explains why, in studying a concrete economy, an apparently irregular and outwardly extreme diversity of production combinations is obtained in each region taken separately. It is especially so when a comparison is made between regions, even those akin to one another in their type of economy. The geographer, accustomed mostly to grasping the characteristic peculiarities and individualities of regions, may at first find it difficult to discover in this diversity of regional production formations any foundation for their scientific classification and the solution of his typological tasks. He must first of all do some analytical work, paying attention to the essence of the production processes themselves.

In all the multiformity of economic phenomena in the present-day industrial economy there are certain constantly recurring elements. Thus, for example, the production of ordinary pig iron in a modern plant is the same in the Donets Basin, the Urals, or the Kuznetsk Basin. This standard process always requires coking coal and iron ore to meet certain specifications, and also reducing agents and ferro-alloys to obtain steel, refractory materials, and fixing agents. Technologically, cement and glass factories are conveniently combined with this process, while coke operations are conveniently combined with coal-tar chemistry, aniline-dye production, some-times nitrogen and artificial liquid-fuel production, sometimes soda production. In certain phases of metallurgical production, com-bination with machine-building is convenient, especially heavy ma-chine-building or metal-assembly production.

The constantly recurring mass production process may be used as a basis to search for the laws of region formation. A particular basic type of production process necessitates the selection of regions processing the best combination of natural resources for it, and coal, ores, and fluxes for the integrated organization of the indus-trial economy.

Examining the whole "portfolio" of present-day industries we can see that, beside the metallurgical cycle, there are also other basic processes for other kinds of mass raw materials and power. Ana-lyzing these basic (leading) mass production processes, we shall in-evitably distinguish other extensive combinations of production pro-cesses that should cause the rise of other types of territorial production combinations (complexes) in certain regions of our country.

Such is the path by which we approach the generalization of the concepts of production processes that is necessary for an analysis of regional combinations (complexes) in economic regionalization. We shall, for the sake of brevity, give the name *energy-production cycle* or simply *cycle* to the typical, constantly existing aggregate of production processes arising in an interrelated manner around a basic process, for a given kind of power or raw material. We note that a change of the basic process from one kind of energy to anoth-er (for example, electric ferrous metallurgy instead of pyrometal-lurgy) affects a change in the composition of the production lines of the cycle. Consequently, the cycle depends not only upon the basic raw materials, but also upon the kind of power used in the process and affecting the technology of the process.

Thus, what is meant by energy-production cycle is the whole ag-gregate of production processes successively evolving in an economic region on the basis of a combination of a given kind of power and raw materials, from the primary forms—extraction and refinement of raw materials—to all kinds of finished products that can be produced on the spot, on the initial premise that the production branch is brought near to the sources of raw materials and power and that a rational use is made of all the components of the raw-material and energy resources. In accord with the actual conditions of the region and the stages of development of the economy as a whole, this cycle may be full (completed) or not full (truncated) in individual regions. Each full cycle, in terms of the assortment of products, may in turn yield either completed finished articles only, or partly finished articles. The proportions among them are fixed in accordance with the economic assignments and the technological requirements on a planned basis. The cycle must be understood as a historical category, evolving in time.

To study the typology of regions and the system of regionalization, it is possible to establish the following generalized cycles.

1. A pyrometallurgical cycle of ferrous metals, including the mining of coking coal and iron ores, their concentration, the coking process, the blast-furnace process, the converting processes for the recovery of steel, the processes of rolling, steel and pig-iron casting, metalworking and machine-building (primarily heavy with some medium). At the same time, it includes the use of coking products (coal-tar chemistry, aniline-dye chemistry, synthetic coal chemistry using gases, and liquid coking distillates), plastics, motor fuel, nitrogen and hydrogen-containing compounds, solvents, alcohols, organic acids, rubber, high-octane additives to liquid fuel, and the like. These often include inorganic chemistry, inorganic acids, alkali, and fertilizers.

2. A pyrometallurgical cycle of nonferrous metals, including the mining and concentrating of nonferrous ores (copper, zinc, lead, gold, silver, and associated minerals), the extraction of coal or gases, the whole complex of pyrometallurgical processes or reduction of ores, the selection of metals and their refinement, the recovery of sulphur, sulphuric acid and chemical products based on the use of sulphuric acid at its place of recovery, the working of nonferrous metals and the recovery of alloys, the production of hardware and machine-building, cable production and the electric industry involving mass consumption of nonferrous metals, and so forth.

3. A petroleum and energy-chemical cycle, including the extraction and refining of petroleum, gases, and salts. This cycle covers: the refining, cracking, and hydrogenation of petroleum and petroleum residues for the purpose of obtaining motor fuel of different grades, lubricating oils, the organic synthesis of high-octane fuel and additives made from gases (natural and cracking gases), or various rubbers, alcohols, esters, organic acids, plastics and solvents, the recovery of hydrogen, nitrogen, compounds, artificial fibers, the refinement of various inorganic salts, the recovery of complex phosphorus-nitrate and potassium-nitrate fertilizers as well as other branches of the chemical industry developed on the basis of petroleum, gases, and salts.

4. The entirety of the hydroelectric-industrial cycles based on cheap mass electric energy with thorough introduction of electric power into both the main and auxiliary (heating) processes. This aggregate of cycles may include simultaneously or separately: elec-

trometallurgy of ferrous, nonferrous, and light metals, electrochemistry (including hydroelectric metallurgy), electrothermics (the production of calcium carbide, electrophosphorus, cyanimides, alumino-cement, and so forth). The modern technology and practice of construction in regions with cheap mass hydroelectric power show that it is possible to construct great energy and production systems over extensive territories (five of the northwestern states of the United States and three provinces of Canada), with very little use of thermal (coal) energy by comparison with the usual norms, with the most varied products, and with a sharp reduction in the use of manpower compared with the coal-and-metallurgy cycle. (These questions have been treated in more detail in the Angarstroy project.) It should be noted that not every hydroelectric station is capable of causing changes of this sort in technology. Hydroelectric stations with seasonal delivery of energy or with a subsidiary role in the amount of power output in thermal power systems are unable to produce such changes in the scale and character of the whole economic region.

5.   The entirety of the cycles of the processing industry developing at a great distance from the sources of raw materials in regions with large populations and internal markets, and in an advantageous economic-geographic and transportation situation, that is, in regions of consumption. These cycles may develop partly on the basis of local kinds of raw materials and energy, but, for the most part, depend on fuel brought from elsewhere, transportable raw materials and semifinished products. This group embraces: the mechanical energy, chemical energy, and thermal energy processing of mineral and organic raw materials and semifinished products including machine-building (medium, light, and precision); the ceramics industry; building-materials production; the food industry; the textile industry and other branches of light industry with all the concomitant activities; and the waste-processing industry. So-called secondary metallurgy often arises here as a satellite. It is impossible to regard this cycle or aggregate of cycles as merely a part of the cycles mentioned above carried out into other regions, since by the very nature of its formation (convenience of economic-geographic location) the regional cycle of the processing industry usually does not depend on any one raw-material region but on several at once (for example, Moscow depends on the Urals, the Donets Basin, Central Asia, and so forth); and yet it forms a usually distinct territorial unity of multiformity, having as its base a highly skilled labor force. In partial forms, these cycles may also develop as concomitant ones in regions where the basic cycles are [already] developed (Donets Basin, Urals, and so forth).

6. The lumber-and-energy cycle has a substantial influence on the formation of the economy in a number of regions. It includes forest exploitation, lumber-processing (mechanical), the process of obtaining wood pulp, cellulose and plywood, the production of shingles, insulating slabs, pressed and glued lumber, and chemical processing of wood, with the recovery of alcohol, varnishes, organic acids, tanning agents, lubricating oils, turpentine, and so forth. The cycle also embraces the production of the necessary thermal and electric energy from production wastes; charcoal metallurgy may be regarded as one of the branches of this cycle.

7. The entirety of the industrial-agrarian cycles include the various branches of socialist agriculture: field cultivation and animal husbandry, as well as the industry processing agricultural raw materials; the working of fiber (vegetable and animal); the leather industry; the processing of food flour and groats, wine, dairy products, sugar, confectionery, tobacco, meat, fish; the soap industry; and so forth. The cycle also includes all rural industry operating on nonlocal and local raw materials and semifinished products and utilizing the seasonal surpluses of agricultural labor. It likewise embraces the respective stationary and mobile energy bases.

8. The hydroreclamational industrial-agrarian cycle is assigned to a separate group in view of its characteristic differences from the preceding ones. It includes hydroreclamation facilities with an accompanying energy base, irrigated agriculture with extensive distribution of industrial crops, and the [related] processing industry.

All these cycles differ from one another in the characteristic relationships between expenditures of energy, raw materials, transportation operations, capital investment, and manpower for all the processes viewed together. Being dependent upon the sufficiently stable technological bases of the leading processes, the full cycles are sufficiently stable, that is, they change relatively slowly over time, owing to improvements in technology, and the balance relationships between the main production processes in the cycle quantitatively characterize their economic efficiency to a sufficient degree. Thus, the economic regions can be characterized not only qualitatively, but also quantitatively by means of the cycle method (full cycles plus truncated cycles).

Since the eight energy-production cycles, by their very design, exhaust all the varieties of processes, it is possible, with their aid, to characterize qualitatively and quantitatively any regional complex in the USSR, however multiform and peculiar it may be.[4] However, here it is necessary to take into consideration a new factor. The point is that in the mutual combination of production branches of different cycles, new ties and new forms of mutual combinations, not previously considered, may arise and actually do arise between them.

Furthermore, the entire natural environment of a real complex and not a [conceptualized] model of it, constructed by the method discussed above (the actual composition of the raw materials, the peculiarities of the energy resources, the course and stage of the historical process of reconstruction of the region, the labor characteristics, and the ethnic-cultural peculiarities of the population) alters the type of structure of the economy that can be logically obtained from the basis of the typical production processes and combinations of them. Thus, something is created that is special, individual, and nonreplicable at any other place, and that is typical for the genuinely living, continually developing economic region.

Hence, considerable differences between a model of the complex and reality are not only possible, but are even bound to exist. Yet it can be said that the above intricate design and analysis of a stable existing complex, having the force of a law for the structure of the economy, have been made to no purpose. We have been interested in the typological problem of regional complexes (combinations) above all else, and less in an individual characterization of each complex taken separately. We have come quite close enough to this typological problem. It must still be noted that not all lines of production of all the cycles are leading ones in the complex. Some of them are of local significance, others of national significance. It is the latter that determine the specialization of the complex in the nationwide geographic division of labor. Moreover, it is positively necessary to take into account the stage of development of the complex, which affects the fullness of the cycles, and the rise of new cycles, which sometimes produces abrupt transition from one qualitative condition to another.

TERRITORIAL-PRODUCTION COMPLEXES AND REGIONALIZATION

Since all the necessary preparatory analysis leading to the definition of the ties and laws governing them within the so-called cycles has already been given, it is possible to develop the notion of territorial-production complexes on a regional scale.

In the 1920s Soviet economic geography had abandoned a one-sided approach and the simple contemplation of such phenomena and facts as can be observed in the past or present. Developing the historical-scientific method in the geography of the socialist society, Soviet geography brought within its purview the study of phenomena in the process of their development, including the future. In the process of battling the bourgeois schools, Soviet geography affirmed that the introduction into science of the evaluation of the prospects of development of regions is, in principle, wholly admissible without violating the objectivity of scientific research, for the reason that under the Soviet system the plan of development of the national economy, built on a scientific foundation, supported by the masses and executed by a centralized economic apparatus organized into a governmental form, becomes an objective force. Hence, with a knowledge of the general laws of socialist development, the process of the formation of economic regions likewise becomes controllable. The study of these regions is the main task of the Soviet regional school of economic geography and, as a science, it has set itself the objective of not only perceiving reality in terms of space and region but also of ascertaining the laws governing its planned transformation.

The teachings of the classics of Marxism-Leninism regarding the basic laws of distribution of the productive forces under socialism have become a permanent part of the economic geography of socialism. The methodology of economic regionalization for the immense territory of the USSR after the works of Gosplan, the Kalinin commission, and the first and subsequent five-year plans, was developed precisely on this basis. However, from the present-day point of view the degree to which the methods of regionalization in its concrete form have been developed leaves something to be desired. The procedures that were suitable at the dawn of Soviet rule for a comparatively uncomplicated economic structure prove to be inadequate twenty-five years after the appearance of the project of economic regionalization of Soviet Russia (1922) in view of the enormous scale and multilateral development of industrialization, as well as in view of the main economic task. This can be seen merely from an analysis of the subject of territorial-production complexes.

Let us show the inadequacy of the methods now employed by means of examples and attempt to find ways of removing the defects. The investigation of the prospects for the development of the economy of any regional complex may be conducted either "from above"—from the general assignments of the Union economy— or "from below"—from the possibilities of developing the productive forces of a given region, striving toward an optimum with respect to integration. The latter course in its extreme form, that is, without the control assignments for the country cannot lead to a positive result, if only for the reason that the number of variants differing in direction and scale can be very large; let us recall that the natural conditions and natural resources themselves still do not decide anything; it is important to refract them through the prism of social relationships and requirements, and the latter are determined not only by internal needs but also by the relationships between the various parts of the state and its external political, economic, and military environment. Investigations from above are conducted by distributing the control assignment figures by separate branches of the economy and, in the subsequent phases, in the planning and investigating organs, also by branches.

The author performed the work originally by the first method (from above). In so doing, the levels of development of the national economy were estimated by the method of per capita norms of 84 products (that is, very broadly). Then the data on the development of the economy of the individual regions were examined and data in research monographs on the major projects for individual regions were evaluated to the extent that was possible. On this foundation, and not mechanically or statistically, the significance of the individual regional complexes in the USSR was evaluated by leading branches of the national economy. In certain cases, the specialization and the type of complexes were determined in a fairly clear manner by this method, especially the complexes of heavy industry operating on coal energy and water power, as they were obtained from the first method (see table 1).

Thus, the long-range specialization and structure of the regional complexes according to the aforementioned method for the group of complexes of heavy industry appear in the following form.

1. Southern Ukrainian Complex

Production branches of interregional significance include: coal mining, smelting of ferrous metals and ferro-alloys, machine-building, basic chemical and coal-tar industry, production of light metals. Among the production branches of intraregional significance are complex-ore metallurgy, mining of rare metals, petroleum refining, and organic synthesis.

2. Urals Complex

Of interregional significance here are: the smelting of ferrous metals (especially high-grade), ferro-alloys, copper, zinc, and light and rare metals; gold mining, machine-building, petroleum extraction and refining, basic chemistry, and the lumber industry; among the industries of local significance for the Urals are coal mining and organic synthesis.

3. Kuznetsk-Yenisey (Middle Siberian) Complex

Among the industries of interregional significance are coal mining, smelting of ferrous metals, ferro-alloys, light and rare metals, gold mining, basic chemistry, organic synthesis and the lumber industry. Hydroelectric power is of great significance. Uncharacteristic of the region are the recovery of copper, zinc, and lead, and petroleum refining.

4. *Eastern Siberian Complex*

The industries of interregional significance are coal mining, use of the energy of the Angara River, recovery of ferrous metals, zinc, lead, rare metals, and gold, basic and organic chemical industry, and machine-building; of especially great significance are the electricity-consuming branches such as the smelting of light metals, electrochemistry, and electrothermics. Copper mining and petroleum refining are not characteristic of the region.

5. *Central Kazakhstan Complex*

The industries of interregional significance are coal mining, copper smelting, zinc, lead, and rare metals. Uncharacteristic are ferro-alloys and nonferrous and rare metals.

6. *Volga-Don Complex*

The industries of interregional significance are ferrous metallurgy, coal mining, machine-building, basic chemistry, petroleum and gases. Uncharacteristic are ferro-alloys and nonferrous and rare metals.

Thus, typical features for all the complexes are coal mining on a mass scale for local use and for shipment elsewhere (an exception is the Urals), the smelting of ferrous metals, heavy machine-building (except for Kazakhstan), basic chemistry, nonferrous metallurgy (in the Donets Basin its significance is more restricted), light metals (to some degree), and organic synthesis (and only for three complexes: the Kuznetsk-Yenisey, the Eastern Siberian and the Volga-Don). Least complicated is the structure of the Southern Ukrainian and Central Kazakhstan complexes. In contrast to this, the Urals, Kuznetsk-Yenisey and Eastern Siberia have an extremely complex outline. These are the conclusions that can be derived by the usual method of analysis. They might have satisfied us eight to ten years ago, not today.

The interrelated and interconnected nature of the industries listed in the table is not shown in any way. In the form in which it is given in the table, a complex appears as a simple "set" of industries arising or existing in a region more or less independently of one another. Still less visible are any of the laws of the socialist economy or any distinctive features of planned regional complexes, setting

TABLE 1

ECONOMIC STRUCTURE OF REGIONAL COMPLEXES OF HEAVY INDUSTRY OPERATING ON COAL
ENERGY AND HYDROELECTRIC POWER BY INDIVIDUAL BRANCHES OF THE ECONOMY

REGIONAL COMPLEXES	COAL MINING AND COKING	HYDROENERGY	FERROUS METAL	FERRO-ALLOYS	COPPER	ZINC AND LEAD	LIGHT METALS	RARE METALS	MACHINE BUILDING	BASIC CHEMISTRY	PETROLEUM EXTRACTION	PETROLEUM REFINING	ORGANIC SYNTHESIS	LUMBER INDUSTRY (RAW MATERIAL)
Southern Ukrainian	+	+	+	+	–	0	+	0	+	+	–	0	0	–
Urals	0	0	+	+	+	+	+	+	+	+	+	+	0	+
Central Kazakhstan	+	–	+	0	+	+	0	+	0	0	–	0	0	–
Kuznetsk-Yenisey[1]	+	+	+	+	0	0	+	+	+	+	–	0	+	+
Eastern Siberian[1]	+	+	+	+	0	+	+	+	+	+	0	0	+	+
Volga-Don[2]	+	+	+	0	–	–	0	–	+	+	–	+	+	–
Frequency of recurrence (+)	5	4	6	4	2	3	4	4	5	5	1	2	3	3
of an industry in (0)	1	1	–	2	2	2	2	1	1	1	1	4	3	–
structure of regions (–)	–	1	–	–	2	1	–	1	–	–	4	–	–	3
Percentage of recurrence of + with respect to total number	82.3	66.5	100.0	61.6	33.4	50.0	66.6	66.6	82.3	82.3	16.7	33.4	50.0	50.0

TABLE 1--Continued

REGIONAL COMPLEXES	WHEAT	MEAT	TEXTILES	GOLD	FREQUENCY OF RECURRENCE OF TYPES OF PRODUCTION OF DIFFERENT SIGNIFICANCE			PERCENTAGE OF RECURRENCE OF + WITH RESPECT TO TOTAL NUMBER OF TYPES OF PRODUCTION
					+	0	−	
Southern Ukranian.	+	0	−	−	8	5	5	44.5
Urals.	0	0	0	+	12	6	−	66.5
Central Kazakhstan	0	+	−	+	7	7	4	38.8
Kuznetsk-Yenisey[1]	+	0	+	+	13	4	1	72.3
Eastern Siberian[1].	0	+	0	+	13	5	−	72.3
Volga-Don[2]	+	+	−	−	9	2	7	50.0
Frequency of recurrence of (+) an industry (0) in structure (−) of regions	3 3 −	3 3 −	1 2 3	4 − 2	62 − −	− 29 −	− − 17	57.4 26.8 15.8
Percentage of recurrence of + with respect to total number . .	50.0	50.0	16.7	66.6	57.4	26.8	15.8	100.0

[1]The Kuznetsk-Yenisey (Middle Siberian) region possesses enormous natural resources of water power: the Ob'Tom', the Upper Yenisey, the Lower Angara. In the fifteen-twenty-year perspective water power may play a most significant role in its energy balance, particularly the Yenisey-Angara station in the region of Yeniseysk. Consequently, for the broader prospect this is a region not only of the coal type, but also of the coal-and-water type. This applies to a still more marked degree to Eastern Siberia with its Angara River resources, which will be the chief basis of productional processes, thus influencing the type of the region.

[2]The Volga-Don region also has two "wellsprings" of energy: Eastern Donbass coal and natural gases.

(+) Industries of unionwide or interregional significance;
(0) Uncharacteristic branches and industries of regional significance;
(−) Absence of the given branch of production.

them off from the capitalist economies of analogous regions such as the Ruhr or the American coal basins.

In the above formulation, only a few branches of the economy developing in the region are, as it were, omitted. Even if the author took them into account, they have disappeared entirely so far as the reader is concerned. An arbitrary element is thereby admitted into the typological characterization in the choice of index industries, and doubt is cast upon the scientific value of the method. If we recall an earlier concern that the totally integrated development of the economy of the regions rightly constitutes a special concern of the socialist state, the importance of this fact will become understandable. At the same time it is quite evident that it is impossible to compile a table that would embrace 85 to 100 production branches by name, let alone every industry.

And, finally, it is quite impossible even roughly to characterize the economics of a complex by a single-dimensional system of quantitative indices under this system of study, without special and unwieldy investigation for each complex. It is all the more impossible to give a comparison between several variants of the development of one and the same complex, as may be required, for example, in an evaluation of given regional complexes in the system of economic regionalization of the Union as a whole.

APPLICATION OF THE CYCLE METHOD

The above defects, in the author's opinion, can be, if not eliminated, at least greatly mitigated with the method of generalized energy-production cycles. For in its very concept are laid the foundations for obligatory combination of interrelated industries and the preferability of arranging them in one complex (with specifically justified exceptions).

A characterization of the economy of the same complexes by the cycle method gives a different picture (see table 2). The following conclusions with regard to the second method will become clear to all who attentively compare tables 1 and 2.

1.   The method is suitable only for studying the development of the economic regions of a socialist state; it is not suited to spontaneously developing capitalist regions.

2.   The number of hypothetical assumptions in the evaluation of the tendencies and trends of development of complexes in the future is sharply reduced by the second procedure; consequently, the reliability of the predictions is increased. The typological features, and likewise the differences, stand out more vividly.

3.   A general coverage of the structure of the economy is achieved instead of a selective one. Table 2 represents a kind of algebraic

formula of the regions, embracing the laws governing each type of tie and the interdependence of the production branches of the regions.

4. By using consolidated indices for the cycles as a whole, it is possible to give approximate typological quantitative evaluations for the elements of a given type of national economy: the balances of requirements for raw materials, fuel, electric power, manpower, water, transportation services, as well as consideration for the demands for industrial land for plants, areas for forest exploitation, agriculture, and the like. Furthermore, with a sufficient knowledge of the general geographic conditions of the region, it is possible to introduce the necessary (consolidated) corrections for local conditions and to obtain as a first approximation in needed cases a certain idea of the differences in the average regional social productivity of labor by comparison with the average conditions throughout the USSR and to know in what sections of the national economy

TABLE 2

ECONOMIC STRUCTURE OF REGIONAL COMPLEXES OF HEAVY
INDUSTRY OPERATING ON COAL ENERGY BY METHOD
OF ENERGY-PRODUCTION CYCLES

REGIONAL COMPLEXES	FERROUS PYROMETALLURGY	NONFERROUS PYROMETALLURGY	PETROLEUM-CHEMICAL	ALL HYDROENERGY-INDUSTRIAL CYCLES	ALL CYCLES OF PROCESSING INDUSTRY	LUMBER-ENERGY	ALL INDUSTRIAL-AGRARIAN CYCLES	HYDRORECLAMATIONAL AGRARIAN-INDUSTRIAL
Southern Ukrainian. . .	+	0	0	+	0	–	+	0
Urals	+	+	+	0	+	+	+	–
Central Kazakhstan. . .	+	+	–	+	0	0	+	0
Kuznetsk-Yenisey. . . .	+	0	–	+	+	+	+	–
Eastern Siberian. . . .	+	+	0	+	+	+	0	–
Volga-Don	+	–	+	0	+	–	+	0
Frequency of (+) . .	6	3	2	4	4	3	5	–
recurrence of (0) . .	–	2	2	2	2	1	1	3
cycles (–) . .	–	1	2	–	–	2	–	3
Percentage of recurrence of + with respect to total number.	100	50.0	30.3	66.7	66.7	50.0	83.4	0

TABLE 2--Continued

REGIONAL COMPLEXES	FREQUENCY OF RECURRENCE			PERCENTAGE OF RECURRENCE OF + WITH RESPECT TO TOTAL NUMBER OF CYCLES
	+	0	-	
Southern Ukrainian. . . .	3	4	1	37.5
Urals	6	1	1	75.0
Central Kazakhstan. . . .	4	3	1	50.0
Kuznetsk-Yenisey.	5	1	2	62.5
Eastern Siberian.	5	2	1	62.5
Volga-Don	4	2	2	50.0
Frequency of (+) . . .	27	-	-	56.2
recurrence of (0) . . .	-	13	-	27.1
cycles (-) . . .	-	-	8	16.7
Percentage of recurrence of + with respect to total number.	56.2	27.1	16.7	100.0

The Kuznetsk-Yenisey (Middle Siberian) region possesses enormous natural resources of water power: the Ob'Tom', the Upper Yenisey, the Lower Angara. In the fifteen-twenty-year perspective water power may play a most significant role in its energy balance, particularly the Yenisey-Angara station in the region of Yeniseysk. Consequently, for the broader prospect this is a region not only of the coal type, but also of the coal-and-water type. This applies to a still more marked degree to Eastern Siberia with its Angara River resources, which will be the chief basis of productional processes, thus influencing the type of the region.

The Volga-Don region also has two "wellsprings" of energy: Eastern Donbass coal and natural gases.

(+) Industries of unionwide or interregional significance;
(0) Uncharacteristic branches and industries of regional significance;
(-) Absence of the given branch of production.

they may be expected. Here the balance estimates can be verified by the method of per capita norms.

5.   Having this sort of characterization, one can, without doing violence to the logic of investigation, proceed to the next and more profound phase of typological study of the structure for the individual territorial components of the regional complex. One can distribute the general dimensions of production of each cycle separately over the territory of a given region without disturbing the integration and

structure of the whole and consistently proceed to study more and more localized complexes (combinations), geographic groupings and, in the course of the investigation, also combines of varying character. There is no doubt that in the second intraregional phase of the investigation the degree of generalization should be less and that the distribution of production over the territory should pursue two objectives. First, to reduce, insofar as possible and advisable, the expenditures for the movement of raw materials, energy, finished products, and other expenditures depending on the geographic distribution of the economy on the territory of the regions. Second, to reduce the total expenditures for the development of the territory, regional planning, and the construction of housing and enterprises. The choice of subregions and nodes should be made primarily in accordance with the sanitary, cultural, and aesthetic requirements with relation to the location of the populated places.

In order to show the application of the cycle method to any type of region, we choose the group of regional complexes developing on the basis of the use of petroleum, gases, and hydroelectric power. An exception here is the Western Ukraine (see table 3). Comparing table 3 with table 2 as to frequency of the recurrence of the cycles, we see that the typical features stand out here no less vividly than in the preceding case. At the same time, one is struck by the diversity of the economy of the complexes having a given trend and by their connection with agriculture. Thus, the long-range specialization and structure of the regional complexes of the petroleum and hydroelectric group are determined as follows by the cycle method.

1. *The Transcaucasian Complex*

The cycles of interregional significance include: the nonferrous metallurgy cycle—copper, rare metals; the petrochemical cycle— extraction and refining of petroleum, petroleum chemistry, including organic synthesis; the hydroelectric cycles—ferro-alloys, aluminum, carbide production and its derivatives, particularly rubber; the industrial-agrarian cycle, represented by the production and processing of subtropical crops, tea, citrus fruits, and so forth; the reclamation cycle—plantings of rice, cotton, and other crops on irrigated lands. Of intraregional significance are the cycles of the processing industry, for example: textiles, ceramics, machine-building, and the lumber-energy cycle.

2. *Northern Caucasian Complex*

This complex stands out vividly as very much akin in structure to the Transcaucasian regional complex, with the difference that zinc and lead appear in place of copper, and northern agricultural crops in place of subtropical ones.

3. *Central Asian Complex*

This [complex] is analagous to the Transcaucasian complex, but with reclamation having much greater importance and petroleum extraction less.

4. *Middle Volga Complex*

Of great importance for this region are the petrochemical cycle and gas, and the extensive development of chemicals from petroleum derivatives and salts (also from potassium salts), including organic synthesis. Furthermore, land reclamation with water resources and hydroelectric power are important in the Transvolga region.

This complex may quite rightly be assigned to the kindred type of regions. Without doubt, the existing differences from the other

TABLE 3

ECONOMIC STRUCTURE OF REGIONAL COMPLEXES DEVELOPING ON
THE BASIS OF UTILIZATION OF OIL, GASES, OR
HYDROELECTRIC POWER BY THE CYCLE METHOD

REGIONAL COMPLEXES	FERROUS PYROMETALLURGY	NONFERROUS PYROMETALLURGY	PETROLEUM-CHEMICAL	ALL HYDROENERGY-INDUSTRIAL CYCLES	ALL CYCLES OF PROCESSING INDUSTRY	LUMBER-ENERGY	ALL INDUSTRIAL-AGRARIAN CYCLES	HYDRORECLAMATIONAL INDUSTRIAL-AGRARIAN
Transcaucasian.	0	+	+	+	0	0	+	+
Northern Caucasian. . .	–	+	+	+	–	0	+	+
Central Asiatic	0	+	+	+	0	–	+	+
Middle Volga.	–	–	+	+	0	0	+	+·
Western Kazakhstan. . .	–	–	+	–	–	–	+	+
Western Ukrainian . . .	–	–	+	0	0	+	+	–
Frequency of (+) . .	–	3	6	4	–	1	6	5
recurrence of (0) . .	2	–	–	1	4	3	–	–
cycles (–) . .	4	3	–	1	2	2	–	1
Percentage of recurrence of + with respect to total number.	0	50	100	66.7	0	16.6	100	83.4

TABLE 3--Continued

REGIONAL COMPLEXES	FREQUENCY OF RECURRENCE OF CYCLES OF DIFFERENT SIGNIFICANCE			PERCENTAGE OF RECURRENCE OF + WITH RESPECT TO TOTAL NUMBER OF CYCLES
	+	0	−	
Transcaucasian.	5	3	−	83.4
Northern Caucasian. . . .	5	1	2	83.4
Central Asiatic	5	2	1	83.4
Middle Volga.	4	2	2	66.7
Western Kazakhstan. . . .	3	−	5	50.0
Western Ukrainian	3	2	3	50.0
Frequency of (+) . . .	25	−	−	52.1
recurrence of (0) . . .	−	10	−	20.8
cycles (−) . . .	−	−	13	27.1
Percentage of recurrence of + with respect to total number.	52.1	20.8	27.1	100.0

NOTE: (+) industries of unionwide or interregional significance; (0) uncharacteristic branches and industries of regional significance; (−) absence of the given branch of production.

more southerly regions are to be explained by two causes: a different geographic situation in the more northerly zones and the different geologic origin; the Volga land is a part of the Russian Platform, whereas the earlier mentioned regions lie in the folded zones of the continent.

5. *Western Kazakhstan Complex*

From the production viewpoint, this is the extreme expression, the accentuation of the qualities peculiar to the Middle Volga region; petroleum and salts determine the great development of the petrochemical cycle and the processing of salts, including potassium salts; irrigation of the land is the only means of cultivation here (see the projects for irrigation with water from the Ural River). The agricultural cycle is for the time being represented by migratory forms of cattle-raising.

6. *Western Ukrainian Complex*

Geographically this would seem to be the exact opposite of the

Western Kazakhstan complex. However, in this representative of the group, there are important forms of kinship with it in the field of industry (petroleum extraction, gas extraction, refining of petroleum and gases, the water resources of the mountain rivers, ore resources, exploitation of salts, with the development of the respective cycles of these bases). There are features similar to the Caucasian regions, attributable to a certain extent to the similarity of the geologic substratum. The economic-geographic differences from Western Kazakhstan are rooted in the conditions for agricultural production, which is represented in diametrically opposite forms (intensive farming versus migratory cattle-raising), affecting the density of settlement and the industrial elite of agriculture—the food and light manufacturing industries of the complexes, which are naturally also very different.

From an examination of this example of descriptions of the structure of the regional complexes of the petroleum and hydroelectric group, as well as on the basis of the earlier example for the complexes of heavy industry, there follows much that is important theoretically about the nature of such complicated formations as are the regional complexes. They appear before our eyes not as the mechanical sum of the productional cycles or processes (let us remember that the concept of cycle has been introduced by us as an auxiliary one); the regional complexes should be examined and studied as a unit of multiformity, as phenomena developing according to all the sociohistorical laws.

Thus far in our construct we have, as it were, ignored the natural geographic environment. It has entered into our arguments in a hidden form: as the resource for industrial raw material obtained at a given place; as the energy of a production process from a nearby natural source; as a resource for industrial-agrarian processes on the basis of agricultural raw material obtained in a given region with all its peculiarities of climate, soils, and vegetation. But the natural environment has thus far not been specifically considered. When we spoke of the heavy industrial complexes, this defect was not so striking, but when we moved to the complexes of the second group, we needed to explain a number of the circumstances that prompted us to regard as legitimate the assignment of all six regions to the given group. The forces of nature, of the geographic environment, broke forcibly into our theoretical construct and compelled us to speak about them.

Earlier, we properly assigned first place in our construct to questions of production (that is, economic and technical questions affecting the structure of the regional complexes), but we neglected such forces of nature as solar energy, wind, soil and so forth, which rendering gratuitous service to man, exert a rather strong influence upon the structure of production of regional complexes. Their organic inclusion in the typological scheme along with the forces of technological processes (electricity, heat, steam, chemical energy, and so forth), is obviously quite indispensable for complete and correct economic-geographic conclusions.

Further, it must be taken into account that the number of different combinations from our eight to ten conventional generalized cycles may be very large. The quantitative relationships between the industries of the different cycles may be almost infinitely varied. Hence, within one and the same grouping of regions (heavy industry, the petroleum and hydroelectric grouping) there may be a fairly broad range of quantitative-qualitative deviations in one direction or another. There can (and do) appear intermediate forms, "hybrids," so to speak, that will be difficult to assign to the one or the other type.

Thus, for example, in our first grouping of the heavy-industrial complex, the Donets (Southern Ukrainian) complex, based on coal energy, stands out quite distinctly as the classical complex. Differing little from it with respect to energy is the Central Kazakhstan complex (tendency toward nonferrous metallurgy); the Kuznetsk-Yenisey complex at this point appears as a "sort of Donbass in Siberia," but in the long run, for example when the hydroelectric station on the Yenisey River, capable of furnishing as much energy as is now furnished by the coal mined in the Kuznetsk Basin, is put into operation, the circumstances must change sharply; new processes will be created, on the basis of hydroelectric and not thermal energy, that is, new cycles will appear on an equal footing; the complex will appear before us as a coal and hydroelectric one.

The Eastern Siberian complex since the initiation of its development has been primarily a hydroelectric one. The Volga-Don heavy industrial complex inclines in another direction from the Donets coal "standard." This is also a transitional type to the complexes of the petroleum-gas-hydroelectric series. On the one hand, it is the same Donets Basin in its eastern part, but, on the other hand, it is the salt, gas, and oil-bearing Volga in its new, present-day appearance. Thus, there are to be several transitional types within the family (group of types) of the heavy-industrial complex and, of course, as we have seen earlier, also within the petroleum and hydroelectric family.

It is possible, with an attentive study of the structures, to arrange the complexes of the two families in a successively changing series forming, in the transition from one regional complex to another, certain jumps from one measure of things (quantity-quality) to another, depending upon the relationships between the dominant forms of energy (and raw materials) and upon the related forms of production cycles having different compositions. The difference in the forms of natural (spontaneous) energy affecting the production processes (for example, in agriculture, lumbering, and hydroelectric power) must be taken into account here.

It turns out that the size of the jumps in the transition from one region to another within one family generally is not greater than in the transition from one extreme member to another extreme member of the two families. Arranging the complexes in accordance with the transitions from one dominant energy source to another with the dominant types of energy symbolically expressed—hydroelectric power (h), coal energy (c), petroleum and gas energy (p)—the following series is obtained.

1. Eastern Siberian: h, transitional to the Far Eastern and Northern Yenisey complexes;
2. Kuznetsk-Yenisey: $c + h$;
3. Central Kazakhstan with the Altay: $c + h$; without the Altay: c;
4. Southern Ukrainian: c;
5. Volga-Don: $c + p$, transitional to the Middle Volga and Western Kazakstan complexes;
7. Western Kazakhstan: p;
8. Middle Volga: $p + h$;
9. Western Ukrainian: $p + h$;
10. Middle Caucasian: $p + h$;
11. Transcaucasian: $p + h$;
12. Central Asian: $p + h$, transitional to the Southern Kazakhstan complex.

In a manner similar to the procedure employed in distinguishing the two families of complexes, heavy-industry and petroleum-hydroelectric, all the other regional complexes can be grouped in three new families marked by the same characteristics of mutual successive transitions of structure and cycle composition. Let us enumerate all the families in full: (1) heavy industry; (2) petroleum-hydroelectric power; (3) industry-farming; (4) processing industry; and (5) northern industry. Let us characterize these families briefly.

1. *Heavy Industry*

These complexes specialize in the mining of coal and the smelting of ferrous and nonferrous metals by pyrometallurgical methods. In the transitional forms of more complex composition, the regions include mass metallurgy operating on hydroelectric power by electrometallurgical methods. The complexes possess a system of machine-building and chemical industries, which arise in them on the basis of the reduction of expenditures for transportation and energy. The complexes have great individual differences depending upon the quality and quantity of the basic raw material being processed.

Within the group there are primarily coal and metallurgy complexes, complexes with the preponderant development of nonferrous metallurgy, complexes in which ferrous and nonferrous metallurgy have approximately equal significance, complexes with the preponderant development of mass hydroelectric power (with regulation provided) and electrical processes. The chemical industry of the complexes varies according to the presence or absence in their territories of petroleum, natural gases, and lumber, as well as according to the degree of diversity of the mineral and chemical raw materials.

2. *Petroleum-Hydroelectric Power*

These complexes specialize in the extraction and refining of petroleum and gases and the production of water power. In the present-day industrial economy petroleum and natural gases occupy one of the leading places as sources of energy and as chemical raw material. All agriculture, motor, and air transportation now have petroleum as a source of motive power. The cost of electric energy obtained from gas occupies a middle place between that of coal and hydroelectric power. Gas is transmitted cheaply for long distances through pipelines. Petroleum and gases yield an immense quantity of organic synthetic products. Hydroelectric power creates an exceptionally favorable combination with petroleum resources, extending the range of electricity-consuming processes both into the field of organic synthesis and into that of the associated electrochemical, electrothermal, and electrometallurgical industries. The ramifications of production are very great, and are still growing.

Under the conditions in the USSR the respective complexes of natural resources (petroleum-hydroelectric) coincide, as a rule, with localities rich in agricultural reserves of national significance requiring land reclamation with water resources. Such reclamation and hydroelectric power require unified water resources as a whole and consequently are closely connected with one another with respect to production, both in the creation of water-resource installations and in their operation. The complexes of this group also differ greatly from one another depending upon the character of agricultural production, the relationships of the cycles, and upon the type and intensity of the physical-geographic processes.

The extreme members of this family come close in type, on the one hand, to the heavy-industry group and, on the other, to the industrial-agricultural group.

3. *Industry-Farming*

A characteristic feature of the complexes is the presence of a large industry processing the foodstuffs and industrial raw materials produced by agriculture (grain, root crops, fruits, fiber, meat, leather, wood, and so forth). In the United States the food industry is first in production cost. Hence, the importance of the branches of this industry in the national economy of industrial-agrarian countries. However, the industry of these complexes is not confined to these forms. In a number of complexes located in the forest-steppe belt, there appear lumber cycles of local significance; in the maritime regions, fish processing appears; in the complexes advantageously located with respect to heavy-industrial complexes, light and medium machine-building; in some maritime regions, shipbuilding and so forth.

Thus, being close to the petroleum-hydroelectric complexes (through agricultural production, hydroelectric cycles, reclamation), the given complexes are transitional (through machine-building) to the processing industry, which also consumes a considerable amount of agricultural raw materials.

4. *Processing Industry*

The group of complexes of the processing industry, which as a rule, are located in the more northerly latitudes with a climate having less intensive energy, is characterized by the high significance of the processing branches of industry, including those producing the means of production from the semimanufactured [items] of the heavy-industrial regions. Here one can cite the most varied *upper levels* of the metallurgy cycles, the cycles of the mechanical, thermal, and chemical processing of articles, as well as the textile, lumber, and food industries.

Processing industries develop with special success in localities with dense population and large local consumption, as well as localities marked by advantages of economic-geographic position, which permit savings in the transportation of the finished products. The modern technology of refining (enriching) certain kinds of raw materials reduces the "ballasts" in the composition of the raw materials, furnishing transportable concentrates. The same object is attained by combining production processes with the recovery of useful products from yesterday's ballast, which also enlarges the sphere of supply. These circumstances determine the expediency and advantageousness of the development of processing cycles in the central regions of the country with their more favorable economic-geographic location, and likewise the advantageousness of developing certain basic industries affording maneuverability (ferrous metallurgy near Moscow and Leningrad and so forth).

It is clear that any change in the economic-geographic position must affect the intensity of the development of these regional complexes. For example, the growth of production in the Urals and Siberia assumes important stimulating significance for the Volga-Kama regions, improving their supply of eastern semifinished products. The establishment of economic ties with the countries of the new democracy in the West improves the economic-geographic situation of Belorussia and the Baltic republics. We note that the excellent moisture supply and dense hydrographic network for the supply of water to industry and large cities is of importance for the successful development of all the regions of this group.

5. Northern Industry

The complexes of this type are located in the forest and even tundra zones of the northern regions. The population here, as a rule, is sparse, owing to the difficult climatic conditions. Under the conditions governing the conquest of territory, only the selective method of developing lands by separate oases and areas is possible, as a rule. It is often necessary to rely not only on permanent cadres of local population but also on the periodic importation (for certain periods of time) of manpower from other, more southerly regions of the country. The principal aim of creating a complex is the working of especially valuable mineral resources, forests, and commercial fish and marine animals. Modern technology and mechanization of the processes of such forms of energy as water and oil power make it possible to establish transitional forms in the regions of the processing industry, on the one hand, and to the heavy-industrial regions, on the other.

We can now compile a synoptic typological table (see table 4) of regional production combinations (complexes) for the stage of the implementation of the main economic task of the USSR. This table embodies as a classificatory feature the above grouping of complexes into five families. In addition, the variability of the complexes of one and the same family brings the different families together so as to produce a sort of continuous series of transitions, including fifty regional complexes of the USSR.

To the left and right of the name of each complex in table 4 are several descriptions of the forces of production processes and forces of the natural processes operating in complexes, viewed in terms of their importance for the economy of complexes. The following have been adopted as indices of the forces of the production processes: coal output (in percentage of Union-wide output), power con-

TABLE 4

REGIONAL TERRITORIAL-PRODUCTION COMBINATIONS (COMPLEXES) (OUTLINE FOR THE
STAGE OF IMPLEMENTATION OF THE BASIC ECONOMIC TASK)

FORCES OF PRODUCTIVE PROCESSES						FORCES OF NATURAL PROCESSES		
Coal Production (Percent of Unionwide Production)	Consumption of Electric Power (Percent of Unionwide Consumption)	Per Capita Norms of Electric-Armament, with 100 Taken as Unionwide Average	Directing Types of Energy*	GROUPS OF COMPLEXES ("FAMILIES")	NAME OF REGIONAL COMBINATION (COMPLEX)	Type of Air Masses	Zonality	Tectonics
...	1.5	250	h	Northern Industry	Karelian-Murmansk (transitional)	ar+at	t+f	pl
2.0	0.5	400	h+c		North-Yenisey	ar	t+f+vz	pl
0.5	0.8	200	h+c+p		Yakut	ar	t+f+vz	pl+fz
4.9	1.2	150	c+p		Dvina-Pechora (transitional)	ar+at	t+f	pl+fz
...	0.3	100	c		North-Ob'	ar+at	t+f	fz
0.5	0.3	100	c+p	Heavy Industry	Chukot-Kamchatkan	ar-p	t+vz	fz
4.2	4.0	200	c+p+h		Far Eastern (transitional)	ar-p	f+vz	fz
6.3	7.3	300	h+c		Eastern Siberian	ar	f-vz	pl+fz
16.6	10.3	250	c+h		Kuznetsk-Yenisey	at-ar	f+fs+vz	fz+pl
6.7	2.5	250	c+h		Central Kazakhastan with Altay	at	s+d+vz	fz
29.2	8.1	150	c		Southern Ukrainian	at	s	pl
6.7	3.9	150	c+p		Volga-Don (transitional)	at	s	pl
6.7	8.7	150	c+p		Urals (transitional)	at	s+fs+f+vz	fz

Table 4--Continued

FORCES OF PRODUCTIVE PROCESSES				GROUPS OF COMPLEXES ("FAMILIES")	NAME OF REGIONAL COMBINATION (COMPLEX)	FORCES OF NATURAL PROCESSES		
Coal Production (Percent of Unionwide Production)	Consumption of Electric Power (Percent of Unionwide Consumption)	Per Capita Norms of Electric-Armament, with 100 Taken as Unionwide Average	Directing Types of Energy*			Type of Air Masses	Zonality	Tectonics
0.5	0.6	100	p	Petroleum-Hydroenergy	Western Kazakhstan	at+tr	s-d	fz
...	3.7	75	p+h		Middle Volga	at	s+fs	pl
1.3	1.9	60	p+h		Western Ukrainian	at	f+fs+vz	pl+fz
1.7	3.5	110	p+h		Northern Caucasian	at	s-vz	fz
1.7	4.4	150	p+h		Transcaucasian	at+tr	vz	fz
2.5	4.8	90	h+p		Central Asiatic	tr	d+vz	fz
0.5	0.8	50	h		Southern Kazakhstan (transitional)	tr+at	d+vz	fz
...	1.2	40	h	Industry-Farming	Black Sea with Moldavia	at	s+vz	fz+pl
1.7	2.8	50	h+c		Central Ukrainian	at	s+fs+f	pl
...	0.4	60	c		Northern Kazakhstan	at	d+s+fs	fz
...	1.9	70	c		Western Siberian	at	s+fs	fz
...	3.1	70	c		Central Chernozem	at	s+fs	pl

TABLE 4--Continued

FORCES OF PRODUCTIVE PROCESSES				GROUPS OF COMPLEXES ("FAMILIES")	NAME OF REGIONAL COMBINATION (COMPLEX)	FORCES OF NATURAL PROCESSES		
Coal Production (Percent of Unionwide Production)	Consumption of Electric Power (Percent of Unionwide Consumption)	Per Capita Norms of Electric-Armament, with 100 Taken as Unionwide Average	Directing Types of Energy*			Type of Air Masses	Zonality	Tectonics
..	5.8	100	c	Processing Industry	Volga-Kama	at	fs+f	pl
5.9	8.5	100	c		Moscow	at	fs-f	pl
..	1.6	60	c		Belorussian	at	f	pl
..	1.6	60	c		Baltic	at	f	pl
..	4.0	100	c+h		Leningrad (transitional)	at	f	pl
100.0	100.0	100	c+h+p					

NOTE: *The dominant types of energy are as follows: c, coal; h, hydroenergy; p, petroleum and gas. The types of air masses are as follows: ar, arctic; at, atlantic; p, pacific; tr, tropical. The types of zonalities are as follows: t, tundra; f, forest; fs, forest-steppe; s, steppe; d, deserts; vz, vertical zonality. The tectonics are as follows: pl, platform; fz, fold zone.

sumption (in percentage of Union-wide consumption), norms of the per capita electrical supply of labor (with 100 as the average for the USSR), and the dominant types of energy (coal, petroleum, hydroelectric power). As indices of the forces of natural processes the following have been adopted: type of air masses and zonality, the main forms of tectonic types (platforms, shields, fold zones), the formation of relief and stratigraphy, the creation of magmatic foci, geochemical processes, the laws governing the concentration and dispersion of the elements in the earth's crust, and so forth. The use of these indices for analysis is based on the following assumption. If the arrangment of our complexes in a successively changing series is founded only on a logical basis, that is, if it is an abstraction having no foundation in reality or the laws governing the regional structure of a socialist economy for such a vast and diversified country as the USSR, then our energy indices for the natural and technological forces should reveal this in the form of complete randomness and inconsistency in the series of these indices. But if we discover a law-governed variation of these auxiliary series in any form, we shall obtain a new and important proof that the structure of the regional complexes is subject to some law of development.

As seen in table 4, there are a number of grounds for believing that such a law does exist, and that it leads to the following conclusions.

1. The highest per capita electric equipment supply for labor (the intensiveness of production processes) is found in the complexes of northern industry (the Northern Ob' and the Chuckchi-Kamchatka complex are not exceptions but only reflections of an early stage in the process of industrialization).

2. As was to be expected, the heavy-industrial regions, along with the new, sparsely settled Siberian regions, occupy second place on the index.

3. The petroleum-hydroelectric complexes—new kinds of heavy and medium industry, combined with agricultural irrigation installations—occupy third place.

4. The fourth and fifth places are occupied by the industrialized farming and processing industry complexes, by nature requiring a considerable expenditure of human labor.

5. The types of air masses change too consistently not to see in this series a confirmation of the correctness of the assumed sequence of complexes. Nor are any substantial contradictions encountered in the zonality column.

6. The tectonics column affords no material for judgement without a parallel examination of the geochemical laws, for which it is as yet difficult to give a generalizing index. However, in a number

of cases, it helps to verify the location of the complexes within a family (it is preferable to group the Eastern Siberian complex along with the Kuznetsk-Yenisey; the Volga-Don complex with the Southern Ukrainian; the Middle Volga complex with the Western Ukrainian; the Black Sea complex with the Central Ukrainian and so forth).

7. It is to be expected that further study of the question from the viewpoint of physical-geographic, geologic, and geochemical regionalization, continuing in the scientific traditions of V. I. Vernadskiy and A. Ye. Fersman, will help to clarify the many problems which remain.

The intention of this outline is to pose in the form of a hypothesis some theoretical problems for geographers. The attentive reader will quite easily detect a number of assumptions in the theoretical construct of the essay that have still not been proven well enough to speak of a complete theory on this question. Today it is only a hypothesis. Theory is the task of the future. However, the author hopes that he has not made arbitrary assumptions, depriving the work of scientific meaning; and if this work stimulates other investigators to new ideas, even different from these statements, the aim of the hypothesis will already have been achieved.

1. N. N. Kolosovskiy, "Concerning the Question of Economic Regionalization" in *Principles of Economic Regionalization* (Moscow, 1958), pp. 60–86

2. Explanations or expansions in brackets were added by the author for the last republication of this article.

3. N. N. Kolosovskiy, "Concerning", p. 69.

4. Theoretically one can foresee in the future the evolution of a ninth cycle based on atomic energy.

YU. G. SAUSHKIN
T. M. KALASHNIKOVA[1]

Basic Economic Regions of the USSR

The question concerning the territorial organization of productive power and the economic regionalization of the USSR has been discussed many times at Party congresses. The Twenty-first Congress of the Communist Party of the Soviet Union again pointed to the necessity of planning by economic regions. "The Control Figures of the Development of the National Economy of the USSR for 1959–1965" states, "the delimiting of large economic-geographic regions (or basic economic regions) assists in the planning of the correct geographic distribution and the most economical spatial organization of the national economy of the Soviet Union."

At the present there is an acute need to work out a new, scientifically based system of basic economic regions of the USSR. This necessity arises from the huge scale of new economic construction in all parts of the Union (especially in the East) related to: (1) new discoveries and achievements in the investigation of the national wealth of the USSR; (2) scientific achievements that place the strongest emphasis on the development of technology and the technological processes of the majority of branches of production; and (3) further development of territorial division of labor within the USSR and between the USSR and the countries of the socialist camp.

Translated from *Voprosy geografii* [Problems of Geography], no. 47 (1959), pp. 42–73.

The reorganization of the administration of industry and construction into territorial units, originating in 1957 at the February plenum of the Central Committee of CPSU and at the Seventh Session of the Supreme Soviet of the USSR, led to the formation, in May 1957, of 105 economic administrative regions.[2] In each of these regions a national-economic council [sovnarkhoz] was formed for the administration of industry and construction.

The delimitation of a large number of economic administrative regions does not eliminate the problem of a scientific general economic regionalization — the investigation of the objectively existing large (basic) economic regions of the country. On the contrary, the transition to territorial forms of administration of industry and construction increases the practical significance of scientific research works in the field of economic regionalization.

The fact is that the administration of industry and construction in territorial units is executed basically according to existing administrative-political units of the country—republics, krays, and oblasts (only in individual cases is the economic administrative region a group of provinces). But the administrative units were created, as a rule, not considering the entirety of their production complex, but rather for other reasons, such as a convenience for administration and direction, mainly agriculture. In many cases the convenience of transport ties with the oblast center was decisive in setting oblast boundaries. The economy must be planned on the basis of knowledge of actually existing territorial production complexes with a specialization on a national scale, and for this a scientifically substantiated economic regionalization of the country is needed, that is, the delimitation of basic economic regions of the USSR. For the planning of the national economy of the USSR by territorial units, Soviet science must delimit these economic regions that actually exist (in one or another stage of development) or will develop in the next fifteen to twenty years.

The basic tenets of economic planning of the socialist state were worked out as early as the first years after the victory of the October Revolution:

> In the GOELRO plan [State Plan for the Electrification of Russia] attention was called to the fact that "individual elements and branches of industry do not represent definite values: their significance changes depending on how these elements and branches are combined. It is, therefore, necessary to study the economy as a whole and to compare the systems of economy and not its links. In connection with this, in developing a rational plan for the economy, the country must be divided into economically independent units—regions—and proceed by comparing the variants of the economic plan that have been worked out in order to implement various measures, in particular electrification."[3]

In developing the ideas of the GOELRO plan, Gosplan [the State Planning Committee in the 1920s] developed a method of economic regionalization, the main tenets of which were as follows:

1. The economic principle must be the basis for regionalization. This principle assumes, first, that the regions must specialize in those types of production that can be developed with the least expenditures. Second, the regions must represent a combined system of the economy with a specialization on a national scale constructed on the principle of its maximum efficiency, that is, on the principle of energy. Thus, specialization as well as organization of the economy of the regions have one goal: the achievement of greatest effectiveness, highest productivity of social labor. Third, the regions must be connected by a well-organized transport network that will facilitate their cooperation.

2. The economic principle of regional division must be apparent in such a form that it contributes to the material and spiritual development of all nationalities . . . applicable to the features of their life, culture, and economic status.

3. It is necessary to consider the process of development of the national economy of a region in the past (history) as well as the future (prospects); for the Soviet Union the latter is of special significance, since no rational regionalization can be accomplished without a long-range plan for the development of the national economy as a whole, and in its turn, a scientifically grounded regionalization. Such a regionalization must simultaneously serve the goals of planning and the goals of administration.[4]

Gosplan held that an economic region is, "as far as possible, a unique, economically complete territory of the country that, due to combinations of natural characteristics, cultural backgrounds, and a population trained for productive activity, would represent one of the links in the total chain of the national economy. This principle of economic completeness makes it possible to build further on a well-selected complex of local resources, outside capital, new technology, and a general state plan based on the best utilization of all possibilities with minimum expenditures."[5] As is known, on the basis of this methodology, Gosplan worked out a system of twenty-one economic regions for the country.

V. I. Lenin, at the Eleventh Party Congress gave high praise to the work of Gosplan on the regionalization of the country. He said: "We now have a division of Russia into regions developed according to scientific principles taking into account economic, climatic and living conditions, fuel availability, and local industry. On the basis of this division, rayon and oblast economic councils have been created. Of course, there will still be specific corrections, but the authority of these economic councils must be increased."[6]

The main theses of the theory of economic regionalization stressed by Gosplan in the 1920s are that the national economy is dynamic and built on the principle of achievement of the largest productivity of social labor; these theses retain their significance even to the present time.

It is natural that the rapid development of productive forces in the USSR demanded changes in the boundaries proposed by Gosplan in 1921. At the Eighteenth Party Congress, the question of a new economic regionalization of the USSR was especially acute. After the congress, the Academy of Sciences of the USSR was commissioned to take up the work of scientific regionalization of the country. However, Gosplan had already begun its work in utilizing first an eleven-unit and then a thirteen-unit division of the country. With the increasing complexity of the national economy and with the development of the territorial division of labor, the number of economic regions increases rather than decreases; moreover, with the increase in territory of the USSR the number of regions could not decrease to thirteen. Thus, each of the thirteen parts into which the country had been subdivided is not an economic region (as defined by Gosplan in the 1920s). The criterion for such a division of the country into parts was unpublished not by chance, of course, but because it cannot be scientifically justified or confirmed. If the thirteen-unit division of the country is still used, it is only because the Academy of Sciences has not fulfilled its assignment, and a network of the country's economic regions has not been provided in practice. N. N. Kolosovskiy discusses the inconsistency of a thirteen-unit division with the Gosplan principles.

Let us briefly examine the thirteen units of Gosplan. Only five of them are territorial-production complexes with a clearly manifested All-Union specialization and represent definite components in the country's national economy (the Urals, the Northwest, the North, the Far East, and Transcaucasia). The other eight units are a group of heterogeneous, unrelated regions.

What kind of economic whole can the Central region represent, when the parts of the country unified there are supposed to fulfill different economic tasks? The oblasts of the nonchernozem zone, led by Moscow, have a highly developed industry. The high concentration of industrial enterprises requires a limitation on further industrial construction. The central oblasts in the chernozem zone have an altogether different type of economy. The economy there is based on the processing of locally produced agricultural output. In the future, however, the iron-ore and metallurgical industries will take on importance, which again is not characteristic of the central oblasts of the nonchernozem zone. The relatively weak industrial development of the Central Chernozem region requires new industrial construction, in contrast to the Central Industrial region. Thus, the specialization and future development of these two territories are completely different.

It is also impossible to agree with the [delimitation of the] Volga region, which stretches from the forests beyond the Volga to the semi-arid lands of Astrakhan' Oblast. The problems of agricultural development in the southern part of the region, with its pasture-oriented animal husbandry and irrigated farming, have nothing in common with the problems of agricultural development in Kuybyshev Oblast and Tataria. The oil industry, which is developing on a gigantic scale in the central part of the Volga region, has achieved only local importance in the southern part so far. On the other hand, high-quality metallurgy distinguishes the southern part of the Volga region from the central part and ties it very closely to the eastern part of Donbass.

Western Siberia is a similar kind of economic conglomerate. What continuous economic intraregional unity can there be between the grain-growing western part of Siberia (with the whole complex of processing industries whose development was generated by the grain specialization of agriculture) and the second coal and metallurgical base—the Kuznetsk Basin?

The thirteen-unit division has Krasnoyarsk Kray, Irkutsk Oblast, and the Yakut ASSR as a single unit. At present the untenability of this union is becoming especially obvious. Powerful territorial-production complexes have developed (and are increasing in complexity) in this territory because of enormous electric-power construction. In the southern part of Eastern Siberia there are two such complexes: the first is taking form on energy from the Angara River and on minerals from the Angara region and the Trans-Baykal region, and the second is based on energy from the Yenisey River and the natural resources of the Yenisey territory. The Yakutsk region stands out quite distinctly in the North, with immense riches and specific ways of developing them.

A number of puzzling questions also arise when one examines the Ukraine-Moldavia region. For example, only four oblasts in the Ukraine—Stalino [Donetsk], Lugansk, Dnepropetrovsk, and Zaporozh'ye—have a coal and metallurgical type of economy; Khar'kov Oblast is related to these oblasts by its industry. But Gosplan, in singling out the Ukraine and Moldavia as a single economic whole, thereby assigns this specialization, completely without foundation, to all 26 oblasts in the Ukraine, and to Moldavia as well. At the same time, specialization in skilled machine-building and sugar-beet processing (industries characteristic of the central part of the Ukraine) is thereby ascribed to the mining region in the southern part of the Ukraine, where metal-consuming heavy machine building is developed and where there is almost no sugar industry.

Gosplan combines the Central Asian republics and Kazakhstan into a single unit. The Central Asian republics in the All-Union division of labor are the main base for cotton-growing, silkworm-breeding, and industries whose development is generated by agricultural specialization: the production of chemical fertilizers, related machinery, and fabrics. Although the Central Asian republics are also distinguished by their large-scale karakul and fine-wool sheep-raising, it was not these branches but cotton growing that was the region-forming branch for their territory. In recent years a metallurgical complex for nonferrous and rare metals has also been taking shape in Central Asia. The cotton complex of production branches that defines the economic character of the Central Asian republics is not at all typical of central, northern, western or eastern Kazakhstan. These parts of Kazakhstan do not have production ties with the Central Asian republics, either. Consequently, the consolidation of the Central Asian republics with the entire vast territory of Kazakhstan is unfounded. Moreover, the territory of Kazakhstan itself, as will be demonstrated below, objectively contains several economic regions, each of which has a pronounced economic specialization. It was these causes that served as the objective basis for the formation in Kazakhstan of several territorial-production complexes (economic regions) on an All-Union scale.

In singling out as a unit the Baltic and Belorussia [as economic regions], as in the preceding cases, there was again a consolidation of territories whose economies were developing in different directions. The economy of the Baltic republics is based not only on skilled cadres but also on their coastal location. Whereas the first element also exists in Belorussia, as it does, incidentally, in northern and central Russia, the second element, an extremely important one for the Baltic, is completely absent. On the other hand, Belorussia has a developed lumber industry, but in the Baltic there is none to speak of.

The further development of Belorussia's agriculture is linked to the solution of complex reclamation problems. This element also sharply distinguishes Belorussia from the Baltic republics.[7]

The analysis of the thirteen-unit division could be continued, but the facts cited make it possible to draw definite conclusions.[8] Basically, such a division does not satisfy one of the main demands of contemporary regionalization theory: it is impossible to judge correctly either the present or the future specialization of the majority of regions. On the other hand, the existing specialization of the territories is hidden, and, as a result, the territories lose their economic character. Therefore, the economic questions and problems of each region cannot be formulated with the necessary clarity.

The unification of economically heterogeneous territories that do not form a production-territorial complex (in the sense of interconnection and interdependence of enterprises) leads to considerable deficiencies in planning. These deficiencies are expressed in errors in locating new industrial enterprises and in additional necessary transportation work. The absence of a scientifically specified regionalization leads also to deficiencies in the planning of agriculture.

Consequently, the task is to develop a scientifically based system of economic regions on the basis of Marxist theory of economic regionalization, principles that were already stated in the works of Gosplan (1921–22).

As already stated, the Academy of Sciences did not fulfill the task of economic regionalization, but science as a whole did not stop investigating in this field. In particular, the Department of Economic Geography of Moscow University in the past twelve years has continually worked on economic regionalization and on the creation of a system of basic economic regions of the country. These investigations were conducted on the basis of the Gosplan methodology of the 1920s, mainly by N. N. Kolosovskiy. He evaluated the natural resources of the USSR from the economic point of view and made a preliminary analysis of the materials and data in monographs concerning large projects of reconstruction in individual regions of the country. Kolosovskiy also calculated the levels of development of the national economy of the USSR by using per capita indices on eighty-four products. On the basis of this work, in 1952, he composed and put forth a project for the system of basic economic regions in the USSR.[9] In the process of creating a network of basic economic regions he worked out a number of new methods for such a regionalization. Some had already been published by Kolosovskiy in 1947.

The Department of Economic Geography of Moscow University wanted to further develop the work on economic regionalization of the USSR by an in-depth analysis of the collected and projected materials and through the discussions of, department proposals with scientific and practical workers on locale. The death (in 1954) of Professor Kolosovskiy, the director, initiator, and primary executor of this, has set back the accomplishment of many undertakings of the department.

At the present time, work on the economic regionalization of the USSR has been revived in the department. The department continued the work of Kolosovskiy using transport, population, agricultural, and industrial statistics, as well as some projected data concerning the future development of the national economy of the USSR as a whole and its separate regions. This work is not yet completed but the part already done confirms the accuracy of Kolosovskiy's proposals on the regionalization of the country. These proposals are used in determining the system of basic economic regions of the USSR. However, some changes that were made in the network will be examined below.

Proceeding from the theory of Gosplan in working on the proposed system of basic economic regions of the country, it was assumed that the formation of such is a sociohistorical phenomenon in which the objective economic processes are reflected. The basic economic regions are formed in actuality according to their own laws of development. The economic regionalization has its own characteristics at each stage of development of society with a material technological base pertinent to it. The basic economic regions must be developed while the country fulfills its basic economic tasks (the next fifteen to twenty years). It is for this period that one must foresee the formation of a production complex of these regions.

In working out economic regionalization, the following questions must be resolved:

1. determination of the primary All-Union specialization for all basic economic regions so that each becomes apparent as a necessary component in the national economy of the USSR as a whole;

2. the specific features of the economic development according to its natural basis of production in each of the basic economic regions must be revealed;

3. establishment of national economic problems, typical and unique, for a given basic economic region, the problems of which must be solved within its boundaries;

4. establishment of the production processes developing on the territory of each of the basic economic regions; the raw material and energy bases of a complex development of productive forces must be ascertained, which combine to form a definite unit (a regional territorial-production complex).

The basic economic regions differ greatly with regard to raw material and fuel-energy resources. Some, with a high population density and a rather small supply of energy resources and raw materials, have a well-developed industry that began and is continuing to grow because of a skilled labor supply. In order to make clear and verify the boundaries of these regions, it is very important to know which regions support them with raw materials and energy resources. A different transport-economic orientation (in its historical development) determines to a considerable extent the specialization of the region and the organization of its productive forces. In establishing an energy base of a region, it is necessary to keep in mind that the basic economic regions develop their economy supported by regional energy systems, which at the same time must be looked upon as parts of an All-Union energy system with interregional electric trunk lines, oil and gas pipelines to supply the regions with types of energy they lack:

5. economic centers of a region and their economic region-forming pivots (main railroad lines, waterways, and the like) must be determined;

6. transport conditions and interregional shipment of goods must be studied since the division of the country into basic economic re-

gions must serve to combat nonrational hauling and excess expenditures for the transportation of goods and to speed up the turnover of valuable items in the country. The boundaries of the basic economic regions must be the criteria for establishing rational interregional hauling, a development that is inseparable from the specialization of these regions.

The economic regions are examined and defined with regard to the prospects for their development. They must have not only scientific-cognitive but, above all, active transformational significance, offering one of the most important tools of long-range planning.

In delimiting a system of basic economic regions, taking into account the experience of national development in the USSR, we assumed that the boundaries of the Union republics must remain inviolate. The largest of them form several economic regions (RSFSR, the Ukraine, Kazakhstan) or an independent region; some republics, smaller in territory, population, and economy, and having a definite economic similarity and unity, comprise an economic group (the republics of Transcaucasia, the Baltic, and Central Asia).

In the future it will be expedient to pose the question concerning the creation of a single management over the economy in an economic region, although in an economic grouping the creation of such supervision can lead to violation of sovereignty of the Union republics. However, the solution of national-economic problems demands the systematic work of regional economic conferences in which the republics could jointly decide how to resolve common economic problems. Such conferences could have a permanent coordinating body.

In the course of delimiting basic economic regions, it was found that the boundaries of administrative districts needed changes, which will be discussed below.

Presently, taking into account the future economic development in the next fifteen to twenty years, the contours of the following twenty-nine basic economic regions and economic groups of the Union republics are sharply defined in the USSR. The basic economic regions of the RSFSR are: (1) Central Industrial; (2) Central Chernozem; (3) Northwestern; (4) Northern; (5) Volga-Vyatka; (6) Middle Volga; (7) Volga-Don; (8) Northern Caucasus; (9) Urals; (10) Western Siberian; (11) Kuznetsk-Altay; (12) Middle Siberian; (13) Eastern Siberian; (14) Yakutia; (15) Southern Far East; (16) Northern Far East (total, 16). The economic regions of the Ukraine are: (17) Southern Mining; (18) Central Ukraine; (19) Western Ukraine; (20) Black Sea Coast (total, 4). (21) Moldavia. The basic economic regions of Kazakhstan are: (22) Central Kazakhstan; (23) Western Kazakhstan; (24) Eastern Kazakhstan; (25) Southern Kazakhstan (total, 4). The economic groups of Union republics include: (26) Central Asia (Uzbekistan, Kirgiziya, Tadzhikistan, Turkmenia); (27) Transcaucasia (Azerbaydzhan, Armenia, Georgia); (28) the Baltic (Latvia, Lithuania, Estonia, and Kaliningrad Oblast of the RSFSR. (29) Belorussia.

Research in economic regionalization at present is characterized by the delimitation of major tasks and problems being resolved by the various territories of the country and, as derivatives of these problems, the boundaries of the economic regions. Consequently, one takes into consideration that the "first stage in delimiting an economic region is to determine the main specialization of the region, and, on this basis, approximate contours of the region are established; it must contain within its boundaries the largest resources necessary for the chosen specialization."[10] The question of territorial organization of the productive forces by regions will be examined in the next stage of the work.

1. *The Central Industrial Region* (comprising the Moscow City, Moscow Oblast, Bryansk, Vladimir, Ivanovo, Kalinin, Kaluga, Kostroma, Ryazan', Smolensk, Tula, and Yaroslavl' economic administrative regions).

This is the oldest and most important region of the processing industry in the USSR, which to a considerable degree, formed and developed as a result of its central position. The main productive forces of the region are its industrial cadres, which have been here a long time, and considerable capital investments accumulated over a long period. Machine construction requiring skilled labor but not metals and a highly developed textile and chemical industry define the production specialization of the region. Although large-scale coal mining and wood-processing are characteristic of the Central Industrial region, these activities play only an intraregional role.

The degree of industrialization of the Central Industrial region is higher than any other region. In the next fifteen to twenty years the main task will not be new industrial construction but the improvement of the economy of the region in order to achieve higher productivity. The Central Industrial region with its dense population and large number of cities is an important consumer of industrial articles and agricultural products; this strongly influences the development of its production complex. Moscow is particularly outstanding, both as a production and consumption center.

The region-forming role of Moscow, the center of a dense network of radial transportation lines, high-voltage electric lines and gas pipelines, is very great. Moscow is located at the focal point of the Central power system, whose electric stations primarily use intraregional resources—coal and peat from outside Moscow. The central power system, encompassing the majority of regional electric stations of the economic region, unites the region. The region uses a huge amount of fuel (oil and coal) brought in from other areas, as well as raw materials (metal, cotton) and food (grain, and other agricultural products); the portion of nonlocal material has considerably increased due to the growth of the processing industry during the last five years. The Central Industrial region receives raw materials and fuel mainly from the southern regions of the country. A reduction in long-distance rail hauls is partially achieved by the construction of power lines, gas pipelines, and oil pipelines from other regions of the country. Modernization of industrial enterprises related to new technological achievements demand greater electrification of these enterprises, that is, the electric consumption by industry in the region increases. An increased demand for electric energy in the production of high-quality materials makes it economically profitable to build atomic power stations in the region. The high price of energy and fuel, compared with many other regions, the high expenditures for transportation of raw materials, and highly skilled and highly paid labor all demand a reorientation of industrial enterprises toward greater specialization and their conversion to the output of high-quality articles. A great deal of work in combining enterprises for complete and complex utilization of raw materials and industrial wastes is also needed. The region is faced with the task of developing intensive suburban agriculture with utilization of industrial wastes and wide use of electricity.

2. *The Central Chernozem Region* (comprising the Belgorod, Voronezh, Kursk, Lipetsk, Orel, and Tambov economic administrative regions). The Central Chernozem is a large region of agriculture and new processing industries, which developed because of a convenient transportation location on transit rail lines connecting the Central Industrial region with the southern economic regions of the country. This region is characterized by the absence of any large local sources of energy and for this reason is noticeably behind the majority of other regions in the country in the development of energy. It would be economically expedient here to develop atomic energy. The prospects for mining the iron ore of the K.M.A. (Kursk Magnetic Anomaly) and developing an iron and steel industry and a building-materials industry are excellent. In addition, one of the important tasks is that of developing the machine-building industry with the utilization of the metallurgical base of the K.M.A. Further industrialization is necessary for a more rational utilization of the large labor reserves available. The region does not have a powerful region-forming industrial or energy center similar to Moscow or Gorkiy. However, Voronezh is becoming an important economic center and is already playing a leading role in the region. Agriculture is of All-Union significance (sugar beets, hemp, sunflowers, corn, rye, potatoes, and livestock). Many agricultural problems must be

solved for all oblasts in the region, including the development of a fodder base, intensification of agriculture in general and, in particular, livestock-raising of All-Union significance with measures taken to raise the fertility of plowed soils and to combat erosion. A very important problem for the region is that of water supply, especially for cities and industry.

3. *The Northwestern Region* (comprising the Leningrad, Murmansk, and Karelian economic administrative regions).[11] The Northwest is one of the largest and oldest coastal regions of industrial processing in the country. It developed on raw materials brought in from other areas but, during the five-year plans, began utilizing local hydroelectric power, peat, timber, and raw minerals concentrated primarily in the Kola Peninsula. The Leningrad industrial and transportation center is the nucleus of the region. The Kirov Railroad, connecting the two seaports of Leningrad and Murmansk, is an important region-forming trunk line. More distant than the Central Industrial region from the Donets Basin, sources of oil, and natural raw material bases, the region was compelled to process and utilize local natural resources to develop intraregional production connections and high-quality industrial products. The uniqueness of the regional production complex lies in the combination of heavy-machine construction, primarily shipbuilding, with precision machine tools, mining, chemical, timber, wood-chemical and textile industry. Fishing, forest exploitation, pulp, paper and cardboard production are of All-Union significance.

4. *The Northern (Dvina-Pechora) Region* (comprising the Arkhangel'sk, Vologda and Komi economic administrative regions). This is a large lumber-industry region in the European part of the USSR with a well-developed lumber-processing industry—sawmills, wood chemicals, pulp, and paper. In absolute terms, the Northern Region is second in the USSR in wood shipments, and the production of lumber, pulp, and paper. The lumber-processing industry is best developed in Arkhangel'sk, the economic center of the region, and in Onega. In recent years, the Northern region has become an important fuel base of the country due to the development of Pechora Basin coal, which goes mainly to Leningrad and Cherepovets. The mining of other types of minerals, especially the extraction of natural gas, is also significant.

The Northern region has great possibilities for further development of its industrial complex based on timber, coal, natural gas, and maritime resources, as well as dairy output of All-Union significance. Energy production, machine construction (shipbuilding), and the chemical industry connect the elements of this complex.

5. *The Volga-Vyatka Region* (comprising the Gork'iy, Kirov, Mari, Udmurt, Chuvash, and Mordvinian economic administrative

regions). This is a large region of industrial processing especially focused on Gork'iy. The city of Gork'iy, fourth in the USSR in industrial production, is an important region-forming industrial and transportation center. Transport and other types of machine construction, timber, wood-chemical and pulp-and-paper industries are of All-Union significance. Lumber and lumber products are shipped to the Central Industrial Region, Middle Volga, Volga-Don and into the mining-industry of the Ukraine. Lumbering on a large scale creates problems of reforestation and transition to high-quality forms of wood-processing and utilization of wastes. As an energy base, the Volga-Vyatka region has a system of three large hydroelectric stations on the Volga and Kama rivers (the Gorkiy, Cheboksary, and Votkinsk stations) and smaller ones on the Vyatka and other rivers. The discovery of coal in the Udmurt ASSR makes it possible to eliminate somewhat the acute shortage of mineral fuel. In contrast to the neighboring Central Industrial region, the Volga-Vyatka gravitates more to the Urals and Middle Volga for raw material and fuel. The favorable geographic location of the region (between the Central, Ural, and Middle Volga regions), a well-developed transportation network, and hydroelectric resources afford great possibilities for further industrialization. Close economic ties and a common lumbering specialization raise the question about incorporating the eastern part of the Kostroma district into the Volga-Vyatka basic economic region.

6. *The Middle Volga Region* (comprising the Ul'yanovsk, Kuybyshev, Penza, Saratov, and Tatar economic administrative districts). This is one of the most important regions of the country for the oil, gas, chemical and machine-building industries. As early as 1955, the Middle Volga was the leading area of the USSR in the extraction of oil. The boundaries of the region are determined, not by its economic tendency toward the Volga alone, but also by the contour of the oil and gas deposits (the eastern wing of the contour is in the Urals, mainly within the Bashkir ASSR.). Grain-growing is of great importance particularly in the trans-Volga region. This economic region developed on the Volga waterway, intersected by a number of railroads. Economic connections with the Urals, Siberia, Kazakhstan and Central Asia also played a great role in forming the industrial complex of the region. Large hydroelectric stations (the Lower Kama, Kuybyshev, and Saratov plants) make up the powerful energy base of the region. Hydroelectric power and the oil- and gas-processing industries form the new, solid foundation for the industrial complex of the region. Electric power, petroleum and petroleum by-products, natural gas, chemical products, machines, and grain are the most important products shipped to other economic regions of the country. There are three industrial centers in the

region: Kazan', Kuybyshev, and Saratov. The most important at present is Kuybyshev, with the most powerful hydroelectric station, many machine-building plants, and the oil- and gas-processing industry. It should be noted that the western, oil-industry section of Orenburg Oblast belongs with the Middle Volga region.

7. *The Volga-Don Region* (comprising the Stalingrad [Volgograd], Astrakhan', and Rostov economic administrative regions and the Kalmyk ASSR, whose industry is directly subject to its Council of Ministers). This is one of the most important basic economic regions in the country, where heavy industry (primarily coal mining, high-quality metallurgy, and machine construction) related in part, to the eastern section of the Donets Basin combines with light industry, fishing of All-Union significance, intensive irrigated agriculture, and well-developed sheep-raising. Although Stalingrad and Astrakhan' oblasts contain oil and natural gas, uniting them with the Middle Volga region would be incorrect, since in the Volga-Don region the oil and gas industry is not the leading one and is in an entirely different combination with the other branches of the economy. This is also true of the timber-processing industry, which is based on outside raw material. The region-forming role is played by the eastern portion of the large Donets Basin with transportation lines, including the Volga-Don Canal, connecting the Donets Basin to the Volga. In the future there will also be electrical-transmission lines linking the Stalingrad Hydroelectric Station, via the Tsimlyansk Hydroelectric Station, with the thermal-power plants of the Donets Basin. The Stalingrad industrial center should be viewed as the foremost eastern outpost of the Donets Basin, as its Volga "gates." The largest industrial centers of the region are Stalingrad and Rostov, the former being the most important in that it is situated near the Volga-Don Canal. Even more important for the future will be the construction of the Stalingrad Hydroelectric Station. The important region-forming role of Stalingrad must be noted as one of the serious objections against singling out a Volga economic region from Kazan to Astrakhan, since the Kuybyshev and Stalingrad industrial centers are the nuclei of different regional production complexes.

8. *The Northern Caucasus Region* (comprising the Krasnodar, Stavropol', Chechen-Ingush, Dagestan, Northern Ossetian and Kabardin-Balkar economic administrative regions). This is an important economic region with an All-Union specialization in the extraction of natural gas and oil, the production of grain, wool, meat, and a number of industrial crops (tobacco and others) and their processing. Some branches of nonferrous metallurgy are of All-Union significance. Hydroelectric power and natural gas are the sources of

energy in this region. The heavily populated belt of the northern foot-hills of the Great Caucausus is the region-forming pivot. Through this region passes the Rostov-Baku main railroad line, which runs parallel to deposits of natural gas, oil, and nonferrous metals. Ir-rigated agriculture, extensive farmland devoted to valuable industrial crops, and the industrial cities of the region are located on it. The natural resources of the region, especially natural gas (one fourth of the USSR reserves), make possible a still more intensive econ-omy. In particular, a high-capacity rural energy system can be created on the basis of the utilization of natural gas. The Northern Caucasus is a multinational region and includes several autonomous republics and oblasts with their own centers. This makes it difficult to develop a large economic center for the whole region; at present Krasnodar stands out, but in the future Armavir will possibly re-ceive heavy development, since it occupies a central position in the region.

9. *The Ural Region* (comprising the Sverdlovsk, Perm', Che-lyabinsk, Bashkir, and Orenburg economic administrative districts). The Ural region, in its defined boundaries, is an integrated, many-sided territorial-production complex, with a clearly manifested All-Union specialization in ferrous and nonferrous metallurgy, heavy metal-consuming machine construction, chemical, timber, nonmetallic mineral industries and, in large measure, an oil in-dustry. Although the coal industry of the Ural region is significant, it supplies only part of the energy needs of the region. The Ural region is compelled to meet its fuel needs by shipping in coal from the Kuznetsk Basin and partially from Karaganda [in Kazakhstan]. In the future, the northern Urals will receive a considerable amount of Pechora coal.

The Ural complex is characterized by its strongly developed in-traregional economic ties. The coordination of these ties is imple-mented to a considerable degree by the leading economic center of the Urals, Sverdlovsk. The development of connections is facilitated by the configuration of the transportation network and a single electro-energy system. The industrial centers of the mining sec-tion of the Ural region cannot be separated from those of the pro-cessing industry—heavy and light industry and food-processing—or from the regions of agricultural production on the cis-Ural and trans-Ural plains.

Presently, the Ural region is sufficiently saturated with indus-trial enterprises with power so great that, in a number of cases, they operate with nonlocal raw materials, particularly iron ore and natural gas. This situation will continue and grow in spite of the abundance of its resources. The complexity of development of the Ural economy will increase steadily, especially with greater specialization and cooperation. Energy production is very significant, including a nuclear power station.

10. *The Western Siberian Region* (comprising the Kurgan, Tyumen, and Omsk economic administrative regions). The Western Siberan region should also include the western parts of the Altay region and the Novosibirsk district, both with single-type economies. Western Siberia is one of the leading grain-producing regions in the USSR and of international significance. The region has approximately 17 percent of the virgin and unused lands that have been developed in recent years (1954–56). Beef and dairy cattle-raising are also of All-Union significance, and a large food-processing industry has been created. The needs of the grain economy have led to the development of farm-machine construction. The new center of machine construction is Omsk, which has become an important center of Western Siberia. The region-forming role of Omsk has increased with the construction of an oil refinery of significance to all of Siberia.

The field-farming of Western Siberia, except for Omsk, is oriented also toward Kurgan and Rubtosovsk centers of farm-machine construction located in the zone of virgin and idle lands.

> In the future, an independent, economic region, the Northern Ob' region, will emerge from the Western Siberian Region; at the present, however, the territory is still undeveloped. In all of the vast Northern Ob' territory there are fewer than 180,000 people. The Northern Ob' territory could not be united with the other territories of Northern Siberia, either, since all its economic development is closely connected with the southern portion of Western Siberia, which already has a highly developed economy and (for Siberia) a high population density.[12]

11. *The Kuznetsk-Altay Region* (comprising the Kemerovo, Tomsk, Altay, and Novosibirsk economic administrative districts). This is a territorial-industrial complex specializing in coal mining, extensive production of chemical products, ferrous and nonferrous metals, and metal-consuming machine construction.

The main economic nucleus of the region is the Kuznetsk Basin (with a yield of 66 million tons of coal in 1956). The problem of how to supply the region with iron ore to satisfy the needs of metallurgy is an acute one, which explains the lag in the output of metal in relation to the extraction of coking coal. Kuznetsk coal does not yet have enough large local consumers so large amounts are shipped to other more distant regions. The discovery of iron ore in Kolpashevo may enable the region to have its own large iron-ore base in the more distant future. Aside from coal mining, the region gets its energy from hydroelectric stations on the Ob' River. The timber industry is very significant, and the grain economy is of All-Union importance.

Novosibirsk, like Khar'kov in the Ukraine, is the center of heavy and electric-power machine construction, based on the nearby coal and metallurgical base. It is also a large center of farm-machine construction for western Siberia, but metal-consuming heavy machine construction is its outstanding characteristic.

> The question concerning the eastern part of the mining Altay (the Zmeinogorsk group) should be discussed, since it gravitates toward the mining Altay situated in Kazakhstan. It is possible that in the future it will be economically expedient to unite the Zmeinogorsk group with the East Kazakhstan economic region.[13]

12. *The Middle Siberian Region* (comprising the Krasnoyarsk economic administrative region and the Tuva Autonomous Oblast [which became an autonomous republic in 1961], whose industry is directly subordinated to the oblast executive committee). The Yenisey River is the main region-forming transport artery; it is not only a waterway but also a colossal source of cheap hydroelectric power. The low-cost open-pit mining of brown coal in the Chulym-Yenisey, Kansk, and other regions makes it possible to build large thermal stations producing cheap electrical energy. A number of deposits of iron ore, light and rare metals of Yenisey ridge and the Sayan Mountains make it possible to develop electrical metallurgy of ferrous, nonferrous, and light metals. The timber, pulp, and paper industries are of greatest All-Union and even international significance.

Krasnoyarsk is becoming one of the main industrial centers of Siberia and is acquiring an important regional significance, especially with the construction of the giant Krasnoyarsk Hydroelectric Station on the Yenisey River.

> The Northern Yenisey area in the north of this region, in the future will form an independent basic economic region. At present, this territory must still be united with the southern portion of Middle Siberia for the same reason as the ties between the Northern Ob' and the southern part of Western Siberia. But continued development of mining and nonferrous metallurgy (in Noril'sk) and the timber industry (in Igarka) will make it possible to form a special territorial-industrial complex of the lower Yenisey.[14]

The question of regionalization of the central part of Siberia is very complex and still unclear. Despite the fact that in our hypothesis two basic economic regions, the Kuznetsk-Altay and Middle Siberian, are distinguished in central Siberia, there are a number of serious arguments for possible unification of the Kuznetsk-Altay and Middle Siberian regions. In recent years, a clear tendency for the expansion of the Kuznetsk metallurgical complex toward the East has been noticed; the industrial ties with the Minusinsk basin, especially with Khakassiya, are becoming even stronger. The nonferrous metals in the southern portion of Krasnoyarsk Kray and the iron ore of the lower Angara will be of great significance. The projected power transmission line (the Krasnoyarsk Hydroelectric Station to the Kuznetsk Basin) will give a definite electric-power unity to these two regions.

13. *The Eastern Siberian Region* (comprising the Irkutsk, Chita, and Buryat economic administrative districts). The extensive and exceptionally low-cost energy of the hydroelectric station on the Angara River is the basis for the future of the Eastern Siberian complex. Selenga, Baykal, and Angara are the energy and economic pivots of the region.

The characteristic features of this region are the common energy source of the Angara River and its tributaries, the common fuel and raw material bases (Cheremkhovo coal, iron ore of the Angara-Ilim deposits and Sosnovyy Bayts, Usol'ye salts, nonferrous metal ores of the Chita district and Buryat-Mongolia), the common lumbering specialization of the region, and a number of common transportation problems with the creation of a continuous waterway along the Selenga, Baykal, and Angara.

Irkutsk is the leading region-forming center of Eastern Siberia, economically uniting the cis-Baykal and trans-Baykal areas.

The production (in the future) of electric energy on a gigantic scale together with ferrous and nonferrous ores and a sparse population necessitates development of consuming industries. The presence of gas, as well as coal, makes it expedient to develop coal chemistry, especially organic synthesis. Large reserves of sodium chloride in Usol'ye allow diversification in the development of the chemical industry. A weak coke base in Eastern Siberia demands a special approach to the question of developing ferrous metallurgy. The solution lies in the extraction of metallurgical coke from local coal, which is possible if hardening components are added.

The proposal to make the Buryat ASSR and Chita Oblast into a separate basic economic region—the Trans-Baykal region—is, in our opinion not yet justified, since the huge energy resources of the Angara will demand the use of the natural resources of the Trans-Baykal region and vice versa. Without the support of Angara energy the development of the productive power of the Trans-Baykal region will be considerably hampered. In the future, when the energy of the Amur river is utilized, energy-consuming industries created, and ferrous metallurgy is developed based on the ores of Borzya, the eastern part of Chita Oblast will be connected with the southern part of the Far East. In that case the boundary between the Eastern Siberian economic region and the southern Far East will shift to the west of the present boundary, or possibly a new Amur economic region will be formed.[15]

14. *The Yakut Region* (the Yakut economic administrative region). The main specialization of the region is the processing of valuable minerals (diamonds, gold, and tin). The territory is still very poorly developed. The low density of population, difficult natural conditions, and lack of roads creates great difficulties for the development of industrial power. Resources of high value (gold, diamonds, tin) are being developed by relatively small industrial centers. The economic life of Yakutia is tied to the Lena River, which is its main transportation trunk line. The transport development of the Lena is an important economic task of Yakutia. The recent discovery of very large iron-ore deposits and coking coal in southern Yakutia (Chul'man, Aldan) has created a new problem—the development of an iron and steel industry. However, the natural conditions of Yakutia make the solution of this problem difficult. It is very likely that Yakutia will not have its own metallurgical industry, but will be a supplier of metallurgical raw material and fuel (or only fuel) for the metallurgical plants of the Far East.

15. *The Southern Far East Region* (comprising the Khabarovsk, Primor'ye, Amur, and Sakhalin economic administrative regions). The geographic and political economic position of the southern part of the Far East (located on the last section of the Siberian rail trunk line, in the lower part of the Amur River on the shore of the Pacific Ocean) and its remoteness from developed basic economic regions of the Soviet Union influence the formation of the economic profile of this region. Its position on the coast, the vast natural resources of the seas, the wealth of various useful minerals (including metals and minerals deficient in the country), the large fuel-energy resources (especially the hydro resources of the Amur), the huge forests, oil, furs, and other natural wealth create favorable conditions for the development of a future complex of industries.

In the All-Union economy, the southern part of the Far East is distinguished by its fishing industry, extraction of nonferrous and precious metals, timber-processing, heavy machine construction, soybean production, extraction and processing of oil, which are leading components of the economic complex. The economic development of the region is closely connected with maritime transport.

16. *The Northern Far East Region* (comprising the Magadan and Kamchatka economic administrative districts). Extraction of rare and precious metals and the fishing industry are the main branches of the economy forming the future territorial-industrial complex of

the region. The abundance of very valuable minerals and the huge resources of the sea represent the potential for strong development of productive power in the region. However, remoteness from the basic economic centers of the country, the absence of rail transportation, and severe natural conditions make it difficult to develop the economy and demand a special approach to the utilization of the territory. The nature of these difficulties is clear in that the development of productive power demands the creation of industrial centers, distributed on undeveloped territory.

The development of the productive force is determined to a large degree by the connections of the region with other regions of the country. Maritime transport plays an important role; its significance for the economic life of the Northern Far East is greater than for any other region of the country.

At the present time the economic development of the Northern Far East has reached such a scale that is has become expedient to distinguish it as an independent basic economic region of the USSR. Gross industrial production is approximately three times as high as in Yakutia. Population growth is related to industrial development. The population is slightly less that that of Yakutia and is two and a half to three times as large as in the Northern Ob' or Yenisey areas, although its area is less than half that of Yakutia and approximately equal to that of the northern Yenisey territory.

Further development of the economic integration of the region during the next two decades will necessitate the creation of a processing industry based on the present extractive industries. An important economic task of the region will be the creation of a vegetable-dairy base to satisfy the needs of the population.

17. *The Southern Mining Region of the Ukraine* (comprising the Stalino [Donetsk], Lugansk, Zaporozh'ye, and Dnepropetrovsk economic administrative regions and Khar'kov Oblast). The Kerch' peninsula of Crimean Oblast should also be included in this region.

The industrial complex of the region is distinguished by the exceptional diversity and great complexity of internal and external economic connections. The natural foundation of the complex include the various coals and salts in the Donets Basin, ores in Krivoy Rog (and Kerch'), manganese in Nikopol', the energy of the Dnieper River, and the fertile chernozem and brown soils. Local natural gases are acquiring significance. On this basis there has been vast development of metallurgy, coal chemistry, and metal-consuming machine construction, not only in the cities of the Donets Basin and the Dnieper region but also in Khar'kov and other cities.

The huge reserves of ores combined with the hydroelectric resources of the Dnieper have made it possible to create industries producing high-quality steel, ferro-alloys, as well as aluminum, magnesium, and zinc (from outside raw material). The transport and electric-power systems enhanced by the unification of the hydro-electric stations on the Dnieper and the thermal-power stations of the Donets are important factors in the creation of an economic entity.

The leading position in the industrial complex is held by the Donets Basin and Khar'kov, the main industrial and transport hub and apex of the industrial complex of the entire economic region.

Further industrial development will proceed by improving present production processes, replacing coking coal with brown coal, introducing new processes of coal exploitation, solving the problem of water supply for the Donets, and developing a number of light industries (especially important for utilization of female labor). The vast scale of industry in the Southern Mining Region of the Ukraine have resulted in a massive industrial population matched by no other single region of cities and worker settlements.

The region is a large producer of grain. Continued growth of agriculture is closely related to the drainage of land inundated by the Dnieper. In interregional terms, this region is a supplier of coal, ore, metal, machines, sodium carbonate, salt, grain, and a large consumer of timber and textiles. In the future, a considerable amount of ore may be brought in from the Kursk Magnetic Anomaly of the Central Chernozem region.

18. *The Central Ukrainian Region* (comprising the Kiev and Vinnitsa economic administrative districts and Poltava and Sumy oblasts). Kiev is the regional hub of the Central Ukraine.

The very favorable geographic position of the Central Ukraine, with its fertile chernozem soils, dense population, and intensive agriculture has had a great influence on the development of the food industry (especially sugar) and non-metal-consuming skilled machine-building, including farm-machine construction.

The main All-Union specialization in agriculture is the production of wheat and industrial crops, primarily sugar beets (about half of the All-Union harvest) and a developed livestock economy. Horticulture plays a considerable role.

Surveying for useful minerals is leading to the discovery of new deposits; especially noteworthy are the results of the investigation of oil-bearing sites in the Central Ukraine. The development of industry on the basis of newly discovered deposits will lead to a more intricate industrial complex in the region.

19. *The Western Ukrainian Region* (comprising the L'vov and Stanislavov economic administrative regions). The Western Ukraine, which began socialist development later than the majority of the

other regions of the country, has already created a foundation for the complex development of the economy. The primary natural resources are oil and natural gas. The development of the Carpathian natural gas supply (with the transfer of the gas to Kiev and Moscow, and in the future to Leningrad) has made the region one of the leading areas in the USSR in natural-gas extraction and has extended considerably the interregional connections of the Western Ukraine. The most important task of the region is the continued development of industry based on the oil-and-gas chemical cycle, the utilization of potassium salts, hydroelectric resources, large local fuel resources (brown coal of the Volyn-L'vov Basin), and its forests, which meet the need for high-value wood. Part of the electric power of the region is transmitted to Hungary.

The agriculture of the Western Ukraine is characterized by the growing role of industrial crops (mainly sugar beets), and consequently, the large requirements of mineral fertilizers. Horticulture and viticulture are also of great significance in the region.

The leading economic center of Western Ukraine, L'vov, is an industrial and transport hub uniting the economy of the region into a single unit.

20. *The Black Sea Coastal Region of the Ukraine* (comprising the Odessa and Kherson economic administrative regions). The largest economic center of the region is Odessa.

The region, situated in the steppe, is a large producer of grain, supplying about two-fifths of the grain of the Ukraine. A developed grain economy, horticulture, and viticulture were the basis for the growth of various branches of the food industry: flour-milling, fruit-canning, and wine-making, and the like.

The region's maritime and border position has played an essential role in the formation of its economy. It shaped the development of an industrial complex of fishing, fish-processing, and whaling industries as well as the specialization in machine construction (shipbuilding and other). The proximity of the Southern Mining region of the Ukraine and the development of international ties via Odessa also contributed to the development of machine construction and other branches of industry.

21. *The Moldavian SSR* (the Moldavian economic administrative region). The development of the productive forces of the Moldavian SSR must be examined in relation to the economy of the Black Sea region of the Ukraine. Although the Moldavian SSR is not as large an economic region as, for example, Belorussia, and its distinction is economically less substantiated, we examine it as an independent region for nationality reasons. Moldavia, a large producer of industrial crops, is similar to the Black Sea Region in its All-Union specialization in agriculture. The economic features of Moldavia

are comparatively one-sided and on an All-Union scale, it is distinguished only by its viticulture. Moldavia's main tasks include the development of the food-processing industry, and the strengthening of ties with the Ukraine to obtain agricultural machines and mineral fertilizers as well as equipment for the food industry. The development of machine construction in the republic must proceed by taking into account the division of labor and cooperation with the regions of the Ukraine. In the future, extraction and processing of oil will be of great significance for the economy of Moldavia.

22. *The Central Kazakhstan Region* (comprising Karaganda, Northern Kazakhstan and Kustanay economic administrative regions). This is a very important coal and grain base of the country and, in the future, should be a very large iron-ore and metallurgical base (ferrous and nonferrous metals). The economic nucleus of the region is the Karaganda coal basin.

A basis for the development of ferrous and nonferrous metallurgy has been created in the form of coal reserves of Karaganda and Ekibastuz, copper ores of Balkhash, Dzhezkazgan, and others, and the recently discovered Kustanay iron ores and bauxite.

Presently, the mining industry lags considerably behind the natural possibilities of the region. Metallurgy lags still further behind the mining industry. These disproportions are due, in particular, to the lack of water in the region. Supplying the internal sections of Central Kazakhstan with water by building a water conduit from the Irtysh River and also utilizing underground water will give an impetus to rapid development of productive power in the region.

The most important economic tasks of the region include the development of the iron-ore industry, providing ore for the Ural region and its own ferrous metallurgical industry, increasing coal production, utilizing new coal deposits, and a further growth of nonferrous metallurgy.

Another very important specialization of Central Kazakhstan, its grain economy, strengthens its ties with Siberia. One of the most important tasks in the economic development of Central Kazakhstan, along with Siberia, is providing grain-producing sovkhozes (state farms) with machines, fertilizer, and equipment. This problem becomes acute if one takes into consideration that Central Kazakhstan contains half of all the virgin and idle lands made useable in the country in the 1954–56 period.

23. *The Western Kazakhstan Region* (comprising the Gur'yev and Aktyubinsk economic administrative regions). The industiral-territorial complex of the region is at the beginning stage of its development. High-quality metallurgy (ferrochrome) and extraction of oil are the All-Union specializations.

The most important problem of further development of the economy is to increase the mining and processing of rare and nonferrous metals. In addition, the region has a developed chemical industry, processing various types of raw materials that are genetically related to deposits of salts and oil. The chemical industry also produces phosphorous fertilizers. This activity, however, depends on nonlocal phosphorites (Kara-Tau) in spite of sufficiently large phosphorite deposits in the Aktyubinsk district.

The food industry is significant for the region, especially fish-processing. The complex of industries connected with the fish industry is rather well developed in Western Kazakhstan—from fishing and processing of fish to shipbuilding.

Agriculture in the region plays a subordinate role. Livestock-raising is of much greater significance than agriculture. As in all of Kazakhstan, sheep make up the largest part of the livestock herds (about 75 percent). A further development of livestock-raising demands improvement of grazing lands, especially their irrigation.

24. *The Eastern Kazakhstan Region* (comprising the Eastern Kazakhstan and Semipalatinsk economic administrative districts). The production complex developed in Eastern Kazakhstan is based on the ore resources of the Altay Mountains (the western part), which are exceptionally rich sources of nonferrous metals in the country (tin, zinc, silver, gold, and others). The complex of nonferrous metallurgy in Eastern Kazakhstan is complete, extending from the primary production process (the mining of ores) to the obtainment of refined nonferrous and rare metals. The development of the metallurgical complex is based on the hydroelectric power of the Irtysh and its tributaries; Ust'-Kamenogorsk, on the Irtysh, is the leading electric-power and industrial hub of the whole region.

Another very important branch of the Eastern Kazakhstan economy having national significance is livestock-raising, which is the basis for the development of the food industry (the Semipalatinsk Meat Combine) and the processing of products of livestock (wool, leathers). The processing of livestock products has resulted in inter-regional and international economic ties in that livestock are shipped in from adjacent regions of Kazakhstan, the Mongolian People's Republic and the Chinese People's Republic. The economic tasks facing the region include the further development of nonferrous metallurgy, strengthening its electric-power base, developing the processing of livestock products and, in particular, the production of wool fabrics, and footwear.

25. *The Southern Kazakhstan Region* (comprising the Alma-Ata and Southern Kazakhstan economic administrative regions). The proximity of Central Asia and the belt of oases, which continues in southern Kazakhstan, have had a definite influence on the formation of the economy in this region. The All-Union specializations in-

clude tin-smelting, mining and processing of phosphorites, rice-growing, and the production of wool and cotton. The nonferrous metallurgical industry also uses the ore resources of Central Asia. The successful development of the Southern Kazakhstan cotton complex is possible under the conditions of the division of labor with the republics of Central Asia, developing first the mineral fertilizer and farm-machine industries. The problem of irrigation of new lands, especially in the Semirech'ye region and along the lower Syr'Dar'ya River, is of great significance. The development of irrigation is related to the construction of hydroelectric facilities, which will make it possible to strengthen the poorly developed electrical power base of the region.

Southern Kazakhstan plays a very important national role by its raising of fine- and semi-fine-wool sheep; this region has about 50 percent of the sheep in all of Kazakhstan.

26. *The Central Asian Region* (comprising the Uzbek, Kirgiz, Tadzhik, and Turkmen republics). The republics of Central Asia are combined into one economic group because they have, and must solve, a number of important common economic problems: irrigation and the hydroelectric operations related to it; production of cotton and silk; mining or ore and smelting of nonferrous and rare metals; extraction of oil and natural gas; livestock-raising; and, in particular, the raising of karakul sheep.

The development of nonferrous metallurgy requires the ore, fuel, and hydroelectric resources of all Central Asia.

Activities important in the All-Union economy include livestock-raising (especially karakul sheep) requiring work in pasture irrigation; silkworm-breeding, which is closely connected with cotton-raising; fruit-and vegetable-growing and viticulture, which are the basis for the canning and wine industries.

A further development of the cotton-industry complex will necessitate a territorial division of labor between the separate parts of Central Asia; some parts are distinguished by their production of mineral fertilizers, others by agricultural machine construction and still others by liquid fuel production.

27. *The Transcaucasian Region* (comprising the Azerbaydzhan, Georgian, and Armenian Republics) has many common problems also. The economic complex of these republics is characterized primarily by a cycle of oil-chemical industries, nonferrous metallurgy, and, in the field of agriculture, production of subtropical crops (citrus fruits, tea, and others) and cotton. The growth of hydroelectric power has resulted in the creation of a number of energy-consuming industries (aluminum, ferroalloys, carbide, calcium, and others).

The most important economic problems of the Transcaucasian

republics include the further development of the oil-chemical cycle in conjunction with the improvement of oil- and gas-refining, further growth of nonferrous metallurgy, and the development of subtropical agriculture. Economic ties between the republics are necessary not only for the development of industries of national significance but also for the intraregional industries. Such a common territorial activity is ferrous metallurgy, the success of which satisfies the needs of all the Transcaucasian republics. The economic unity of this group of Union republics is also due to the integrity of the transport system and the unity of the fuel-energy base.

28. *The Baltic Region* (comprising the Latvian, Lithuanian, and Estonian republics and Kaliningrad Oblast of the RSFSR). The national-economic complex of the Baltic Region is characterized by its outstanding role in sea transport, the combination of a diverse processing industry based mainly on outside raw materials, highly intensive agriculture in dairy products and swine-breeding, and maritime fishing. In the past decade, industrial growth has been especially intensive; the pace of the Baltic exceeded that of many other basic economic regions of the USSR. By combining the utilization of shale, peat, and hydro-resources, a considerable energy base is being developed.

The industrial All-Union specializations of the Baltic include electrical-machine construction, shipbuilding, instrument-making, wood-processing, paper, and light industry oriented to the output of labor-consuming industries of high quality, great variety, and wide assortment. The essential element of All-Union specialization is the food industry, particularly the processing of fish, milk, and meat products.

The economic unity of the Baltic republics is based on their common maritime position and the tasks within the system of the All-Union division of labor, as well as on the communality of economic problems facing them (especially in land reclamation and the development of agriculture). It is important to note the intraregional ties in shipbuilding, the use of fish resources, the acquisition of certain types of agricultural machines and fertilizers, the utilization of forest resources, and the interrelated work of the ports and dense railway network as a single system.

The economy of Kaliningrad Oblast of the RSFSR is similar in most features to the economy of the Baltic republics (the important role of maritime transport and transit shipments, development of fishing and fish-processing, ship repair, pulp and paper industry, development of the dairy industries). Kaliningrad is an important Baltic port, a large industrial hub, and one of the nationally significant focal points of maritime fishing. Similar problems of development of both industry and agriculture demand joint planning of the economies of the Kaliningrad economic administrative region and the Baltic republics. From an economic point of view, it would be artificial to tie the Kaliningrad district to the Northwestern region, since it is territorially isolated from, and has little participation in, the intraregional connections of the Northwest; it does not reflect the objective regularities of formation of the basic economic regions and the economic tasks facing each of them.

29. *The Belorussian SSR* (Belorussian economic administrative region) is a basic economic region with a population of more than eight million, almost equal in population to all the republics of Transcaucasia. The economic specialization of Belorussia is defined by a combination of light industry (especially linen, based on local raw material, wool, and knitted fabrics), many types of wood-processing and timber-processing (matches, plywood, furniture, timber-chemical industry) and the increasingly important machine construction based on nonlocal metals (motor vehicles, tractors, machine tools, and radios). The agricultural specializations are potato-growing, swine-breeding and dairy cattle-raising, which has led to the creation of a food-processing industry of national importance. Local energy sources include peat and hydroelectric resources (the utilization of which is still in its initial stage). In the future, the complex energy-chemical utilization of peat will occupy an important place in Belorussia's industry.

The region has a distinct center: Minsk, a very large industrial city and an important transport hub. The role of Minsk is enhanced by its geographic position, the configuration of the territory of the republic and its railroad network.

The most important future economic tasks of the region include land-reclamation work, especially in the forest zone [Pripet Marsh], with the improvement and enlargement of agricultural lands, the intensification of agriculture with corresponding expansion of the food industry, the stabilization of the energy base of the republic, the creation of a chemical industry based on local resources (sodium chloride, potassium, peat, and wood) and the further industrialization and reconstruction of the economy of the western oblasts of Belorussia.

The Department of Economic Geography of Moscow State University is continuing the work on economic regionalization. The work will progress in three directions: (1) determining boundaries and compiling characteristics of the basic economic regions; 2) establishing systems and connections of economic regions of different scales and types; (3) further regionalization of the basic economic regions of Central Russia and also some of the other regions.

Economic regionalization is inseparably tied to the planning of the national economy of the USSR. Improvement in planning and the necessity of raising local initiative and the development of local economic life demand that not only Gosplan of the USSR and the republic level planning groups plan the economy but also the local groups such as the sovnarkhozy of the basic economic regions.

The achievement of the maximum economic effect in the further development of productive forces demands a remedy for two deficiencies in planning and building. First, as pointed out at the Twenty-first Party Congress, "It is necessary to overcome the defective practice of the former departments in planning and distributing the enterprises so that each construction in a given region was independent, isolated from the others, and had a local building base; and each enterprise was planned separately with its own subsidiary repair plants, which led to a rise in construction costs and to an unjustified increase in expenditures for the exploitation of the enterprises." And, second, in-depth specialization and cooperation between the basic economic regions demands "decisively fighting against a localist understanding of a complex economy as a closed economy."[16] Planning by basic economic regions makes it possible to overcome departmental narrowmindedness as well as localist restrictiveness.

The proposed project of economic regionalization of the USSR is a working hypothesis that can be adopted for working out scientific problems and for the solution of a number of practical problems, in particular for the compilation of a general future plan of development of the national economy of the USSR in the next fifteen to twenty years. Of course, in the process of drawing up a general plan much can change. But without a preliminary working hypothesis of regionalization, all this work cannot be done with sufficient accuracy.

1. With participation of P. N. Stepanov, S. A. Kovalev, I. V. Nikol'skiy, and V. P. Lebedeva.

2. In the following months a partial consolidation of oblasts occurred, and in November 1957, the number of these regions decreased to 103, and then increased to 104.

3. Plan GOELRO [Plan of the State Commission for the Electrification of Russia], 2d ed. (Moscow, 1955), p. 185.

4. Theses worked out by the Commission of the All-Union Central Executive Committee on the question of economic regionalization of Russia in Voprosy ekonomicheskogo rayonirovaniya [Problems of Economic Regionalization] (Moscow, 1957), p. 103.

5. Ibid.

6. V. I. Lenin, Sochineniya [Works], 4th ed., 33:275–76.

7. N. N. Kolosovskiy, "Voprosy ekonomicheskogo rayonirovaniya SSSR" [Problems of Economic Regionalization of the USSR], Voprosy geografii [Problems of Geography], no. 47 (1959).

8. See Yu. G. Saushkin, T. M. Kalashnikova, and V. P. Lebedeva, "Economic Regionalization as a Method of Investigation of Economic Phenomena," in Nauchnyye doklady vysshey shkoly (geologo-geograficheskie nauki)[Scientific Reports of the School of Higher Learning (geological-geographic sciences)], no. 41 (1958), pp. 101–13.

9. This was published by I. I. Belousov in 1957. See I. I. Belousov, "On the Project of Economic Regionalization of the USSR Drawn up by N. N. Kolosovskiy," *Voprosy geografii*, no. 41 (1957), pp. 29–46.

10. "Economic Regionalization of Russia: Report of Gosplan to the Third Session of the All-Union Central Executive Committee," *Voprosy ekonomicheskogo rayonirovaniya* (Moscow, 1957), p. 128.

11. In essence, the western (Sheksna) part of Vologda Oblast also belongs here; its inclusion in the Northwest is due to the fact that Cherepovets metallurgy is based on iron ore from the Northwest; the metal produced, however, is almost all used by Leningrad. Moreover, the plants being built here at present are oriented to Leningrad as its consumer. The timber industry of the Sheksna territory is weakly connected with the Dvina-Pechora area. Agriculture has also more common features and connections with Leningrad Oblast than with Vologda Oblast.

12. Kolosovskiy, "Voprosy ekonomicheskogo rayonirovaniya SSSR."

13. Ibid.

14. Ibid.

15. Ibid.

16. *Kontrolnyye tsifry razvitiya narodnogo khozyaystva SSSR na 1959–1965 gg* [Control Figures of the Development of the National Economy of the USSR for 1959–1965].

V. S. VARLAMOV

Geographic Features of Interregional Ties Between the Territorial- Production Complexes of the Western and Eastern Regions of the USSR

Transport and economic ties provide the most descriptive indices of the rationality of the distribution of productive forces.[1] The study of such ties for the purpose of analyzing the industrial structure requires that one distinguish between interregional and intraregional ties. This division, important also for the understanding of the ties themselves, is based on the idea of regional production complexes. Interregional ties represent an exchange of the products of specialization of different regional complexes. The intraregional ties are of a different nature: they provide for the exchange between branches of specialization and supplementary branches within the regional complex and thus assure its material unity. The distinction between interregional and intraregional ties is important also as a method for working out future economic ties in transportation by means of consolidated accounts, making it possible to generalize proportionately (in regionalization, reflecting the objectively formed territorial-industrial complexes) the infinite number of individual connections of enterprises, stations, junctions, cities, and villages, and reducing them to ties between and within complexes.[2] Such consolidated accounting opens the way for the utilization of mathematical methods and the newest computers.

Translated from *Voprosy geografii* [Problems of Geography], no. 65 (1964), pp. 38–52.

"The study of interregional connections is necessary to indicate the geographic division of labor, which is the specialization of regions, and to establish the place of each region in the economy of the country and its interaction with other economic regions."[3]

Posing the problem in this manner demands an investigation of the ties in several ways. With respect to both theory and practice, the study of interregional ties by individual branches and individual products is very important. However, another aspect of the study of interregional ties is also very important. The focus of investigation must be the region, not the branch or industrial sector. Moreover, the ties of the region are not examined in terms of one product or the output of one [industrial] branch, but in terms of all basic products moving into and out of the region. This approach is important because it allows examination of the regional ties in terms of various products *together in mutual association*, thus promoting a better explanation of the mechanism of interaction of a given region with other regions and the discovery of those links (surpluses and shortages of various products) by which the regions are connected into a single chain of interregional specializations.

In practice, such a study is important because an integrated analysis of regional ties in their totality also makes it possible to discover the deficiencies in interregional connections not discerned by a sectoral approach. From the standpoint of one sector, for example, it is rational to export paper, plywood, and matches from Belorussia, which has a surplus of these products. However, if one considers the importation of pulp from Kaliningrad Oblast, which imports wood for the pulp industry and timber from other regions of the USSR, then the export of the above products from Belorussia shows up in quite a different light. The national economy suffers considerable losses from such nonrational transport. Such an analysis leads to the conclusion that the scale of development of the timber, wood-processing, and paper industries in Belorussia are insufficiently coordinated and need correction in order to guarantee rational proportions between all industries in the economic complex of the republic.

It is important to study not only the freight structure but also the territorial structure of connections between economic regions, that is, to study which regions have better or worse ties and why, what territory is encompassed by the interregional ties of a given region, and the distances of interregional ties. Such a study plays an inherent role in clarifying the picture of the interregional geographic division of labor. By studying the territorial structure of connections, we can discover the defect in their geography. For example, lumber from the European part of the USSR must not be shipped to Central Asia in order to avoid cross hauls of lumber from Siberia, but this occurs even now; and Karelian lumber must be sent to the western half of the European part of the country, while Arkhangel'sk, Komi and Vologda lumber is sent to the eastern half.[4]

The study of ties between complexes means clarifying their interaction. This makes it possible to characterize the economy of the industrial branches of the USSR as an interconnected system of complexes and not as a sum of the industrial branches.

In this article, only the most common features of the interregional connections that distinguish the western and eastern regions of the USSR are examined. The interregional ties are calculated in conformity with the network used in the works of the IKTP [Institute of Combined Transport Problems] and include thirty-one regional production complexes.

What questions must be subject to clarification in a geographic study of ties, particularly interregional ties? Obviously, these questions must deal with the density of the ties, their direction (or territorial structure), the dynamics of development (including the dynamics of their territorial structure), the relationship between intra- and inter-regional ties, freight structure, and evaluation of the efficiency of the ties. In examining each of these questions, primary attention must be given to the dependence of the connections on the industrial and territorial structure of the complexes that give rise to them. So, for example, the density of ties is dependent on the balances between production and consumption by complexes and subregions.

In terms of absolute volume of interregional communication, the eastern regions clearly yield to the western.[5] In the East, the Kuznetsk-Altay complex has the greatest interregional freight turnover. However, in the European part of the USSR, the Middle Volga region has approximately the same freight turnover [as the Kuznetsk-Altay]; the Donets-Dnieper complex, however, has twice the volume of interregional movement of the Kuznetsk Basin. The total volume of interregional transport of the Siberian and Far East regions (arrivals and departures) in 1960 was more than 230 million tons, or 12 percent of the all-Union total.

The European regions of the USSR (excluding the Urals) have 65 percent of the total interregional freight turnover. This is determined first by the fact that in these regions, occupying somewhat more than one-fifth of the national area, are concentrated 70 percent of the population, about two-thirds of the fixed assets of industry, about two-thirds of all the grain harvested, more than two-thirds of the cattle, more than four-fifths of pigs, and almost half of the sheep. Thus, the European territory plays the predominant part in both All-Union production and consumption.

In terms of ton-kilometers, the interregional ties of the eastern regions have a higher coefficient (20–22 percent as against 12 percent in terms of tonnage), which indicates greater average distances of interregional hauls in Siberian complexes. Comparatively small in general, the scale of interregional hauls is due to the lower economic development of the Siberian regions. However, these rates are increasing rapidly: in the course of five years, 1956–60, the interregional hauls of Siberia and the Far East have increased 60 percent (the average national growth rate of shipments was 53 per cent).

Until recently, the coefficient of interregional transport of Siberia and the Far East, taken as a whole, hardly differed from that of the European regions and the Urals; 41.1 percent for European and 41 percent for Siberian regions. In 1960, this index for Siberian regions was noticeably lower than for the European (35.5 percent and 41.9 percent, respectively).

However, if one examines the interregional ties not in isolation but with respect to the development of the regions, their population, and the size of industrial and agricultural production, then a much higher degree of development of interregional connections in the Siberian regions becomes evident. Indeed, in 1960, the per capita interregional transport average in the USSR was 9 tons; in the European regions (excluding the Urals), 8.5 tons (with the Urals, 9.3 tons); in Kazakhstan and Central Asia, 6.9 tons; and in Siberia and the Far East, 9.8 tons; the latter having the highest index for any group of regions. Considering the individual regions, of course, there will be higher indices, but at present we are interested in the characteristics common to large groups of regions. Among the Siberian complexes, first place (using this index) is occupied by Western Siberia (12 tons) and the Kuznetsk-Altay region (12.6 tons), which is related, first, to their geographic situation in the far western part of Siberia close to the European regions, and, second, to the transit situation of Western Siberia and the significance of the Kuznetsk-Altay as an all-Union coal base. As one moves eastward, the index decreases (the Yenisey region, 8.8 tons; Eastern Siberia, 10.4 tons; the Far East, 3.7 tons; Yakutia, 2.4 tons), illustrating the increase in difficulties and expenditures in overcoming the greater distances (in the interregional ties of Siberian complexes, the ties with the western regions are more important than those with each other). Only the Yenisey region differs, owing, first, to the intraregional character of its coal mining (coal mining and consumption balances in the Yenisey region indicate that production exceeds consumption only by 17 percent, whereas in the Kuznetsk-Altay region the ratio is 165 percent), and second, to the absence of imported crude oil (there is no oil-processing in the region). Eastern Siberia, however, in addition to oil products, ships in crude oil for processing and export.

If the data on the interregional connections are related to the magnitudes of gross industrial and agricultural production, then this index also shows that the group of regions in Siberia and the Far East exceeds all other groups of regions in the USSR by a considerable margin: setting the national average index at 100, the European regions, including the Urals, have an index of 98, Kazakhstan and Central Asia, 101, and Siberia and the Far East, 123.

The indices of the great development of interregional ties can be attributed to the lack of complexity in the economy of Siberia and the Far East. The weakly developed economy leads to a large volume of shipments into and out of the region. Many types of products are not produced or are produced in insufficient quantity within the region and must be shipped in; many goods produced within the region have not yet found consumers and are shipped out of the region. Moreover, insufficient integration and development of the various stages of processing and spatial disunity result in a predominance of shipments from the region of unprocessed products and unrefined raw materials (unenriched coal and timber), containing many potential waste products, thus increasing the work of transport by exploiting interregional freight lines.

The weak development of the economic complexes of Siberia and the Far East and the inadequate development of the processing industries are reflected also in the interregional freight structure. Thus, in the interregional shipments out of the western regions of the USSR, the share of mass freight (basically raw materials and semifinished products) is 60.5 percent; in the Urals, 66.8 percent; in Kazakhstan with Central Asia, 66.9 percent; and in Siberia and the Far East 82.8 percent.[6] Considering the receipt of freight in the interregional exchange, the picture reverses: Siberia and the Far East have the lowest share of mass freight (53.7 percent) and the highest share of the so-called other freight (other freight includes processed products). For the western regions the share of mass freight received from within their boundaries is 63.6 percent; for the Urals, 74 percent; and for Kazakhstan and Central Asia, 62.6 percent. Coal accounts for 42.6 percent of interregional shipping of Siberia and the Far East. In the interregional shipping of the European regions, coal accounts for only 18.1 percent, although the largest coal basin of the USSR (the Donets Basin) is here. The absolute volume of coal shipment, however, is almost twice as great as in the Siberian regions. The Donets-Dnieper region, however, uses 64 percent of its coal and ships out a processed, more valuable product created with energy obtained from its own coal, whereas the Kuznetsk-Altay region, because of a much smaller development of the upper levels of its complex (chemistry, machine-building), uses only 38 percent of its mined coal, shipping the rest to other regions. The interregional ties in other regions of Siberia and the Far East are similar, with a lower degree of development of the industrial complex than the Kuznetsk-Altay region.

The differences in the territorial structure of their interregional connections also attest to the varying degrees of development of the production complexes in the eastern and western regions.

Indeed, an analysis of tables of interregional flows indicates that if one divides the territory of the USSR into several groups of regions or zones—the European section (without the Urals), the Urals, Siberia with the Far East, and Kazakhstan with Central Asia—then for the European zones a great predominance of intrazonal ties or interregional ties will be very characteristic. The European regions exchange freight mainly among themselves. Their interregional ties form a rather closed system of intrazonal exchange: of the total shipments out of the European regions only 11.7 percent (about 79 million tons) are sent into other zones, whereas the shipping from other zones comprise 19.3 percent (about 125 million tons) of the total interregional imports of the European zone.

This feature of the system of interregional ties in the European part of the USSR is a direct result of the territorial structure of its economy. The highly developed, specialized regions of the European part of the USSR are tightly connected by the geographic division of labor. Here several groups of regions can be distinguished, varying by type of interregional ties, and determined largely by the type of economy: (1) The Central, Leningrad, Baltic, Belorussian, and Volga-Vyatka regions;[7] (2) The Karelian-Murmansk and Dvina-Pechora regions; (3) The Donets-Dnieper region; (4) The Middle Volga, Volga-Don, Northern Caucasus, and Transcausian regions; (5) The Central Chernozem, Central Ukrainian, Southern Ukrainian, and Moldavian regions; (6) Western Ukrainian region. The exchanges among these groups of regions form the basis for the whole system of interregional ties of the European part of the country.

The Central, Volga-Vyatka, Leningrad, Baltic, and Belorussian regions belong to the typical complex of manufacturing industries. The specialization of these regions is of historical origin, based on the various branches of the manufacturing industry, working with shipped-in raw materials and supplying almost all regions of the country with their products. In terms of shipments out of these regions, products of the manufacturing industry predominate: machines, metal wares, fabrics, fertilizers, clothing, footwear, various types of chemical products, and paper. For these regions, a shortage of all mass raw material freight is characteristic: coal, oil, oil products, ferrous metals, iron ore, timber,[8] grain, and other agricultural freight. All have a freight-turnover balance deficit; imports greatly exceed exports (for the Central region, by a factor of 1.9; for the Leningrad region, 2.3; for the Baltic 3.4; for Belorussia, 4.4; for the Volga-Vyatka region, 1.4). These regions are especially deficient in energy resources. Their coal supply is based mainly on ties with the Donets, where 71.3 percent of its shipments originate. The Leningrad region receives a considerable amount of Pechora coal. Petroleum comes from the Middle Volga (44.8 percent), the Urals (24.4 percent), and Northern Caucasus (29.8 percent, although the percentage is sometimes greater).

Thus, the sources of all mass products, in which the five described regions are deficient, lie in the European part of the country. This is why most of the interregional ties of the Central, Baltic, Belorussian, Volga-Vyatka and Leningrad regions are locked within the European territory of the country (excluding the Urals). Although the products of the manufacturing industries go from these regions to all parts of the country, the general picture is hardly altered: the material shipped out is less by total weight than the raw materials and fuel brought in. Moreover, a large part of this production is absorbed in the European regions of the USSR. The Central region is especially distinguished by the concentration of a number of manufacturing industries whose products are needed by the whole country. This shows up in the higher coefficient of ties with the East in the total volume of interregional ties of this region—20 percent (in the interregional ties of the other European regions the share of the East is 15 percent).

The specialization of the Karelian-Murmansk and Dvina-Pechora regions is based on the utilization of its timber and mineral (ore, coal) wealth. Most important items shipped out are timber, coal (to the Leningrad region and the Baltic) and ore (to the Cherepovets plant from the Karelian-Murmansk region). All regions of the European part of the USSR except the Volga-Vyatka, Middle Volga, and Transcaucasus regions, are supplied with Karelian and Dvina-Pechora timber.

The Donets-Dnieper complex is second in the country (after the Central) in volume of the gross industrial product. This region is a coal-metallurgical base for the whole European part of the Union (without the Urals), supplying it with coal, pig iron, rolled metal, iron ore, products of heavy machine-building and chemistry, light metals and salt. In terms of surplus iron ore, pig iron, rolled metal and coal, the Donets-Dnieper region is first in the country. It also has a surplus of grain. The Donets Basin and the Dnieper region have the best-developed connections with the remaining regions of the European zone. Of the other zones, the one best connected to it is the Ural. Ties with the regions of Siberia, Far East, Kazakhstan, and Central Asia are not great. Timber, oil, and oil products are brought in to the Donets-Dnieper region.

The eastern edge of the western zone (Middle Volga, Volga-Don, North Caucasus, and Transcaucasia) is distinguished by its specialization in the extraction and refining of oil. Oil freight comprises from 50 to 75 percent of all shipments from these regions (excluding the Volga-Don region). Oil freight is shipped from here to all regions of the USSR (including 40.8 percent to the western regions, 30.3 percent to the eastern, 28.8 percent exchanged among the four regions of the zone mentioned above). Most important is the connection of these regions with the remaining part of the western zone. But for them, in contrast to the other regions of the zone, an increased development of ties with other zones is also characteristic, particularly with the Urals (27.2 percent of total interregional shipments). For Transcaucasia and the Northern Caucasus the development of ties with Central Asia is characteristic.

In addition to shipments of oil freight, common to all Volga-Caucasus regions, each exports products of other branches of specializations: the Middle Volga and the Northern Caucasus send grain; the Volga-Don region, coal and grain; Transcaucasia, manganese ore. From these regions to the Industrial Center and, to a lesser degree, to the Urals, gas and electric energy from the Volga hydroelectric stations is transmitted. Coal and metal are brought in from other regions (to satisfy the needs of one of the branches for specialized machine construction).

The main supplier of metal for the Transcaucasian, Northern Caucasus, and Volga-Don regions is the southern coal-metallurgical base (Donets-Dnieper). The Urals and the southern metallurgical base play analogous roles in supplying metal to the Middle Volga (in recent years Urals metal has begun to predominate).

The Central Chernozem region, the Central Ukraine, and Moldavia are united around local agriculture in the formation of their economic complex. These regions ship out grain and other agricultural materials and products, derived from processing agricultural raw material. The development of machine construction, determining the main branches of the economy, demands the importation of rolled iron from the Donets-Dnieper region and the Urals (pig iron is sent out from the Central Chernozem region). Timber and petroleum are also imported. Although the economic development of these regions is based mainly on local raw materials, they have a transport balance deficit because of the need to bring in considerable amounts of Donets coal.

The Western Ukraine, with its specialization in the extractive industry (oil, gas, salt) and intensive lumber and agricultural economy, ships out oil products, processed wood products, agricultural products, and other freight, and receives coal, metal, and grain, from the Donets-Dnieper and other regions of the Ukraine.

Thus, each of these regions of the western zones is connected to a much higher degree with the other regions of their zone than with regions of other zones in the country. The intrazonal connections (by volume) are distinctly significant here.

The ties of the western zone with other zones of the USSR are determined mainly by surpluses of oil and oil products, iron, and manganese ore, which are shipped to other zones or exported to other countries (iron ore, oil freight) and by shortages of coal, steel, rolled metal, timber, lumber, grain, and cement.

The territorial structure of interregional ties of industrial complexes in the Siberian–Far East zone has a completely different character than in the western zone. Here *interzonal* exchange predominates. Of the total shipments out of the Siberian and Far East regions, 67.4 percent are sent to other zones, to the Urals, the European part of the USSR, Kazakhstan, and Central Asia. From these zones Siberia and the Far East receive 54.4 percent of total interregional freight arrivals. Thus, the territorial-production complexes of Siberia and the Far East enter into an interregional exchange mainly with regions of other zones and not with their neighbors within the Siberian and Far Eastern zone, which is to say the *interregional* ties of Siberian and Far East complexes are primarily *interzonal*. Such a territorial structure of interregional ties is explained to a considerable degree by the recency and incomplete formation of Siberian-Far East complexes and the dependence of their industrialization mainly on the more developed adjacent western regions. The other Siberian regions often produce the same materials found in all the other Siberian regions. Almost all of them, for example, have a surplus in timber and coal, constituting 70 percent of the total shipping from these regions, so that they cannot exchange these products. Such a situation is the result of the fact that the development of newly assimilated, colonized territories, as a rule, begins in the interest, and for the needs, of more developed regions, and not in the interest of the pioneer territories. The various products of these young regions are shipped to the more developed regions to satisfy the general national demands, leading initially to the development of ties, not among the pioneer territories themselves, but with the developed regions. In the succeeding stages of development, the branches of specialization in pioneer territories create new consumption needs, causing the creation and growth of auxiliary, supplementary industrial branches. Thus, the volume, structure, and direction of interregional connections can serve as one of the indices of the stage of economic development of regions.

The diverse territorial structure of the interregional ties of complexes in the various zones can be clearly seen from table 1, where the Donets-Dnieper complex exemplifies the European zone,

TABLE 1

INTERREGIONAL TIES OF THE DONETS-DNIEPER, WESTERN SIBERIAN, AND
SOUTHERN URALS PRODUCTION COMPLEXES IN 1960
(In Percent)

	Total	With the European Part of USSR	With Urals	With Siberia and Far East	With Kazakhstan and Central Asia
Donets-Dnieper					
Shipments in	100	85.8	8.8	2.4	3.2
Shipments out.	100	96.3	1.7	0.9	1.1
Western Siberian					
Shipments in	100	24.5	47.6	21.7	6.2
Shipments out.	100	25.8	18.2	39.6	16.4
Southern Urals					
Shipments in	100	35.8	8.8	29.8	25.6
Shipments out.	100	46.8	17.4	22.1	13.7

the Southern Urals, the Ural zone, and Western Siberia, the Siberian-Far East zone.

In the interregional freight exchange of the Donets-Dnieper complex, one feature becomes apparent that is very characteristic for all regions of the western zone: the intensive development of ties mainly with regions of their own zone (especially neighboring ones) and the much smaller volume of exchange with the Urals and the Asian part of the country. The Donets-Dnieper region, the main base of heavy industry for the whole European part of the USSR, has a transport-balance surplus in ties with the majority of its regions.

Western Siberia illustrates another type of territorial structure of interregional connections. Most of its interregional connections are with other zones. It is connected only somewhat better to the Siberian regions than to the European regions, but much less so with the Urals.

The Urals, lying between the European and Asian parts of the USSR, is equally well connected to both. This situation is the result of its economic-geographic position and to the fact that the Urals is still the main base of industrialization for the Asian section of the USSR. For the European part of the country, it is an important source of metal (second only to the Donets-Dnieper), the most important source of petroleum (along withe the Volga region) and the largest supplier of timber. The Asian part of the USSR and the Urals have a fuel-energy shortage in common, necessitating the import of coking coal and, recently, iron ore. Therefore, in its ties with the regions of the European part of the USSR the Urals has a transport balance surplus with a powerful raw material base, whereas in its ties with the Asiatic part of the country it has a balance deficit in which raw materials are brought into the region (coal from the Kuznetsk Basin, Central Kazakhstan, and the Virgin Lands, and ore from the Virgin Lands and Western Kazakhstan and Siberia) and oil freight, metal, and various industrial products are sent out.

The differences in the territorial structure of the interregional ties of the European and Siberian regions determine the sharp difference in the average distance of the interregional shipment. The average distance of a shipment for the country is 1,492 kilometers, whereas for the European regions (for goods sent out of the regions) it is 1,214 kilometers; for the Urals, 1,816 kilometers; for Kazakhstan and Central Asia, 1,770 kilometers; and for Siberia and the Far East, 2,180 kilometers. The average distance of interregional transport for goods brought into the regions are, for the European region, 1,330 kilometers; for the Urals, 1,540 kilometers; for Kazakhstan and Central Asia, 2,030 kilometers; and for Siberia and the Far East, 2,088 kilometers.

Finally, the insufficient integration of the economy of the Siberian and Far East regions, which becomes apparent when comparing the interregional connections of western and eastern regions of the country, is also one of the reasons why the interregional freight flows to the West predominate over those to the East.[9] The great development of interregional connections, characteristic for the eastern regions, and the predominance of interzonal connections explain the formation of the most powerful freight flow in the country. Since the shipments out of the eastern regions are predominantly bulk materials, and much lighter other freight comes from the European part of the USSR and the Urals (manufactured and processed items), there is a greater freight flow westward than eastward. As can be seen in table 2, this excess is characteristic for all interzonal ties; in the ties of the European part of the USSR with the Urals, Siberia, and with the Kazakhstan-Central Asia regions, between the Urals and the Siberian-Far Eastern regions, between Kazakhstan and Central Asia on one hand, and Siberia and the Far East on the other: in all cases the freight flows coming from the East are greater than those coming from the West. In the connections of the Urals with Kazakhstan and Central Asia, the freight flows into the Urals are larger than those in the return directions. The excessive freight flow to the west in relation to the return flow causes huge and steady flows of empty rolling stock.[10]

It is known that empty runs of rolling stock cause unproductive expenditures and that the utilization of empty rail lines lowers transport costs by comparison with costs on busy lines. Ye. D. Khanukov has shown that the most effective way to solve the important national-economic problem of reducing unproductive runs of rolling stock is the rationalization of the distribution of productive forces, taking into account the effective utilization of transport.[11]

The insufficient integration (development) of the Siberian and Far Eastern economy is one of the main causes of unequal freight loads on the eastern and western lines and must be overcome to reduce the unproductive runs between the two parts of the country. The change in the character of their (Siberia and the Far East) economy, the advancement of these regions into new and more mature stages of economic development, and the addition of new components to the existing production complexes will lead to a change in the structure and directions of interregional ties; a decrease in the proportion of raw materials shipped out of Siberia and the Far East and the increase in the products of industrial processing, the replacement of less transportable products with more transportable ones (for example, the export of raw timber will be gradually replaced by lumber, plywood, paper, cardboard, and other products).

TABLE 2

INTERZONAL TIES IN THE USSR IN 1960
(Millions of Tons)

Shipments in	Total Shipped to Other Regions	European Part of the USSR (without the Ural Region)	Ural Region	Siberia and the Far East	Kazakhstan and Central Asia
European part of the USSR (without the Ural region).	593.0	523.8	39.7	18.0	11.5
Ural region.	145.8	81.9	19.9	26.6	17.4
Siberia and Far East	134.9	22.7	41.5	44.0	26.7
Kazakhstan and Central Asia	78.5	19.5	25.9	7.9	25.2
Total shipped in.	952.2	647.9	127.0	96.5	80.8

The development of new methods of energy transfer will play an important role in reducing empty runs, exemplified by the replacement of coal shipments from Siberia and Kazakhstan with the transfer of electric power by long-distance electric powerlines. Its significance is clear from the fact that timber and coal are the main freight presently determining the dimensions of the uneven freight loadings in western and eastern directions.

The unevenness of shipments in the western and eastern directions is stable in character, which creates the necessary conditions for the utilization of empty flows in the distribution of new enterprises (for example, it is possible to utilize the empty runs to transport aluminum raw materials from the European part of the USSR or even from Hungary, to Siberia).

Thus, even the most general examination of geographic features in the interregional ties of various territorial groups of industrial complexes allows one to conclude that, in the economy of the Siberian and Far East regions, integration has not been nearly sufficiently developed. The huge growth of industry planned here for the future will demand the growth of transportation also.

The question arises of how to solve this problem most economically, taking into account the present structure of the transport-economic ties of Siberian regions with their strongly developed interregional and interzonal ties. It is evident that a complete, economically substantiated plan is needed for the optimum territorial organization of industry and transport in the future. .

Rational construction of the economy in the economic regions and the creation of economic regions where the various branches of production (from extraction to the final stages of processing) would be reasonably and expediently combined and where many freight hauls would constitute a closed circuit because of the developed structure of the economy within the regions, can be a formidable lever with which one can assure that growth of transport would be considerably slower than the growth of industry. The growth and formation of regional industrial complexes must mean, simultaneously, the growth and development of regional systems of communication, in which the heavy flow of raw materials (now going in an unprocessed form to other regions) is utilized internally, and the bringing together of various stages of processing, enclosing within the regions comparatively small areas. If possible, however, the finished product must be sent from the regions, resulting in more economical transport costs, which are added to savings in production costs obtained by the economically expedient combination of industry in regional complexes.

1. This article puts forth the main propositions of a paper given by the author at the Second Scientific Conference of Geographers of Siberia and the Far East (Vladivostok, September 1962).

2. This does not exclude but, on the contrary, assumes the presence of a following stage of investigation that must produce more detailed connections after the most common interregional proportions have been established. Of course, this detail makes it possible to correct the established interregional connections.

3. I. I. Belousov, "Problems of the Geography of Transport," in *Sovetskaya geografiya: Itogi i zadachi* [Soviet Geography: Accomplishments and Tasks] (Moscow, 1960), p. 384.

4. Ye. D. Khanukov, *Transport i razmeshcheniye proizvodstva* [Transport and Distribution of Industry] (Moscow, 1956), pp. 290–91.

5. See, for example, the map "Interregional Connections of the Economic Regions of the USSR," *Voprosy geografii* [Problems of Geography], no. 57, p. 156.

6. The proportion of production of the processing industries in all of the freight shipments by rail (both out of the region and within the region) follow: average for the USSR is 31.4 percent; the regions of the European part of the country (without the Urals), 31 percent; the Urals, 41.4 percent; Kazkhstan and Central Asia, 24.4 percent; Siberia and the Far East, 25.6 percent.

7. It is true that the Volga-Vyatka region differs considerably from the other regions of the group due to the lower level of its industrial development and not to qualitative differences in the type of economy.

8. Except for the Volga-Vyatka region, which has a surplus of timber.

9. Other causes lie in the distribution of natural resources and population, historical-geographical factors, and the like.

10. It must be kept in mind that table 2 shows ties of all types of transport, including the pumping of oil and oil products from west to east by pipelines. If one excludes this, then the difference between loaded and empty rail movements becomes still greater (this applies also to empty rolling stock). In addition, table 2 indicates the directions of the ties between regions within zones and the ties within regions, which also changes the relationship between loaded and empty flows on specific sections of the railroad network.

11. Ye. D. Khanukov, *Transport i razmeshcheniye proizvodstva*, p. 303.

N. M. BUDTOLAYEV
V. P. NOVIKOV
YU. G. SAUSHKIN

On Methods of Drawing up a

Territorial (Spatial) Model of the

National Economy of the USSR

The national economy of a country can be compared to a chain in which the individual links of production and consumption are joined consecutively. This system of mutally connected links begins with the exploitation of natural resources (including their investigation and survey), extends to the production of the end product, and finally, to the consumption of this product. In the system of mutually connected links, consumption gives rise to new (reverse) links and new chain reactions. The planning of the USSR national economy urgently demands the modeling of a single chain of connections of the various links of production and consumption; despite its great significance, an interbranch balance [approach] cannot replace such a model.

The interbranch balance characterizes the mutual relationships selected for statistical calculation of the branches and not their real productive connections; it is limited to production and does not include consumption by the population and in many cases combines completely different industries and separates related ones.

The model can be structural and territorial (spatial). A structural model reflects concrete productive connections of various links of

Translated from *Vestnik Moskovskogo Universiteta,* seriya geografiya [Herald of Moscow University, Geography series], no. 6 (1964), pp. 7–16.

the national economy (including the utilization of natural resources and labor resources, and consumption by population) in the country as a whole. The territorial or spatial model analyzes production ties in territorial cross section according to areas, regions, and industrial units, relating them to a specific area and to definite distances between centers of concentration of resources, production, and consumption.

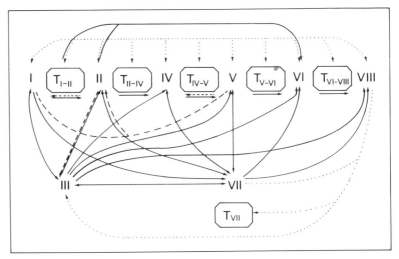

Fig.1. Scheme of a Structural Model of the Economy

Legend

I,II,...VIII	Structural links of the economy
——→	Direct relations
←— — —	Reverse relationships (feedback)
←——→	Mutual relationships of equal intensity
←··········	Relationships involving the use of manpower resources and the effect of consumer demand on production
T_{I-VIII}	Material and labour inputs in transportation between links

The creation of a structural model is an indispensable stage in the construction of a territorial model. Therefore, the first step of the authors in modeling the national economy of the country was to draw up a diagram of a structural model and show its basic links. From a philosophical point of view, the diagram of a structural model of the national economy reflects the interchange between nature and man through a series of intermediate links; this differs greatly from an interbranch [intersectoral] balance that does not analyze the entire system of links, nature-production-man.

Figure 1 illustrates the direct exploitation of natural resources, that is, the structural link of the national economy in which the earth is the main productive force (mining, agriculture, fishing, forestry). The exploitation of water resources (irrigation, drainage of land, water supply) is also considered. Link I is the foundation of the national economy: the broader the base, the more possibilities for the development of the following structural links in the chain of the economy. Our country is still behind the United States in per capita production of goods in mining, agriculture, forestry, and fishing. According to data for 1959, in the USSR per capita production in mining was 6.0 tons; agriculture and forestry, 4.4 tons; and in all extractive industries, 10.4 tons. In the United States these figures[1] were 12.5, 5.3, and 18.8, respectively. To this mass of primary raw material, fuel and water used in the economy and by the population must be added.

Link II is the structural link of the national economy representing the stages of the simplest (coarse) processing of primary raw material and fuel obtained (extracted) from nature: the concentration of ores and coal, the desulphurization and dehydration of oil, the coking of coal and distillation of oil, lumber production, cotton-cleaning, the grinding of grain, and the like. It is characteristic of this link that the types of raw materials and fuel extracted from nature are processed individually, *vertically*, mainly mechanically without any considerable combination of various types of raw material and without extensive processing.

Link III is electric power, supplying all other links in the chain of the national economy with electricity and heat, and is therefore separated from the straight line of connections of these links. It is directly connected with natural resources, first of all coal, utilizing the "natural storehouse of energy," but also utilizing the wastes of industries of the second link (dotted line).

Link IV includes mass production branches based on a combination of several or even many types of raw materials, fuel (energy), semi-finished products, and industrial by-products: ferrous metallurgy, the basic chemicals and others. In this structural link the combining of industries acquires great significance and horizontal ties come to the fore.

Link V is the processing industry, which produces the finished (end) output in the form of means of production, the consumer of which is not the population but the various branches of the economy. The production ties of this link of the chain of the national economy (predominantly the branches of machine construction and metalworking) are very complex. Connections of cooperating enterprises predominate.

Link VI includes the processing industry, whose finished (end) output goes directly to satisfy the diverse needs of the population, individual as well as collective.

Link VII is the building of cities and other settlements (residential and industrial buildings), roads, bridges, dams, canals, ports, and many other facilities without which no material production can exist and develop. This branch on the model diagram, like electric power, is placed apart from the main series of ties since it too, and to the same degree, serves all links in the single chain of the economy.

Finally, Link VIII is man himself, that is, the population of the country analyzed from the point of view of its distribution (urban and rural), labor resources, productive capacity of social labor and consumption. In Soviet economic-geographic investigations little attention has been paid to consumption by the population. Yet of the 1961 national income of the USSR amounting to 152.9 billion rubles, consumption constituted 71.3 percent, including personal consumption of 63.6 percent of the entire national income.[2]

A large consumer of material wealth and labor is transport, and this is reflected in the diagram of the structural model. In it the transport expenditures and the use of material wealth and labor connected with them are shown not by one structural link but are located on the main directions of movement (transporting raw materials, fuel, and semifinished products from one link in the chain of the economy to another). The loads of construction are unified on the model diagram.

In the chain of the national economy represented on the diagram of the structural model, the system has diverse tendencies in the various links: at maximum I, II, III, VI, maximum utilization of labor resources and maximum growth of consumption by the public (VIII) the transport expenditures (I–VIII) must be minimized and must be limited to the investments in IV, V, and VII that would be necessary to guarantee the development of the entire system.

We offer examples of such chain reactions in order to make more tangible the connections of the links in the chain of the economy, which in a general form are shown in the diagram of the structural model.

The mass extraction of high-quality iron ore, ferrous metallurgy, production of pipes, preponderant development of prospecting and extraction of oil and gas, development of the petrochemical and chemical industry, increased production of motor fuel, synthetic materials, nitrogen fertilizers and means of plant protection, growth of agricultural production, expansion of food and other light industries guarantee the needs of the population [as will] an increase in oil and gas extraction, wide use of petroleum residue and natural gas for energy, a decrease in transporting coal over great distances, a decrease in the demand for metal in the mass transport of coal, utilization of coal at the place of mining for the production of electric energy, the creation of large electric energy centers, and the development in them of branches of the metallurgical and chemical industries requiring high electric capacity.

These chain reactions connect all branches of industry. These connections are very complex; and the reduced use of metal, for example, in coal transport results in an increase of consumption of metal, in mining, oil and gas transport, and the development of the chemical industry, including the production of nitrogen fertilizers, based on oil processing; it leads to an increase of grain production and to an increased need for motor fuel, trucks, and railroad cars for transporting grain.

The task is to compose a structural model in which the chain reactions would satisfy the following demands: (1) rapid involvement in production of the largest and most economically profitable natural sources of raw materials and energy, expansion of the natural base of production; (2) establishment of the most progressive ways of developing the productive forces of the country, primarily with the help of a rational combination of electrification, chemicalization, and mechanization of the economy that would enable the USSR in the shortest possible time to surpass the level of development of productive forces in the United States; (3) the choice of sequentially linked progressive technological systems under which the totality of all enterprises would be utilizing fully and cheaply the raw materials and energy, yielding the largest quantity of necessary end products, which in turn would guarantee expanded reproduction of the national economy and which furthermore would satisfy the needs of the population at the given stage of development; (4) full utilization of the labor resources of the population, taking into account the complex mechanization and automation of industry and the increased skill-levels of the workers. A greater interregional and international socialist division of labor and rational territorial organization of productive forces are of prime importance for the solution of this problem.

The possibilities of constructing and utilizing a structural model of the USSR national economy are very great. Yet in the construction of a model one must consider the limitations that determine: (1) the possible size of investments; (2) the period of construction and introduction into operation of new facilities; (3) the degree of development of science and technology; and (4) the degree of development of a new modern technology. For example, the dynamics of chemical technology is such that in the United States where one-fifth of the workers in the chemical industry do research work, by 1970 (in the Americans' opinion) three-fifths of the chemical products sold will consist of still unknown or little-known products.[3]

The structural model expresses the main directions of development of the economy and the rational relationships between its links, but it cannot show the optimum territorial organization of the productive forces and the rational movement of material wealth within the country. The structural model does not visualize how the whole chain of various links of the national economy is distributed in the territory of the country nor how the flow of material wealth and people, being in constant motion, moves within this territory.

The territorial model of the national economy must provide an answer to these questions. It is this model that makes it possible to accelerate the development of productive forces of the USSR with the same investments, by means of their most rational territorial organization, that is, by means of the mobilization of reserves still insufficiently utilized in the economic competition of two systems. A parallel composition of a structural model and a spatial model with all variants is possible. We shall try to proceed in this manner.

For a spatial model it is necessary first of all to divide the country into several large parts that differ greatly in natural and economic conditions, levels of development of their productive forces, and the living patterns of the people. In other words, the irregularity of the distribution within the territory of the country of natural bases, natural conditions of production, and productive forces that actually exist must be reflected in the model. In 1963 the authors reported on their search for methods of analyzing the relationships between the West and East of the USSR.[4] Now, with the more complex work of constructing a spatial model, six groups of economic districts (*rayony*), called regions (*regiony*), have been distinguished on the territory of the USSR: (1) the Central region (Central Russia, the Northwest, Belorussia, the Baltic); (2) the Southern region (the Ukraine, Moldavia, the Northern Caucasus and Transcaucasia); (3) the Volga-Urals region (the Volga and the Urals); (4) the Central Asian region (Kazakhstan, Central Asia); (5) the Siberian region (Western, Central, and Eastern Siberia); (6) the Far Eastern region (the Far East with the Yakut ASSR). Each of these regions possesses a unique combination of geographic situation, population, natural resources, economic conditions, and consumption patterns.

Each of the regions has its own combinations of resources and its own limitations for economic development, both of which are very important as a basis for the division of labor between the largest parts of the country. The energy resources of the six regions can be compared in table 1.

The division of labor between the regions is greatly influenced not only by the presence of particular resources but also by the lack of others. Therefore, in composing a spatial model it is necessary to determine also the quantitative limits of the particular resources

TABLE 1

POWER RESOURCES OF THE REGIONS OF THE USSR

TYPES OF NATURAL RESOURCES	REGIONS					
	Central	Southern	Volga-Urals	Central Asian	Siberian	Far Eastern
Hydroelectric resources. . . .	0	+	+	X	+	++
Coal	X	+	0	+	++	0
Petroleum.	-	+	++	+	X	0
Natural gas.	-	++	+	+	0	-

NOTE: -, represents the absence of or a small quantity of a given type of resource; 0, the intradistrict significance (within a large economic district); X, the intraregional significance; +, the All-Union significance; and ++, the international significance (in the system of socialist countries).

in the region (labor resources, water, and fuel). Table 2 shows the combinations of the most important limitations in the development of the economy and of living standards of the population in the six regions of the country. As table 2 shows, every region has some shortage in local resources that must be remedied.

The most important economic indices need to be compared with the labor and natural resources, investments in the national economy, economic effectiveness of the investments (the increase in the gross production of the industry in kopeks per ruble of all investments); the size and structure of the fixed assets (including the relationship between the productive and nonproductive assets); the shares of the regions in the industry of the USSR by personnel, assets, gross production; the branch structure of industry; the gross production of industrial branches per ruble of industrial production assets and per capita of industrial production personnel; the main economic indices of agriculture, and the like.

These comparisons make it possible to determine the economic effectiveness of the utilization of the resources and the combination of branches of the economy in each of the regions. Such a comparison is of special significance for the determination of the most economically profitable means of developing the productive forces of the eastern regions of the country. The huge resources of these regions are evident. In order to accelerate the rate of national economic growth in the USSR, it is necessary to utilize now the most profitable resources from the vast natural wealth of the east. But a strict selection of branches of the national economy is necessary, and these must be developed primarily in the eastern regions without using traditional planning, as has been done in the past.

During the period 1918–60 the effectiveness of investments in the economy varied greatly (in terms of the increase of gross production of industry per ruble of all investments) in the various regions of the country: from 68 kopeks in the Central region to 36.5 kopeks in the Siberian region, 32.4 kopeks in the Central Asian region and 24.2 in the Far Eastern region. The lesser effectiveness of investments in the Eastern regions (in the form of industrial production) can be explained to a considerable degree by the fact that, together with specific sectors successfully utilizing the natural wealth in these regions, there were created certain labor- and asset-consuming branches that were especially expensive in the East and demanded huge additional investments. These branches could be developed much more successfully in the western regions of the country. For the most part, this refers to machine construction and metallurgy. These calculations have not taken into account the huge additional expenditures for the nonproductive and productive, but nonindustrial, assets that were needed in the East and that, of course, must be considered in composing a spatial model.

In an earlier report we noted the sharp imbalance of railroad flow from east to west (loaded direction) and back (enormous flow of empty cars). Hence the concept of economic distance introduced in our model differs greatly from the actual number of kilometers. Furthermore, we compiled and analyzed the matrix of departures and arrivals of the basic loads for the six regions, and we examined the basic principle of territorial division of labor within the country. Similar matrices are only one of the sources of information for the composition of a spatial model. These are merely threads, and not the only ones, from which the rope has to be woven. Having clarified the actual picture of the mutual connections between regions, that is, the territorial division of labor between them that is expressed in the exchange of production (raw material, fuel, semifinished and finished products), we proceed to the analysis of their production costs in the various regions and to the cost of transporting them to the regions of consumption with a consideration of the established economic distances and the necessity of hauling in the empty directions, equalizing as much as possible the unbalanced shipments. For example, calculations have shown that it will be economically profitable to send a considerable amount of cheap iron ore from the Kursk Magnetic Anomaly to the Urals and Kustanay iron ore to Siberia, to create a complete huge flow of iron ore going from west to east in the empty direction, to greatly develop metallurgy in the Urals, the Center and the South by limiting the further growth of metallurgy in Siberia, Kazakhstan, and Central Asia.

A most complicated fact is that for the spatial model, as well as for the structural model, the basic units are links and not economic sectors. The essence of this complication lies in the interlocking of the various industries. Therefore, the ferrous metallurgy of the Urals, the Center, the South and other parts of the country must be examined not only from the point of view of production of cast iron, steel, and sheet iron but also of its consumption of coke, ore, scrap metal, and from the point of view of its most profitable peripheral connections: with the chemical industry, production of building materials, and the like. For example, the metallurgy of the Center has an advantage over a number of other regions of ferrous metallurgy in that the chemical industry connected with it is situated in one of the main regions of consumption of mineral fertilizers in the Central Chernozem zone of great agricultural intensity. The development of metallurgy in the Center leads directly to growth in machine construction and production of bread, meat, and sugar there. This, however, does not lead to its economic autonomy. On the con-

TABLE 2

LIMITATIONS IN THE ECONOMIC DEVELOPMENT OF THE REGIONS OF THE USSR

TYPES OF LIMITATIONS	REGIONS					
	Central	Southern	Volga-Urals	Central Asian	Siberian	Far Eastern
Shortage of labor resources. . .	−	−	−	+−	+	++
Insufficient transport network .	−	−	−	+	++	++
Shortage of water resources. . .	++−	+	+−	++	−	−
Shortage of fuel:						
coal	++	−	+	+−	−	+
oil.	++	−	−	−	++	+
natural gas.	++	−	−	−	++	+
Shortage of electric energy. . .	++	+	+	+	−	+
Shortage of metals:						
ferrous metals	+−	−	+−	+	−	+
nonferrous metals.	++	+−	−	+ −	− −	+−
Shortage of raw materials for the production of mineral fertilizers.	−	+−	−	+	+	+

TABLE 2--Continued

| | REGIONS | | | | | |
TYPES OF LIMITATIONS	Central	Southern	Volga-Urals	Central Asian	Siberian	Far Eastern
Shortage of timber	±	++	±	++	−	−
Shortage of salt	+	−	−	−	+	++
Shortage of local grains	±	−	±	±	±	++

NOTE: −, means that at present there is no limitation of the given type in the region; +, shows that the given limitation exists; ++, emphasizes an acute deficit of the given kind of resource (product); ± and ±, specify that within the region are areas with directly opposite values of resources (products), both with an excess and an acute deficit.

trary, the region will be in still greater need of oil, gas, motor fuel, electric energy, fuel-consuming products of the chemical industry, nonferrous metallurgy, and other forms of production obtained from other regions. The interregional division of labor increases, and becomes more varied. It is the matrices that make it possible to carry over the chain reactions between the links (and their concrete elements, the industries) of the structural model onto the plane of the territorial division of labor between the regions.

The spatial model examines the chain reactions in the sphere of production and consumption in a spatial aspect, in a concrete geographical environment, under actual demographic conditions of the territory and the security of the economic development by the labor resources. In comparing, for example, the bases of ferrous metallurgy, one must consider the water supply and the cost of industrial water. Calculations by M. M. Davydov show that the creation of an artificial flow of water from northern rivers to the south (via Moscow, the Chernozem Center, and the Donets Basin, that is, through the largest and most promising modern industrial regions) is economically expedient, it guarantees the development of a gigantic new region of ferrous metallurgy in the zone of the greatest iron-ore beds and the greatest consumption of metals. In these calculations the productive capacities are compared with the labor resources, the provision for a network of city and rural settlements, living space, transportation and other facilities in the cross section of the regions of the country. All this requires a vast amount of information, which is introduced into the model.

In further stages of the work of constructing a spatial model it will be necessary to proceed from the regions to district territorial complexes and from there to the industrial hubs (agglomerations), that is, from 6 units to 33, and from 33 district complexes to the 120 to 150 largest agglomerations. If the calculations are focused on the industrial agglomerations, it will then be possible to put the distribution of the productive forces of the country on a firm scientific basis.

1. N. M. Budtolayev, V. P. Novikov, and Yu. G. Saushkin, "Problems of the Economic Development of the Western and Eastern Parts of the Soviet Union," *Vestnik Moskovskogo Universiteta,* seriya geografiya [Herald of Moscow University, Geography series], no. 4 (1963).

2. L. I. Vasilevskiy, "The U.S. Transport System: A Comparative Analysis," in *Sorevnovaniye dvukh sistem, Problemy ekonomicheskoy nauki* [The Competition of Two Systems: Economic Problems] (Moscow, 1963), pp. 185–86.

3. G. T. Grishin, S. P. Ivanov, V. P. Novikov, Yu. G. Saushkin, and M. N. Stepanov, "Paths to the Economic-Geographic Modeling of Regional Territorial Production Complexes," in

Materialy k IV s'yesdu Geograficheskogo obshchestva SSSR, Simpozium "B." Osnovniye voprosy ekonomicheskoy geografii SSSR [Material on the Fourth Congress of the USSR Geographical Society, Symposium B: Basic Questions of the Economic Geography of the USSR], Reports, part 2 (Moscow, 1964).

4.   *Narodnoye khozyaystvo SSSR v 1961 g.* [The USSR National Economy in 1961] (Moscow, 1962), p. 599.

5.   *Prirodnyye resursy SSSR, ikh ispol'zovaniye i vosproizvodstvo* [Natural Resources of the USSR, Their Utilization and Reproduction] (Moscow, 1963).

6.   *Sel'skoye khozyaystvo SSSR* [Agriculture of the USSR] (Moscow, 1960).

7.   Richard S. Thoman, *The Geography of Economic Activity* (New York, 1962).

B. B. RODOMAN

Mathematical Aspects of

The Formalization of

Regional Geographical

Characteristics

The theories of spatial systems worked out by geographers and the attempts to solve concrete practical problems with their help do not release geography from an obligation to compile detailed characteristics of a territory that can be utlized for various purposes. In this age of cybernetics, formalization of the descriptive sciences, and the automation of mental work, the traditional task of geography —the description of the earth—must expand in important new directions: (1) compilation and constant renewal of land cadasters— the technical specifications of the geographic environment, which include cartographic representation and wide use of the laconic language of tables, formulas, numbers, and special symbols; (2) improvement of the content and form of geographical information. Each of these two approaches is closely related to the other. For the first, automation seems inevitable. The second does not yet require automation, but a critical reexamination of present methods of geographical description from the viewpoint of mathematics and logic could help eliminate from geographical work excessive and useless information, increase verification [of hypotheses], and improve the presentation of interrelations between reported facts.

Translated from *Vestnik Moskovskogo universiteta,* seriya geografiya [Herald of Moscow University, Geography series], no. 2 (1967) pp. 28–44.

The most precise and prevalent form of geographic description is the regional characterization of a territory. Regionalization by geographic characterization is a special method of grouping information, which consists of dividing the entire territory into sections, to each of which is ascribed a definite term or expression. The automation of the process of geographical description is unthinkable without a mathematical approach to the process of regionalization.

The object of this article is to show how, in the study of regionalization, the basic concepts of mathematics are reflected and of what significance the various branches of mathematical science can be for the study of regions.

THE FORMS AND DIMENSIONS OF GEOGRAPHIC SPACE AND ITS REGIONS

The object of geographical study in most cases is a horizontal layer within the earth, which includes at least one surface divided among geospheres, differentiated by the predominant aggregate state of matter (on the earth it is the lithosphere, atmosphere, and hydrosphere), and extends a certain distance upward, downward, or in both directions from this surface. Digressing from the specific content of the concepts *geographic envelope, landscape sphere, geographic environment, anthroposphere,* and similar terms designating the complex (multicomponent) geospheres, we may call this layer a *geographic space* (*geospace*) and any finite part of this space a *geographic region* (*georegion*). Georegions are in fact three-dimensional; they not only have length, width, and area, but also height and volume. Linear boundaries between regions pass through physical land surfaces, water bodies, and through the bases of each. But the true lateral boundaries of the regions are surfaces that pass through the linear boundaries.

An essential aspect of complex (multicomponent) georegions is their multilevel or layered quality. Thus, a common natural region on a plain includes the basement rocks, soils, vegetation, and lower layers of the atmosphere. The upper and lower limits of natural regions evidently do not coincide with the limits of the geographic envelope, but are arranged as shown in figure 1. The lateral boundaries of natural regions related to tectonic bodies or mountain relief may deviate considerably from the vertical. In economic regions the upper and lower boundaries are usually not distinguished, but these regions too are three dimensional, which is especially evident if one notes the exploitation of mineral resources and their role in the economy of the regions, which are distinguished on the basis of the existence of extractive industries.

The question of boundaries of three-dimensional regions is far

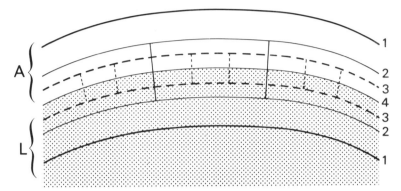

Fig.1. Distribution of the Three-Levels of
Natural Regions Within the
Geographic Envelope (profile).
Boundaries   1 – geographic envelope,
2 – regions, 3 – subregions,
4 – land cover
A – Atmosphere, L – Lithosphere

from simple. It is possible that for an approximate description of the true form of each georegion an arbitrary geometrical figure will have to be introduced—a *georegionoid* (similar to the *geoid* for the description of the form of the earth).

Although emphasizing the need for a three-dimensional approach to the study of regions we cannot forget that the vertical and any of the multitude of horizontal directions in the geospace are never equivalent and that they are incomparable. Due to the great difference between the vertical and horizontal dimensions of many geographical objects and the corresponding scale representations, our macro-geography remains essentially two-dimensional. Man's image of objects situated under and above him is invariably projected onto the surface of the lithosphere or sea on which he moves. All horizontal mathematical surfaces used in geodesy and cartography are close to that surface and related to it. Our orientation on the earth's surface is planar and projectional in character. However high we may fly or however deep we may penetrate the interior of the earth or the ocean, in answering the question of where we are, we will first refer to a point on the land or water surface, and only then indicate at what height or depth we are.

The *anisotropic* character of geospace and projectional character of human conceptions of the earth are reflected in the concept *territory.* This term, the exact nature of which is difficult to define, evidently designates a certain arbitrary two-dimensional space into which the three-dimensional geospace is transformed in the ordering of geographic information. By drawing the boundaries of a region in a territory we usually include all levels of natural and man-made features without special reference to the upper or lower limits of the region. The territory is measured in terms of area, not its volume. Thus, macrogeoregions can be simply regarded as two-dimensional.

By disregarding the vertical, we obtain an unused dimension, and can again move to the three-dimensional model in which the vertical component will no longer depict the actual relief of the earth's surface but will characterize the territory in some quantitative terms. A generally known example is the barometric pressure map of the atmosphere. Similarly the relief of temperatures or population densities can be depicted by isolines, modeled plaster, or metal casts.

A territory serves as the base for the construction of many, more arbitrary geospaces. For example, an economic-geographical space is possible in the representation of actual distances in terms of travel or cost. Just as the substitution for usual rectangular coordinates by a logarithmic or semilogarithmic grid simplified the representation of some empirical relationship because of straightening of the lines of the graph, the replacement of a "normal" territory by a more arbitrary space with different dimensions, such as non-Euclidean ones, reveals and makes obvious new spatial relationships between geographic objects. It would be interesting to work out algorithms for machine translation of ordinary cartographic representatives into a system of arbitrary measurement.

DISCRETENESS AND THE UNFOLDING OF A TERRITORY

In the process of geographical characterization, correspondence is established between a territory and statements concerning it. If this correspondence is set down on paper (or similar material), then one of the *areas of correspondence* is a cartographic representation, another is a text. In a geography book, there are few statements in which the name of the entire described territory is the subject. The task of geography is to describe differences from place to place. Therefore, it will be sufficient if hereafter we will examine a number of statements in which the subject, implicitly or explicitly, is the name of a place located in the territory being described, but not covering it entirely.

In a text there are, of course, a multitude of statements. In order to relate statements to various places of a two-dimensional territorial continuum, this continuum must be rendered discrete, that is, presented in the form of a finite set of places. There are two means of rendering a territory discrete.

1. *Point method*, where places are objects whose dimensions and forms are not essential compared with the scale of geographic description. The position of a point-like object is analytically determined by two geographic or geodetic coordinates, for example, latitude and longitude. Graphically it can be shown as a point symbol.

2. *Area method*, where places are objects whose dimensions and form are essential and therefore must be communicated to the reader. The position of an area-like object may be given analytically as a set of pairs of coordinates of points through which the boundary of an area passes, but it usually is shown only graphically. To depict a given object on a map in area-like terms means to plot the boundary and to indicate *what* this boundary outlines.

The area method is preferable to the point method, since it allows continuous description of a territory in which no place remains for points not contained in one or another area-like object. The achievement of continous areal discreteness of a territory is regionalization.

The point method and regional geographic description have two common elements: (1) the *center* of a nodal socioeconomic region; 2) the *nucleus* or representative *focus* of a natural or economic region, delimited on the basis of maximum internal homogeneity. In a number of cases the geographical description may be transformed without significant change by appropriate substitution of centers (nuclei) for regions and vice versa.

An oral statement is linear (unidimensional) and unidirectional. Sound by sound, letter by letter, words are formed, word by word into phrases, and all are arranged on one line, the beginning and end of which cannot be interchanged and similar to a mathematical line, a text has only length, and no width. Letter configuration, type size, arrangement of lines, and so forth, have no relation to the volume of the text, or more exactly, to its length. Time, too, is unidimensional and unidirectional and, therefore a text is well adapted to historical exposition to relate a succession of events.

It is quite different when objects are enumerated that exist simultaneously but in different places. Here the problem of sequence of exposition arises. To establish a correspondence between places and statements mean to project a two-dimensional topological grid of regions or points onto a unidimensional line of text or, in other words, to choose some kind of linear order for a path taking in all regions, points, or foci. This ordering is expressed by numbering or alphabetical-letter indexing of regions on a map and a corresponding arrangement of textual regional characterization. The unfolding of a regionalized territory is like disassembling a mosaic and then arranging all the pieces in one row.

If the alignment of the path is not suggested by the spatial distribution of the regions, as, for example, in the case of natural zones, then the territory may be unfolded arbitrarily by horizontal rows, vertical columns, on the diagonal, or along a spiral. It is sometimes possible to arrange the regions in the order of increase or decrease using some objective criterion (for example, elevation) or for methodological reasons to examine regions proceeding from simple to complex, from the important to the unimportant, and so forth.

Thus, in the process of geographical description, the following steps are distinguished: (1) *achieving discreteness* of a territory, that is, representing it in the form of a finite set of places; (2) *establishment of correspondence* between places and statements; (3) *unfolding* of the territory by (a) sequential numbering of places, (b) numbering the corresponding statements, (c) arranging statements in the order of the numbers.

The sum of propositions that can be combined in a single concept and placed before these propositions as a title—as well as the concept itself, used instead of a proposition—will be considered a single statement. In particular, the entire characterization of a region may sometimes be expressed by its name alone, which more frequently consists of several words rather than one word.

CORRESPONDENCE BETWEEN REGIONS AND STATEMENTS

If the task of geographical description consists in characterizing existing regions, then, to apply mathematical terminology, we can say that the set of regions is the point of origin and the set of statements is the zone of closure of correspondence between regions and statements. In regionalization: (1) one and only one statement must correspond to each region; (2) identical statements may correspond to different regions. If the second condition is not applicable, the regionalization is called an *individual* regionalization; if it does apply it is a *typological* regionalization (see figure 3). Thus, an individual regionalization exists when, and only when, there is a *one-to-one* correspondence between regions and nonidentical statements.

Individual regions usually bear proper names, for example, Volga-Shoshinsk lowland or the Volga-Vyatka economic region. Typological regions do not receive proper names, but use the name of a type of territory: outwash plains; grain, potato, and dairy-and-meat regions. In individual regionalization, in most cases it is enough to draw the boundaries of the regions on a map and within these outlines place indices. In typological regionalization one must, in addition, clearly depict the distribution of types, that is, color or shade identically regions of the same type.

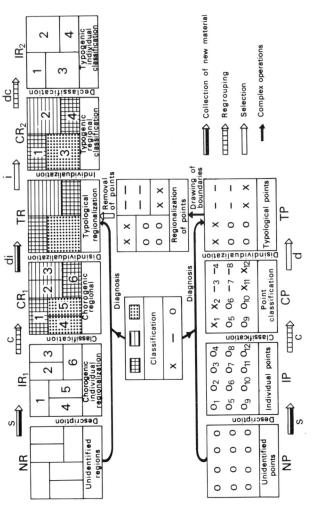

Fig. 3. The Simplest Techiques of Single-level
Individual and Typological Regionalization

The method of regionalization based on the description of "unidentified" outlines and their subsequent grouping we shall call *chorogenic*. The opposite course is also possible, where the point of origin of the correspondence is a previously given classification of geographic objects, and the point of closure is the set of areas of distribution of the classes within a territory. Let us assume that the concepts "tropical climate" and "red-earth soil" are given. Regions must be found that correspond to these concepts by plotting this climate and soil type on a map. This type of regionalization we call *typogenic*. The obtained regionalization is, for the most part, typological, but may also be individual.

In the chorogenic approach, the volume of the concept is given (the territorial region) and the content must be determined. In the typogenic approach, the content is given and the volume must be determined. In practice, both approaches are combined, of course. In the process of accumulating geographic knowledge, the point of origin and the point of closure of the above-mentioned correspondence are constantly interchanged. How can we set down on paper the correspondence between regions and statements? In principle one can write or print a textual characterization of a region directly within the outline on the map, but it is evident that in most cases this is very inconvenient. Therefore, within the regional outline a compact symbol is placed, such as a number or a letter index, and then this symbol is repeated in the text followed with the statement. The index thus serves as an intermediary between the regional outline and its characterization.

In a typological regionalization the types (classes) of regions are differentiated by a colored or shaded background, which cannot be introduced into the text without requiring in the type-setting a set of miniature printing plates. For typographical and other reasons the chain of relationships between cartographic representation of a region and the statement about it becomes even longer. So, for example, in a monochromatic (black and white) map of a typological regionalization included in a book, the following chain results: (1) in the cartographic description, the boundaries of regions and the shaded background; (2) in the legend, printed as part of the linecut—samples of the shadings and the corresponding class indices; (3) in a typeset map caption, the class indices and names explaining them; (4) in the text, the class names and their characterizations. In the language of mathematics, such a chain is called a *composition of correspondences*. The number of possible compositions may be calculated by combinatorial analysis; the concrete forms of the chains described by using the theory of graphs, and the distribution of background colors. Indices, regional names in the cartographic description, legend, caption, and text can be expressed by a matrix.

REGIONALIZATION AND A DIAGRAM OF INTERREGIONAL CONNECTIONS AS A DISCRETE IMAGE OF A FIELD

Let us return to the territorial continuum and assume that any point within it may be characterized by numbers. This facilitates the mathematical modeling of a geographic characterization. The potential correspondence between all points of the territory and any numbers means that the territory may be regarded as a *field*. A layered tinting of the bands between the isolines of the field would be the simplest model, or quantitative variant of typogenic typological regionalization, in which the numerical intervals serve as names and characterizations of types (classes) of territory.

If, for some reason, the field cannot be depicted by isolines, then the territory can be divided into regions and, for each region, the average value of the variable under study could be used. So, for example, one can depict on a map the mean temperature or mean elevation of any kind of region. A model, or quantitative variant, of a chorogenic individual regionalization is obtained. If one then groups the regions according to intervals of the values for a given variable, and colors or shades the regions according to their assignment to a given group, we obtain a method of representation called a cartogram, which is a model, or quantitative variant, of a chorogenic typological regionalization.

The concept of territory as an aggregate of fixed points and regions having certain defined properties is only one half of geography, *static* geography. The other half, the *dynamic*, is the study of space permeated by flows of matter, energy, and information. The velocities, intensities, and directions of these flows form a *vector* field.

Any place is intersected by transit flows that have no significant influence on it. Therefore, for the description of the dynamic state of the geospace, especially in socioeconomic geography, in many cases specific trajectories of the flows are not important but only their general schematic direction between the places of origin, transformation, and destination. The infinite set of points of origin and destination can be reduced to a finite set if the centers of geographic regions are taken as these points. Instead of the actual centers (many regions, for example, the natural regions, do not have such centers) one can take arbitrary points and compute all converging and diverging flows from the places where these flows intersect the boundary of the region. But this means that entire regions instead of localized points will function as places of origin and destination of flows. Instead of a continuous vector field, a finite set of isolated vectors assigned to regions at their "centers of gravity" is obtained.

The intensity and direction of incoming and outgoing flows of each region will be the expression of its external relations, and analogous data on flows between subregions will express internal relations. Thus, the horizontal flows of matter, energy, and information on the surface of the earth can be generalized with any given degree of exactness and with the necessary selectivity in the form of a hierarchical scheme of interregional links based on the multilevel regionalization.

When generalized flows are localized and related to their territories not by means of points, but by regionalization, the composite description of interregional connections has no need of graphic geometrical illustration in the form of cartographic flow diagrams, but can be given topologically in the form of a block diagram or in tabular form as matrices of interregional flows. The scheme of interregional flows is an inherent part of a regionalization map. Both schemes are the two sides of a coin. Regionalization can be looked upon as a discrete representation of a *scalar* field, and the scheme of interregional flows as a discrete representation of a *vector* field.

The vector sum of homogeneous, commensurate horizontal flows not subject to transformation and not leaving or entering the limits of the territory under study, must evidently be equal to zero, that is, the interregional flows in this case will be balanced. The matrix of interregional flows expressed quantitatively in identical units of measurement is called an interregional balance. The compilation of such balances is an important method of geographic research and applied geography.

REGIONALIZATION AND SET THEORY

There is hardly a branch of mathematics more important for regionalization than set theory. Geographic regions can be interpreted as a set of points or subregions, and regionalization as the disjointing of sets.

Looking upon a region as a set of subregions, we can translate into the language of geography the theorem of the *five alternative relationships* contained in set theory. According to this theorem, two regions A and B may have the following relationship to each other: (1) they have no common territory; (2) they intersect; (3) *A* is part of *B*; (4) *B* is part of *A*; (5) they are identical (figure 2; the left vertical row of figures). This is more precisely expressed by symbols:

(1) $A \cap B = \phi$
(2) $A \cap B \neq \phi, A \cap B \neq A, A \cap B \neq B$
(3) $A \cap B \neq \phi, A \cap B = A, A \cap B \neq B, A \subset B$
(4) $A \cap B \neq \phi, A \cap B \neq A, A \cap B = B, B \subset A$
(5) $A \cap B \neq \phi, A \cap B = A, A \cap B = B, A = B$

(The symbols: $A \cap B$—a set of subregions that are members of both region A and region B; ϕ—non-existent territory; $A \subset B$—A is part of B without covering it entirely.)

Regarding the boundaries of the region as sets of points, we can break down this listing further and, within the possibilities (1), (3) and (4), distinguish cases in which the boundaries of two regions: (a) do not have a single common point; (b) have one common point; (c) have many points in common (see figure 2).

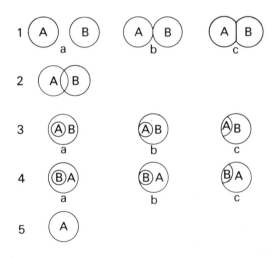

Figure 2. Views of the Spatial Relations Between two Regions, grouped by Point Theorems About Five Possibilities

Applying these cases to relationships between the entire regionalized territory and of the regions within it, as well as to the relationships between various sets of regions, we obtain various formalized kinds of modes of regionalization: (1) simple one-level regionalization, the result of a single continuous division of the territory into nonoverlapping parts; (2) simple multilevel regionalization in which a region of any rank, except the lowest, contains a whole number of designated subregions; (3) the same as the previous but complicated by the presence of subregions not encompassing the entire territory, or by regions coinciding with their single subregions; (4) the coexistence of several independent systems of regionalization within a given territory; 5) a *complex* regionalization in which hybrid regions, formed by the intersection of two original equivalent sets A and B, are looked upon as a new, third regionalization C; (6) *complex-subordinate* regionalization in which the hybrid regions C are considered subregions of A or B.

Regarding regionalization as the disjointing of a set of regions, we can say that set M of regions R is a complete, simple, single-level regionalization of country L, if:

$(\forall R \in M)$ $[R \subseteq L]$ Any region of the set M is entirely
within territory L;   (1)

$(\forall R \in M)$ $[R \neq \phi]$ Any region of set M has an area
different from zero;   (2)

$(\forall R_i \in M)$ $(\forall R_k \in M)$ $[R_i \neq R_k \rightarrow R_i \cap R_k \neq \phi]$.
No two non-identical regions of set
M contain subregions common to both;   (3)

$L \subseteq \cup R$
$R \in M$ The territory of country L is
contained in the territory that
combines all regions of set M.   (4)

From conditions (1) and (4) it follows that:

$\cup R = L$
$R \in M$ The union of all regions R contained
in set M coincides with the territory
of country L (we have attempted to
translate into geographic language
the rules for disjointing sets).   (5)

(Symbols; $\forall$ —any; $\in$ is an element or member of; $\subseteq$ is contained in; $\rightarrow$—hence it follows that; $\cup$—union.)

In a similar manner the distinctive features of other forms of regionalization can be designated.

Of still greater interest to the geographer are the relationships between a set of places contained within a region and a set of regional properties. In textbooks on logic a distinction is drawn between the differentiation of an individual and the division of the volume of the concept about that individual; the law of the reverse relationship between the volume and the content of a concept is cited. However, for the needs of geography this is not sufficient. We think that for the solution of the question concerning the relationship between regionalization and classification it is necessary to construct set-theory models of geographic objects, that is, examining them as fields of intersection, union, and complements of sets of places and properties. Interesting steps in this direction have been made by Golledge and Amadeo.

Having designated a set of places contained within a region as X, and the class of objects encompassed by the geographic characterization as Y, and the time of existence of the region as T, we find that the various headings of geographic description can be formalized as one of the following three phrases: (1) the Y characterization of territory X at time T; (2) a characterization of the class of phenomena Y within territory X at time T; (3) characterization of time T within territory X from the point of view of the class of phenomena Y. In (1) above X is the name of the object under study, and Y and T simply restrictions limiting the content of the monograph; in (2) and (3) the name of the object is Y or T, respectively; however, the real object of study in all three cases is neither X nor Y, nor T, but the intersection $X \cap Y \cap T$. Taken as a continuum, it represents an infinite set of objects, properties, places, and moments of time; but if rendered completely discrete, it looks like a finite set of subclasses, subregions, and subperiods. Examining a geographical object as the intersection of sets that are amenable to disjointing, we can better picture the structure of geographic description.

REPRESENTATION OF GEOGRAPHIC DESCRIPTION IN THE FORM OF MATRICES

The assignment of an object simultaneously to three sets converts the discrete model of geographic description into a spatial grid, a three-dimensional *matrix*.

The majority of geographical work is written to characterize the present; the time component enters in only as a more or less reduced introduction to the description of individual phenomena and regions, and appears again at the end of the work in the form of a section on "prospects," frequently the only reference in the entire monograph.

If one ignores the chronological aspect then, instead of a cumbersome parallelepiped we can obtain a two-dimensional matrix that is easily represented on paper. This could be a table in which the horizontal rows represent regions and the vertical columns the natural components of the geographical envelope, branches of economy, or types of human activity, or, conversely, rows represent components, and columns, regions. In a multilevel division, columns of subregions and subcomponents are combined under the names of regions and components into long blocks divided by double or heavy lines.

If all components change in time at approximately equal rates, then the historical dimension can be given by preparing several region-component matrices. If the components do not change at the same rate, then they may be characterized by supplementary regional-chronological and component-chronological matrices, including matrices of periodic changes of some object by seasons of the year or times of day. All such matrices we will call *structural*.

The spatial component of a structural geographical matrix is by its nature three-dimensional, but we have already simplified it twice: we substituted three-dimensional space for a two-dimensional territory and a territory was unfolded into a line of text. Thus, space has been entered into the matrix as one-dimensional. Analogously the classifications of economic sectors and components of the geographial environment that are initially multidimensional are reduced to a unidimensional form.

In addition to structural matrices there exist matrices of interrelationships. These are tables in which the same elements are enumerated in identical order (for example, only economic sectors or only regions) as headings of columns and rows: first as senders of goods, energy, or information, that is, "agents"; and then as receivers or as elements influenced by the agents, that is, "the patients".

In contrast to the structural matrix, a complete matrix of interrelationships is always square: the number of rows equals the number of columns. The process of creating a structural matrix is a logical multiplication of m components by n regions; the process of constructing a matrix of interrelationships is the multiplication of n elements by the same n elements, that is, logical squaring (powering).

In both matrices the same elements can be taken as subjects (the same regions or the same components). Two such matrices can be fused into one, that is, have common rows or common columns.

The construction of the model of the structure of a textual geographical description can be regarded as the unfolding of a geographical matrix, that is, the ordering of its elements into one line. The unfolding of a structural matrix in a direction parallel to the economic branches (components) yields a topical survey in which each topic is characterized first as a whole and then by regions. The unfolding of the structural matrix in a direction parallel to the regions yields a regional survey in which each region is characterized at first as a whole and then by topics. Thus, the two main traditional divisions of geographical monographs are obtained.

The unfolding of a matrix of interrelationships in the two different directions yields two different orders of exposition: (1) *active*, when each component is characterized as an agent affecting all the others, as a cause of a set of effects, and each region acts as a sender; (2) *passive*, when each component is characterized as a receipient experiencing the influence of all others, as the effect of a set of causes, and each region functions as receiver.

Let us designate the references to components or regions contained in the section titles of a geographical work with capital letters A, B, C, and the reference in the text of sections with small letters a, b, c. For simplicity's sake we will limit ourselves to three components. With arrows we designate the direction of the described relationships with polynomials in parentheses representing the content of chapters, and with the letters before the parentheses representing the titles. Then, an active sequence of exposition would be designated as

$$A\,(b \leftarrow a \rightarrow c),\, B(a \leftarrow b \rightarrow c),\, C\,(a \leftarrow c \rightarrow b),$$

and a passive sequence as:

$$A\,(b \rightarrow a \leftarrow c),\, B(a \rightarrow b \leftarrow c),\, C\,(a \rightarrow c \leftarrow b).$$

For any number of components $m > 3$, the polynomials in the parentheses become $(m - 1)$—pointed stars with centripetally or centrifugally directed arrows.

The fact that the same things are regarded as causes and also effects does not contradict the principle of causality, since cause and effect are produced by different aspects, properties, and components of objects, usually at different times.

The time has come to define more accurately the differences between the concepts of description and characterization, which, up to now, were used as synonyms. A *description* should be restricted to a simple enumeration of component parts and properties of an object, whereas a characterization requires an enumeration in which the interrelationships between the parts and properties as well as the external relationships of the object are indicated. Geographical description may be represented as the result of the unfolding of a structural geographic matrix, and a geographical characterization as the result of the unfolding of a matrix of interrelationships or of a *complex* matrix (an agglomerate of fused matrices) in which there is both a structural submatrix and a submatrix of interrelationships.

Table 1 shows a complex matrix that includes four submatrices— *quadrants*. The upper left quadrant shows the intercomponent relationships and the lower right, interregional relationships. The elements of the column and row headings corresponding to general textual surveys of components and regions participate in the unfold-

TABLE 1

A COMPLEX MATRIX MODEL OF THE STRUCTURE OF GEOGRAPHIC CHARACTERIZATION (WITHOUT TAKING INTO ACCOUNT HISTORY AND EXTERNAL RELATIONSHIPS)

A \ P	C_1	C_2	$...C_m$	R_1	R_2	$...R_n$
C_1	*	$C_1 \rightarrow C_2$	$...C_1 \rightarrow c_m$	$C_1 \cap R_1$	$C_1 \cap R_2$	$...C_1 \cap R_n$
C_2	$C_2 \rightarrow C_1$	*	$...C_2 \rightarrow c_m$	$C_2 \cap R_1$	$C_2 \cap R_2$	$...C_2 \cap R_n$
$.\ .\ .$	$...$	$...$	$...\ ...$	$...$	$...$	$...\ ...$
C_m	$C_m \rightarrow C_1$	$C_m \rightarrow C_2$	*	$C_m \cap R_1$	$C_m \cap R_2$	$...C_m \cap R_n$
R_1	$R_1 \cap C_1$	$R_1 \cap C_2$	$...R_1 \cap C_m$	X	$R_1 \rightarrow R_2$	$...R_1 \rightarrow R_n$
R_2	$R_2 \cap C_1$	$R_2 \cap C_2$	$...R_2 \cap C_m$	$R_2 \rightarrow R_1$	X	$...R_2 \rightarrow R_n$
$.\ .\ .$	$...$	$...$	$...\ ...$	$...$	$...$	$...\ ...$
R_n	$R_n \cap C_1$	$R_n \cap C_2$	$...R_n \cap C_m$	$R_n \rightarrow R_1$	$R_n \rightarrow R_2$	$...$ X

NOTE: A, agents or senders; P, patients or receivers; C_1, $C_2...C_m$, components or sectors; R_1, $R_2...R_n$, regions; $\rightarrow$, direction of the relationship; $\cap$, intersection of sets; *, characterization of intracomponent relationships or absence of a statement; X, characterization of intraregional relationships or absence of statement (* and X, main diagonals).

ing in the same manner as the elements of the quadrants. In an unfolding of the matrix by rows, the upper right quadrant yields a regional description of components and the lower left, a topical description of regions. In the unfolding of the matrix by columns, the upper right quadrant yields a topical description of the regions and the lower left, a regional description of the topical components. For more completeness, the four quadrants can be supplemented by submatrices of external relationships between components or subregions and objects, situated outside the territory being studied.

This matrix form is not numerical; the symbols do not represent numbers but names of regions and components. These names are the *subjects* of statements, titles of sections of geographical description. The column and row headings represent the initial *elementary* subjects; in the cells of the table are the *hybrid* subjects formed by the combinations of the elementary subjects. If in the structural matrix C_i is population, R_j is the Ukraine, then the hybrid subject $(C_i \cap R_j)$ represents the population of the Ukraine. If in a matrix of interregional relationships of component C_i, the symbol R_k designates Belorussia then the hybrid subject $(C_i \cap R_j) \rightarrow (C_i \cap R_k)$ reads "the movement of population from the Ukraine to Belorussia." In order to obtain a statement, the hybrid subject must be followed (in the text) or be replaced (in the table) by the predicate. It may be a word or a number. In the latter case we would have in the structural matrix: $C_i \cap R_j = 46$ million people.

We could decide, for example, that in a horizontal unfolding, the transition from one cell of the table to another in a predetermined order, would mean the beginning of a new paragraph, the transition from quadrant to quadrant within a single row would indicate the beginning of a new section, the transition from row to line, the beginning of a new chapter, and so forth. The process of modeling the structure of geographical characterization involves repeated alternation of two types of steps: (1) the logical multiplication of linear (unidimensional) classifications and schemes of regionalization; (2) the unfolding of matrices obtained as the result of multiplication (see figure 5).

THE PROCESS OF REGIONALIZATION AND OPERATIONS RESEARCH

Regionalization can be understood in two ways: statically, as a state of information; and dynamically, as the process leading to it. The formalized types of regionalization states we have called regionalization *modes* and the stages of the process, regionalization *operations*. To define an operation means to name two modes and point out which has been derived from which. The description of the process of regionalization can be represented as a sequence of modes similar to frames in a motion picture.

Figure 3 shows some of the simplest techniques of regionalization. The frames are cartographic representations of modes and their names; symbols over the arrows designate operations and long rec-

FIGURE 5

THE PROCESS OF MODELING THE STRUCTURE OF GEOGRAPHIC CHARACTERIZATION

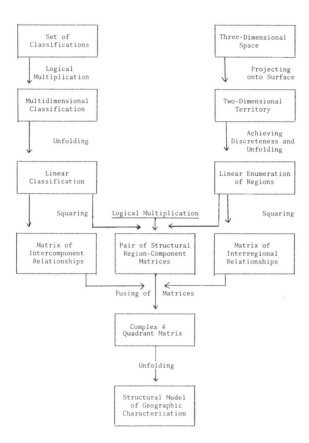

tangles within frames are regions. Numbers are given for regions instead of names and characteristics of the individuals; shadings and symbols of various forms designate different classes. In addition, modes are designated by capital Latin letters and operations by small letters.

This does not, of course, exhaust the techniques of regionalization. For example, Soviet physical geographers often obtain *IR* (individual regionalization) by combining small typological regions. In figure 3, $TP \rightarrow IR_2$ would represent such an operation with microregions substituted for point symbols in mode TR.

The mode CR (classification of regions) is the simplest form of combining IR (individual regionalization) and TR (typological regionalization) on one map. It can be taken as $IR \cup TR$ if one keeps in mind the set of points that make up the lines of a drawing. Having printed the maps on transparent paper, we obtain CR by superimposing IR on TR, and by superimposing IR, TR, or both modes on CR we do not change the external appearance of CR.

The enumeration of operations can be ordered and simplified if one regards them as sequential or parallel combinations of such operations as the disjointing of sets, the union of subsets, the combination of a place symbol with a property symbol, the removal of symbols, and so forth.

By designating the volume of phenomena in the objective world as reflected by mode M as v, and those reflected by mode M' as v', we may define three kinds of operations $M \rightarrow M'$ and group the operations shown in figure 4 according to these categories:

1. collection of new information $(v < v')$: s (description), i (individualization)
2. regrouping of existing information $(v = v')$: c (classification), de (declassification)
3. discarding unnecessary information $(v < v')$: di (disindividualization)

The sequence of different types of operations forms a cycle characteristic of any type of scientific research. These cycles can be seen in figure 3 (the periodic alternation of the three types of arrows).

Operations (1) and (2), corresponding to condition $v \leq v'$ we shall call *reversible* because they can be followed by the reverse movement $M \leftarrow M'$ without using information not already contained in M. Operations of types (3) are *irreversible*, as are all processes of cartographic generalization. One can also distinguish pairs of mutually reversible operations: i and di, c and dc.

Discarding the last letters in the abbreviated designations of point and regional modes we can designate with arrows all operations that can occur without using information not already contained in the transformed mode:

$$N \leftarrow I_1 \rightleftharpoons C_1 \rightarrow T \leftarrow C_2 \rightleftharpoons I_2.$$

These arrows indicate paths in which material can be regrouped and reduced; a procedure for those who compile a map at their desk and cannot supplement the available sources.

Diagnosis of objects, that is, assigning them to classes of previous classification, may yield the same results as developing a new classification by comparing the identified individuums. Consequently, operations *s* (description), *c* (classification), and *di* (disindividualization) form a *group* (in the mathematical sense of the word), that is, $g = s + c = di$. The plus in this formula may be interpreted as the symbol of the vector addition of paths in a discrete space, the points of which correspond to various modes. In the process of formalization of geography, the theory of groups, elaborated by the outstanding geographer, mathematician and Arctic explorer O. Yu. Shmidt, can be applied. The mathematical method of logical and map-compiling operations and the application of the algebraic methods would help us find means to mechanize and automate regionalization.

REGIONALIZATION AND THE THEORY OF GRAPHS

The theory of graphs is as close to questions of regionalization as set theory. With regard to regionalization, graph theory has significance in at least the following case.

1.   A network of boundaries between regions is a planar multiangular graph. Consequently, different concrete networks of regions can be studied topologically and described in terms of the theory of graphs.

2.   If a point is placed within each edge of graph G and points of adjoining edges are connected by lines, a graph G^* is obtained which is the dual of the original graph. This operation, called *inverting*, has an interesting economic-geographic interpretation: by inverting a graph of economic nodal regions we obtain a topological scheme of communications; and by inverting the graph of the transport network we obtain a scheme of regions oriented toward transport nodes.

3.   Any graph can be replaced by an *adjacent matrix* showing which vertices are connected and which are not. For a *directed* graph this would be a complete square matrix, and for a *nondirected* graph it would be half a matrix lying on one side of the main diagonal. On the other hand, the presence or absence of connections between the regions or points, as well as interrelationship between regionalization and a matrix of interregional flows is a geographic interpretation of these facts.

4.   A multilevel classification or regionalization scheme can be graphically depicted by a graph in which the vertices correspond to taxons (regions or classes of a certain rank) and the arcs show the relation of subordination and inclusion of taxons. Such a graph we call *taxonomic*. The following taxons can be distinguished: (1) *abstract* taxons—the province in general or a species in general; (2) *concrete* taxons—such as the province Yenisey Ridge, and the

species Siberian larch. A graph depicting abstract taxons would be a taxonomic *ladder*, and a graph of concrete taxons a taxonomic *pyramid*. In contrast to the traditional forms of classification the taxonomic graph of regionalization can have cycles (see figure 4).

5.   A block diagram of the process of regionalization corresponds to a directed graph in which the vertices designate nodes and the arcs operations, or vice versa. The scheme of the process of regionalization, and the entire process of compiling a geographical characterization may take the form of a grid graph in which the most time-consuming operations would be indicated, and, consequently, the critical path could be selected.

6.   The cartographic representation of regions is related to the classical unsolved problem of the theory of graphs concerning the *four colors*; to demonstrate or refute that any geographic map can be colored by using no more than four colors in such a way that regions having a common boundary other than a common point will carry different colors. For five colors this possibility has already been demonstrated.

7.   As indicated earlier in the section discussing the correspondence between regions and statements, graphs can depict forms of correspondence between regions and statements (two-stage graph) and means to combine a cartographic representation of a region; with its textual characterization.

8.   Having described the relationships between landscape components with a directed graph and expressed the force of mutual influence of the components in some unit of measurement, we can find a rational order of exposition of the characterization of a region, proceeding primarily from cause to effect. For this purpose the components must be arranged in the order of the increasing sum of incoming (+) and outgoing (−) influences calculated on a graph or matrix; or by extending the network of relationships along a path of maximum influences creating a model of the resistance of material, of the strain in strong arcs and breaks in weak arcs; or, finally, to find, if possible the suitable *Hamiltonian* path, that is, a line passing through all the vertices of the graph once and only once. (These are only suggestions by the author that have not yet been verified in practice.)

9.   By coding a graph of regionalization or intercomponent relationships with a linear sequence of symbols similar to physical-chemical diagrams of the composition and properties of metal alloys, one can obtain the fundamental characteristics of the morphological structure of natural and agricultural landscapes suitable for machine classification and rapid retrieval in a cadaster library, or as standards for automatic diagnosing of landscape types.

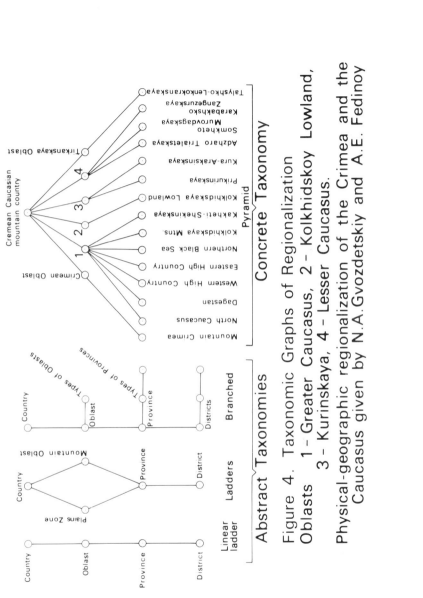

Figure 4. Taxonomic Graphs of Regionalization
Oblasts   1 – Greater Caucasus, 2 – Kolkhidskoy Lowland,
3 – Kurinskaya, 4 – Lesser Caucasus.
Physical-geographic regionalization of the Crimea and the
Caucasus given by N.A.Gvozdetskiy and A.E. Fedinoy

Evidently one can find an infinite number of similar examples of application of graph theory to regionalization. The topological properties of the material world are more important and fundamental than the metrical properties. Topology, discreteness, and whole numbers dominate in schemes describing the structure of scientific knowledge. Therefore, the presentation of regionalization schemes, as well as schemes of relationships between regions and components of the geographic envelope in the form of graphs may become one of the promising directions of mathematical geography.

1. B. L. Gurevich, and Yu. G. Saushkin, "The Mathematical Method in Geography," *Vestnik Moskovskogo Universiteta,* seriya geografiya [Herald of Moscow University, Geography series], no. 1 (1966).

2. O. Ore, *Grafy i ikh primeneniye* [Graphs and Their Applications] (Moscow, 1965).

3. B. B. Rodoman, "Methods of Individual and Typological Regionalization and Their Representation on Maps," *Voprosy geografii* [Problems of Geography], no. 39 (1956).

4. B. B. Rodoman, "Logical and Cartographic Forms of Regionalization and Problems of Their Study," *Izvestiya Akademii Nauk SSSR,* seriya geografiya [Proceedings of the Academy of Sciences of the USSR, Geography series], no. 4 (1965).

5. Yu. A. Shikhanovich, *Vvedeniye v sovremennuyu matematiku* [Introduction to Modern Mathematics] (Moscow, 1965).

6. V. S. Shteyn, *Resheniye nekotorykh informatsionnykh zadach fiziko-khimicheskogo analiza s pomoshch'yu elektronnykh tsifrovykh vychislitel'nykh mashin* [Solution of Some Information Problems of Physical-Chemical Analysis With the Aid of Electronic Digital Computers], Summary of a dissertation for the degree of candidate of technical sciences, Moscow Chemical Technology Institute, 1966.

7. B. J. L. Berry "Approaches to Regional Analysis: A Synthesis," *Annals of the Association of American Geographers* 54, no. 1 (1964).

8. R. G. Golledge, and D. M. Amadeo, "Some Introductory Notes on Regional Division and Set Theory," *The Professional Geographer,* no. 1 (1956).

SECTION III

Resource Management

Introductory Note

The study of land and other natural resources and their utilization has long occupied a prominent place in Soviet geography.[1] An inventory of Russia's natural wealth was indeed the first major assignment given by the young Soviet government to the Academy of Sciences in 1918.[2] The significance of the study of resources has been reinforced by a number of factors: the large territorial base of the nation and its diversified natural conditions and resources; a socialist political-economic structure with national ownership of all land and other natural resources; a system of integrated long-range national planning; and a prevailing *engineering*, transformative philosophy with regard to the natural environment.

Among the major concerns in resource studies historically have been questions of (1) the discovery, inventory, and study of natural resources; (2) the evaluation and proper utilization of resources in the context of integrated regional planning; and (3) the conservation, renewal, and transformation of resources and the environment.

Geographers have been involved in resource management planning at scales that range from that of the individual economic unit (such as the sovkhoz, or state farm) to the regional and national level. Their work involves the following tasks:[3]

1. *Collection of information about natural conditions and resources.* Included here are such activities as data collection on the location, and dynamics of the natural environment and resources and development of a cadastral system for purposes of data comparability. Much of the work carried on under this heading has been by physical geographers engaged in studies of climatic and water resources, mineral resources, land and soils resources, and biotic resources. In addition to practical inventories, these studies have resulted in a number of important theoretical contributions to the world literature in physical geography.

2. *Reduction and presentation of information* in the form of special purpose maps, atlases, and reference compendiums and monographs. Both physical and economic geographers, as well as cartographers and technicians, participate in this aspect of resource activity and have contributed to the impressively large volume of materials already produced.

3. *The formulation and analysis of technical and economic alternatives* for the solution of resource utilization and renewal problems. Less developed than the categories above, this area is one of increasing concern and is one in which economic geographers, economists, and engineers are most likely to make a special contribution.

Among the recent trends evident in resource management studies is a shift away from the previous emphasis upon the approaches of physical geography to a greater emphasis on economic-geographic approaches; that is, a shift from discovering, studying, and inventorying the natural environment and its resources to a concern with proper economic evaluation and utilization. To some degree this can be viewed as a historical phenomenon reflecting the past success in discovery, study, and inventory of resources; however, it is also a reflection of the increased national political concern with economic efficiency in managing the Soviet economy. The necessity of a more precise economic evaluation of land and environmental differences under an economic system that levies no direct charge on land has resulted in a system of regional prices for agricultural and other commodities, which in effect incorporates some degree of environmental evaluation.[4]

It is not yet completely clear how the new concern with economic evaluation and efficiency will affect the traditional emphasis on large-scale national and regional environmental transformation schemes. Past schemes of this sort, for example, Stalin's nature-transformation plan of 1948–50, are today criticized strongly on an economic basis: "the plan did not give proper attention to the economic factor, it failed to take account of the then existing level of collective farming, it did not calculate the economic effectiveness of large capital investments in the transformation of nature."[5] Although it has not dampened continued advocacy by geographers of such major regional transformation schemes as the diversion of northern rivers to the Volga basin, artificial regulation of the Caspian Sea level, full development of irrigation and water use in Central Asia based on an overall transformation of the hydrological cycle, and so forth, it does appear certain that in the future such schemes must be guided by the principle enumerated at the Twenty-second Party Congress: "Priority in resource development should be given to those resources that are accessible for immediate development and will yield maximum economic benefit." As interpreted by Saushkin this means that "before an attack on . . . unfavorable natural conditions is contemplated, the following economic calculation should be made: How much additional production will be obtained and after how much time; as a result of such a reclamation project when and how will the investment in the nature-transformation project be recovered?"[6]

A recent theme associated with that of a more economic evaluation of resource development is the increasing concern with integrated regional resource planning; in the future resource development is to be studied as part of the process of regional and national economic planning and "not simply within the framework of . . . 'a natural-economic complex.'"[7] It is recognized that this integrated evaluation will present substantial methodological problems for though "physical and economic geographers can name a large number of factors affecting such an evaluation . . . they still do not know how to weigh them, how to generalize them in one or several integrated indices, or how to evaluate them in terms of cost of labor and materials or in monetary terms."[8] The use of mathematical methods and computers is seen as essential to solution of the problem.

A theme, increasingly reflected in literature of recent years, is that of conservation and environmental quality. This concern is evident not only in professional literature but in items that have appeared in the Soviet press revealing the considerable public interest in this issue.[9] The concern for conservation, of course, poses a potential conflict given the development orientation of Soviet economic geography. For example, the same authors who warn that "the time has come to give up the naive faith in the inexhaustibility of natural resources" are also at pains to stress that the "problem of the rational utilization of natural resources is far broader and more complicated than "mere conservation of nature"; and "that while such conservation is undoubtedly important, the main objective of a Socialist economy is the rational utilization and expanded reproduction of natural resources."[10] The problem of pursuing conservation goals within an overall philosophy of rapid economic development provides some interesting parallels to the contemporary American dilemma and is an additional topic of common interest to geographers of both societies.

In the initial article, "Rational Utilization of Natural Resources and Problems of Soviet Geography," Armand presents a strong plea for conservation. A socialist economy has the possibilities to avoid depletion and qualitative decline of natural resources, but to turn possibilities into reality "the government, scientific organizations, and all the people must show a constant and untiring concern for the preservation and regeneration of natural resources." Serious current problems are noted and traced to "the power of the myth of inexhaustible natural resources." Current Soviet conservation efforts including recently passed conservation laws are noted, but Armand points out they "are only the first stage"; in their implementation there "must be created an effective state and public apparatus for estimating resources and planning their utilization." He proposes creation of unions of resource producers and consumers as closed systems; these unions would thus be forced to be concerned with resource regeneration; he also asks for subordination of the narrow planning of individual enterprises to integrated regional planning. He concludes with a discussion of the contributions and deficiences of the present Soviet geographic effort in resource management.

The second article, by Komar, is essentially a substantive contribution in which regional patterns of raw material extraction, use, and mobility over the present territory of the USSR are compared for 1913 and 1960. Raw material extraction is shown to have shifted dramatically from the 1913 dominance of agricultural and forest products to the present emphasis on mineral raw materials and fuels; eastern regions have come to enjoy a much higher rank although older, densely settled areas with no major resource deposits remain surprisingly high in absolute production. Although fifty to sixty percent of raw materials are consumed at site or within a limited distance of production, interregional shipments of raw materials have grown faster than extraction, which is interpreted as a testimony to the "deepening territorial division of labor."

An economic approach to the problem of a natural hazard is demonstrated in the third paper of this section, "Evaluation of the Effectiveness of Flood Control." Flooding is shown to be a major problem in urban areas located along certain rivers. Borisov then presents a form of cost-benefit analysis that can be employed to reach a decision as to whether it is economically preferable to (1) bear the loss, (2) remove the endangered objects from the flood zone, or (3) construct alternative engineering projects for purpose of protection. Central to the evaluation procedure is a method proposed for estimating direct and indirect losses from flooding.

The final selection, "On the Problem of the Economic Evaluation of Lands," by Gurevich and Landa, illustrates the more rigorous economic approaches to agricultural land management developed in recent years. The first question treated is that of the optimal distribution of labor and capital expenditures among different land or soil types under cultivation of a single crop; the authors show that it is necessary to take into account not only the productivity of each soil associated with its characteristic yield capacity but also the productivity of each in accordance with changes in yield capacities brought about by changes in expenditures. Under certain limiting assumptions, equations, indices, and graphs are then developed to solve optimization problems. The procedure is extended to the problem of a single soil type with a combination of crops. An illustration is given employing actual data form a sovkhoz.

1. See in this respect I. P. Gerasimov, "The Study, Rational Utilization, and Preservation of Natural Resources," in *Soviet Geography: Accomplishments and Tasks*, ed. Chauncy D. Harris (American Geographical Society, 1962), pp. 261–65. Also a number of papers on the topic have appeared in *Soviet Geography: Review and Translation*.

2. Yu. G. Saushkin, "A History of Soviet Economic Geography," *Soviet Geography: Review and Translation* 7, no. 8 (October 1966):5.

3. I. P. Gerasimov, D. L. Armand, and V. S. Preobrazhenskiy, "Natural Resources of the Soviet Union, Their Study and Utilization," *Soviet Geography: Review and Translation* 5, no. 8 (October 1964).

4. Robert G. Jensen, "Land Evaluation and Regional Pricing in the Soviet Union," *Soviet Geography: Review and Translation* 9, no. 3 (March 1968):145–49. The entire issue is devoted to this particular problem.

5. Saushkin, "A History of Soviet Economic Geography," p. 45.

6. Ibid., p. 62

7. I. P. Gerasimov, et al., "Natural Resources of the Soviet Union," p. 12.

8. Ibid.

9. Again a number of articles on the topic of resource conservation have appeared in *Soviet Geography: Review and Translation*. See particularly the issues of June, September, and December 1960. See also, I. P. Gerasimov, "The Past and Future of Geography," *Soviet Geography: Review and Translation* 7, no. 7 (September 1966):9. Translation of Soviet press items concerned with conservation can be found in *The Current Digest of the Soviet Press*.

10. D. L. Armand, et al., "The Role of Geographers in the Study, Mapping, Economic Appraisal, Utilization, Conservation and Renewal of the Natural Resources of the USSR," *Soviet Geography: Review and Translation* 1, no. 6 (June 1960):5.

D. L. ARMAND

Rational Utilization of

Natural Resources and

Problems of Soviet Geography

In addition to human labor, the sources of created material values are the natural resources used in agriculture, industry, and energy production, that is, raw minerals and fuels, water, solar energy, fertile soils, vegetation, and useful animals. Therefore, growth in production is inevitably associated with an increased consumption of natural resources.

According to the decisions of the Twenty-first Congress of the CPSU [Communist Party of the Soviet Union], in the course of the next fifteen years basic branches of industry will increase their production by more than two or three times. Moreover, some new branches, which will use many times as much energy and water per ton of manufactured production as the old ones, will develop most rapidly.

The fulfillment of these plans of industrial development will be possible only with increased utilization of useful minerals, sources of energy, water, and building materials. The growth of agriculture, together with the increased productivity that is its main task at the present time, demands further utilization of unplowed lands and a search for new feed resources. The question arises: Can we end-

Translated from *Izvestiya Akademii Nauk SSSR*, seriya geografiya [Proceedings of the Academy of Sciences of the USSR, Geography series], no. 1 (1961), pp. 48–56.

lessly continue to remove the wealth in the storehouses of nature or will there come a time when the reserves of some important resources will be exhausted? An if this danger becomes real, what can be done to prevent it?

Historical experience shows that in the early stages of the development of productive forces natural riches always seem inexhaustible. Gradually, however, in connection with population growth and its needs, with industrial development and also with an increase in international trade, in one country after another the demand for natural riches begins to exceed the supply. The crisis accelerates if the economy has a spontaneous or rapacious character, as can be observed in capitalistic and especially in colonial countries. Then forests are destroyed over wide areas, reserves of fish and furs are exhausted, the fertility of the soil declines under the assault of erosion and dust storms, rivers become shallow or the water in them becomes unusable due to pollution with industrial wastes, and even the atmosphere becomes polluted and turbid.

It is necessary to distinguish clearly the specific features of individual natural resources. With respect to the group of exhaustible and nonrenewable resources, to which belong, above all, useful minerals, the task is one of rational exploitation so that not only the richest deposits but also the poorest ones are utilized and all valuable components are extracted from complex ores. With respect to the exhaustible and renewable resources, that is, fertile soil, useful vegetation, and fauna, the task must be to balance their expenditure with the possibilities of their natural or artificial renewal leaving a constant positive balance that will guarantee increased production. Finally, for rational utilization of the third group—the inexhaustible resources of solar radiation, wind energy, and atmospheric precipitation—the task amounts to one of delivering them to the consumer at the right place and time and in a state of high, natural quality. A qualification must be made here: water is an inexhaustible resource only on a global scale. Due to its uneven distribution in space and time, it sometimes becomes least available where it is most needed. Therefore, an important problem arises in its redistribution; the preservation of its purity and its economic utilization.

As a rule, the depletion and qualitative decline of natural resources are avoidable. In capitalist countries they arise as the result of anarchy in production and the pursuit of high profits. In a planned socialist economy the socioeconomic preconditions are created for intelligent utilization and constant reproduction of natural riches, an increase of its reserves, or a search for substitutes for those that cannot be renewed by nature. However, in order for the possibilities of a socialist structure to turn into reality, the government, scientific organizations, and all the people must show a constant and untiring concern for the preservation and regeneration of natural resources. Planning of rational resource utilization and renewal must be accepted as one of the most important problems of planning organizations at all levels.

The government and the Party are making all possible efforts to preserve natural wealth. Accompanying the control figures on development of the national economy for 1959–65 was the statement: "An important task of the coming seven-year period is the intensive involvement of the rich natural resources of our country in economic circulation." It is very important to emphasize that the primary concern is not the consumption of natural resources but their circulation, that is consumption and receipt, utilization and renewal. This idea is still more clearly imparted in the case of forestry where the problem of "preservation and renewal of forests" is directly referred to.

In April of last year, an important resolution concerning the regulation of water resources was passed by the USSR Council of Ministers. Its main purpose is to bring planning to the utilization of water; to replace departmental [branch or sectoral] planning, in which the interests of individual water users often came into conflict, with cooperative planning; to create a network of Union and republic agencies studying water resources and directing their use by intelligent distribution.

In the majority of Union republics during recent years, laws have been enacted to preserve nature. Of greatest interest in the content of the law of the Russian Federation passed 27 October 1960 by the RSFSR Supreme Soviet (*On the Preservation of Nature in the RSFSR*, 1960).[1] The distinguishing feature of this law is that, while paying sufficient attention to the preservation of natural areas, its emphasis is placed mainly on regulating the exploitation of natural resources. The following must be noted as the most important aspects of the law:

1. The enumeration of elements of nature to be preserved; among them are both individual natural resources and resource complexes having scientific, cultural, and sanitary significance.

2. The enumeration of the properties of each natural resource that determine its usefulness for the national economy, and also the means by which it can be kept in an acceptable state; the law includes also the prohibition of destructive methods of utilizing natural wealth.

3. The establishment of a principle of expanded regeneration of renewable natural resources guaranteeing the satisfaction not only of current needs of the country but also its growing future needs; in essence this points to the close connection of the problem of natural resources with the problem of the transformation of nature.

4. A directive on the importance of taking into account the interconnections existing in nature and in the national economy, which would permit us, while developing one branch of the economy, to meet the interests of all other branches and would allow us to foresee the effect of the exploitation of one resource on the others; this directive must be looked upon as confirmation of the principle of comprehensive planning.

5. The division of resource conservation into three major stages: the quantitative and qualitative calculation of resources, the planning of utilization, and the attainment of control over the implementation of the laws and plans.

In regard to conservation by the removal of areas from exploitation, the most important aspects of the law are, first, the multiplicity in the forms of reserves, which range from large state reservations to individual reserved landmarks and natural monuments; and, second, the clause stating that the territory of preserves is excluded forever from utilization.

In addition to the RSFSR law on the conservation of nature, the outstanding legislative accomplishments in the Ukrainian SSR must be noted. Recently laws and decrees were issued concerning natural conservation, especially the conservation of the Dnieper basin, field-protecting foresty, and the prevention of erosion. Minimum erosion-prevention measures obligatory for each collective and state farm have been officially approved. Important legislative documents have been passed in the Moldavian, Baltic, Transcaucasian, and other Union republics.

One cannot underestimate the importance of such resolutions. However, they are only the first stage of transition from the planning of natural resource utilization to the planning of a resource replenishment cycle. The needs for further growth of the material well-being and cultural level of the Soviet people and the needs of peaceful economic competition with the capitalist countries make it imperative to give this matter the highest priority. In order to visualize more clearly how much remains to be done, we will mention some unfavorable aspects of the topic under discussion. Lately they have frequently served as the subjects of troubled articles in the press.

In such hilly regions as the central Russian and Volga highlands, the Right Bank [of the Dnieper in the] Ukraine, the foothills of the Caucasus and other regions, accelerated erosion is developing. According to rough calculations, the USSR loses annually about 50,000 hectares of arable land and natural pastures due to the growth of gullies, and considerably more as the result of soil erosion. How serious these processes are can be seen from the directives of the chairman of the Presidium of the RSFSR Supreme Soviet, N. N. Organov, who in his report at a session of the Supreme Soviet said that many collective farms in erosion areas have lost up to 10 percent of their agricultural land in the last thirty years as the result of erosion. Dust storms in the steppe belt of the Ukraine, in the Northern Caucasus, Transvolga region, and Kazakhstan annually damage hundreds of thousands of hectares of young crops. In 1960 young crops on several million hectares were destroyed or heavily damaged. Several hundred thousand hectares of recently developed lands have already changed into barren ground as the result of weathering of light soils. About one-third of the lands with an irrigation network are not used for their prescribed purpose; a portion of these in Central Asia and Transcaucasia have been subjected to secondary salination. Drained lands in a number of western and northern regions are not well used and become swamps again.

These figures for losses may seem small compared with the 36 million hectares of virgin land recently developed. Many people are reassured by the fact that in the USSR only 10 percent of its territory has yet been plowed. However, there is no reason for complacency. The remaining 90 percent is mainly tundra, taiga, swamps, deserts, mountains, and also unarable, but useful, agricultural and forest land, the conservation of which is necessary for maintaining a favorable climate and water supply and to serve the needs of a diversified economy. It must be remembered that we also have large reserves of virgin land, but to make them arable there must first be serious improvements and, consequently, huge expenditures of labor and resources. Therefore, even small losses of developed lands are intolerable.

The total water intake from rivers for the needs of industry, agriculture, and the urban economy presently reaches 500 cubic kilometers a year, that is, 30–40 percent of the stable (minus flood water) annual river flow in the USSR. If one takes into account that the rivers of largest volume flow in poorly developed regions, it becomes understandable why at present the water shortage creates difficulties for the development of industry and agriculture, and for raising the cultural and sanitary level of life in such important regions as the Donets Basin, Krivoy Rog, the Urals, Karaganda, and others. The water shortage is aggravated by the pollution of many rivers by industrial wastes, city sewers, and driftwood. Pollution makes the water unsuitable for many types of utilization and, above all, is detrimental to the fishing industry. Valuable species of fresh-water fish are disappearing from our market, and fishing is increasingly concentrated in distant regions of the world ocean.

The yearly growth of wood in the forests of the USSR exceeds the annual cutting. It would seem that forest wealth is continually increasing. In fact, however, the increase in growth occurs in outlying taiga regions where the forests stand too long and rot while standing. In more developed regions, close to transport routes and centers of consumption, forests are felled much faster than they can grow back. The Ukraine, in a number of regions, is already cutting forests originally planned to be cut in 1975–80. And although a great deal of forest planting is going on it will take many years until a normal percentage of mature forest is reestablished. In wide areas the clearings are poorly replanted, which leads to a deterioration of river regimes and local climate; or the clearings are reforested with poor varieties of trees.

The general reason for the deficiencies enumerated, as well as many others, is that we are still under the power of the myth of inexhaustibile natural resources; we continue to believe that the reproductive ability of nature will compensate for the damage. Being in need of forests, fur-bearing animals, and fish, we speak more often about lumbering, hunting, and fishing than we do of a foresty, animal, and fishing economy. However, as reality shows us, such an attitude towards natural resources is becoming a less permissible luxury with each passing day.

In economically developed countries there are hardly any such types of natural resources that can be taken without calculating whether one is using up the legal percentage or cutting too deeply into basic capital. The development of productive forces has reached such a stage that it is necessary in each of the regions to assist nature to restore its wealth, to consciously create favorable conditions, and to expend labor and resources for the maintenance of soil fertility, for the replanting of high quality forests in cleared areas, for the increase of fish, useful animals and birds, and for the regulation of the soil water balance.

What then must be done for further rationalization of the involvement of natural wealth in the economic cycle? First of all, in the implementation of laws on conservation passed in the Union republics, an effective state and public apparatus for estimating resources and planning their utilization, regeneration, and control must be created. Let us discuss each of these tasks in more detail.

Without a single, scientifically based, official estimate of all types of natural resources, the most profitable norms for the use of each cannot be established, and the best distribution of industrial organizations that exploit the resources cannot be achieved. As pointed out in the above-mentioned RSFSR law, such an estimate includes the classification of each resource by types (for example, lands by relief and character of soils; water, by salinity, and flow), division according to economic value, and the compilation of a cadastral survey, that is, a system of special maps and descriptions making it possible to determine the distribution of resources over the whole area of the country, as well as the type, economic value, and reserves at each point. The single system of estimates must include all enumerated resources and be conducted within the framework of the same regions set aside in the system of national regional division. A single system of estimates will enable each sovnarkhoz and oblast to know exactly its natural wealth, the prospects for its development and the possibilities of utilization.

Planning, in keeping with the principle of expanding resource regeneration, must lead to a situation in which the utilization and regeneration of the raw materials consumed by some branch of industry will become an inherent part of the production cycle—in some cases a special department of the enterprise, or more often of the combine. For example, to a combination of woodworking, cellulose, and wood chemical plants, a forest economy of an area large enough so that the yearly growth covers the annual need of wood supplies, can be delimited. This combination must be concerned with avoiding overcutting and with growing an amount of forest equal to or larger than that felled. Work under conditions of a closed cycle (which of course, does not preclude expansion of production) will contribute to a maximum utilization of wastes and the transition to a cultivated forest economy. Such a system will sharply lower the need for developing new forest clearings, for building expensive roads that are abandoned after cutting the forest, and for building temporary villages for the lumbermen. The transition from the system of constantly shifting logging operations to territorially stable forest establishments makes it possible to avoid unproductive expenses and to create for the employed workers cultural and living conditions resembling the conditions of industrial centers.

A corresponding system of a closed cycle for agricultural lands implies that choice of an economic profile, crops, and agrotechnical methods by which the depletion of soil fertility is excluded and, on the contrary, a constant increase in soil fertility is guaranteed. It is necessary to introduce a personal responsibility on the part of the managers for damage to lands in cases where measures were not taken in time to combat the growth of gullies, of washouts, blowouts, salinization, marshes, shrub growth, and the overgrazing of pastures. Carrying out a single qualitative evaluation of lands makes it possible to establish the amount of damage in each case. Similar principles can be applied in the utilization of water, fish resources, land fauna, and the like.

A task no less important than an expanded regeneration of natural resources is that of the integration of planning. Due to the presence in nature of dialectic interconnections, the exploitation of any natural resource involves the change of many natural conditions. The economic significance of indirect effects grows with an increase in the intensity of utilization of the resources and the scale of new projects. Each accomplished project uncovers new wealth but also hides many dangers.

The building of large hydroelectric stations provides industry with ever increasing sources of electric energy, creates the preconditions for the creation of new energy-consuming industries and, around them, new cities, opens the possibility of irrigating arid land, and eliminates shallows and sandbanks in rivers. But at the same time difficulties arise: the best agricultural lands are inundated, and their destruction continues after the reservoir has filled due to wave erosion of the banks, the filled-in gullies cut into the territories of nearby collective farms, dams prevent access of fish to spawning places; roughness in the reservoir and ill-timed opening of flood gates impede shipping. The same occurs when virgin land is being developed. The productivity of the plowed steppes increases immeasurably, the country receives torrents of grain, and there arise numerous new opportunities for labor and culture. But a careless choice of virgin lands is sometimes the cause of the light soil blowing away, and it leads to dust storms. The cutting and digging out of steppe plants causes rivers to run low and lakes to dry out, a deterioration of sanitary and hygienic living conditions, a disappearance of game animals and birds, and a multiplication of harmful insects.

Experience shows that the organizations directing construction often approach this from a narrow view and do not consider indirect aftereffects of their activity and thereby inadvertently bring harm to associated branches of the national economy. And yet negative aftereffects connected with economic transformation and new projects can always be avoided if one is able to foresee them and take them into account.

It is evident that in the future there must be a quite different approach to planning; it will consist of the subordination of narrow planning of individual enterprises to integrated regional planning. What must be planned is not a hydroelectric station nor a virgin-land sovkhoz but rather a new region of socialist economy gravitating to these large enterprises. The plan must encompass energy supplies, industrial enterprises, forestry, the development of transport and communication, new and reconstructed settlements with forest-park zones, water supply, agriculture including land-use arrangement, poultry farming and fishing in the new reservoirs and in the rivers. In the elaboration of a single project, not only should all the necessary technical specialists participate, but also scientists who know the local natural resources and their interconnections and are capable of showing ways of utilizing and enriching the whole natural complex that are most profitable for the entire country.

A rational utilization of natural resources, and the responsibility not only for fulfilling the plan but also for providing for the needs of coming generations, which will be much greater than the present, will demand from planners, designers, and economists much more exact work and also initiative, innovation, the rejection of traditional, primitive methods, and a conversion to more complex and modern methods.

Soviet legislation provides for measures to suppress the activity of various kinds of poachers, plunderers, and simply negligent managers. The public is ever more involved in the struggle against them. It is this that identifies the idea of the conservation of nature in the narrow sense. It must be supported and intensified. However, it is clear that errors in planning and in the technology of the extraction of natural wealth by state organizations equipped with a powerful technology can, and do, cause incomparably greater harm than individual poaching when these organizations disregard their responsibility concerning the regeneration of natural resources.

Therefore, a very important element in the struggle for intelligent utilization of natural resources is control. It must begin at the first stage of developing national-economic plans concerned with the use of natural wealth and then extend to their design and implementation.

The maintenance of the necessary level of reserves of all natural resources and the uninterrupted satisfaction of the constantly growing demands of the national economy pose a very complicated problem. At the same time its solution is absolutely necessary. Without it the building of a socialist economy will meet with ever-growing difficulties in the form of insufficient raw materials, water, and land parcels of a certain quality, which will arise now here, now there, and perhaps in crucial points and in the most important centers of the country. To cope with this problem by simply imposing new functions on the existing planning organs, drafting institutes, ministries, and sovnarkhozes is clearly impossible. Responsibility for natural resources must be concentrated in a special, very authoritative state agency. It is these considerations that compelled the Soviet scientific community to come forward with the proposal that there be created a USSR Council of Ministers' Committee on Natural Resources (General Resolution of the Third Congress of the USSR Geographical Society, 1960).[2]

The character of problems that arise in the solution of the problem of preserving natural resources—the necessity of integration, spatial differentiation, and thorough study of natural processes—places high demands upon Soviet science. It is difficult to find a branch of knowledge that should not make a contribution to their solution. However, geography bears the greatest responsibility. Modern geography is a whole system of sciences, including physical and economic geography and their branches. The presence of two complex disciplines as the basis of the geographic system of sciences and the close cooperation in geographic institutes of representatives of their more specialized branches make geography a science directly designated for the elaboration of both the theoretical bases and the solutions to a number of practical problems entering into the problem of resources. It is the professional duty of geographers to know how to analyze the interactions of the various facets of nature and branches of the economy and to predict their consequences.

Soviet geographers and specialists of allied disciplines are doing a great deal in this respect. The most important stage in the detailed study of natural resources is the physical-geographic division into regions. With sufficient emphasis on questions of the national economy, it can and must be the basis for a systematization of the diverse types of localities in our country, each of which possesses its own special set of natural resources and, therefore, must be developed according to the special economic profile most profitable for the interests of the state.

In the postwar years Soviet geographers raised the question of a change to a more precise form of inventorying land resources and of evaluating them qualitatively, and creating in the USSR a land cadaster. Since then, the participation of geographers, together with specialists in allied disciplines, has become a tradition in experimental works in land inventory and the rational organization of territory on this basis.

The participation of geographers in the study of water, vegetation, climate, and other resources is essential. The development of measures to combat erosion of soils, and sand and snow drifts on roads and canals are traditional branches of geographic investigations and have a copious literature. Between 1948 and 1953 many geographers participated in the work on the shelter belts and in creating a network of hydrometeorological observatories.

Methods of quantitative and qualitative inventory of the value of pasture land and also the recently developed methods of quantitative inventory of wild fauna are important contributions to the study of natural biological resources.

The numerous but uncoordinated works of geographers on the study and improvement of utilization of individual natural resources have necessitated establishing in geographic institutes the study of the problems of the rational utilization of natural resources as a whole. The main tasks of a detailed study must be:

1. The working out of a common system of cadastral survey of all types of natural resources, a system based on general principles and a national network of geographic regions. Such a system

must serve as the scientific basis for a general State inventory of natural resources carried on by the central statistical administrations and the corresponding ministries of the union republics.

2. The publication on an increasingly larger scale of national and regional maps and atlases of natural resources drawn up on the basis of common principles (later on the basis of cadastral surveys) and methods of cartographic representation of the material.

3. Participation of geographers in regional planning. In this work, which unites various specialists, the geographers must play the role of specialists in natural interrelationships and in the prognosis of the consequences of human interference in the course of natural processes and also, to a considerable degree, the role of uniting the experts of related fields.

The problems outlined above, in their full extent are still too difficult for the existing geographic organizations and cadres. We outlined them as a goal to which one must strive. We noted above existing deficiencies in the conservation of nature and in the utilization of natural resources together with achievements of the state. Similarly, with regard to the question of the role of geographers, it is necessary, together with their successes, to point out failures and weaknesses, which in some respects are very important for the topic under discussion. These failures and weaknesses depend partly on the geographers themselves, and partly on the institutions they must serve. At the present time geographers participate too little in the solution of the problem of natural resources, and their knowledge is not utilized to the proper extent. One of the reasons for this is that the geographers themselves do not realize the practical value of their science. Many of them are afraid to take on the solution of national-economic problems. In a number of cases they are, indeed, poorly prepared for them; their education is too theoretical and they are not proficient in the modern physical-technical methods of influencing nature. While Soviet geography long ago became a science capable of providing an enormous "yield," the national economy, up to now, still has the opinion that geography is a purely general, educational subject, not even included in the plan of polytechnical subjects in secondary schools. It is natural that this underestimation negatively affects the school programs in geography. The programs of geographic faculties of higher schools of learning are often not aimed at solving national economic problems, in particular, the most important of them—the problem of natural resources. Only in a few faculties, and then only on their own initiative, are such subjects as land and forest improvement, qualitative evaluation of lands, and the organization of territory considered as faculty disciplines or offered as courses. There are no textbooks for any of these courses.

In order to create the necessary pool of trained personnel, better preparation in physics and mathematics should be required of all geographic faculties; practical courses that bring geography closer to the natural spheres and stress its economic application should be offered; and special groups of students specializing in the problems of natural resources should be formed. It will then become increasingly possible to place the very important state problem of inventory, rational utilization, and expanded regeneration of natural resources on a solid scientific basis.

1. "On the Preservation of Nature in the RSFSR," *Pravda*, no. 302, October, 1960.

2. "General Resolution of the Third Congress of the Geographical Society of the USSR," in *Izvestiya Vsesoyuznogo Geograficheskogo Obshchestva* 92 [Proceedings of the All-Union Geographical Society], no. 3 (1960).

I. V. KOMAR

Territorial Improvement in the

Utilization of the Natural Resources

Of the USSR in the Soviet Period

The scientific analysis of the interaction of man and nature in the process of production includes, as its major element, the geographical approach to this problem. It is conditioned, first and foremost, by a profound territorial differentiation in the natural environment and by a tremendous diversity in the socioeconomic conditions of production in different countries and regions. In the USSR such an approach is closely related to the problems of the systematic, rational distribution of productive forces, the effective specialization and comprehensive development of economic regions, the ensuring of a full and accurate forecasting of the profound alterations in the natural environment under the influence of man's activity, and the development of economically advisable projects for transforming nature.

All that has been said fully applies to an analysis of one of the major forms of social and natural environmental interaction: the appropriation of natural resources by man, that is, resource consumption. We have previously undertaken to give a summary evaluation of the scale of natural resource use throughout the USSR; to inves-

Translated and abridged from *Izvestiya Akademii Nauk SSSR*, seriya geografiya [Proceedings of the Academy of Sciences of the USSR, Geography series], no. 5 (1967), pp. 38–52.

tigate the dynamics of the extraction of primary natural materials and products in this country over nearly half a century (1913–60); and, finally, to make certain predictions in the area of resource consumption.[1]

The aim of the present article is to examine the questions listed above from a regional standpoint and to reveal characteristic traits in the territorial structure of natural resource utilization in the USSR, using for this purpose the existing grid of major economic regions but not being limited to it.

An account of the regional aspects of extraction of primary natural materials and products requires more detailed data for each region, and this brings us once again, and this time with great urgency, to the problem of the incompleteness of statistical accounting in this area of man's economic activity. Insufficiencies and incompleteness in the accounting of the utilization of different types of materials are much more telling for some regions than for others. The method of calculating for each region the extraction of these materials is considerably more intricate. This method entails the application of regional conversion coefficients (for instance, for converting hay to grass, or for converting the volume indices employed in the extraction of gas and certain building materials to units of weight); of differential territorial norms in utilization (for example, the quantity of firewood used for day-to-day needs); and the carrying out of special reevaluations in order to achieve maximal comparability of data in time and space.

The data introduced below on the dynamics and the pattern of primary natural raw material extraction by large economic regions of the USSR, cannot, therefore, make claim to a high degree of precision; however, in our opinion, they do reflect the chief regularities in use of the country's natural resources from a regional standpoint.

The summary calculation by regions has been rendered in expressions of weight and according to the major groups of primary materials and products. In general, because full data are unavailable, the dynamics of natural resource utilization by districts is given, not for the whole Soviet period, but only through 1960.

From the data in table 1 the absolute predominance of the products of agriculture in the overall balance of primary materials extracted in prerevolutionary Russia, both for the country as a whole and for its regions, is strikingly obvious. In an overwhelming number of regions, agriculture yielded more than three-fourths of all primary materials and products extracted; in the vast outlying southeastern regions, Central Asia and Kazakhstan, in spite of their wealth in mineral resources, this proportion rose to nearly 100 percent.

Forest resources were another major type of primary material derived from nature in the majority of regions in Russia. Especially

TABLE 1

EXTRACTION OF PRIMARY NATURAL MATERIALS AND PRODUCTS
BY MAJOR ECONOMIC REGIONS OF THE USSR AT 1913 LEVEL

TYPES OF PRIMARY MATERIALS AND PRODUCTS	UNIT OF MEASURE	RSFSR				
		Northwestern	Central	Volga-Vyatka	Central Chernozem	Volga
Mineral fuel	In millions of tons	...	1.5	0.3	0.1	...
	%	...	1.9	1.2	0.3	...
Ferrous metallic ores (iron, manganese). . . .	In millions of tons	...	0.4	0.1	...	0.2
	%	...	0.5	0.3	...	0.2
Useful nonmetallic minerals	In millions of tons	3.3	4.2	0.7	0.6	2.6
	%	6.7	5.2	3.0	1.7	2.6
Timber (freshly cut)	In millions of tons	15.7	8.5	1.3	4.2	6.8
	%	32.1	10.5	5.5	11.9	6.8
Agricultural products. . . .	In millions of tons	29.7	65.4	20.9	30.3	90.2
	%	61.2	81.9	90.0	86.1	90.4
Totals	In millions of tons	48.7	80.0	23.3	35.2	99.8
	%	100.0	100.0	100.0	100.0	100.0
Additional: other metallic ores. . . .	In millions of tons	...	...	...	...	...

TABLE 1--Continued

TYPES OF PRIMARY MATERIALS AND PRODUCTS	UNIT OF MEASURE	RSFSR				
		Northern Caucasus	Urals	Western Siberian	Eastern Siberian	Far Eastern
Mineral fuel	In millions of tons %	3.8 5.1	1.3 2.0	0.8 1.1	0.8 2.3	0.4 1.8
Ferrous metallic ores (iron, manganese). . . .	In millions of tons %	... ...	1.6 2.5	... ...	... ...	... ...
Useful nonmetallic minerals	In millions of tons %	3.0 4.1	4.9 7.6	0.2 0.3	0.3 0.8	0.2 0.9
Timber (freshly cut) . . .	In millions of tons %	0.4 0.5	5.5 8.5	2.1 3.0	0.9 2.6	0.9 4.1
Agricultural products. . . .	In millions of tons %	67.1 90.3	51.4 79.4	66.4 95.6	31.7 94.3	20.3 93.2
Totals.	In millions of tons %	74.3 100.0	64.7 100.0	69.5 100.0	33.7 100.0	21.8 100.0
Additional: other metallic ores.	In millions of tons	...	21.6	1.4	12.9	5.1

TABLE 1--Continued

TYPES OF PRIMARY MATERIALS AND PRODUCTS	UNIT OF MEASURE	UKRAINIAN SSR AND MOLDAVIAN SSR			Baltic	Transcaucasia
		Donets-Dnieper	Southwestern	Southern		
Mineral fuel	In millions of tons %	22.8 34.0	1.1 1.5	... ...	... ...	7.8 29.8
Ferrous metallic ores (iron, manganese). . . .	In millions of tons %	6.7 10.0	... ...	0.5 1.6	... ...	1.0 3.9
Useful nonmetallic minerals	In millions of tons %	4.2 6.3	1.5 2.1	5.8 18.2	2.1 5.4	0.3 1.1
Timber (freshly cut)	In millions of tons %	0.5 0.7	3.0 4.2	0.5 1.6	2.5 6.4	0.2 0.8
Agricultural products.	In millions of tons %	32.9 49.0	65.7 92.2	25.0 78.6	34.4 88.2	16.7 64.4
Totals.	In millions of tons %	67.1 100.0	71.3 100.0	31.8 100.0	39.0 100.0	26.0 100.0
Additional: other metallic ores.	In millions of tons	0.4	`	...	...	1.7

TABLE 1--Continued

TYPES OF PRIMARY MATERIALS AND PRODUCTS	UNIT OF MEASURE	Central Asia	Kazakhstan	Belorussian	USSR
Mineral fuel	In millions of tons %	0.3 0.7	0.1 0.1	... ...	41.1 4.2
Ferrous metallic ores (iron, manganese). . .	In millions of tons %	... ...	... ...	... ...	10.5 1.1
Useful nonmetallic minerals.	In millions of tons %	0.4 0.9	0.1 0.1	0.3 0.8	34.7 3.5
Timber (freshly cut)	In millions of tons %	0.1 0.3	0.3 0.3	3.6 10.2	57.0 5.8
Agricultural products. . . .	In millions of tons %	39.7 98.1	115.6 99.5	31.2 89.0	834.6 85.4
Totals	In millions of tons %	40.5 100.0	116.1 100.0	35.1 100.0	977.9 100.0
Additional: other metallic ores. . .	In millions of tons	...	0.4	...	43.5

noticeable was the extraction of timber, which, however, for the most part went to filling the daily needs of the population (for household heating, cooking, buildings, and the like) and to a considerably lesser extent was used commercially and for export.[2] Taken together the primary products of agriculture and forestry amounted, to more than 90 percent of all primary materials extracted in 14 regions out of 18. Even in those few regions where the most important centers and areas of mining were located—on the territory of the present Donets-Dnieper region (the Donets Basin coal fields, extraction of ferrous metallic ores), the Ural region (extraction of ores and of usable rare nonmetallic minerals), the Transcaucasian region (Baku oilfields), the Eastern Siberian and the Far Eastern regions (gold fields)—primary agricultural products and timber accounted for one-half to four-fifths and more of the overall weight of primary materials extracted.

Such a nonindustrial pattern in the natural resource utilization in prerevolutionary Russia was consistent with the country's economy and the prevalence in most regions of a general farming, farming-forestry, and forestry economic profile. Related to these structural peculiarities was the coincidence of basic extractive industries with corresponding natural zones; within these zones, because of economic, historic, and other factors, natural resources were in general more widely used from their western sectors.

However, in spite of the low degree of resource exploitation in the eastern regions then, it would be improper to underestimate their importance in the extraction of primary natural materials and products in prerevolutionary Russia. Out of the general volume of this type of extraction in 1913, the eastern regions situated beyond the Urals provided more than one fourth. One must note that it was a matter of specific types of natural resources that, especially in the initial stages, could be exploited in a primitive and predatory fashion and without significant investments. In addition, it was economically quite justified to transform these usable natural resources into a highly transportable product on the spot, in a simple manner without large outlays for the general development of the territory.

Resources of this nature included, above all: the auriferous gravels and ores processed by miners workers in Siberia and the Far East, amounting to 20 million tons a year; and forage resources in the vast natural pastures and hay fields, which fed livestock in the amount of about 200 million tons a year. Let us recall that in prerevolutionary Russia the eastern regions lying beyond the Urals yielded nearly 80 percent of all the gold mined and up to 30 percent of all the cattle (converted to conventional livestock). Beginning with the end of the nineteenth and early twentieth centuries, these regions played a growing role in terms of agricultural products, in particular in cereal grains (producing more than 10 percent of all grain in 1913).[3]

The immense changes that have taken place in Soviet times in the scale, structure, and the very character of natural resource utilization in the country have also found expression in the territorial distribution of extraction of primary natural materials and products (table 2).

During the period under review rapid growth took place in the total utilization of natural resources throughout all the major regions of the country. In this way, the characteristic wide-scale, systematic involvement of the population from all parts of the country in actively participating in raising productive powers has found striking manifestation. This process has been totally rational from the strictly economic standpoint as well.

Another major improvement brought about by the socialist transformation of the country and its conversion into a mighty industrial power was a sharp change in the natural resources utilized in all major regions of the USSR. The dominant position in the majority of these regions was taken over by the extraction of materials in the mineral-fuel and mineral raw-material groups. From table 2 it is apparent that in 1960, in 9 out of 18 of the regions under investigation this extraction accounted for more than half of all primary materials and products extracted (counting extraction of nonferrous ores, there were 10 of these districts). In a number of remaining regions the proportion for the mineral group of materials amounted to 40–50 percent (Kazakhstan, Central Asia, the Baltic, and others).

The proportion of timber in the overall mass of primary materials extracted decreased, however, for the most part through lower demand for it as fuel. At the same time, the felling of commercial timber, to be processed for varied and complex industrial needs, increased sharply.[4]

In summary, in 1960, in 12 out of 18 major regions, mineral raw material and fuel, as well as timber, comprised a large portion of the total mass of primary materials extracted from nature, and in three districts, nearly one-half. Nevertheless, the proportion made up of products of agriculture remains considerable. In 1960, in the majority of regions, they amounted to not less than one-third, ranging among the regions from between one-tenth to two-thirds. One change in the pattern of natural resource exploitation has also entailed its enrichment, both through overcoming the onesidedness in exploitation characteristic of the past and by broadening the scope of initial materials involved in economic turnover.

Growth in the extraction of natural materials was experienced in different regions of the country with varying degrees of intensity

TABLE 2

EXTRACTION OF PRIMARY NATURAL MATERIALS AND PRODUCTS
BY MAJOR ECONOMIC REGIONS OF THE USSR IN 1960

TYPES OF PRIMARY MATERIALS AND PRODUCTS	UNIT OF MEASURE	RSFSR				
		Northwestern	Central	Volga-Vyatka	Central Chernozem	Volga
Mineral fuel	In millions of tons %	33.0 16.0	83.8 26.0	9.8 11.3	3.7 3.4	115.2 38.0
Ferrous metallic ores (iron, manganese). . . .	In millions of tons %	5.6 2.7	0.6 0.2	... ...	5.2 4.7	0.8 0.3
Useful nonmetallic minerals	In millions of tons %	68.2 33.0	128.4 40.0	12.4 14.5	26.7 24.4	56.6 18.5
Timber (freshly-cut) . . .	In millions of tons %	76.3 37.0	27.7 8.7	32.6 38.1	1.1 1.0	10.7 3.5
Agricultural products. . . .	In millions of tons %	23.0 11.3	80.8 25.1	73.1 36.1	120.9 66.5	108.5 39.7
Totals.	In millions of tons %	206.1 100.0	321.3 100.0	85.8 100.0	109.8 100.0	304.2 100.0
Population figures as of 1 January 1961	In millions of persons	11.1	26.0	8.3	7.9	16.5

TABLE 2--Continued

TYPES OF PRIMARY MATERIALS AND PRODUCTS	UNIT OF MEASURE	RSFSR				
		Northern Caucasus	Urals	Western Siberian	Eastern Siberian	Far Eastern
Mineral fuel	In millions of tons %	55.3 23.5	65.5 19.5	84.1 35.4	36.9 23.8	23.6 34.5
Ferrous metallic ores (iron, manganese) . .	In millions of tons %	... ...	36.7 11.0	5.6 2.3	1.8 1.2	... ...
Useful nonmetallic minerals.	In millions of tons %	66.8 28.3	106.9 32.1	33.6 14.2	34.1 21.9	14.3 20.8
Timber (freshly cut). . .	In millions of tons %	4.3 1.8	55.3 16.4	21.3 9.1	41.8 26.9	14.4 21.0
Agricultural products . .	In millions of tons %	... 46.4	70.7 21.0	93.3 39.0	40.6 26.2	16.2 23.7
Totals	In millions of tons %	234.9 100.0	335.1 100.0	237.9 100.0	155.2 100.0	68.5 100.0
Population figures as of 1 January 1961. . . .	In millions of persons	12.3	14.6	11.6	6.7	4.9

TABLE 2—Continued

TYPES OF PRIMARY MATERIALS AND PRODUCTS	UNIT OF MEASURE	UKRAINIAN SSR AND MOLDAVIAN SSR			Baltic	Transcaucasia
		Donets-Dnieper	Southwestern	Southern		
Mineral fuel.	In Millions of tons %	175.5 38.5	32.6 12.4	0.1 10.1	16.5 18.4	25.4 29.6
Ferrous metallic ores (iron, manganese) . . .	In Millions of tons %	68.0 14.8	... ...	6.5 5.9	... ...	6.0 6.8
Useful nonmetallic minerals.	In millions of tons %	108.2 23.7	50.9 19.4	47.5 43.2	25.2 28.0	25.3 29.5
Timber (freshly cut). . . .	In millions of tons %	0.6 0.1	10.8 4.1	0.3 0.3	7.3 8.1	1.4 1.6
Agricultural products . . .	In millions of tons %	104.6 22.9	168.6 64.1	55.0 50.5	40.6 45.5	28.1 32.5
Totals	In millions of tons	456.9	262.9	109.4	89.6	86.2
Population figures as of 1 January 1961	In millions of persons	18.4	19.4	8.3	6.8	10.0

TABLE 2--Continued

TYPES OF PRIMARY MATERIALS AND PRODUCTS	UNIT OF MEASURE	Central Asian	Kazakhstan	Belorussian	USSR
Mineral fuel.	In millions of tons %	15.6 14.2	34.0 15.0	25.7 26.3	836.3 23.8
Ferrous metallic ores (iron, manganese) . .	In millions of tons %	... ...	5.8 2.5	... ...	142.6 4.1
Useful nonmetallic minerals	In millions of tons %	28.7 26.1	38.5 16.8	15.5 15.9	887.8 25.3
Timber (freshly cut). . .	In millions of tons %	0.3 0.3	1.7 0.7	6.2 6.3	314.1 9.0
Agricultural products . . .	In millions of tons %	65.3 59.4	146.9 65.0	50.2 51.5	1,317.4 37.8
Totals	In millions of tons %	109.9 100.0	226.9 100.0	97.6 100.0	3,498.2 100.0
Population figures as of 1 January 1961. . . .	In millions of persons	15.0	10.0	8.3	216.1

between 1913 and 1960. On the average, extraction in the majority of districts more than tripled, with a minimum increase of 90 percent and a maximum of nearly 600 percent. The new pattern in utilization of natural resources took shape for each region in a unique fashion. In the intricate complex of conditions that had given rise to these differences, an important role was played by the presence of some particular natural resource or combination of resources that could be extracted economically, and that permitted creation in the region of a specialized production base of national or interregional importance.

Over the years, during the building of socialism, previously existing fuel and iron-ore bases have received strong development and in addition many new and important ones have been created; the same applies to bases of nonferrous ore extraction, the exploitation of large tracts of forests, output of grain, cotton, and the like. Such bases constitute a network of much larger-scale productive units at present than in the prerevolutionary past. At the same time, they have been systematically redistributed over the territory of the USSR thereby fostering the economic growth of all parts of the country and a more rational utilization of its natural resources. Because there was such an abundance of natural wealth as yet untouched at the time of the first five-year plans it has been feasible to exploit highly economical resources (the cheap coal of the Kuznetsk Basin, iron ores of Magnitogorsk, copper ores of Kazakhstan, and oil of the Volga-Ural region).

Taking place alongside the strengthening of specialization in economic districts, the comprehensive development of their economies had a different kind of influence on the pattern of resource exploitation. By stimulating the exploitation of resources of local importance in all regions with the aim of limiting interregional exchange to truly necessary products, this development leveled out the structure of the primary natural raw-materials extraction to a certain degree. Tables 1 and 2 offer a visual representation of the large growth and wide distribution over major regions of the exploitation of local fuel resources as well as mineral building materials. The latter show a rather well delineated correlation with the size of urban population in these regions.

Channeling locally important resources into the economy may lead to the possibility that the district, even though poorly provided with or lacking major types of resources, will stand out in total quantities of primary materials and products extracted. An example is the densely populated Center with its manufacturing industry of high capacity, which stands out distinctly in overall extraction of primary materials, alongside the Donets-Dnieper region, the Urals, and the Volga region, with their rich natural resources.

In total quantities of initial materials extracted, the large regions may be broken down into three groups: (1) those with extraction of more than 300 million tons—the Donets-Dnieper region, the Urals, the Volga, and the Central region; (2) those with extraction of 200–300 million tons—the Southwestern Ukraine, the Northern Caucasus, the Northwestern region, Western Siberia, and Kazakhstan; (3) all other regions. In extraction of primary materials per capita, however, the most notable are the majority of the eastern regions (including the Urals) and the Donets-Dnieper region. This index is also above the average for the country in the Northwestern, Volga, and Northern Caucasus regions.

For the majority of the country's regions, it is still characteristic that there be present large, specialized production branches for the extraction of a certain kind of primary material so that its own needs, as well as those of other regions, may be provided for. This leads us to problems of the interregional exchange of primary materials and products, and to an evaluation of the character and scale of such exchanges.

Practically speaking, all, or nearly all, the materials and products extracted from nature by man's efforts may be redistributed territorially, and this fact alone accounts for the possibility of divorcing processing industries and centers of consumption from the place of their extraction. Only certain types of products because of their natural properties, need be consumed or given over to processing on the spot (they, however, are rarely extracted in large quantities).

The expansion of the interregional exchange of primary products, which depends essentially on the unevenness in territorial distribution of various natural resources and on the differing quality and conditions of exploitation, is taking place under the powerful influence of the growing social division of labor, of the concentration, cooperation and combination of various areas of production, of progress in transportation, and of the gravitation of many types of industrial processing to areas where their products are consumed. Opposite tendencies are simultaneously at work: the great efficiency, in a number of instances, of developing the productive powers of regions rich in natural resources; the gravitation of a large number of manufacturing branches to regions where primary materials originate; and a host of other factors.

Out of all primary natural materials and products obtained in the USSR and accounted for in table 2, those consumed locally, according to our estimates, amount to only 10–15 percent, chiefly fodders and certain types of nonmetalllic resources for the building-materials industry, such as brick clays.[5] However, this percentage ought to be increased several times (to approximately 50–60 percent) if regard is taken of the transfer of the primary materials over small distances (from several 100 meters to several tens of kilometers).

A judgement of the scale of territorial redistribution of primary materials and products over more considerable distances may be made on the basis of quantities shipped by general transportation means. In 1960 these shipments amounted to 1.4–1.5 billion tons, or approximately 40 percent of the total extraction of primary materials.[6] It is characteristic that in comparison with 1913, the total volume of such shipments rose at a far faster rate than the total extraction of primary materials and products. In 1960, seven to eight times as many of these shipments were made compared with the total shipments in 1913, and at least ten times as many as of comparable types of freight. These figures testify to the deepening territorial division of labor that has taken place in the USSR during the building of socialism and are intimately connected with the new structure of natural resource utilization.

Interest arises in an evaluation of the quantities of primary materials and products exchanged between regions by general transportation and calculated for regions in various levels of the hierachy: for the administrative-economic districts delineated in 1957, and for the major economic regions of the 1960 structure. By our calculations these figures in 1960 amounted to about 600 million tons for the former category and about 360 million tons for the latter.[7]

Thus the territorial redistribution of primary materials has a clearly manifested pyramidal structure, shaped entirely by the natural and economic conditions of development and distribution of the country's productive powers. Perfecting this structure, ought, however, to be the object of constant attention; the more so because the total production of the extractive sectors constitute two-thirds of all freight shipped in the USSR by rail, and water routes for general use, and also by the major pipelines.

The roles of the various types of primary materials and products in the general material cycle are far from uniform, of course. This may be seen from the data in table 3, which take in the overwhelming proportion (over 95 percent) of the total production of extractive industries transported by the means cited above.

First, the great differences should be noted in the relations between overall freight volume and that of various types of raw materials. Thus, in the fuel group (principally coal, oil, and gas) the relationship is 87 percent, for raw lumber, 66 percent, for grain, 50 percent, and for nonmetallic mineral raw materials (building materials), 42 percent. Greatest mobility is exhibited, therefore, by the major contemporary types of fuel.

As for the exchange between the administrative-economic regions as delineated in 1957, the proportion of all these materials was considerably less but also varied. At the present stage, turnover among these regions accounts for 40–50 percent of all fuel and iron ore extracted, up to 30 percent of all prepared timber, approximately 20 percent of all grain harvested, and somewhat more than 10 percent of all nonmetallic mineral raw materials. Finally, into the exchange among major regions goes approximately 25 percent of all fuel extracted, 17–23 percent of the iron ore, grain, and timber, and only 5 percent of all nonmetallic mineral raw materials.

TABLE 3

INTERREGIONAL FREIGHT SHIPMENT OF
PRIMARY MATERIALS AND PRODUCTS IN 1960
(Rail, Water, and Pipelines)

TYPES OF PRIMARY MATERIALS AND PRODUCTS	OUT OF ALL FREIGHT SHIPPED		FREIGHT SHIPPED			
			Between Administrative Economic Regions		Between Major Regions	
	Percentage of Total	Percentage of Extraction	Percentage of Total	Percentage of Extraction	Percentage of Total	Percentage of Extraction
Mineral fuel.	48.0	87	53.1	40	55.6	25
Iron ore.	7.0	74	10.0	44	6.3	17
Building materials. .	25.6	42	16.7	11	11.5	5
Timber.	14.9	66	15.7	29	20.5	23
Products of agriculture (grain) .	4.5	50	4.5	21	6.1	17
Total.	100.0	64	100.0	27	100.0	16

The ratio among the principal types of primary natural materials and products exchanged between regions differs sharply from the ratio characteristic of prerevolutionary Russia, where the lead was taken by grain and forestry shipments and shipments of fuel occupied a subordinate position. It reflects the utilization of the country's natural wealth at the present stage of development, as well as the geographical peculiarities of the natural resources channeled into the economic cycle.[8]

1. I. V. Komar, "The Dynamics and Structure of Natural Resource Utilization in the USSR," *Izvestiya Akademii Nauk SSSR,* seriya geografiya, no. 3, (1966).

2. The available statistical data on utilization of forest resources are particularly incomplete. Figures for timber output, which amounted in 1913 to 67 million cubic meters (or 57 million tons), in fact reflect only the timber removed from public forests. The general quantity of timber felled in prerevolutionary Russia reached values of 250–300 million cubic meters per year (G. M. Benenson, *Drevesina v narodnom khozyaystve SSSR* [Timber in the Economy of the USSR] [Moscow-Leningrad, 1947]; idem, *Lesozagotovitel'naya promyshlennost' v narodnom khozyaystve SSSR* [The Timber Industry in the USSR] [Moscow-Leningrad, 1952]). Taking the average figure for timber felled in 1913 as 275 million cubic meters (or 235 million tons), we have undertaken the task of distributing it by districts, by taking account of information by regions on numbers of population, norms of timber use for everyday needs, tracts of forest land under all types of ownership, official indices on timber output from public forests, and other data. As a result we have the following, very approximate values for quantities of timber felled in 1913 per economic regions (in millions of tons of newly cut timber): Northwestern, 40.2; Central, 42.2; Volga-Vyatka, 8.5; Central Chernozem, 10.2; Volga, 17.0; Northern Caucasus, 1.3; Urals, 34.0; Western Siberian, 17.8; Eastern Siberian, 9.3; Far Eastern, 6.4; Donets-Dnieper, 0.7; Southwestern, 20.4; Southern (including the Moldavian SSR), 0.9; Baltic, 9.4; Transcaucasian, 3.0; Central Asian, 0.5; Kazakhstan, 0.5; Belorussian, 12.7.

The values cited, despite being conditional, indicate that the role of timber in the overall mass of primary natural materials extracted in the country and in its regions in 1913 was in fact considerably higher than is shown in table 1.

3. According to the data offered by R. S. Livshits, the eastern districts furnished 13.3 percent of the total evaluated production of Russia's mining industry. Even greater was their share in the evaluated products of agriculture. However, the proportion of the eastern part of the country beyond the Urals amounted to only 3.9 percent, a telling indication of the agrarian raw-material character of the economy of the eastern lands at that time, and true of a number of other outlying areas of Tsarist Russia (*Razmeshcheniye promyshlennosti v dorevolyutsionny Rossii* [Distribution of Industry in Prerevolutionary Russia] [Moscow, 1955]).

4. Data on the quantities of timber felled in 1960 are also incomplete, for the most part because of a lack of accounting for the quantities utilized from the forests of collective farms. However, the completeness of the tally has risen greatly through the growing role of centralized operations in the state forests.

5. Many nonferrous ores should be considered also, as their characteristically low metallic content requires that they be processed extensively where they are extracted. We may recall that in prerevolutionary Russia, for every pood of gold obtained in 1911, 1.2 million poods of auriferous gravels were washed, and in the Urals this amounted to as much as 7 million poods (*Sbornik statisticheskikh svedeniy o gornozavodskoy promyshlennosti Rossii za 1911 g.* [Collection of Statistical Data on the Mining Industry of Russia for 1911] [Petrograd, 1918]). And the total volume of auriferous materials processed in this year exceeded 2.4 billion poods.

6. In view of the lack of corresponding statistical data this calculation does not include freight shipped by automobile and air transport, but they do not significantly participate in such shipments of primary materials. Another notation relates to the inclusion in these freight figures of partially processed materials.

7. We might add that in foreign-trade turnover of these types of raw materials and products in 1960, the USSR exported 60 million tons and imported 4 million tons (*Vneshnyaya torgovlya SSSR 1960 god. Statisticheskiy obzor* [Foreign Trade of the USSR in 1960: Statistical Report] [Moscow, 1961]).

8. The degree to which different types of primary materials share in exchange between districts is determined by a host of factors in combination. In particular, for the chief types of fuel, significance lies in the high concentration of resources in a few deposits that may be efficiently exploited and, simultaneously, in the general overall need for fuel; also important is the continual transport cost reduction of the chief types of fuel (particularly oil and gas). Of course, it is characteristic of the mineral building materials that their resources are widely scattered and that it is not economical to ship them; but a number of these materials, especially those of high quality, are unevely distributed even in the major regions such as Western Siberia and Belorussia.

A. P. BORISOV

Evaluation of the Effectiveness

Of Flood Control

Catastrophic floods causing territorial inundations entail great losses to the national economy. The largest floods occur on the Volga, Neva, Ural, Severnaya Dvina, Amur, Yenisey, and Dnieper rivers.

Usually the floods originate from a sharp increase in the amount of water in the river due to melting snow and glaciers, heavy rainfall, and the blockage of the river bed by ice. At the mouth of the rivers, where they empty into the sea, floods often arise because of winds that cause the water level to rise. Floods of this type occur in Leningrad. The rise of water in the river Neva is, as a rule, a yearly occurrence, and sometimes it occurs several times a year. A flood occurs when the water level in the river rises more than 1.4 meters above normal.

Since the founding of Leningrad, there have been more than 200 floods with a water rise of more than 150 centimeters. By the level of water rise, these floods are distributed in the following manner:[1]

Translated from *Izvestiya Vsesoyuznogo geograficheskogo Obshchestva* [Proceedings of the All-Union Geographical Society], no. 4 (1966), pp. 310–16.

Height of water level above normal (in centimeters)	Number of floods
150–200	162
201–250	42
251–300	7
301–350	1
351–375	2

Since the elevation of the territory of Leningrad is low, considerable areas are flooded by the rising water. For example, with a rise of 1.5 meters, the flooded area of the city and suburbs is more than 50 square kilometers. Such loss by floods reaches serious proportions when a considerable number of valuable units of the national economy are concentrated on comparatively limited areas, as in the case of urban and interurban lands with a high degree of economic development.

The floods along the middle course of the Amur in the basin of the river Zeya are also catastrophic; during the last sixty years there have been twenty-four small, five average, and eleven large floods. In the course of the development of the Amur region, frequent floods were one of the main reasons for the limited development of the riverine territories.

Floods cause great harm to such large cities as Arkhangel'sk, Orsk, Novokuznets, and others. In Arkhangel'sk, for example, about half of the city's territories are periodically flooded. Floods cause not only considerable material loss, which thereby impedes the further economic development of the regions adjacent to the rivers, but they also disrupt the compactness of areas being developed by cities because the flooded localities, which usually are not built-up, often border on built-up districts. In the city of Novokuznets, for example, such territories comprise more than 1,000 hectares. They number hundreds of hectares in Novosibirsk, Yaroslavl', Omsk, Kursk, Orsk, and other cities.

By causing the discontinuous development of a city, the flooded territories are a great inconvenience for the population and result in higher costs for city engineering installations and transportation communications. It is natural, therefore, under conditions of rapid development, that the favorable positions of these flooded territories with respect to the existing developed areas suggest these territories as a reserve for new housing in the cities. The potential role of these territories in the city plan raises the practical necessity of engineering their protection.

The basic measures to ensure the protection of certain areas from flooding are embankment, filling-in, and regulation of river flow. Embankment is the most prevalent method, especially under urban conditions when built up riverine districts are in question. It is accompanied, as a rule, by measures for drainage and, in many cases, the installation of pumping stations for pumping out the water from the embanked areas. Filling-in land is a reliable protective measure, especially for the development of a riverine belt with a low level of flooding. A complete fill-in of land reduces the problems of architectural planning. The regulation of flow by creating reservoirs above the city is an effective means of flood control. Creating single reservoirs or systems of reservoirs makes it possible to hold back part of the water during high water and then gradually to let the water into the rivers during the following period. Thus, for example, the danger posed by catastrophic flooding on the river Moskva was considerably diminished by the creation of the Istrin reservoir. Similarly, a system of dams almost fully eliminated the danger of floods on the Volga. The Mingechaursk reservoir played a large role in eliminating the devastating floods on the lower Kura. The Chardar'insk reservoir on the river Syr-Dar'ya stopped the flooding of a section of the important railroad artery connecting Central Asia with the European part of the country.

The protection of territories from flooding by creating reservoirs is especially economically expedient in the comprehensive utilization of the reservoir (for energy, transport, irrigation, and flood control). Thus, to eliminate the effects of catastrophic floods in the basin of the river Zeya, the construction of the Zeya power installation is being planned to create a large reservoir (full volume, 96 cubic kilometers; effective volume, 60 cubic kilometers). After this reservoir is built, the most devastating floods will cease. Floods that occurred in this basin every two to four years can occur only once in ten to fifty years. Calculations show that the average yearly loss from floods will thereby decrease by 87 percent. In addition to flood control the Zeya installation will in the future annually produce 4.9 billion kilowatt hours of electric energy and will make it possible to bring 90,000 hectares of the most fertile bottom land into economic use as well as improve conditions for navigation. We now have many positive examples of the planning and building of comprehensive protective projects for important industrial units and urban residential districts.

The comparative economic effectiveness of an engineering protection project is determined by comparing the expenditures for its completion with other possible measures of avoiding loss, such as removal of structures from the flood zone. The comparison of alternatives is conducted on the basis of capital investments and operating expenditures.

The equating of capital investment with operating expenditures is accomplished by an accounting of the period of investment return. That alternative is considered optimal which guarantees a minimum of total expenditures, while taking into account capital investments, in the yearly operating outlays during the standard period of return. Capital investments and operating expenditures for alternative protective projects are compared with the average yearly loss, calculated for the period of return, from the formula:[2]

$$\frac{K_z}{L - D_z} \gtreqless T_n$$

or in a converted form from the formula

$$K_z \gtreqless (L - D_z) \, T_n \tag{2}$$

where K_z is the initial investment in the building of the complex of installations of engineering protection; D_z, the yearly expense for operation of the engineering installation; T_n, the standard period of return of investments; L, the average yearly loss from floods.

As a result of these comparisons the economic effectiveness of engineering protection becomes apparent; it is expressed in the value of prevented loss from flooding. This means that in removing the danger of floods by means of engineering protection, the annual loss from floods to the national economy covers the expenditures for protection and yields a difference in the form of a saving. Subsequent comparisons establish the comparative economic effectiveness of the possible alternative protective measures. The expenditures on the variant with the best economic indices, as obtained by comparisons derived from formulas (1) and (2), are compared with expenditures for the alternative of removing the structures subject to floods, based on the following formula:

$$K_v + K_{ov} \gtreqless K_z + D_z T_n + K_{oz} \tag{3}$$

where K_v is the lump-sum expenditure associated with the removal of the affected installation as part of the expenses for the demolition and transfer of residential and public buildings and structures; the adaptation of residential and public buildings and structures; the moving of the inhabitants from residences; the compensation for unused investment of labor and resources in land parcels and; the preparation and organization of lands after the removal of structures;

K_{ov} are the losses caused by additional premature investments required for installations being removed (renovation, reconstruction, and expansion); K_z are the losses for additional capital investments connected with the premature expenditures required over a period of time for the renovation of the protected objects.

Depending on specific conditions, the expenditures for the variant of removing structures can be compared with either the magnitude of the loss (if engineering protection is not economical) or with the expenditures of the alternative variant of engineering protection. The results of these comparisons make it possible to come to one of three possible solutions: (1) rejection of protection from loss is found to be bearable for a certain period (that is, if protection and removal of structures is considerably more expensive); (2) removal of structures from the flood zone; (3) construction of engineering protective projects.

In the case where the comparative economic effectiveness of engineering protection is established, further research is conducted to discover the most economical variant of protection. The comparisons are made among the alternative forms of protection, which may be differentiated as to the most rational and economical grouping of engineering installations within the whole complex of the chosen scheme of protection. The selection of the most economical installation of engineering protection (types of protection, distribution of installations, types of constructions, and the like) can be made by means of alternative comparisons of expenditures from the formula

$$K_z + D_z T_n \gtrless K_{z1} + D_{z1} T_n \tag{4}$$

where K_z and K_{z1} are the initial investments in building the engineering protection installations at each variant of the designated solution; and D_z and D_{z1} are the annual operating costs of the engineering installations for the alternative variants.

If, as a result of comparison, the economic expedience of engineering protection is established, then, in the process of further technical elaboration of the general scheme of protection, its separate elements must be worked out by calculating the most advantageous combinations of the various types of engineering protection. It is necessary here to consider the economics of the individual, interchangeable types of engineering installations (a dam or the filling in of the land) and to combine them correctly under specific conditions.

The establishment of the magnitude of flood loss is of great significance in making apparent the economic effectiveness of flood control measures. In practice, the computation of the actual loss from floods of varied recurrence is very difficult. Defects in calculations and the length of time since the flood make it impossible to have exact data on the loss. In order to determine data on flood loss, it is necessary to classify losses. All types of loss must be subdivided into property (material) and nonproperty losses (moral).

The property loss includes the loss of all types of material valuables and also expenditures for emergency measures and the cleaning up of destruction. It must be subdivided into direct and indirect loss.

The direct loss includes: (1) major productive and nonproductive capital investments, removal and destruction by water, damages calling for capital repair, damages calling for current repair, and depreciation of investments due to the flood; (2) working funds (finished products, raw materials, fuel, semifinished products, and the like), removal by water and irreparable damage, loss in value, and expenses associated with re-sorting; (3) expenditures arising from emergency work, eliminating the flood consequences, wages to workers during the period they are unemployed due to the flood, the evacuation of the population, losses and damage to personal property, and government help to organizations and persons who have suffered from the flood (aid, exemptions, and the like); (4) loss and damage to agricultural crops.

In the case of the complete destruction of basic productive and nonproductive capital investments, damage is calculated on the balance of cost minus depreciation at the moment of the flood. The damage from partial destruction is determined by the expenditures for the repair of damage or by compensation for restoration of material values. The damage caused by the loss of finished products is calculated in wholesale prices. For the products delivered by the enterprises to the commercial-marketing organizations, the cost and profit of market and commercial organizations are included as damage. For semifinished products the damage is determined by wholesale prices minus the production expenses necessary for the completion of the production cycle. All expenditures related to the realization of emergency work and the elimination of flood after-effects are calculated from actual expenses.

Indirect damage includes: (1) losses associated with the disruption of work at the enterprises (underproduction); (2) losses from disruption in the functioning of cultural-service establishments; (3) losses from the disruption of water pipes, sewerage, energy systems, transport and other communications; (4) losses associated with the impossibility of rational land use.

Enterprises, not directly subject to flooding, experience indirect loss through the disruption of normal production ties with enterprises subject to flooding. In view of the complexity of calculating the indirect losses, their value in calculations are often taken as approximately twenty-five to fifty per cent of the direct loss.

The nonproperty loss, although not expressed in monetary evaluation (damage to health, destruction of unique historical or artistic monuments, and the like) is an important factor in the qualitative evaluation of the negative influence of floods.

The size of the loss from floods depends directly on the rise of the water level, rate of rise, recurrence and duration of floods and also on the dimensions of the inundated areas and the level of their economic development. In projecting the usefulness of engineering protection, it is expedient, for economic comparison, to determine the average annual loss of the basis of the observed magnitude of the loss from floods of various probability. Graphs, showing the dependence of magnitude of loss on the rise of water levels for floods of different probability, must be the initial data for the calculation of the average annual loss. As an example, one of the possible methods for calculating the average annual loss is shown in table 1.

Considering the provisional character of this calculation and also the defects of the initial data, corrections must be introduced into the calculation, since the loss is not correctly estimated in certain cases. The results obtained must be corrected also for the increase in loss associated with the future development of installations in the flood zone. Collected data on floods observed in the last forty years served as initial data for calculating the average annual loss in the example cited. All these floods, depending on the degree of inundation, were classified in terms of periods of their recurrence. The portion related to the average annual loss (table 1) was calculated from the total sum of loss from floods of the corresponding recurrence. For the sake of completeness of computation the value of the indirect loss was added to the sum of direct loss obtained in this manner. In the example cited, the indirect loss was accepted as a quarter of the sum of direct loss. The total sum of the average annual loss in this calculation with some adjustments came to 1,462,500 rubles ($1,170,000 \times 1.25$ [plus an unspecified adjustment factor]).

In order to illustrate the application of the proposed methods for the economic evaluation of alternative methods of flood protection, the resultant indices of comparison made during planning of a flood protection work of one of the cities are shown in table 2.

For the purpose of protecting the territory of this city two variants of flood protection were worked out: the first alternative [in table 2] projected the protection of the city by damming; the second alternative [in table 2] provided for regulation of the river flow by means of a reservoir. As a result of calculations making possible a comprehensive utilization of the reservoir (energy, transport, and so forth), the summary expenditures for this alternative [alternative 2] proved to be the smallest.

TABLE 1

CALCULATION OF AVERAGE ANNUAL LOSS FROM FLOODS

RISE OF WATER LEVEL IN A RIVER DURING FLOODS (cm.)	RECURRENCE OF FLOODS (yrs.)	RECURRENCE OF FLOODS CALCULATED FOR 1 YEAR (%)	DIRECT LOSS FROM 1 FLOOD (Thous. rbls.)	AVERAGE ANNUAL LOSS (Thous. rbls.)	INCREASE IN AVERAGE ANNUAL LOSS (Thous. rbls.)
0–100	3	33.3	0	...	...
101–200	4	25.0	302.5	75.6	75.6
201–300	7	14.3	597.4	85.4	161.0
301–400	9	11.1	2,480.0	275.5	436.5
401–500	20	5.0	6,600.9	330.0	766.5
501–600	33	3.3	9,614.6	291.4	1,057.9
601–700	100	1.0	11,164.6	111.6	1,169.5

Average annual loss. 1,170.0

TABLE 2

COMPARISON OF EXPENDITURES ACCORDING TO ALTERNATIVES
OF PROTECTION OF THE CITY FROM FLOODS
(In Thousands of Rubles)

	Loss from Flood	Alternative Removal of Affected Objects	Alternatives of Engineering Protection	
			1	2
One-time capital expenditures.	...	17,500.0	18,910.0	10,200.0
Annual operating expenditures.	1,462.5	...	140.0	65.0
Operating expenditures over ten years.	14,625.0	...	1,400.0	650.0
Total corrected expenditures. . .	14,625.0	17,500.0	20,310.0	10,850.0

For the determination of the economic effectiveness of flood protection projects, the alternatives were compared, but they were also compared with the total value of average annual loss, accepted for comparison, during the standard period of return. The variants of flood protection works under comparison were also evaluated and compared with the amount of expenditure for removal of structures from the flood territory. Such comprehensive examination guarantees the choice of the most economic alternative of flood protection.

In determining the economic effectiveness of measures for flood control, the comparison for the choice of the most rational alternative must be done not only from the standpoint of cost but also from the standpoint of a number of technical-economic indices making it possible to reveal more fully the rational alternative of engineering protection. Possible indices include: the proportional capital investments in engineering protection per unit of cost of building volume, or the area of protected basic investment per unit; the proportional capital investments per linear meter of flood protection installations (dams and embankments), or per hectare of protected territory and others.

The quantitative data must be augmented and defined more accurately by qualitative analysis of the measures under examination.

1. R. A. Nezhikhovskiy, *Reka Neva* [The Neva River] (Leningrad, 1957).

2. A. P. Borisov, *Ekonomicheskoye obosnovaniye inzhenernoy zashchity gorodskikh territoriy ot zatopleniya* [The Economic Basis of Engineering Works Protecting City Areas from Flooding] (Leningrad, 1963).

B. L. GUREVICH
I. M. LANDA

On the Problem of the

Economic Evaluation of Lands

We feel that in constructing a scale for the evaluation of land (soils), one must proceed from some general principles. The most natural and the simplest principles, in our opinion, are the following.

1. The construction of evaluation scales must be based on the soils of those agricultural enterprises where the technological level [sposob vedeniya khozyaistva] is both up to date and similar enough to be comparable. These would be the lands of modern collective and state farms typical of various zones.

2. The evaluation of the above-mentioned soils must be based on statistical-economic materials referring to a time interval during which the technological level can, without noticeable errors, be considered stable.

3. We shall consider the technological level relatively stable if the following conditions are fulfilled for the lands of modern and representative agricultural enterprises:

a. There exist for the given technological level a characteristic level p_o of the production costs (rubles per hectare) that, de-

Translated from *Vestnik Moskovskogo Universiteta*, seriya geografiya [Herald of Moscow University, Geography series], no. 2 (1962), pp. 16–24.

pending on the crop cultivated, is the same for all allotted lands and that changes with a change in the technological level.

b. At a given technological level the changes in production costs (rubles per hectare) are small compared to the characteristic level p_o. In other words, if one plots on the axis p the characteristic point p_o, the expenditures p, permitted by the technological level cannot got beyond the limits δ of point p_o (see figure 1).

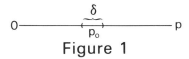

Figure 1

The characteristics of the technological level described above follows directly, in our opinion, from the directives of Lenin, who wrote in 1901, criticizing the law of decreasing fertility of the soils: "Of course, in comparatively small dimensions the 'additional investment of labor and capital' can also occur (and does occur) on the basis of a given, invariable level of technology. . . . The invariable state of technology puts comparatively *very narrow* limits on the additional investments of labor and capital."[1]

Let us examine a soil selected in accordance with the first principle. Different yield capacities y (centner per hectare) of a given cultivated crop correspond to the various expenditures p on this soil. The yield capacity y is a function of p:

$$y = f(p). \tag{1}$$

c. Under a change in the technological level, the character of the dependence of yield capacity on expenditures becomes different, that is, the function y itself changes. But whatever the technological level, y is a single-valued, continuous, and progressive function of p.

Thus, it is assumed that (1) at a given technological level a completely determined yield capacity corresponds to each admissible level of expenditures per hectare (single-valued dependence of y on p), and (2) that small changes in yield capacity correspond to small changes in investments per hectare (continuous dependence of y on p).

Finally, assuming y to be a progressive function of p we can thereby record the apparently trivial fact of an increase in yield capacity with additional investments permissible under the given technological level.

The graph of the function $y = f(p)$, for each given soil and a particular crop that is cultivated on it, represents at this technological level a small arc, as shown on figure 2. Such is the geometric characteristic of a given, invariable technological level. *The examination of a specific technological level must be local.*

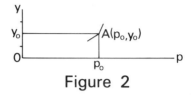

Figure 2

It is well known that a small arc can be replaced by a straight line without any noticeable error (as a first approximation). We shall draw this section across the characteristic point A (p_o, y_o). We designate the yield capacity of a certain crop at the given technological level characteristic for this soil with $y_o = f(p_o)$ (see figure 3).

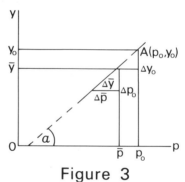

Figure 3

Therefore, the study of the dependence of the yield capacity y on the expenditures p is reduced to the study of the linear function,

$$y - y_o = \lambda (p - p_o), \qquad (2)$$

here

$$\lambda = \tan \alpha = \frac{y - y_o}{p - p_o} = \frac{\Delta y_o}{\Delta p_o}, \qquad (3)$$

or

$$\lambda = \frac{y - \bar{y}}{p - \bar{p}} = \frac{\Delta \bar{y}}{\Delta \bar{p}}, \qquad (4)$$

where p represents any admissible expenditure level (per hectare); $\bar{y}$ is the yield capacity; λ is the rate of change in the yield of the crop on this soil (the rate is the same for all admissible levels of expenditure).

Returning to equation (2), p_o does not depend on the soil, since it is the characteristic level of expenditure for a certain crop at a given technological level. For p_o it is best to take an expenditure level averaged from all selected soils (in accordance with the first principle), pertaining to the time interval pointed out in the second principle.

The yield capacity $y_o = f(p_o)$ is the characteristic yield capacity. For different soils it varies with the given technological level and the specific crop under cultivation. The characteristic yield capacity introduced here coincides with the "normal yield capacity" of V. V. Dokuchayev. The characteristic yield y describes a given soil (its productivity for a particular crop at a given technological level). The relationship

$$x_1 = \frac{y_o}{p_o} \tag{5}$$

we shall call the first productivity index of the soil or the soil productivity at the characteristic yield capacity. It stands to reason that for different crops the productivity x_1 of the same soil will differ. But the productivity of the soil according to the characteristic yield capacity is not its only (and total) characteristic. We have already pointed out that the rate of change in yield capacity is the same for a given soil but changes with transition to another soil. Consequently a second characteristic feature of the soil is λ.

Let us call y the second productivity index of the soil and designate it by x_2. From equation (4) we obtain

$$x_2 = \frac{\Delta y}{\Delta p} . \tag{6}$$

Consequently, this index of soil productivity depends on the change in yield capacity.

The indices of productivity, x_1 and x_2, are dimensional values measured in their own units:

$$D = \frac{\text{centners}}{\text{rubles}} . \tag{7}$$

Equations (5) and (6) can be presented in the following form:

$$y_o = x_1 p_0; \Delta y = x_2 \Delta p . \tag{8}$$

From figure 4 it can be seen that $x_1 = \tan \alpha_1$ and that $x_2 = \tan \alpha_2$.

If the productivity index x_1 according to the characteristic yield capacity is directly proportional to the latter, then the productivity

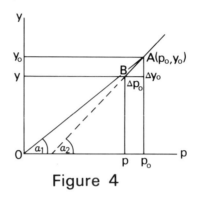

Figure 4

index x_2 according to the change of yield capacity is proportional to this change (at a given additional expenditure) and is inversely proportional to the additional expenditures (at a given change of yield capacity).

The results obtained can be stated in this manner: the *economic evaluation of each soil demands not one but two productivity indices,* x_1 and x_2.

To assign a soil economically to a given technological level means to assign indices of its productivity y_o [this is an error and should read y_o/p_o, Eds.] and λ, that is, [to determine] the graph of a straight line according to equation (2). Consequently, in the first approximation the graph itself is a description of the soil. And in practice, in comparing various soils, it is expedient to turn to their graphs. Let there be, for example, three soils sown to grain crops, soils I, II, III to which graphs I, II, and III correspond (see

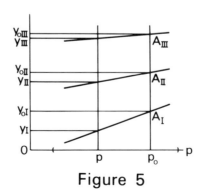

Figure 5

figure 5). The limits of the characteristic level of p_o are marked in parentheses. Expenditure levels other than these are not examined, since they are inadmissible under the given technological level.

From figure 5 it can be seen that according to the yield capacity (not only the characteristic y_o but also any factual y) soil III is more productive than soil II, which in turn is more productive than soil I.

If we now evaluate the *changes* of yield related to additional expenditures, we obtain a different result: $\lambda_I > \lambda_{II} > \lambda_{III}$ (see figure 5). Consequently, with the same additional expenditures $\triangle p$ (addition to level p) the greatest increase in yield comes from soil I. Soil III has nearly zero productivity according the change in yield capacity; the additional expenditures on soil III are almost without effect. Soil II is an intermediate soil. It is clear that additional investments will be most effective on soil I (in spite of the fact that the yield capacity is higher on soil III).

In planning the distribution of expenditures for all three soils, it is necessary to take into account not only the productivity of each soil corresponding to the characteristic yield y_{oI}, y_{oII}, or y_{oIII} but also the productivity of each of these soils corresponding to changes in yield capacity $\triangle y_I$, $\triangle y_{II}$, $\triangle y_{III}$, which are due to changes in expenditure $\triangle p_I$, $\triangle p_{II}$, $\triangle p_{III}$.

By using the indices of productivity x_1 and x_2 and the graphs, one can qualitatively and quantitatively solve a number of important practical questions, such as the optimization of the distribution of expenditures among the various soils.

Equation (7) introduced the units of productivity D = centners/rubles. If it concerned only one crop, the value D would be fully determined. However, various crops can be cultivated on the same soil. It stands to reason that the value D for grain, as opposed to that of some special crop, is essentially different. Therefore, the introduced units of productivities x_1 and x_2 are indeterminate. The resolution of this uncertainty can be realized in the following manner: we refer to each crop by its own unit of productivity (D) and then, finding the conversion coefficient from the value for one crop to that for another, we select one of these Ds as standard.

Let g and s be two crops (grain and special). Let

$$D_g = \frac{\text{centners}_g}{\text{rubles}}, D_s = \frac{\text{centners}_s}{\text{rubles}}.$$

In such a case

$$\frac{D_s}{D_g} = \frac{\text{centners}_s}{\text{centners}_g}.$$

It is common in economic terms to consider the ratio of one centner of crop s to one centner of crop g as equal to the ratio of their costs at a given technological level and the ratio of the latter can be taken as equal to the ratio of the standard expenditures (at a given technological level) needed for the production of each of the centners under consideration, that is,

$$\frac{D_s}{D_g} = \frac{\bar{z}_s}{\bar{z}_g}$$

from which

$$D_s = \frac{\bar{z}_s}{\bar{z}_g} D_g \tag{9}$$

where $\bar{z}_s$ and $\bar{z}_g$ are the standard expenditures per centner of crops s and g.

According to equation (5), the productivity index $x_{1s} = y_{os} / p_{os}$ is expressed in D_s, but $x_{1g} = y_{og} / p_{og}$ in D_g. Therefore, considering equation (9), we have

$$x_{1s} = \frac{y_{os}}{p_{os}} \cdot \frac{\bar{z}_s}{\bar{z}_g} D_g$$

that is,

$$x_{1s} = \frac{y_{os}}{p_{os}} \cdot \frac{\bar{z}_s}{\bar{z}_g} D, \tag{10}$$

where

$$D = D_g \tag{11}$$

is the unit of productivity chosen as standard. Since

$$p = yz \tag{12}$$

where z is the expenditure per centner, then $p_o = y_o z_o$. We designate the *characteristic* expenditures per centner with z_o. It must be emphasized that the standard expenditures $\bar{z}$ per centner and the characteristic expenditures z_o are determined by the relation $z_o = p_o / y_o$. And since, at a given technological level the characteristics yield capacity y_o depends on the soil, so too z_o, at the same technological level also depends on the soil. Yet the standard expenditures $\bar{z}$ are standard for a given technological level and do not depend on the soil.

Now one can rewrite formula (10) in the following form

$$x_{1s} = \frac{y_{os}}{y_{os} z_{os}} \cdot \frac{\bar{z}_s}{\bar{z}_g} D = \frac{1}{z_{os}} \cdot \frac{\bar{z}_s}{\bar{z}_g} D.$$

Dividing what is obtained by

$$x_{1g} = \frac{y_{og}}{p_{og}} D = \frac{y_{og}}{y_{og} z_{og}} D = \frac{1}{z_{og}} D,$$

we find

$$x_{1s} = \frac{z_{og}}{z_{os}} \cdot \frac{\bar{z}_s}{\bar{z}_g} x_{1g} \ D. \tag{13}$$

We turn to the second index of productivity pointed out in equation (6), that is $x_2 = \Delta y / \Delta p$. For crop s we have

$$x_{2s} = \frac{\Delta y_s}{\Delta p_s} D_s$$

or, considering equations (9) and (11),

$$x_{2s} = \frac{\Delta y_s}{\Delta p_s} \cdot \frac{\bar{z}_s}{\bar{z}_g} \ D.$$

Dividing by $x_{2g} = \dfrac{\Delta y_g}{\Delta p_g} D$ we find

$$x_{2s} = \frac{\Delta y_s}{\Delta p_s} \cdot \frac{\Delta y_g}{\Delta y_g} \cdot \frac{\bar{z}_s}{\bar{z}_g} \ x_{2g} D. \tag{14}$$

And so, the first and second indices of the productivity of the soil for crop s are expressed respectively by analogous indices of productivity of the same soil for crop g in units of D.

Formulas (13) and (14) show that the soils sown to a special crop are equivalent to the soil sown to grain crops. Up to now we examined a soil sown completely to one crop (grain or special); however, the actual agricultural situation is more complex. Therefore, to construct a scale for the evaluation of land it is necessary to give formulas equating soil under a complex of crops to soil under only grain.

Let the area sown to a special crop be k percent of the total sown to grain and the special crop. In that case the productivity indices for the combination of these crops, x_1 and x_2, are the arithmetic means weighted from x_{1s} and x_{1g} for x_1, and from x_{2s} and x_{2g} for x_2. Therefore the weights are k for x_{1s} and x_{2s}; $100 - k$ for x_{1g} and x_{2g}. And so

$$x_1 = \frac{kx_{1s} + (100 - k)x_{1g}}{100} \ D$$

and

$$x_2 = \frac{kx_{2s} + (100 - k)x_{2g}}{100} \ D.$$

Substituting here x_{1s} and x_{2s} with the values pointed out in equations (13) and (14), we obtain:

$$x_1 = \left(\frac{\dfrac{z_{og}}{z_{os}} \cdot \dfrac{\bar{z}_s}{\bar{z}_g} \, k + 100 - k}{100} \right) x_{1g} D.$$

$$x_2 = \left(\frac{\dfrac{\Delta y_s}{\Delta p_s} \cdot \dfrac{\Delta p_g}{\Delta y_g} \cdot \dfrac{\bar{z}_s}{\bar{z}_g} \, k + 100 - k}{100} \right) x_{2g} D$$

that is,

$$x_1 = \left[1 + \frac{\Sigma(A_1 - 1)\, k}{100} \right] x_{1g} D, \tag{15}$$

$$x_2 = \left[1 + \frac{\Sigma(A_2 - 1)\, k}{100} \right] x_{2g} D,$$

where

$$A_1 = \frac{z_{og}}{z_{os}} \cdot \frac{\bar{z}_s}{\bar{z}_g},$$

$$A_2 = \frac{\Delta y_s}{\Delta p_s} \cdot \frac{\Delta p_g}{\Delta y_g} \cdot \frac{\bar{z}_s}{\bar{z}_g}. \tag{16}$$

The summation sign (Σ) in equation (15) is there because not one but several different special crops may be found together with grain in the complex of crops.

Thus, the problem of indices of productivity of soil under a complex of crops is solved: all lands indicated above can be evaluated on a single scale of evaluation.

The relationships between (6) and (8) make it possible to rewrite equation (2) in the following form:

$$y - x_1 p_o = x_2 (p - p_o)$$

or

$$y = x_2 p + (x_1 - x_2) p_o. \tag{17}$$

It is natural to call equation (17) the equation of productivity of the soil or the soil equation having in view the definite aspect of the soil at a given technological level.

Equation (17) together with equation (12) made it possible (as a first approximation, fully adequate for local examination) to clarify in economic terms the processes of change in the yield capacity and production expenditure per centner of product with a change of investments per hectare.

Below is a calculation of the productivity indices x_1 and x_2 from corrected data collected by A. I. Polosa, a student in the geography faculty of Odessa University, on the Berezino State Farm in Tarutin region of Odessa Oblast. Since the goal is essentially to illustrate the method of calculation, we can limit ourselves to the examination of four basic crops of the crop rotation for 1958–60 at the state farm: (1) winter wheat, (2) corn grown for grain, (3) winter barley, and (4) oats.

The initial values needed for the calculation, harvested sown areas (s), yield (y), and direct investments (p) are given by crop in table 1.[2]

Using the method of least squares we find the soil (land) equation for each of the four crops:

$$y = ap + b. \tag{18}$$

In order to determine a and b we work out the normal equations

$$a\Sigma p^2 + b\Sigma p = \Sigma yp, \; a\Sigma p + 3b = \Sigma y. \tag{19}$$

The values Σy and Σp are in the last line of table 1 (columns 2 and 3, 5 and 6, 8 and 9, 11 and 12). The values Σp^2 are obtained by squaring the values of columns 3, 6, 9 and 12. The values Σyp are found by multiplying the pairs of indicated columns, 2 and 3, 5 and 6, 8 and 9, 11 and 12. Substituting the obtained values in the system of equations (19), for each of the four crops, and working out the corresponding systems, we find the following four land equations.

Sown to wheat: $y = 0.156 p - 43.3$.

Sown to corn: $y = 0.053 p - 15.2$.

Sown to barley: $y = 0.115 p - 19.4$.

Sown to oats: $y = 0.099 p - 34.1$.

(20)

Taking the corresponding values p_o one can easily work out from the coefficients of equations (20) the following table characterizing the land study for each separate group (see table 2).

In order to determine the productivity indices x_1 and x_2 of the land under the four crops *as a whole*, we point out the proportions (k percent) of each of the crops in the sown area, taking into account

TABLE 1

YEARS	WHEAT			CORN			BARLEY			OATS		
	$\frac{s}{1}$	$\frac{y}{2}$	$\frac{p}{3}$	$\frac{s}{4}$	$\frac{y}{5}$	$\frac{p}{6}$	$\frac{s}{7}$	$\frac{y}{8}$	$\frac{p}{9}$	$\frac{s}{10}$	$\frac{y}{11}$	$\frac{p}{12}$
1958	745	20.0	406.3	240	27.0	794.0	100	20.9	389.0	80	17.7	515.0
1959	1,166	22.8	423.3	150	18.3	630.8	200	26.8	388.8	38	15.1	503.0
1960	733	20.2	405.8	150	25.8	770.0	100	30.0	410.0	50	13.5	476.0
TOTAL	2,644	63.0	1,235.4	540	71.1	2,194.0	400	77.7	1,187.8	168	46.3	1,494.0

NOTE: s is area in hectares; y is yield in centners per hectare; p is direct investment in rubles per hectare.

in table 1 columns 1, 4, 7 and 10, namely: for wheat, 70.5 percent; corn 14.4 percent; barley, 10.7 percent; and oats, 4.4 percent.

TABLE 2

INDEX		WHEAT	CORN	BARLEY	OATS
Typical investment Rubles per hectare	$(\underline{p}_o)$	420	700	390	490
First productivity index Centners per ruble	$(\underline{x}_1)$	0.05	0.03	0.06	0.03
Second productivity index Centners per ruble	$(\underline{x}_2)$	0.16	0.05	0.11	0.09
Characteristic yield capacity Centners per hectare	$(\underline{y}_o)$	22.3	22.0	23.8	14.7
Characteristic net cost Rubles per centner	$(\underline{z}_o)$	18.8	31.8	16.3	33.3

NOTE: The data on the characteristic investment $(\underline{p}_o)$ as well as the zonal net cost values $\bar{z}$ were arbitrarily estimated.

Taking wheat as the standard we find the values A_1 and A_2 from formulas

$$A_1 = \frac{x_{1i}}{x_{1l}} \cdot \frac{\bar{z}_i}{\bar{z}_l} \; ; A_2 = \frac{x_{2i}}{x_{2l}} \cdot \frac{\bar{z}_i}{\bar{z}_l} , \qquad (21)$$

where $i = 2, 3, 4$ are the indices for corn, barley and oats, and the values $\bar{z}_2 = 30$ rubles / centners, $\bar{z}_3 = 20$ rubles / centners, $\bar{z}_4 = 25$ rubles / centners, are taken for $\bar{z}$. The zonal net cost of a centner of wheat is $z_1 = 20$ rubles/ centners. The first and second productivity indices of wheat are designated by x_{11} and x_{21}. Finding A_1 and A_2 we calculate x_1 and x_2 of the land as a whole in accordance with the formulas

$$x_1 = \left[1 + \frac{\Sigma(A_1 - 1)k}{100}\right] x_{11}D; \; x_2 = \left[1 + \frac{\Sigma(A_2 - 1)k}{100}\right] x_{21}D \qquad (22)$$

and obtain

$$x_1 = 0.0498 \frac{\text{centners of wheat}}{\text{ruble}}; x_2 = 0.1152 \frac{\text{centners of wheat}}{\text{ruble}}.$$

And so in spending 1 ruble for the land under examination at the given technological level, one can obtain 0.05 centners of wheat, 0.03 centners of corn,[3] 0.06 centners of barley, or 0.03 centners of oats, which at the given technological level and the given combination of crops in the rotation is equivalent to 0.0498 centners of wheat. Correspondingly, with x_2, the land under examination is evaluated by its response to 1 ruble of additional expenditure.

1.   V. I. Lenin, *Sochineniya* [Works], 5:93.

2.   The investment and cost indices of production are calculated in the old monetary units.

3.   The lower productivity of corn is explained by the insufficient attention given this remarkable crop before the Plenum of Central Committee of the CPSU in January, 1961. It is evident that the high yields of corn are associated with the change in the technological level.

SECTION IV

Agricultural Geography

Introductory Note

Agriculture has long been an important object of study in both pre-revolutionary and Soviet geography. Such an interest is not suprising given the extent and importance of agriculture as well as the magnitude of the problems associated with this branch of the economy in both periods. The evolution of Soviet geographic research displays a number of interesting facets that bear a striking resemblance to the various shifts in governmental economic policy regarding agriculture. In the Stalin period, when the emphasis was on the ability of man to transform nature, a large proportion of research in agricultural geography was done by physical geographers. In the Krushchev period agricultural policy veered away from the grandiose schemes for the transformation of nature but continued to stress extensification of agriculture, which led to considerable expansion of farming into areas previously considered marginal for cultivation. The Virgin and Idle Lands Program, initiated in 1954, serves as an excellent example of extending and expanding the agricultural base of the USSR. Correspondingly, the bulk of the geographic research on agriculture stressed the physical and technological potential of areas for development.

Soviet agricultural problems persisted in the post-Krushchev period, and again major policy changes were discussed and, in part, implemented. The new concerns voiced by the Party and government emphasized that the measurement of performance should be related to production per hectare and that the dependence of agriculture on the natural elements be reduced to a minimum. Whereas the earlier programs stressed expansion of the agricultural land base, the new emphasis was placed on more intensive, effective utilization of existing agricultural lands. Recent geographic research clearly reflects these new directions. Studies by geographers concerned with agricultural pricing and precise regionalization schemes that account for land productivity have become numerous.[1] Among the new, evaluative disciplines emerging in Soviet geography is a branch involved with the economic evaluation of agricultural lands.[2] Since the 1950s geographers have been the leading proponents of an all-Union land cadaster that would provide a basis for judging the economic effectiveness of land according to cost and yield data. Similarly, agricultural regionalization, a persistent theme of geographic research from the earliest years of the Soviet regime, has been affected inasmuch as the current call for more precise planning and pricing procedures relate to a need for more precise regionalization schemes.[3]

The main content of agricultural geography involves land-use and systems of farming. The two broad problems of the field include: (1) discernment of the existing distribution of agriculture and the economic evaluation of the existing distribution (the first or inventory stage); and (2) the solutions to questions of a *more rational* distribution of agriculture.[4] The goal of such research is to bring about a proper distribution of agriculture by natural, economic zones or regions. It is this goal that, for agricultural geography more than many other branches of Soviet geography, brings together physical and economic geographers.

The articles selected for inclusion in this section represent work done by Soviet geographers in recent years and represent a cross section of the field. With the exception of the lead article, the selections represent the application of mathematical-statistical techniques in the study of agricultural problems and reflect the increasing rigor and sophistication in Soviet economic geography. (Other articles relating to agriculture are included in Section 3, Resource Management, and Section 9, Historical Geography).

The first article by Chelintsev is an example of the early traditional approach to agriculture in that it provides a descriptive overview of the existing agrarian system. The main theme is a regionalization scheme that attempts to identify spatial variations in agriculture in order to understand the conditions that give rise to such differences. Regions are distinguished by selecting indices for farming and livestock raising and then mapping them within the framework of five natural zones: (1) the nonchernozem forest, (2) the chernozem forest-steppe, (3) the chernozen steppe, (4) the arid, unirrigated zone, and (5) the arid, irrigated zone and piedmont. Two oblasts of the forest-steppe zone are then used as case studies of intensive and extensive agriculture. The most useful content of the article lies in the extensive tabular data.

The article by Mikheyeva is an elaboration and extension of a model outlined in an earlier study by the author.[5] In this study a model is developed in four stages, each of which allows an examination of the quantitative relationships between resources and production. The first stage yields a monocultural solution that determines the land capacities. The second stage provides quantitative indicators for the size and structure of the sown areas in the regions, whereas the third stage yields the necessary information regarding the structure of the livestock economy. The last stage is a corrective one that allows for an evaluation of labor resources and the detection of production deficiencies. The remainder of the study focuses on the meaning and significance of the estimates derived from the modeling procedure and delves into an interesting comparison of land evaluations and differential rent.

Sheynin's article on price zones for agricultural products examines the criteria used for combining areas into price regions. Grains and sugar beets in the Ukraine are used as an example of delimiting price zones on what is termed the production principle, whereas vegetables and dairy products are utilized to outline regions based on the transport principle. Both the production and transport principles of pricing are explained and their corresponding advantages and disadvantages noted. The transportation principle is suggested as the more efficient principle, providing an economic advantage over the production principle that has been most commonly used in price setting. The similarity of Sheynin's ideas with the classic Von Thunen model, so well known in the West, is worthy of note.

The final article by Gorokhova, Kamenetskiy, and Shvartsberg is an application of linear programming to the problem of allocating machinery to agricultural enterprises. They examine the question of the optimal size and distribution of tractor parks in a case study set in Irkutsk oblast. The solution arrived at indicates that a more rational and economical operation is feasible, but the authors caution their readers that more complex problems require even more sophisticated techniques of analysis.

1. See Robert G. Jensen, "Land Evaluation and Regional Pricing in the Soviet Union," *Soviet Geography: Review and Translation* 9, no. 3 (March 1968); see also the translated articles included in this number of *Soviet Geography*.

2. Examples of this emphasis are demonstrated in two issues of *Voprosy Geografii* [Problems of Geography], no. 43 (1958), entitled "The Quantitative Inventory and Evaluation of Lands"; and no. 54 (1961), entitled "Geography and Land Resources." See also B. L. Gurevich and I. M. Landa, "On the Problem of Economic Evaluation of Lands," in Section 3.

3. N. V. Gvozdetskiy, "Physical-Geographic Regionalization for Agricultural Purposes," *Soviet Geography: Review and Translation* 5, no. 7 (September 1964): 3-9; N. V. Gvozdetskiy, "Physical-Geographic Regionalization of the USSR for Agricultural Purposes," *Soviet Geography: Review and Translation* 1, no. 9 (November 1960): 5-19.

4. A. M. Rakitnikov, "Economic Geographic Research on Agriculture," in *Soviet Geography: Accomplishments and Tasks*, trans. Lawrence Ecker and ed. Chauncy Harris, American Geographical Society Occasional Publication, no. 1 (New York, 1962), pp. 233-34.

5. V. S. Mikheyeva, "An Economic-Mathematical Model of the Distribution of Agricultural Production by Regions of the Country," *Vestnik Moskovskogo Universiteta seriya geografiya* [Herald of Moscow University, Geography series], no. 6 (1962), pp. 12-17.

A. N. CHELINTSEV

Agricultural Regions of the USSR

The Program of the Communist Party of the Soviet Union, adopted at the Twenty-second Congress of the CPSU, emphasizes that one of the main ways to promote a sharp rise in agricultural production is the intensification of farming. At present a truly revolutionary change in farming in the collective and state farms of the country is occuring. The extensive and inefficient utilization of land resources, which are the most important part of national wealth and the main means of agricultural production, is being reorganized into a highly intensive agriculture, primarily by replacing extensive and unproductive crops by intensive and highly productive ones such as corn, sugar beets, and legumes.

In order to differentiate scientifically, not dogmatically, those regions where there have been sharp increases in the rates of intensification, the most careful scientific analysis of the level of intensity and productivity of agriculture by natural-economic zones and regions of our country is needed. Such an analysis is presented in this article

The division of the USSR into agricultural regions using 1956–59 data clarifies the factual specialization of agriculture in the

Translated from *Voprosy narodnogo khozyaystva SSSR* [Questions on the National Economy of the USSR] (Moscow, 1962), pp. 233–46.

country. In some parts of the Union agriculture differs in productive organization as well as in the yield per unit of agricultural area. Agricultural regionalization has not only the task of illustrating these spatial differences but also of determining the influence under which these differences arise. This article is devoted in large measure to clarifying the conditions under which the various forms of agricultural production and differences in productivity occur. In general, such conditions include the development of productive forces of the country and the national economic environment under the effect of which agriculture is formed. These forces do not act uniformly on the territory of the Union.

In solving the problem of agricultural regionalization a number of factors are considered: population density (number of inhabitants per square kilometer), development of a domestic market for agricultural products, densities of urban and rural populations, supply of agricultural labor (the number of collective farmers per 100 hectares of plowed land), and the sufficiency of the means of production (the latter could be represented as an index of the quantity of plowed land per tractor).

The choice of indicators characterizing agricultural production has two conditions: (1) they must show the component parts of agriculture and livestock-raising; (2) they must be criteria that change in the process of agricultural development, but the achieved level of production characterized by these criteria must show some degree of intensity. Consequently, the choice of criteria is based on theoretical positions of the agricultural economy concerning the successive change of indicators during the increase in agricultural intensity. For example, the percentage of sown area shows the composition of agriculture together with that index related to the greater or lesser potential intensity of managing agriculture.

To distinguish the agricultural regions the following indices are taken:

1. For farming: the percentage of plowed land in the total agricultural area, the percentage of sown land, the percentage of area in cereal, fodder, and commercial crops (total and by individual types), in vegetables, potatoes, and grain yields.
2. For livestock raising: density of all large livestock per 100 hectares of plowed land, density of cows, pigs, and sheep per 100 hectares of sown area, percentage of cows in a herd of large cattle, milk yield per cow, production of milk and meat, and wool per 100 hectares of agricultural land.

Every criterion as well as the production indices were plotted on a map at the appropriate location for each. The quantitative characteristics of these indices determine what level of development an area has achieved. The development of an area cannot be characterized by individual indices but rather by interrelated indices that are also linked with the premises of regionalization. Above all, the two basic components of agriculture—farming and livestock-raising—must be linked up.

In view of the great difference in natural conditions for agriculture production, it is necessary to distinguish agricultural regions within the limits of each basic natural zone, namely: (1) nonchernozem forest, (2) chernozem forest-steppe, (3) chernozem steppe, (4) arid unirrigated, and (5) arid irrigated and piedmont.

Without as yet characterizing each of the twenty-five agricultural regions we have delimited, let us compare one of the most intensive regions of the forest-steppe belt, Khmel'nitskiy Oblast, with the most extensive in this belt, Chita Oblast.

The population density of the Khmel'nitskiy Oblast is 80 per square kilometer; farming is of a crop-rotation type with more than 20 percent in sown crops, almost without clean fallow, and a minimum of unplowed land (plowed land is no less than 80 per cent of the entire agricultural area). In livestock-raising it has one of the highest densities—49 head of cattle, 58 pigs, and 20 milk cows per 100 hectares of agricultural land. The oblast has one of the highest milk yields per 100 hectares—356 centners—and meat of all kinds—66 centners, of which pork accounts for 46 centners per 100 hectares of plowed land. There are very few sheep—23 head [per 100 hectares]. Due to the low density of urban population the local market is small and production is directed to the national market. The labor force supply is high—54 [per 100 hectares].

The population density of Chita Oblast is only 2.4 [per square kilometer]. Only 37.5 percent of the agricultural land is under cultivation, the other land is being used as natural meadows and pasture; plowed land is mainly under cereal crops (70 percent), whereas in Khmel'nitskiy Oblast 58 percent is in grain. In Chita Oblast 4 percent of the area is in labor-consuming nongrain crops. In spite of the large quantity of natural fodder land, there are 50 head of cattle and 321 sheep per 100 hectares of sown area, and a limited number of workers per 100 hectares of plowed land (on the collective farms—8 workers). Milk yield in 1957 was only 44 centners; meat, 7 centners, and pork, 7 centners.

The degree of density is measured by the density of the entire population per square kilometer and it is compared with the oblast supply of collective farm workers calculated per 100 hectares of plowed land, whereas the volume of the local agricultural market is characterized by the percentage and density of the urban population.

Let us compare now the agricultural effectiveness indices in each of the natural belts with the factors of intensity. The data by regions for each of the natural belts are shown in table 1. As indicated in the table, the level of development of productive forces increases from east to west, and there appear conditions for greater intensity of agricultural production. Agricultural effectiveness increases correspondingly. Thus, in the nonchernozem belt, milk yield rises from 62 to 383 centners per 100 hectares of agricultural land, and meat production from 8 to 43 centners including pork, which increases from 7 to 49 centners. In the chernozem-forest belt milk yield increases from 62 to 467 centners; and meat production from 8 to 53 centners, including pork (7 to 51 centners). In the chernozem-steppe milk production increases from 25 to 242 centners, meat from 3.3 to 34.6 centners (including pork which increases from 2.4 to 24.1 centners).

Upon further analysis characteristic differences are discovered between the three natural belts in the production increases (meat and milk) from east to west. The production of meat in the forest-steppe as in the steppe is higher than that of milk, but in the nonchernozem belt it is the opposite: the indices of meat yield are in most cases lower than the indices of milk yield. At the base of the interregional differences lies the population density of their territories and the dissimilar supply of labor resources. From the cited data (see table 1) one can note the increase in density of the urban population from east to west.

With the rise in density and the higher percentage of urban population, the density of agricultural population drops (smaller supply of workers per 100 hectares of plowed land); and conversely, with the lowering of the urban population density (smaller influence of the agricultural market) the density of agricultural population rises (larger number of workers per 100 hectares). An example of the first case is Kemerovo Oblast, and of the second, the Central Chernozem oblasts and the Chuvash ASSR.

The extent to which both these conditions create differences in agricultural intensity between agricultural regions can be seen by comparing the structure of production in all agricultural regions.

Besides the density indices, the size of the plowed area per able-bodied person, and the indices of the volume of the local agricultural market, that is, the density of urban population, in the table the volumes of the leading branches of livestock-raising are shown, including the density of cattle per 100 hectares of plowed land and per 100 hectares of sown area. The supply of the most important means of production—tractors—is characterized by the area of plowed land per 1 tractor.

TABLE 1

INDICES OF INTENSITY OF AGRICULTURAL PRODUCTION BY NATURAL ZONES
1956-1959

Region	Population Density	Plowed Land Per Able-Bodied Collective Farmer	Density of Urban Population	Percentage of Plowed Land	Cattle Density (Per 100 Hectares of Plowed Land)	Cattle Density (Per 100 Hectares Sown Area)	Plowed Land (Hectares Per 1 Tractor)	Percentage of Sown to Plowed Land	All Fodder Crops
1	2	3	4	5	6	7	8	9	10
Nonchernozem Forest Belt									
1a	3.0	...	...	12.6	18.3	24.1	16	100	43.2
1b	0.2	...	1.5	8.5	(24.4)	(43.5)	64	75	11.1
2a	4.9	15.3	3.3	42.3	24.0	113.0	67	100.3	18.2
2b	3.2	10.1	1.1	43.7	10.0	52.0	118	74.6	21.9
3	1.9	10.1	1.1	60.0	18.0	38.0	113	83.6	29.6
4	1.6	...	0.6	42.6	17.7	54.5	126	85.6	20.9
5	21.0	6.2	16.4	56.9	25.3	56.0	122	91.2	25.0
6	21.7	4.5	26.0	71.1	19.3	32.0	135	84.5	23.2
7a	71.2	3.7	57.9	64.0	36.0	47.4	98	91.1	37.7
7b	26.2	2.2	15.7	48.2	25.1	60.0	65	88.1	41.7
8	47.1	2.34	19.6	69.4	41.1	55.0	120	96.3	19.6
9	35.9	...	19.5	56.6	30.1	58.0	82	92.3	36.0
10	4.3	2.2	2.0	28.1	25.3	107.8	59	84.7	41.2

TABLE 1—Continued

Region [1]	Population Density [2]	Plowed Land Per Able-Bodied Collective Farmer [3]	Density of Urban Population [4]	Percentage of Plowed Land [5]	Cattle Density (Per 100 Hectares of Plowed Land) [6]	Cattle Density (Per 100 Hectares of Sown Area) [7]	Plowed Land (Hectares Per 1 Tractor) [8]	Percentage of Sown to Plowed Land [9]	All Fodder Crops [10]
Chernozem Forest-Steppe Belt									
11	18.3	9.3	9.5	59.4	16.6	30.1	120.0	93.3	19.8
12	23.8	10.45	15.4	65.8	16.5	28.8	134.0	82.9	23.5
13	41.1	3.7	15.8	77.6	22.1	34.2	157.0	84.5	19.6
14	35.4	6.2	15.7	81.8	17.1	24.3	169.0	83.0	20.3
15a	55.4	3.7	28.8	79.3	20.9	29.8	129.0	84.8	23.5
15b	44.8	3.65	11.2	83.4	21.8	33.0	144.0	86.3	20.7
16a	67.6	2.5	24.2	81.3	45.1	50.7	108.0	100.8	21.0
16b	82.6	...	24.2	54.3	56.3	102.0	100.0	105.4	29.0
17	43.3	3.2	18.9	58.0	27.9	44.2	91.0	102.3	21.5
Chernozem Steppe Belt									
18	4.5	15.3	1.8	30.9	4.6	17.4	105	90.2	16.8
19	17.5	11.4	12.5	65.0	13.9	16.5	158	88.8	19.1
20	32.9	9.4	17.6	66.1	15.7	27.0	161	89.0	21.5
21	70.4	5.48	41.3	80.5	32.6	35.7	122	94.9	18.7

TABLE 1--Continued

Region	Population Density	Plowed Land Per Able-Bodied Collective Farmer	Density of Urban Population	Percentage of Plowed Land	Cattle Density (Per 100 Hectares of Plowed Land)	Cattle Density (Per 100 Hectares Sown Area)	Plowed Land (Hectares Per 1 Tractor)	Percentage of Sown to Plowed Land	All Fodder Crops
1	2	3	4	5	6	7	8	9	10
Arid Unirrigated Belt									
22a	1.8	8.6	11.3	6.3	3.2	31.2	69	87.3	36.5
22b	19.7	7.5	7.8	22.6	11.5	32.1	92	94.0	23.9
Arid Irrigated and Piedmont Belt									
23	4.8	3.4	1.9	8.1	3.1	60.9	80	89.4	23.0
24	11.9	3.1	4.2	12.6	9.3	68.0	51	96.3	18.0
25	45.0	1.7	20.0	32.1	39.6	132.0	79	89.5	17.4

TABLE 1--Continued

Region	Grain Crops	Commercial Crops	Potatoes	Vegetables	Corn	Flax	Mushrooms	Sugar Beets	Sunflower	Hemp
1	11	12	13	14	15	16	17	18	19	20
Nonchernozem Forest Belt										
1a	5.8	...	39.9	10.6	...	...	...	...	...	...
1b	80.5	...	5.0	1.2	...	...	...	...	...	...
2a	62.6	1.0	13.7	4.7	...	...	...	...	...	...
2b	68.7	0.39	3.8	0.6	...	...	0.46	0.41	0.02	0.005
3	72.8	1.2	4.6	0.06	...	...	0.7	0.09	0.02	0.07
4	72.0	4.5	5.1	0.4	...	3.1				
5	60.0	0.2	9.2	1.6	...	0.07				
6	65.0	4.7	6.6	0.7	...	5.4				
7a	48.0	3.8	12.1	1.9	...	3.7				
7b	38.1	16.7	11.5	1.9	...	12.7				
8	55.1	8.3	13.9	1.2	...	4.3		2.6		0.5
9	46.1	3.0	9.7	1.5	...	2.1		0.9		
10	44.5	3.1	11.7	1.6	...	3.0				

TABLE 1--Continued

Region	Grain Crops	Commercial Crops	Potatoes	Vegetables	Corn	Flax	Mushrooms	Sugar Beets	Sunflower	Hemp
1	11	12	13	14	15	16	17	18	19	20
Chernozem Forest-Steppe Belt										
11	71.0	3.20	2.8	0.55	...	0.2	0.02	...	0.4	...
12	72.1	1.10	2.6	0.50	...	0.4	...	...	...	...
13	67.6	3.4	8.0	0.6	...	...	...	0.1	1.4	0.9
14	68.0	6.6	3.4	0.6	...	...	...	0.3	4.6	...
15a	65.0	1.8	9.9	1.25	0.2	0.15	...	3.0	0.0	0.09
15b	62.1	9.6	5.4	0.5	4.3	...	...	...	4.7	0.63
16a	156.0	12.7	9.4	2.3	15.4	0.5	...	5.2	3.8	...
16b	48.0	6.3	16.1	1.7	13.7	...	...	3.2	1.2	...
17	62.4	10.3	2.7	4.2	16.7	...	...	0.2	7.9	2.3
Chernozem Steppe Belt										
18	88.7	2.1	0.77	0.4	...	0.2	...	...	1.9	...
19	74.0	74.0	0.9	0.7	6.5	0.2	...	...	3.1	...
20	68.0	88.3	1.0	1.8	15.0	0.3	...	...	6.9	...
21	62.5	7.8	2.2	2.9	20.0	...	...	...	5.8	0.9

TABLE 1--Continued

Region	Grain Crops	Commercial Crops	Potatoes	Vegetables	Corn	Flax	Mushrooms	Sugar Beets	Sunflower	Hemp
1	11	12	13	14	15	16	17	18	19	20
Arid Unirrigated Belt										
22a	62.5(80)	0.65	1.7	1.2	2.2	...	...	...	0.4	...
22b	66.7	4.45	2.5	4.1	10.2	...	...	...	...	...
Arid Irrigated Piedmont Belt										
23	69.8	5.0	1.3	0.9	4.8	...	...	0.01	0.31	...
24	60.7	35.9	3.5	2.3	5.5	...	...	0.5	...	0.7
25	70.1	4.7	5.5	2.6	12.5	...	...	0.1	...	...

TABLE 1—Continued

Nonchernozem Forest Belt

Region	Tobacco	Makhorka	Cotton	Milk (Centners)	Meat (Centners)	Wool (Centners)	Pork (Centners)	Grain Yield (Centners)	Percentage of Cows in Herd	Milk Yield Per Cow (Thousand Liters)
	21	22	23	24	25	26	27	28	29	30
1a	...	...	...	...	...	...	...	...	48	2.12
1b	...	...	...	...	...	...	...	5.2	49	0.87
2a	...	...	...	155	16.5	13.0	25.0	7.8	34	1.72
2b	...	...	...	62	8.0	11.0	17.0	7.5	40.5	1.19
3	...	...	...	114	17.0	38.1	13.0	9.2	42.5	1.48
4	...	...	...	141	13.5	12.0	26.5	9.0	44.5	1.69
5	...	...	...	244	25.5	33.0	19.0	11.0	77.0	1.99
6	...	...	...	160	16.0	39.0	10.0	7.0	46.8	1.71
7a	...	...	...	321	20.0	43.0	19.0	6.0	52.6	2.23
7b	0.04	...	...	270	26.7	22.3	48.0	6.1	55.0	2.14
8	...	0.02	...	295	48.0	34.6	87.0	9.9	50.1	1.73
9	...	...	...	383	26.5	39.8	47.0	8.9	51.5	2.29
10	...	...	...	249	15.0	14.2	23.5	6.25	51.3	1.73

TABLE 1--Continued

Region	Tobacco	Makhorka	Cotton	Milk (Centners)	Meat (Centners)	Wool (Centners)	Pork (Centners)	Grain Yields (Centners)	Percentage of Cows in Herd	Milk Yield Per Cow (Thousand Liters)
1	21	22	23	24	25	26	27	28	29	30
Chernozem Forest-Steppe Belt										
11	...	0.72	...	123.7	13.0	42.0	8.5	9.35	48.0	1.82
12	...	...	...	115.3	13.3	40.0	7.0	11.25	42.5	1.73
13	...	...	...	143.0	20.0	65.0	17.0	8.5	46.6	1.61
14	...	...	...	181.0	26.5	85.0	12.3	9.1	45.0	1.83
15a	...	...	...	246.0	29.0	69.0	20.0	6.5	47.5	2.43
15b	...	...	...	219.1	28.3	68.5	15.0	9.8	43.4	2.01
16a	...	...	...	354.0	53.0	47.5	39.0	15.8	45.1	1.92
16b	...	0.05	...	467.0	51.7	51.3	51.0	14.3	49.3	1.69
17	0.3	...	...	185.5	25.0	114.7	16.7	16.2	40.5	1.86
Chernozem Steppe Belt										
18	...	...	...	24.9	3.3	26.3	2.4	8.0	35.6	1.33
19	...	0.02	...	67.0	10.4	50.0	4.3	8.1	41.7	1.7
20	...	...	...	95.0	15.0	128.0	10.0	7.8	37.0	1.64
21	0.07	...	...	242.0	84.6	76.6	24.1	16.1	43.5	2.02

TABLE 1--Continued

Region	Tobacco	Makhorka	Cotton	Milk (Centners)	Meat (Centners)	Wool (Centners)	Pork (Centners)	Grain Yield (Centners)	Percentage of Cows In Herd	Milk Yield Per Cow (Thousand Liters)
1	21	22	23	24	25	26	27	28	29	30
Arid Unirrigated Belt										
22a	...	...	...	6.0	4.5	22.0	4.0	6.0	36.2	0.84
22b	...	...	...	43.0	8.3	169.0	13.7	7.6	34.0	1.19
Arid Irrigated Piedmont Belt										
23	...	...	2.9	14.8	3.4	69.0	3.6	8.3	34.9	1.24
24	0.05	...	30.0	29.0	6.0	94.5	5.7	6.9	34.8	0.76
25	1.0	...	6.6	123.0	19.0	180.5	13.2	9.7	32.7	0.89

TABLE 1--Continued

Region	Cows Per 100 Hectares Sown Area	Pigs Per 100 Hectares Sown Area	Sheep Per 100 Hectares Sown Area
1	31	32	33
Nonchernozem Forest Belt			
1a	119	399	...
1b	117	32	...
2a	30	56	25
2b	21	23	254
3	16	39	47
4	24	35	59
5	25	43	55
6	17	34	65
7a	25	56	62
7b	34	57	73
8	29	60	38
9	30	55	34 (46)
10	57	49	67
Chernozem Forest-Steppe Belt			
11	15	20	38
12	12	15	31
13	16	31	88
14	11	20	63
15a	15	51	67
15b	15	34	67
16a	26	61	32
16b	49	94	37
17	19/16	45/15	121/38

TABLE 1--Continued

Region	Cows Per 100 Hectares Sown Area	Pigs Per 100 Hectares Sown Area	Sheep Per 100 Hectares Sown Area
1	31	32	33

Chernozem Steppe Belt			
18	6.9	6.7	75.0
19	6.0	5.9	50.0
20	9.9	28.0	66.7
21	15.9	42.0	46.0

Arid Unirrigated Belt			
22a	9.5	6.0	53.0
22b	19.3	20.0	361.0

Arid Irrigated Piedmont Belt			
23	19.0	5.6	201.0
24	24.0	4.2	73.0
25	33.0	49.0	198.0

There are also indices of the volume of agricultural production per 100 hectares of agricultural land. A comparison of these indices with livestock indicates that the size of the herd and its structure is closely connected with the degree of intensity of the entire agricultural operation.

The comparative numerical data that characterize agricultural production according to the factors and level of its intensity makes it possible to determine the system (trend) of agriculture in each region. The significance and composition of the indices of factors for agricultural intensity need no special explanation. The indices of the system (trend) of agricultural production, however, demand an explanation.

First, the higher livestock density indicates a livestock-raising trend. In this case the system is called livestock-raising, which is confirmed not only by the ratio of cattle to sown area but also by the ratio of cattle to plowed land.

Second, the meaning of the term livestock-raising is supplemented by the term "large horned livestock raising" if other types of cattle such as pigs and sheep have a low density; otherwise the terms pig-raising or sheep-raising is added to cattle-raising.

Cattle-raising is characterized as a meat type if the percentage of cows in the herd is low (30–35 percent) or as a dairy type if cows constitute 50 percent or more. If meat predominates the trend is called meat-dairy; if milk predominates—dairy-meat. Meat livestock-raising is the most extensive whereas dairy, the most intensive.

Third, the intensive livestock-raising trend corresponds to the presence of dairy cattle and is accompanied by the development of swine-raising (in terms of density and yield of total meat and pork), whereas sheep-raising, other conditions being equal, is less developed. On the other hand, the extensive livestock-raising trend means a predominance of meat cattle and the development of sheep-raising.

Fourth, the degree of intensity of livestock-raising is an index of the degree of intensity of the entire agricultural operation, therefore, extensive or intensive livestock-raising is also accompanied by corresponding indices in farming. Thus, extensive livestock-raising is accompanied by a higher percentage of grains and a lower percentage of nongrains (potatoes, vegetables, commercial). Intensive livestock-raising, however, is accompanied by a lower percentage of grains and higher percentage of nongrains. The transition to intensive livestock-raising is accompanied by a transition in farming also.

Fifth, parallel with the degree of intensity of livestock-raising and the magnitude of the percentage of nongrains the name of the appropriate commercial crop is also included. Each of the five natural zones has its own set of these [commercial] crops. Thus, in the nonchernozem belt it is flax, potatoes, and, at times, vegetables; in the forest-steppe chernozem belt—sugar beets, potatoes, sunflower, corn and other cultivated crops, and tobacco; in the steppe belt—sunflower, corn for cereal, cultivated fodder; in the irrigated arid belt—cotton, tobacco, and others.

Let us now name the agricultural systems (trends) of each of the 25 agricultural regions and their subregions in the five natural zones.

NONCHERNOZEM ZONE

Region 1a (Magadan, Sakhalin, and Kamchatka Oblast): extensive livestock-raising, beef cattle, and intensive vegetable-potato system.

Region 1b (Yakutsk ASSR): extensive livestock-raising, beef-cattle-cereal crops.

Region 2a (Primor'ye Khabarovsk Kray, Amur Oblast): grain, soybean, potatoes-vegetables, meat, and dairy products.

Region 2b (Chita Oblast, Buryat and Tura ASSR): extensive livestock-raising, beef cattle-raising, sheep-raising, grains.

Region 3 (Irkutsk Oblast, Krasnoyarsk Kray): grains, meat-dairy cattle, average sheep-raising economy.

Region 4 (Tyumen', Tomsk Oblast): grain, newly developed flax cultivation, dairy-beef cattle, newly developed swine-raising.

Region 5 (Sverdlovsk Oblast): grain-potatoes, dairy-beef cattle, newly developed swine-raising.

Region 6 (Perm' and Kirov oblasts, Udmurt and Mari ASSRs, Kostroma Oblast, and the nonchernozem part of Gor'kiy Oblast): grain-flax, beef cattle, average sheep-raising economy.

Region 7a (Moscow, Ivanovo, Vladimir, and Yaroslavl' oblasts): dairy cattle, swine-raising, potatoes (in some oblasts vegetables).

Region 7b is divided into two parts. The first (Leningrad Oblast), coincides in direction with Region 7a; the second (Pskov, Novgorod, Kalinin, and Smolensk oblasts) has dairy cattle, swine-raising, and flax-raising.

Region 8 (Belorussian SSR, Bryansk, Volyn, Rovno, Zhitomir, Kiev, and Chernigov oblasts): grain, sugar beets, dairy-beef cattle, swine-raising, and lupine seeds. To assign this whole region to the nonchernozem belt is somewhat arbitrary since part of Kiev and Chernigov oblasts have chernozem soils.

Region 9 (Baltic republics and Kaliningrad Oblast): dairy cattle, swine-raising; newly developing sugar beet agriculture in Lithuania and Latvia, and flax-growing in Estonia.

Region 10 (Vologda and Arkhangel'sk oblasts, Karelian and Komi ASSRs) nonfarming: dairy cattle, swine-raising, and potatoes in Karelia; livestock-raising, dairy-beef cattle in Arkhangel'sk

Oblast and Komi ASSRs; beef-dairy cattle, flax-growing, newly developed swine-raising in Vologda Oblast.

FOREST-STEPPE (LESOSTEPPE) CHERNOZEM BELT

Region 11   (Kemerovo, Novosibirsk, and Omsk oblasts and Altay Kray): grain, extensive livestock-raising, beef-dairy cattle, newly developed swine-raising, some sheep-raising.

Region 12   (Kurgan and Chelyabinsk oblasts): grain, extensive livestock-raising, beef-dairy cattle, newly developed swine-raising, and some sheep-raising.

Region 13   (Tartar, Chuvash, Mordvin, and Bashkir ASSRs, and the chernozem part of Gor'kiy Oblast): grain, average intensity livestock-raising, beef-dairy cattle, newly developed swine-raising, sheep-raising, some potatoes; in the Chuvash ASSR, grain, dairy cattle, swine-raising, and potatoes.

Region 14   (Kuybyshev, Ul'yanov, and Penza oblasts): grain, beef-dairy cattle, some swine-raising, and vegetables.

Region 15a   (Tula and Ryazan' oblasts): grain, potatoes, dairy cattle, vegetables (Ryazan' Oblast).

Region 15b   (Orel, Kursk, Belgorod, Tambov, Lipetsk, and Voronezh oblasts): grain, potatoes, dairy-beef cattle, swine-raising, and sugar beets (Kursk and Lipetsk oblasts), sunflowers, corn (Tambov, Voronezh, and Belgorod oblasts), hemp, sugar beets (Orel Oblast); corn, sunflowers, and potatoes.

Region 16a   (the nine forest-steppe oblasts of the Ukraine and the Moldavian SSR): intensive livestock-raising, dairy-beef cattle, swine-raising with planting of commercial crops, potatoes (except in the Moldavian SSR, L'vov, Stanislav, and Transcarpathian oblasts), tobacco [Nicotiana rustica] (Poltava and Cherkassy oblasts), tobacco (Moldavian SSR, Khmel'nitskiy, Vinnitsa, Ternopol', and Chernovtsy oblasts), flax (L'vov Oblast), and sheep-raising (Moldavian SSR).

Region 16b   (Stanislav and Transcarpathian oblasts): the trend is the same as Region 16a.

Region 17a   (Krasnodar Kray): grain, corn, sunflowers, dairy-beef cattle, and sheep-raising.

Region 17b   (Kabardino-Balkar, North Ossetian, and Chechen-Ingush ASSRs): corn, sunflowers, hemp, beef cattle, sheep-raising.

STEPPE CHERNOZEM BELT

Region 18   (the nine northern oblasts of Kazakhstan): grain, extensive livestock-raising, beef cattle, sheep-raising.

Region 19 (Orenburg, Saratov, and Volgograd oblasts): grain, some sunflowers and corn, beef cattle, sheep-raising.

Region 20 (Rostov and former Kamensk oblasts): grain, sunflowers, corn, beef-dairy cattle, sheep-raising (wool), and some swine-raising.

Region 21 (the nine Ukrainian steppe oblasts): grain, corn, sunflowers, transitional fruit crop rotation, dairy-beef cattle, swine-raising (Crimean Oblast).

ARID NONIRRIGATED BELT

Region 22a (Karaganda and Gur'yev oblasts): extensive livestock-raising, beef cattle, sheep, and grain (Karaganda Oblast).

Region 22b (Astrakhan' Oblast, Kalmyk ASSR, eastern part of Stavropol' Oblast and the area north of the Dagestan Lowland): extensive livestock-raising, beef cattle, sheep, and grain.

ARID IRRIGATED AND PIEDMONT BELT

Region 23a (Kzyl-Orda, South Kazakhstan, Dzhambul, Alma-Ata, and Taldy-Kurgan oblasts): cotton, grain, extensive livestock-raising, beef cattle, sheep.

Region 23b (the oblasts of southern Kazakhstan [Dzhambul, Alma-Ata, and Taldy-Kurgan]): grain, sugar beet, extensive livestock-raising, beef cattle, sheep.

Region 23c (Kzyl-Orda Oblast): extensive livestock-raising, beef cattle, and sheep.

Region 24 (four Central Asian republics): cotton, grain, extensive livestock-raising, beef cattle, sheep (wool); in the Turkmenian SSR, Karakul sheep-raising.

Region 25 (Transcaucasian SSR and southern Dagestan): cotton (Azerbaydzhan and Armenian SSR), cotton with other commercial crops, fruit, grapes, subtropical crops, beef-dairy cattle, sheep (Georgia and Armenia), swine-raising, subordinated grain.

The significance of the changes that occur from east to west lies in the increase of national economic factors (reasons) determining the increase in level of intensity in agricultural production. Simultaneously and in accordance with this, the indices of organization and production effectiveness change.

There is a parallelism concerning the population density of a territory, the supply of agricultural labor and the indices of intensity of agricultural production that allows one to think of a direct connection of national economic factors with the manifestation of intensity (and effectiveness) in agriculture. The influence of the size of the agricultural market on the intensity of agricultural production of the region (accounting for the geographic position of the region) is illustrated in the following manner: with an increase in the size of the market there occurs an increase in the indices of production intensity compared to its neighbors not only in the west but also in the east, whereas with a considerable decrease in size of the market one similarly observes a decrease in the indices of intensity. These are the general manifestations of the regularity of agricultural production intensity noticed in comparing the agricultural regions from east to west in the three main natural agricultural zones: forest, forest-steppe, and steppe. In order to obtain a more complete picture we shall examine these zones.

In the nonchernozem belt, an increase in the influence of the intensity factors of agriculture is associated with an increase in the portion of sown land from Regions 1 and 2 to Regions 6 and 7a, and livestock-raising (number of cattle per 100 hectares plowed land and per 100 hectares of sown area) drops initially with some fluctuations up to Regions 6 and 7 and then begins to rise. Thus, an increase of farming and a still more rapid growth of livestock-raising is observed. In the general balance of production then, livestock-raising begins to replace farming (which contines to grow), which up to then increased more rapidly than livestock raising.

With a decrease in the proportion of plowed land to total agricultural land, farming becomes more skilled, according to the cited data. It is conducted with greater expenditures: beginning with Region 6, the significance of mechanization increases (from 65–82 hectares to 128–135 hectares per tractor).

On the plowed land the proportion of sown area increases; at first it is 74–84 percent and beginning with Region 7a, 82 percent and even 96 percent. This indicates that plowed land is used with almost no provision for clean fallow with compensation in the form of fertilizers. The use of other local fertilizers also increases.

Up to Region 6 grains occupy 60 percent or more of the crop area and then decrease to 55–44 percent and in Region 7b as low as 38 percent.

With the decrease of grains in Region 7a the area under labor-consuming commercial crops, potatoes and vegetables, increases. Together they comprise 17 percent of the area. In Region 7b this portion rises to 29.5 percent and in Region 8 to 23.4 percent. This growth in farming proceeds within a framework of increasing livestock significance as indicated above. In Region 9 the proportion of labor-consuming crops decreases (15.2 percent). However, the decreased proportion of farming in the two preceding regions is balanced by an increase in livestock-raising, its intensification here (in Region 7*a*) is represented by the high percentage of cows in a herd (more than 50 percent) and a high milk yield (more than 2,000 liters). The productivity, that is, milk yield, increases and the yield of other livestock products per 100 hectares of agricultural lands also increases.

In moving from east to west, from the extensive regimes to the more intensive, the less labor-consuming types of livestock-raising (sheep-raising) attain a special significance. This varies in different regions judging from the yield of wool per 100 hectares of agricultural land (frequently from 30–38 kilograms), but the increase does not occur from the moment of change from extensive grain-farming to livestock-raising (starting with Regions 7*a* and 7*b*). The wool yield is at a maximum in the most extensive regions (from Region 1*a* to Region 4). This is an indirect confirmation of the increasing intensity of agricultural production from the Far East to the Baltic.

Such are the organizational-economic stages of agricultural intensification with the increasing influence of its factors in the nonchernozem belt, that is, a higher degree of population density, greater labor force, and density of urban population.

In the forest-steppe chernozem belt the relationships between the leading sectors of agricultural production—farming and livestock-raising—is characterized by an increase of plowed agricultural land from 43.7 percent in Region 2*a* and, further westward reaching 80 percent in Region 15*b*. The smaller proportions in Regions 17 and 16*a* can be explained by hilly terrain (in Region 17—North Caucasus autonomous republic, and in Region 16*b*—Transcarpathia). Thus, judging from this index, the role of farming grows rapidly from east to west.

The presence of a higher percentage of unplowed land from Region 2*b* to the cis-Urals chernozem area does not give rise to a higher density of large-horned cattle and, only with respect to sown areas, can one note that in these regions and in the eastern chernozem area (14) the density of cattle and the resultant proportion of the large-horned cattle are higher. In the other regions the increase in cattle and the significance of livestock-raising continues at the expense of grazing on plowed land. In this case a characteristic of the chernozem forest-steppe belt is the similar proportion of unsown fallow land kept at a level of 18–14 percent. In the southernmost (North Caucasus and Ukraine) zone there is no similar practice. On the contrary, the sown area here occupies more than 100 percent of plowed land, which is explained by two crops in one year.

The transition to a noticeable change in farming intensity and livestock-raising begins with Region 15 and includes Regions 16*a* and 16*b* (Ukraine) where there is a considerably higher population density and, similarly, an increased number of workers per 100 hectares of plowed land. This is characteristic also for livestock-raising; a higher density of cattle per 100 hectares of plowed land.

With respect to farming, a decrease is noted in the percentage of grains from Region 15a (65 percent) to Region 16b (48 percent). At the same time the proportion of labor-consuming crops increases; commercial crops, potatoes and vegetables (beginning with Regions 15a and 15b to Region 16b) from 19 percent to 31 percent, and it must be noted that in the last two regions 15–14 percent is added in corn for cereal. With a somewhat lower population density and labor supply the North Caucasus (17a) region resembles the more intensive forest-steppe regions with regard to the number of labor-consuming crops as a result of its favorable climate and soil. This together with the great yielding capacity of the grains that provide fodder (especially corn), leads to an ever greater intensification of agriculture from east to west, that is, livestock-raising increases not only in the form of cattle per 100 hectares of plowed land, but also in the proportion of the more intensive branches of livestock-raising. This is directly reflected by the fact that beginning with Region 15a the yield of agricultural products is greatest: milk from 246 to 467 centners per 100 hectares of agricultural land, meat from 29 to 53 centners, including pork (from 20 to 51 centners per 100 hectares of agricultural area).

With regard to the other, more extensive regions (the eastern regions), beginning with Regions 13 and 14, they have higher indices in the percentage of plowed land and in livestock density per 100 hectares of plowed land, a higher percentage of grains (68 and 72 percent, whereas labor-consuming crops have an index of 12 and 10 percent. With a considerable plowed area and high cattle density in these regions, a high milk yield is obtained per 100 hectares of plowed land (from 143 to 181 centners). It is the same with meat production, which is higher than in the extreme eastern regions.

With regard to the degree of organizational improvement and agronomic measures it must be noted that the average grain yield in the eastern extensive regions is not lower, and in Region 12 even higher, than in the others: Regions 17, 16a and 16b have the highest yields—14 to 16 centners per hectare, which is due to the large proportion of sown corn.

With respect to livestock-raising, a higher percentage of cows per herd (49.3 percent) must be noted only in the Region 16b (Carpathian); in the other regions it fluctuates between 40–48 percent. The most obvious evidence of imperfect techniques of animal husbandry in all regions is the low milk yield, it is below 2 thousand liters per cow.

Comparing the increased indices of livestock-raising and farming an alternation of regions from east to west is not noticeable as in the case of the nonchernozem belt. However, with the increase in intensification of agriculture as a whole from east to west, an ever increasing livestock emphasis is noticeable beginning with the central chernozem oblasts (15a, 15b, 16a and 16b).

The North Caucasus (18) is the most extensive region in the chernozem-steppe belt with regard to farming structure and livestock-raising. It is a grain cultivation region with a low cattle density (4.6 head per 100 hectares of plowed land), milk yield (29.4 centners), and meat yield (3.3 centners per 100 hectares of agricultural land). Located in the center of immense developed areas of virgin lands, with considerable additional means of production (technology), the region has such an increased percentage of plowed area, chiefly in grains, that the proportion of livestock-raising at the present time is very low. With regard to farming, with its predominance of grains (88.7 percent), the region has an insignificant number of labor-consuming crops (3.0–3.5 percent).

In the region to the west and south, the Middle Volga (19), these indices increase; but milk yields are only 67 centners, and meat, 10.4 centners.

Region 20 (Rostov Oblast) with a greater labor force supply has, by all indications, higher indices, especially for commercial crops; in connection with this, the proportion of grain crops is only 68 percent and farming predominates. Milk yield is 95 centners and meat, 15 centners per 100 hectares of agricultural lands.

Only Region 21 (Ukrainian steppe) with a considerably larger number of workers per 100 hectares of plowed land (but less than in the forest-steppe Ukraine) and with the greater role of local and national agricultural markets and being mainly in grains (62.5 percent of the sown area including 20.6 percent in corn for cereal), has about 13 percent in commercial crops and as a result, higher grain yields (16.1 centners per hectare) and fodder yields. The plowed area of the region is very great (80.5 percent). The region has a developed livestock economy: milk production is 242 centners, meat, 84.6 centners per 100 hectares of agricultural land. This makes it a grain-corn region, but to a considerable extent also an intensive livestock-raising region (yield per cow is 2.02 thousand liters).

The preceding account of the factual state of agricultural production in the regions of the USSR encompasses the period up to 1959. Zonal conferences of agricultural workers have shown that, to some extent in all specific groups of oblasts, krays and republics, agricultural production has continued to rise as a result of measures taken by the Communist party and the Soviet government. These conferences have shown that the development of production in the collective and state farms was impeded by not fully taking into account the important role of such crop giants as corn for cereal (in the south) and for silage (in the south and north), as well as such fodder crops as sugar beets (for fodder) and broad beans; also such valuable legumes as peas, soybeans, lupine, and millet were disregarded. This limited the growth of fodder supplies and, correspondingly, the number of agricultural animals of all kinds and their productivity.

The change in the structure of sowings and, in connection with this, the sharp rise in the intensity and productivity of farming has become one of the decisive levers for a radical improvement of agricultural production. Herein lies the great merit of Khrushchev, who gave the impetus to the revolutionary change in the level of intensity of collective and state farm production.

Thus, based firmly on the great achievements of our socialist agricultural production, especially after the September 1953 plenary session of the CPSU Central Committee, further development of intensity and specialization continues on an economic-revolutionary basis.

V. S. MIKHEYEVA

Methods and Methodology in the Application of Optimal Programming To Problems of the Location of Agricultural Production and Regional Agricultural Specialization

CRITERIA OF OPTIMALITY AND THE POSING OF PROBLEMS

The application of mathematical methods in combination with the use of computers opens great possibilities in national-economic planning for the solution of one of the most complex economic-geographical problems—the problem of agricultural regionalization of the country. In order to find the most rational plan for the distribution of agricultural production and to determine the specialization of agricultural regions, criteria of economic effectiveness, an optimal plan, must first be established. The Program of the CPSU states: "Achievement of the greatest results at minimum cost in the interest of society—this is the immutable law of economic development."

Indices of labor productivity are the criteria of economic effectiveness of material production. The more consumer values created per unit cost of socially required labor, the more fully realized is the final production objective—satisfaction of the needs of society. One of the factors in the growth of labor productivity and development of material production is the social division of labor. Studying the

Translated from *Kolichestvennyye metody issledovaniya v ekonomicheskoy geografii* [Quantitative Methods of Research in Economic Geography] (Moscow, 1964), pp. 30–61.

economy of Russia by regions, Lenin explains the teaching of Marx on the significance of the division of labor in production in general and distinguishes its particular form—the territorial division of labor. "In direct relation to the division of labor in general, there is, as already mentioned, the *territorial* division of labor, the *specialization of individual regions* in the production of one product, sometimes one type of product, and even a part of a product."[1]

The successive development of various types of divisions of labor is accompaned by significant growth in the productivity of labor. Territorial division of labor is important for a number of reasons. It permits each type of production to be developed where there are the most advantageous conditions for its development, from the point of view of achievement of high labor productivity. There is no doubt that the criterion of optimality (economic effectiveness) in the distribution of agricultural production is the achievement of optimal (highest) labor productivity in agriculture in the country as a whole. This means that such a pattern of distribution must be determined in which either a unit of goods is produced at a minimum labor cost or a maximum quantity of goods is produced per unit of cost. On this basis, it is possible to suggest two problems concerning the distribution of agricultural production by regions of the country.

With given resources to find such an optimal pattern at which: (1) the gross volume of agricultural production by types, planned for the country, will be produced with minimum costs; (2) within the limits of the fixed sum of expenditures for the production of all agricultural production its maximum yield will be achieved in the necessary variety.

The choice of one or the other formulation of the problem in defining the optimal pattern for the distribution of agriculture depends on the actual economic situation and the character of planning, long-range or current.

Of course, the desire to obtain maximum agricultural production is not a goal in itself. Maximum for the sake of maximum is unnecessary in any sector of the national economy. As much production is demanded as is necessary for the complete satisfaction of the population in material goods, and the required time saved in production must be utilized for the satisfaction of the spiritual needs of society. Considering the continuous growth of material needs of society it must be borne in mind that, although these needs outrun the possibility of production, they stimulate its development, but at the same time correspond to it and are determined by it. We cannot desire that which we do not know or that which it is not possible to produce at the present time. Therefore, with regard to the present, the requisite needs for complete satisfaction of the material goods for society are known. This maximum of need may be fixed in indices of the long-range plan of the national economy. In formulating problems concerning the distribution of agriculture, it may find expression in the form of an assigned yield of agricultural production for the country as a whole.

Undoubtedly, there is also the desire to decrease the required time expenditure in the sphere of material production, since its further development depends on the growth of technological progress, which demands ever greater expenditures of time for the intellectual development of society.

Thus, *the first formulation of the problem* assumes that the planned volume of agricultural production fully satisfies the needs of the country with regard to internal consumption as well as for export, but demands, if possible, the achievement of greater economies in outlay. Such a formulation of the problem is necessary in drawing up a plan in general long-range terms (fifteen to twenty years), but is also feasible for the five- or seven-year plans.

In the work of planning agencies conducted on the basis of many years of experience, the size of the gross agricultural harvest is established by proceeding from a rigorous calculation of the possible size of the arable land area and crop yield for the given level of agrotechnology. Therefore, in applying the problem to the determination of the optimal pattern for the distribution of agriculture, there cannot be, in practice, a situation where there are inadequate land resources to produce the planned volume of production. Moreover, the calculations of the planners cannot disagree with the exact computer calculations, which, on the basis of mathematical methods, examine hundreds of thousands of variants of the desired plan until the most economical of them is found, the optimal plan. Therefore, with the initial data (yielding capacity, costs, dimensions of arable area, magnitude of gross production, and so forth) that are utilized in ordinary planning, the use of mathematical methods of optimal programming and computers allow an assigned volume of gross production to be obtained not only at considerable economy in expenditures but also at considerable economy of resources (arable land, labor resources, means of mechanization, fertilizers, and the like).

Since in this problem expenditures are brought to a minimum, it is natural that the poorest lands are not fully utilized, that is, those of low productivity that produce less per unit of expenditure than the lands of average and higher quality.

At the present time, the problem of providing the country with agricultural products is evidently such that not only the better and average lands must be utilized but all arable land in the country.

In connection with this, by utilizing the poorer lands some surplus of agricultural products can be obtained.

The proposed problem can be expanded by formulating it in the following manner: at minimum expenditures to provide for the planned volume of agricultural products by types and to produce additionally that product which is most needed by the government. Hereby the limitation of resources must be dropped, with the exception of the condition of maximum dimension of the possible arable area. Indeed, in the first formulation of the problem it was assumed that there are sufficient resources for the production of the planned volume of agricultural products. If the purpose is to utilize the entire arable land fund and produce an additional product, it is not known whether there are enough other resources for this. In connection with this it must be noted that in solving the problem of agricultural distribution in the first formulation it may prove expedient in general not to take account of resource limitations (except for arable land). The achieved optimal pattern will provide the answer to the question of how much and what resources must be in each region of the country in order to produce the planned agricultural volume at minimum cost.

The second formulation of the problem assumes that the country either experiences a sharp deficiency in agricultural products for internal consumption or tries to expand the export structure with agricultural products. Such a formulation is characteristic for current planning or planning for the near future.

With the goal of producing a maximum of agricultural products under given conditions the reservation must be made that this goal concerns the maximum production of a *required* variety. Otherwise a huge output of one or two items can be obtained that demand the least expenditures in resources for their production and have a low cost.

Further, there arises the question of the expediency in introducing into the formulation of the problem conditions for a possible sum of expenditures for production. Solving the problem of maximum agricultural output within the limits of the established sum for expenditures, an attempt is made to obtain it at a minimum production cost, that is, a high labor productivity in agriculture for the country as a whole is sought. If this condition is removed, then the question in essence becomes, what is the maximum volume of agricultural products attainable with a certain amount of resources of various types available? The criterion of optimality in such a formulation will be the maximum yield of production per unit of resources. Whether it will be identical with the criterion of optimal labor productivity depends on whether one can take into account all the conditions of agricultural production, of which the cost of the product is composed. In all probability this is impossible at the present time.

We shall dwell on the formulation of the problem of maximum production within the limits of the established sum of expenditures and assigned resources *with obligatory utilization of the entire land fund,* rather often proposed by the workers of planning organizations. The demand of *obligatory* utilization of *all* land for obtaining maximum production under conditions of limited means and resources is economically incorrect. The available limited volume of means and resources must be utilized for obtaining maximum production on better and average land. If, however, they (these means) are stretched also in utilizing poor land, then the total gross yield or agricultural production will be lower.

Concerning the achievement of maximum production with full utilization of the land, it is possible only if the expenditures for production are not taken into account and if unlimited resources are available. In such a case the criterion for optimality is the maximum yield of products per unit of land area, that is, the productive capacity of the land at a given level of development of productive forces in agriculture.

Any formulation of the problem that sets forth the initial positions and the final goal of its solution is the basis for developing an economic-mathematical model. The latter reflects the quantitative dependencies between the various factors and conditions of the problem to be solved in mathematical form. The more precisely and fully they are expressed in the economic-mathematical model, the higher will be the quality of the results of the solution.

Unfortunately, the most important problems of economic geography having a problematic character are problems of regionalization, general economic as well as special, distinguished by their many factors, complexity of quantitative interrelationships, and the large dimensionality of the initial data. Thus, for example, at the present time it is impossible economically and technologically to reflect, within the framework of some single economic-mathematical model, the whole array of complex interrelationships of the many factors of different plans (natural-geographical, socioeconomic, organizational-technological) that form the basis of the distribution and specialization of agricultural production, that is, to depict the complex process of region formation, which forms the territorial-production complexes of agricultural regions.

It is difficult to say whether all factors are known that affect the process of agricultural region formation. If one can disregard the unknown, however, and it does not affect the course of our thinking, it is much more difficult to avoid a consideration of those factors that are known only qualitatively but cannot be introduced in the problem since no means to express them in quantitative form have been found.

For both the economic geographers and mathematicians, questions of particular difficulty are connected (1) with the dynamics of the economic process under investigation (for example, reproductivity of resources of arable land in the development of new lands), and (2) with the nonlinear dependence that exists between the quantitative initial indices used in the problem and the factors that determine them. However, this is no basis for refusing to apply mathematical methods in solving complex economic problems, since there are other ways to approach the economic principles of the formulated problem within the framework of economic-mathematical models.

But even the mass of normative material that is available is so huge (tens of thousands of variables) that it is hardly possible to solve such a problem with the help of existing techniques. Undoubtedly, it is a fact that for the reflection of the complex dependencies existing among the factors under consideration there are as yet no sufficiently developed mathematical methods.

THE METHOD OF MODELING BY STAGES

The method of modeling by stages consists of combining methods of linear programming and methods of economic analysis. It has nothing in common with the mathematical method of block solution of problems of large dimensions that attempts to simplify the development of algorithms, compose programs, and reduce working time of the computers.

The method of modeling by stages must also not be identified with the variance methods calculation—the solution of a problem with various criteria of optimization and different initial information. Thus, in the problem concerning the distribution of agricultural crops various cost indices may be taken into account (cost of production on collective and state farms, the average for all categories of the economy), various limits on the size of sown areas calculated by a different method, and so forth. The problem may be solved for a maximum of production of agricultural products, for a minimum of expenditures, or for a minimum of labor expenditures. As a result, optimal variants of solution to the posed problems are obtained that correspond to the given conditions, the information utilized, and the limits of established mathematical accuracy.

The method of modeling by stages is an *economic* method of consecutively intermittent reflection of the economic process in economic-mathematical models of mathematical programming. Like any economic method it is a calculation designed to penetrate deeper into the essence of the economic process under study, attempting to utilize mathematical methods not only as a tool for the solution of problems but also as a means of obtaining additional economic information. Since the realization of the algorithm in the course of solving the problem imitates the development of the economic process, the consecutive discontinuity of the solution makes it possible to show at each stage the quantitative interrelationships between resources, factors and production.

For us a good algorithm is one that obtains the most additional information, sufficiently clear in an economic sense. In this connection, mathematical methods based on the solution of problems by means of approximating an optimal plan from conditionally optimal plans have a considerable advantage over methods based on permissable plans. The basic idea of this method is the selection of the most important factors for each stage of the mathematical solution of the problem. In this connection it must be said that *most important* in this case is the factor that to the highest degree overlaps the demands of the other factors, acting in the same direction on the results of the solution. Thus, despite the multiplan principles of agricultural distribution, they all assume a certain heterogeneity of agriculture. Instead of introducing the suitable conditions in the economic-mathematical model for each factor it is expedient to take into account the demands of crop rotation. The structure of scientifically specified crop rotations opens considerably the possibilities for complex development of agriculture as well as specialization in individual branches within reasonable limits. A characteristic feature of this method is the possibility of searching via models for further ways to reflect factors not considered at the phase of interstage economic analysis.

The results of the optimal solution obtained at each preceding stage are subjected to economic analysis, after which it can be established in which direction the further solution must proceed, what new factors must be accounted for in the problem, and what information must be utilized in the following stage. A scheme for the stage method of solution to the problem of regional distribution of agriculture is presented below with the following formulation: *to find a pattern for the distribution of agricultural production that will satisfy the national requirements for plant products by types of crops (including feed crops), at minimum cost and that will give maximum yield, and livestock production per unit of feed resources.* The solution of this problem is conducted in four stages.

The first stage is realized within the limits of the economic-mathematical model for best utilization of productive capacities.[2] In conformance with its conditions the following factors are taken into account:

1. national requirements for plant products by type;
2. size of arable area under cultivation possible for each region;
3. yield capacity of each crop in each region;
4. expenditures for the production of a unit of each type of plant product in each region.

The optimal pattern obtained at the first stage of solution of the problem is called the monoculture pattern. It points out which one-two crop sequence each region should specialize in to be profitable for the country as a whole.

The system of optimal evaluations obtained in the solution of the problem corresponds to the optimal pattern of the first stage. On the basis of this system of evaluation one can see (1) the tendency of the most profitable sequence for all regions in the production of each crop and (2) the tendency of the most profitable crop sequence in each region. In connection with this, the evaluation can be utilized for developing limits for the size of the sown area of each crop in each region, which will be introduced in the solution of the problem at the second stage.

The second stage is realized within the limits of the economic-mathematical model of distribution of agricultural crops.[3] *The designations of this model are*: the sought (unknown) values—X_{ik} is the size size of the sown area with kth crop in the ith region; assigned values—

n	number of regions
i	the number of the region
m	number of agricultural crops
k	the number of the crop
l	number of groups of crops
j	the number of the group
S_i	amount of arable land under cultivation in the ith region
λ	yielding capacity of the kth crop in the ith region
P_k	the magnitude of gross production of kth crop in the country as a whole
C_{ik}	cost of production of the kth crop in the ith region
ϵ_j	the last number of crop in the jth group (crops entering into groups have consecutive numbers)
$\bar{B}$	limit to the size of the sown area under jth group of crops in the ith region
d_{ik}	limit to the size of the sown area of kth crop in the ith region.

Conditions of the model

$X_{ik} \geq 0$ —The size of sown area cannot be negative   (1)

$$\sum_{k=1}^{m} X_{ik} \leq S_i$$

—The area in the ith region used for sowing must not exceed the amount of arable land. (2)

$$\sum_{i=1}^{n} \lambda_{ik} X_{ik} = P_k$$

—The gross harvest of kth crop in all regions must be equal to the assigned volume of its production in the country as a whole. (3)

$$B_{ij} \leq \sum_{k=\epsilon j-1}^{\epsilon t} +1 X_{ik} \leq \bar{B}_{ij}$$

—On the sown area in the ith region limitations may be imposed of the type. (4)

$$d_{ik} \leq X_{ik} \leq \bar{d}_{ik}$$

—On the sown area in ith region under separate crops limitations may be imposed of the type. (5)

It is necessary to minimize the linear function

$$\sum_{i=1}^{n} \sum_{k=1}^{m} C_{ik} \lambda_{ik} X_{ik},$$

the total expenditures for the production of all agricultural products in all regions.

Conditions 4 and 5 make it possible to take into account new factors: (a) the minimum required volume of production of difficult to transport products (potatoes, vegetables, milk, commercial factory beets) in each region; (b) the maximum possible size of sown area for groups of crops that are similar agrobiologically, taking into account the system of farming practiced in each region; (c) the maximum size of sown areas in each region under individual crops, corresponding to the geographic area they occupy in the region; (d) production capacity of factories processing agricultural raw material, from the point of view of loading the equipment with a specific type of product.

The model does not take into account the regional labor resources to satisfy agricultural production. This demand may be introduced into the model in the form of an inequality of the type:

$$\sum_k p_{ik} \lambda_{ik} X_{ik} \leq F_i,$$

where λ_{ik} is the expenditures of direct labor in man-days for the production of one unit of production of the kth crop in the ith region, F_i, the quantity of labor resources at hand (fund of working time in man-days) in the ith region occupied in crop cultivation. Such a condition (regional labor resources) is not introduced in the model because, a priori, one cannot say in what manner the farm labor must be distributed in each region and sector without having solved the problem of the distribution of agricultural production.

This model does not take into account the regional provision of agricultural production with mechanized means and fertilizers. Their distribution is realized by the planning organs in a centralized manner and must be conducted on the basis of a precalculated optimal pattern of the distribution of agriculture that establishes the needs of the regions in terms of technology and fertilizers.

Thus, as the result of the solution of the second stage we obtain an optimal pattern that can be called the *optimal pattern of the distribution of agricultural crops*. In contrast to the monocultural variant it yields practically suitable quantitative indicators on the size and the structure of the sown areas in the regions.

Correspondingly, with the volume and the structure of the feed base, which are determined on this basis, one can resolve the question concerning the distribution of livestock branches in the regions of the country. Conditions 4 and 5 of the model make it possible to project a feed base of a size sufficient for the number of cattle which it is expedient to have in each region in order to produce the minimum required volume of perishable products—milk and some milk products.

The third stage is for the purpose of solving the problem of development of livestock-raising in each region of the country and is formulated in the following manner: to determine the optimal structure of livestock branches in each region of the country at which a maximum yield of livestock production will be achieved with given feed resources. For this, the results obtained at the second stage concerning the volume of the feed base and the structure of feed resources in each region are utilized, taking into consideration the size and productivity of existing natural hay fields and pastures. Hereby the system of optimal estimates obtained from the solution of the second stage of the problem is utilized. On the basis of estimates of production for each region, the type of feed is determined for each group of cattle.

The solution is realized within the limits of the economic-mathematical model on selections and mixtures, the modifications of which have been developed and have received sufficient explanation in the Soviet and foreign literature.[4] The result of the solution of the third stage is quantitative data on the optimal structure of livestock branches and on the maximum yield of livestock production in each region.

After conducting stages 2 and 3 of the solution, the results of the problem can be called the *optimal pattern of the distribution of agriculture production*.

The economic analysis of the results of stages 2 and 3 of the problem basically amounts to the solution of two problems: (a) to what extent is the proposed pattern of the distribution of agricultural crops and livestock branches guaranteed by labor resources existing in the regions; (b) what quantity of agricultural products is produced in excess in each region and which items were deficient in each region.

In order to answer the first question one must calculate the proposed distribution pattern of agricultural production by using normative labor costs in man-days per unit for each type of production, and compare this with the present amount of labor resources of the region.

To answer the second question a comparison must be made between the need of the regions for each type of agricultural product and the quantity (of the product) according to the optimal pattern of the distribution of agricultural production. The economic analysis amounts to a determination of excesses and deficiences of the regions by types of agricultural production and to a determination of freight flow lines, that is, to the establishment of a plan of interregional transfer of agricultural products.

The fourth stage, the corrective stage, has as its purpose the introduction into the optimal pattern of the distribution of agricultural production the necessary corrections related to the insufficiency of labor resources in some regions (if this factor was not taken into account at stages 2 and 3) and with the insufficiency of individual types of agricultural production in some regions.

The essence of these corrections amounts to comparisons of (1) expenditures for attracting a labor force to the deficient region with the expenditures that represent the difference in reducing the production of labor-consuming products in that region and increasing their production in other regions, and (2) transport outlays for shipment of agricultural products to the deficient region with the difference in the expenditures arising as a result of expansion of the production of the deficient product in that region and lowering its production in other regions. Further necessary corrections are introduced in the optimal pattern of distribution of agricultural production.

The comparison of expenditures for attracting a working force and for transport outlays for shipment of agricultural products to deficient regions is realized with the use of methods of mathematical programming for individual links of the economic analysis with a wide use of a system of optimal estimates obtained at the preceding stages of the problem.[5]

We think it is impossible to take into account the transport factor in any other way when solving the problem of the distribution of agricultural production for the following reason. Transport expenditures for shipment of freight depend on the volume of shipment and length of haul. At the beginning, when providing the model with initial indices these values are unknown. There are also reasons of a practical and technical nature. In order to consider directly the influence of the transport factor in this problem it is necessary to have an immense amount of information concerning transport costs for shipment of each type of agricultural product in all possible directions.

In a solution of the problem by stages there is the preliminary determination of *what* freight, in *what* volume, and *from where to where* they will be transported. Experience in experimental calculations shows that a considerable portion of the agricultural products is not at all amenable to shipment.

All these factors sharply restrict the volume of initial information utilized in the solution of the problem, and this is of great practical significance since the collection of initial material is exceedingly difficult and the quality of normative data is poor.

The foregoing clearly indicates the great possibilities the method of modeling by stages holds for the solution of economic problems that are large and have complex interrelationships. The positive significance (of such modeling) consists of the following:

1. At each stage of the solution of the problem one can obtain additional economic information concerning the economic process under study.

2. The results of the preceding solutions and the system of optimal estimates accompanying them make it possible to formulate the problem more clearly and to select accurate information for the following stages.

3. It allows one to proceed toward the solution of the problem with a minimum volume of initial data.

4. It makes it possible to take into account factors for which there are no initial indices since their quantitative value depends on the end result of the problem.

Inasmuch as the method of modeling by stages consists in combining optimal methods of mathematical solution with methods of economic analysis, the final solution of the problem is only an approximation toward an optimal plan that is considerably better than a plan developed without the use of mathematical methods.

Despite the imperfect method of modeling by stages, its application to the solution of large and complex economic problems is of considerably greater economic effect than purely mathematical methods of optimal programming for the solution of particular technical production problems. A small economy achieved in solving a large problem surpasses by many times in absolute terms the larger relative economies in solving a partial economic problem. It must be noted that practical planners in developing a plan for the solution of minor economic problems with ordinary methods commit significantly fewer errors in relation to the possible optimum than when working out large economic problems.

The method of modeling by stages, developed two years ago for the problem of distribution of agricultural production, has found wide application in solving many problems of sectoral planning of the distribution and specialization of individual branches conducted by the laboratory of economic-mathematical research at the Siberian section of the Academy of Sciences, where it has been further modified.

The role that is played by the system of optimal estimates in the process of modeling by stages in the solution of a problem was discussed above. It is expedient to examine in more detail their (optimal estimates) economic content that makes it possible to point out the possibility of utilizing optimal estimates when working out other economic questions.

THE ECONOMIC MEANING OF OBJECTIVELY STIPULATED ESTIMATES OF THE OPTIMAL PATTERN OF THE DISTRIBUTION OF AGRICULTURAL CROPS

The duality theorems of problems of optimal programming confirm that, in the solution of any problem of optimal planning, for each optimal pattern of resource utilization there is a corresponding system of objectively stipulated estimates of factors, resources, and production.[6] The form of the derived optimal pattern depends on the actual economic situation taken into account in the construction of the economic-mathematical model, within the limits of which the solution of the problem was conducted.

Each of the contributing factors and resources affects the formation of the optimal end result differently, which finds a concentrated expression in the value of the special purpose function. There exist definite quantitative dependencies between the factors and the *special purpose* functions, which at the stage of completion of the optimal solution received the name of objectively stipulated estimates (factors, resources, production). The estimates characterize the measure of influence of the individual factors on the magnitude of the *special purpose* function. Thus, the estimates of land in the regions corresponding to the optimal pattern of the distribution of agricultural crops indicate how much the expenditures decrease (or increase) for the production of agricultural products if the sown area of the regions is increased (or decreased) by one unit.

The use of mathematical methods in a planning-economics calculation makes it possible to obtain definite information concerning the economic process under study in the form of quantitative indices, not only in the final solution of the economic problem but also at every step leading to it.

For the economist these indices are new and their economic meaning is not always clear, since, for the most part, they express quantitative regularities in the economic *process*, and not its results. Some economists attempt to define these unusual indices using the familiar terminology of political economy, but in most cases this is unsuccessful. The concern, of course, is not with definitions. In time, with our deeper knowledge of the new field of economic science, planometrics, they will be found. At the present time it is more important to determine precisely their (indices) economic content and to find ways to utilize them in planning.

For example, let us examine the land and production estimates that accompany the optimal pattern of the distribution of agricultural crops obtained as a result of solving the problem within the limits of the economic-mathematical model of the best utilization of productive capacities with the help of the delta method of conditionally optimal plans.

In drawing up the initial conditionally optimal plan the requirements of the country in agricultural production are fully satisfied. The production of each crop was carried out on the best lands, that is, in regions where the expenditures for its production were minimal. The best lands (precisely lands, not soils) were considered to be not those having the best naturally fertile soils but those that, due to the generalized economic-geographic conditions, are characterized by the least expenditures for the production of a unit of product.

Of course, there was not enough of this type of land, and this limitation was discovered. As the problem was being solved there occurred a successive shift in sown crops from the best land to somewhat poorer land, and in connection with this the total expenditures for the production of agricultural products in the country as a whole increased in comparison with the expenditures that characterized the initial conditionally optimal plan of crop distribution. Thus, the calculations at each step indicated by quantitative index estimates of the land in the regions illustrated how the shortage of the highest-quality land influenced the rise in cost of production of the total agricultural production in the country.

In the optimal pattern of distribution of agricultural crops, the regions having the best and average lands proved to be completely under cultivation. In regions with poorer lands the arable land remained either completely unused or was only partially under cultivation. From the point of view of familiar economic concepts these poorer lands vary, that is, they should be divided into the categories of worst (not utilized at all), and poorer lands (not fully utilized). The estimates of the land, however, do not reflect this, since they quantitatively characterize only the shortage of land. From this point of view it makes no difference whether all the land is uncultivated or is only partially used.

It is of interest to examine the attempt to apply the term differential rent to the land estimates. In their economic meaning they are quite similar. Let us examine how a land estimate is formed in the course of calculations in solving a problem. Let us assume that in region 1 the cost of grain is 3 rubles per centner, yield is 20 centners per hectare, and production cost of these 20 centners correspondingly equals 60 rubles. Because there is not enough land in region 1, this product must be produced in region 2, where the cost of grain is 5 rubles per centner, that is, expenditures for its production is 100 rubles. The difference in expenditures for the production of the same amount of grain (20 centners) in these two regions equals 40 rubles. These are estimates for one hectare of land in region 1, which in a sense represents the amount of differential rent from a unit of sown area.

Let us examine the formation of differential rent 1, using an example from the textbook *Politicheskaya ekonomiya* [Political Economy].[7] The differential rent region 2 is obtained from the difference in individual production price for one centner in these two regions, multiplied by the volume of production of the product in region 2. The differential region 1 has a zero value (see table 1).

Let us assume that the sown area of region 2 must be decreased by 1 hectare and a product equivalent to 6 centners produced in region 1: since in region 1 the capital expenditures per centner of product equal 25 rubles (200:8), then, for the production of an additional 6 centners 150 more rubles are needed. At 20 percent normal profit, the average profit for the expenditures of the total capital (350 rubles) is 70 rubles.

We composed table 2 in the manner in which we calculated the land estimate: (1) products grown in region 1 were assigned the yielding capacity of region 2; (2) we considered that expenditures per centner were not dependent on the volume of production.

Other conditions being equal, we obtained the same differential rent in both cases: the poorer land of region 1 has a differential rent equal to zero; the better land of region 2 has a differential rent equal to 60. In reality this is not so, because a change in the volume of production leads either to a decrease or to an increase in expenditures for each centner and, consequently, to a change in the size of differential rent.

If only this were the essence of the matter one could hope that in solving the problems, removing the nonlinear dependencies of expenditures from the volume of production, land estimations would be obtained that would then be the differential rent. Unfortunately, the land estimates and differential rent, similar in an economic sense, have a different economic content.

Differential rent is created in continuous connection with the conditions of realization of that additional income that it represents. It depends on the social expenditures for production (or social cost) of agricultural products stipulated by the level of prices that must be calculated on the basis of full national economic expenditures.

Land estimates as well as differential rent, arise as a result of land resource limitations; they are formed in its image and likeness, but, because within the limits of the model the whole economic situation cannot be reflected, their economic content is determined by the actual situation in a particular problem.

One can only say that land estimates have the nature of rent. Their magnitude is determined by that economy in expenditures that is obtained from utilizing the best and average lands in the optimal pattern of the distribution of agricultural crops. It must be taken into account in finding the production estimates.

Before examining the question of possible practical applications of optimal estimates their concrete character must once more be emphasized. In showing the extent of influence of individual factors on the magnitude of the special purpose function, the estimates possess only a relative stability directly following from the economic situation that was taken into account in the problem.

Thus, for example, the land estimates in a certain region, amounting to 2,000 rubles per 100 hectares of sown area, shows by how much the total expenditures for the production of an agricultural product decrease if the sown area in this region is increased by 100 hectares. Using this estimate one cannot calculate the saving in expenditures that the country will obtain by increasing the sown area of this region by 5,000 to 10,000 hectares. It is possible that the significance of land estimate is real only to a certain magnitude of expansion of sown area (1–3 thousand hectares), but further influence of the land factor on the special purpose function will have a different character.

The estimates are reliable only within a specific confidence interval. It must be stated that the actual magnitude of any of the economic indices (labor productivity, economic yield, differential rent) is also not constant but varies depending on the degree of change in its determining conditions. Estimates can serve as an important aid in setting up an economic cadaster of lands in the USSR. Since in our country land is not an object of buying or selling but only an object for application of labor in agricultural production, the development of an economic cadaster of lands attempts to obtain criteria for the economically effective (most rational) distribution of agriculture.

TABLE 1

REGION	SIZE OF SOWN AREA (In Hectares)	CAPITAL OUTLAY (In Dollars)	AVERAGE PROFIT (In Dollars)	PRODUCED (In Centners)	INDIVIDUAL COST OF PRODUCTION		TOTAL COST OF PRODUCTION		DIFFERENTIAL RENT 1 (In Dollars)	
					Of All Production (In Dollars)	Of One Centner (In Dollars)	Of One Centner (In Dollars)	Of All Production (In Dollars)	Per 2 Hectares	Per 1 Hectare
1	2 Hectares	200	40	8	240	30	30	240	0	0
2	2 Hectares	200	40	12	240	20	30	360	120	60

TABLE 2

REGION	SIZE OF SOWN AREA (In Hectares)	CAPITAL OUTLAY (In Dollars)	AVERAGE PROFIT (In Dollars)	PRODUCED (In Centners)	INDIVIDUAL COST OF PRODUCTION		TOTAL COST OF PRODUCTION		DIFFERENTIAL REND 1 PER HECTARE (In Dollars)
					Of All Production (In Dollars)	Of One Centner (In Dollars)	Of One Centner (In Dollars)	Of All Production (In Dollars)	
1	3.5 Hectares	350	70	14	420	30	30	420	0
2	1 Hectare	100	20	6	120	20	30	180	60

Land estimates most fully satisfy this purpose. They are the indices only of the optimal plan. Knowing the land estimates it is possible to determine immediately the optimal plan; knowing the optimal plan, one can find the system of estimates corresponding to it from the formulas:

$$q_{ik} = \frac{Q_i}{\lambda_{ik}} + C_{ik}; \tag{1}$$

$$Q_i = \lambda_{ik}(q_{ik} - C_{ik}) \tag{2}$$

where q_{ik} is the estimate of production of kth crop in the ith region; Q_i, the estimate of the land in the ith region; λ_{ik}, yield capacity of kth crop in the ith region; C_{ik}, expenditures for the production of a unit of produce of kth crop in the ith region.

Since each hectare of land in the ith region has an estimate equal to Q_i, the estimate of each centner of production of any (k) crop in this region is composed of the expenditures for the production of a centner of the product (C_{ik}) plus that part of the estimate of one hectare of land (rent estimate), which produces each centner

$$Q_i / \lambda_{ik}.$$

For the examination of the interdependence between the land estimates and production estimates we turn to tables 3, 4, and 5. An analysis of these tables shows that:

1. The region with the poorest land (3) has a zero land estimate, its area is deficient and remains as excess.

2. Estimates of production in the region with the poorest land (3) are equal to individual expenditures for the production of each centner (see tables 3 and 5). They have no rent addition per centner.

3. The estimate of production of each crop in the regions where its cultivation is provided for is determined by the conditions of production on the poorest land.

Thus, expenditures for the production of a centner of crop B in region 1 equal 4, but the estimate of production will be equal to the estimate of production in region 3, and, as a result, region 1 has a rent per hectare equal to (5-4) 50=50.

The production of crop A in region 1 receives an estimate equal to $50/15 = 3.3$; but the production of the same crop in region 2 will be the same in spite of the fact that the individual expenditures for it are less (8 rubles per centner) than in region 1 (10 rubles per centner). Hence region 2 has a rent per hectare equal to (13.3-8) 20=106.

TABLE 3

INITIAL DATA FOR THE SOLUTION OF THE PROBLEM
OF THE DISTRIBUTION OF AGRICULTURAL CROPS

DIMENSION OF SOWN AREA BY REGION		A	B	C
		3,000 Centners	4,000 Centners	5,000 Centners
Region	Dimension	Yield in Centners/Hectare in the Upper Corner Expenditures in Rubles/Centner in the Lower Corner		
1	150 Hectares	15 10	8 35	50 4
2	100 Hectares	20 8	10 30	40 5
3	100 Hectares	12 14	6 45	40 5

TABLE 4

OPTIMAL PATTERN OF THE DISTRIBUTION OF AGRICULTURAL CROPS

DIMENSION OF SOWN AREA BY REGION		A	B	C	
		3,000 Centners	4,000 Centners	5,000 Centners	EXCESS (+) OR DEFICIT OF AREA
Region	Dimension	Sown Area in Hectares			
1	150	120		30	0
2	100	60	40		0
3	100			88	+12

TABLE 5

ESTIMATES OF LAND AND ESTIMATES OF PRODUCTION
ACCORDING TO THE OPTIMAL PATTERN

		A	B	C
Region	Estimate of Land (Rubles per hectare)	Estimated Value of Production In Rubles Per Centner		
1	50	13.3	41	5
2	106	13.3	40.6	7.5
3	0	14	45	5

TABLE 6

		A	B	C
Region	Estimate of Land (Rubles per hectare)	Rent for a Centner in the Upper Corner, Rent for Hectare in the Lower Corner (In Rubles/Centner)		
1	50	3.3 50	5.6 45	1 50
2	106	5.3 106	10.6 106	0 0
3	0	-0.7 -91	-4.4 -26	0 0

4. Only crops that must be produced in regions conforming to the optimal plan give a rent from a hectare of sown area equal to the estimate of its land. The highest rent from the sown area in a region can be obtained only if the crops are distributed according to the optimal plan.

5. The estimates of production of each crop in those regions where cultivation is provided for by the optimal plan are equal and minimal, compared to the estimates of the production of this crop in those regions where its cultivation is not provided for.

The correlation of these estimates shows what the increase in total expenditures will be, if a given crop is sown on land where it must not be cultivated. Thus if crop A is cultivated in region 3, then each centner of the product produced there will cause an increase in expenditures equal to 0.7 (14-13.3=0.7) (see table 6).

Since the prices for agricultural production must stimulate an economically effective distribution of agriculture, the optimal system of estimates of production can serve as assistance to price formation.

The optimal plan of the distribution of agriculture is dual: it is characterized by a minimum of expenditures for the production of the product and a maximum of their rental estimates. This confirms the validity of the thesis that the most rational plan of the distribution of agricultural production is an economic cadaster of lands in the USSR.

1. V. I. Lenin, "The Development of Capitalism in Russia," in *Sobraniya sochineniya* [Collected Works], 5th ed. (Moscow, 1958), 3:431.

2. V. S. Mikheyeva, "A Mathematical-Economic Model of the Distribution of Agricultural Production Among the Regions of the Country," *Vestnik Moskovskogo Universiteta, seriya geografiya* [Herald of Moscow University, Geography series], no. 5 (1962).

3. A. G. Aganbegyan and V. S. Mikheyeva, "On the Determination of an Economically Efficient Distribution of Agricultural Production Among the Natural-Economic Regions of the Country," in *Optimal'noye planirovaniye* [Optimal Planning], Mathematical-economic series, no. 3 (Novosibirsk, 1962).

4. Ye. M. Chetyrkin, "Normative Models in the Economics of Animal Husbandry," in *Primeneniye matematiki v ekonomicheskikh issledovaniyakh* [The Use of Mathematics in Economic Research], ed. V. S. Nemchinov, vol. 2 (Moscow, 1961).

5. Ibid.

6. L. V. Kantorovich, *Ekonomicheskiy raschet nailuchshego ispol'zovaniya resursov* [Economic Calculation of the Best Utilization of Resources] (Moscow, 1960).

7. Moscow, 1954.

L. B. SHEYNIN

The Zones of Purchase Prices for Agricultural Products and Two Principles for the Delimitation of Such Zones

The specialization of agriculture depends on a whole host of factors, among which natural conditions and price level are of decisive significance. Natural conditions are reflected in production costs, that is, the cost per unit of output. Depending on the price levels, however, with the same production costs an economy can be profitable or unprofitable. Thus, a planned differentiation of prices can have an influence on the specialization and intensiveness of agriculture just as significant as the differences in natural conditions.

In practice, the price of a product extends throughout a given territory. A territory with the same price for a similar product forms a zone. In some cases the price for all producing regions is fixed at one level, and then all producing regions would form one zone. In regions where a commodity is not produced (or if produced, not purchased by the government), a price is not fixed at all.

What then are the principles that unite certain localities into one zone and that differentiate prices by zones? It is easier to explain this with a specific example. This is how the oblasts and rayons of the Ukraine were divided into purchase price zones (for collective farms) in the last three years (the zones are given in ascending order of purchase price).[1]

Translated from *Voprosy geografii* [Problems of Geography], no. 54 (1961), pp. 143–49.

Grain Crops

1. Oblasts: Stalino [now Donetsk], Zaporozh'ye, Dnepropetrovsk, Poltava, Odessa, Kherson, Kirovograd, Nikolayev, Crimea, part of Lugansk, Cherkassy, Vinnitsa, Khmel'nitsky, Kiev, and Khar'kov.

2. Part of the rayons of Lugansk and Khar'kov oblasts (adjacent to Voronezh and Kursk oblasts of the RSFSR).

3. Ternopol' Oblast; and also part of the rayons of Chernovtsy, Chernigov, Sumy, Zhitomir, Volyn', L'vov, Rovno, Kiev, the plains region of Stanislav, and the northern region of Khmel'nitsky, Vinnitsa, and Cherkassy oblasts.

4. Drogobych and Transcarpathian oblasts, the mountainous and foothill regions of Stanislav, Chernovtsy, northern rayons of Sumy Oblast, the wooded regions of Chernigov, Zhitomir, L'vov, Kiev, Volyn', and Rovno oblasts.

Sugar Beets

1. Olbasts: Kiev, Cherkassy, Vinnitsa, Khmel'nitsky, Zhitomir, Chernigov, Sumy, Khar'kov, Poltava, Dnepropetrovsk, Lugansk, Kirovograd, Odessa, Nikolayev.

2. Oblasts: Volyn', Rovno, Ternopol', L'vov, Chernovtsy.

3. Oblasts: Drogobych, Stanislav.

Vegetables (Except Tomatoes)

1. All Oblasts except those below;
2. Oblasts: Stalino [Donetsk], Lugansk.

Tomatoes

1. Oblasts: Dnepropetrovsk, Zaporozh'ye, Kherson, Nikolayev, Odessa, Crimean.

2. All others.

Cattle and Sheep

1. All oblasts except those below.
2. All oblasts of the Western Ukraine (except Chernovtsy).

Milk and Cream

1. All oblasts and regions except those below.

2. Regions of whole-milk delivery, zones of cheese and canned milk plants (according to a special list).

Calves

1. All regions except those listed below.
2. Regions of dairy cattle raising (according to a special list).

The prices of such products as potatoes, sunflowers, buckwheat, corn, grass seeds, volume feeds (hay, straw), pork, eggs, poultry, butter, and some other products are not differentiated and are uniform in the entire republic.

Thus, considering the zones of grain crops, each of them unites territories with approximately the same natural conditions, where production costs per unit of product are relatively similar. In zones with better natural conditions, where the cost price per unit of production is relatively small, the price is lower; in zones with a higher level of production costs, the price is higher. In a similar manner zones are distinguished for sugar beets, tomatoes, cattle, and sheep. Each of these types of product territories are united in one zone, which has similar production conditions. This unification of territories is called unification by production principle, and the zones themselves can be called production zones.

The formation of vegetable zones is based on a different principle. In oblasts with a very dense population engaged in industry (Donets Basin) the prices are higher than in the rest of the Ukraine. Milk zones are also established so that near large cities and industrial centers the prices for milk would be relatively higher than in more distant regions. It is obvious that in such a case the zones are distinguished on the principle of transport proximity to the marketing points. Such zones in contrast to those formed on the production principle, can be called *transport zones*. The transport zones together with the center attracting their produce form a closed complex.

The differentiation of prices by production zones has the following positive features:

1. The government can guarantee the profitability of production in a given region by adjusting costs and preventing excessive earnings or losses. Uniform remuneration to producers is thus achieved, since in more fertile oblasts (where less labor per unit of output is required) the price is lower, and in less fertile oblasts (where more labor is expended per unit of production) the price is higher.

2. By raising prices in some zones the government can regulate the intensity of production and the yield per unit of area. This can prove profitable if, for example, raised prices lead to such an increase in the amount of goods that it can prevent its production in other zones where, because of local conditions, production is too expensive. (It is true, however, that the uniformity of remuneration to producers disappears, since in zones with raised prices the profits will be above normal: in other words, there will no longer be the convenience foreseen in paragraph 1 above.) Similarly, the government can lower prices in individual zones where production is too expensive if this product is being produced in other zones where production is cheaper.

3. The government can selectively compensate for the increased costs per unit of production when such cost increases are unavoidable (for example, unfavorable weather conditions that caused lower yields), and also selectively withdraw excess profits if they were accumulated without human participation (for example, increased yields under favorable weather conditions).

Price-zone formation based on a production principle, however, has serious faults also. First, the uniformity of compensation to producers is possible only in an ideal case, since natural conditions vary not only from oblast to oblast and from region to region but also from economy to economy. Detailed soil investigations show clearly that large zones (as those mentioned above) include land of the most diverse fertility, which frequently requires dissimilar expenditures for cultivation. A single price for such a zone means, of course, that the different economies will have different incomes even with similar levels of management skill.

Thus, only the prices in general and total correspond to the production costs in a given zone, and this means that some operations have an insufficient profit level or even losses, while the others have increased profits—both without sufficient reasons for it.

Second, differentiation of prices aiming at an approximately equal profitability in the zones takes on the function of redistribution of revenues not natural for prices. The function of redistribution of revenues is the task of the tax system, and it is therefore hardly expedient to impose it on the price apparatus.

Third, prices oriented to production zones do not take into account differences in the level of transport costs that must be considered in order to deliver the product to the points of consumption or processing. This leads to the fact that similar products at the same marketing point are evaluated differently, depending on the point of origin. Thus, for example, cattle received at the meat-packing plant from more distant points is more expensive to the plant than cattle from nearby, although there is no other difference in the cattle received. This disrupts the economic accounting at the plant and makes the results of its work (profit) dependent on originating points of cattle shipment. The same thing occurs in the dairy industry (in connection with dissimilar milk-delivery distances) and in other enterprises that process agricultural raw material.

Setting of prices while ignoring the transport situation results in the remote farms not being induced to send out a finished or semifinished product instead of raw material, because they want to decrease the mass of shipped output and thereby compensate for increased long-distance transport costs of each unit of freight that the processing organizations must bear. Thus, for example, a single price for sunflowers in all of the Ukraine does not induce those collective farms far from processing mills to process sunflowers locally with their own means.

Let us return to an examination of the price zones based on the transport principle. The main merits of this principle are listed below.

1. The transport zones may extend over the whole territory of the country making it possible for all interested enterprises to market their product at a single price in the consumption center (or at a lower price at home, if transport costs are considered). In planning the distribution of agricultural production all agricultural enterprises are thus involved.

2. Distinguishing transport zones facilitates a more rational distribution of production with respect to transport since the higher prices near the consumption centers stimulate a corresponding specialization and intensification of production with a yield increase near the consumption points, as it should be from the point of view of the national economy.

Similarly, the remote enterprises, because of low local price levels of raw material, are induced to process this raw material locally into a more compact commodity that can be transported more cheaply.

3. In differentiating prices according to transport zones, where the price differences correspond strictly to the freight charge, speculation and all kinds of income of a speculative character are impossible (for example, selling the product to the government in another zone where the price level is higher).

At the same time, a systematic division of the territory of the country into transport zones gives rise to serious difficulties. It is necessary to specify beforehand what price at the end point of consumption and processing will evoke a corresponding influx of goods from what localities and in what quantities. It is necessary to know (if only roughly) the production costs and the transport situation of typical enterprises within the areal supply limits of individual centers of agricultural consumption and processing. The inadequacy of this information is demonstrated by the fact that only since 1958 have all collective farms in the Soviet Union begun to calculate their own production costs, but the transport costs of each enterprise for delivery of the product to procurement points and the costs of the governmental procurement organizations for the further transportation of products have still not received the proper attention on the part of the enterprises themselves and on the part of the planning organizations. And yet for each unit of production it is necessary to know the point of origin, production costs, place of destination and transport costs, whoever bears the costs.

A second difficulty is connected with the establishment of a minimum price difference between neighboring zones, keeping in mind that the zones themselves must not be too small and that there be no unwarranted price jumps in the transition from one zone to the other. The solution of this question touches upon the question of transport tariffs.

A third difficulty centers on the fact that the transport principle of zone formation demands constant attention to changes occurring in production (including transport) as well as consumption in all the territories of the country, and presumes a corresponding price regulation.

A fourth difficulty lies in the necessity to equalize the differences in the incomes of individual farms since the purchase prices set according to transport zones will no longer serve as the instrument for equalization of incomes of farms.

It must, however, be noted that all such difficulties are temporary: as planning and the taxation systems are perfected they will disappear.

In conclusion, comparing the zones established on the production and transport principle, the differences between them can be generalized in the following manner. In a production zone a single price is applied to the enterprises operating under similar production conditions. In a transport zone a single price is applied to enterprises operating under the same transport conditions; the differentiation of prices by zones is done in such a manner that any unit of product regardless of the point of origin acquires a single price in the consumption center. Thus, the transport zones guarantee for all commodity units of a given kind, wherever they have been produced, a single common measure, whereas the production zones distort this common measure by adapting it to the production peculiarities in each zone. This explains the economic advantages of the transport zones over the production zones.

Nevertheless, if the principal rule for zone formation is the production and not the transport principle, this is not the result of some advantages inherent in this system but to the difficulty in changing to a more efficient territorial delimitation with a provision for a transport pattern. This difficulty lies in an unclear representation of those agricultural production-consumption complexes into which the territory of the country is divided, and in the lack of factual knowledge of transport costs that fall on each commodity unit of a given kind, depending on the distance between the place of production and the place of consumption in each such complex. In connection with this it would be expedient to study the experience of foreign geographers working on this problem.

1.   By decree of the Council of Ministers of the Ukrainian SSR, no. 1366, of 26 September 1958.

A. V. GOROKHOVA
M. R. KAMENETSKIY
A. I. SHVARTSBERG

The Use of Linear Programming in

Resolving Certain Problems in

Agricultural Geography

The considerable increase in the volume of production on the collective and state farms and the growth of their technical equipment greatly complicate the internal production linkages and demand constant improvement in the methods of planning. In planning for the demands in mechanization and the utilization of machine-tractor fleets on the collective or state farms, it is quite possible to use mathematical methods and computers. In this article an attempt is made to solve a problem of optimal planning for the completion of a tractor park and its rational utilization.

The problem concerns a complex of agricultural tasks of the Ust'-Ordynskiy State Farm of Ekhirit-Bulagat Rayon in Irkutsk Oblast from April to October. For each type of work the most economical types of machines were chosen. Simultaneous with the choice of procedure, the optimal graph of its utilization was determined that would guarantee the fulfillment of the entire volume of work within the assigned agrotechnical time limits with minimal expenditure of labor and means. In addition to a practical purpose (determining the optimal size of tractor parks and their distribution by types of work), the solution of the problem had methodological significance.

Translated from *Doklady Instituta geografii Sibiri i Dal'nego vostoka* [Papers of the Institute of Geography of Siberia and the Far East], no. 19 (1965).

The problem of optimal planning of the structure and distribution of a tractor park was solved by the simplex method of linear programming on a BESM-2M computer with a program developed at the Energetics Institute of the Siberian section of the USSR Academy of Sciences.

The work on the optimal planning of a machine-tractor fleet consisted of two sections: (1) the preparation of the initial technico-economic data; (2) the development of an appropriate algorithm and the writing of a program for the computer.

For the solution of the problem the following original data are used: (1) the plan of sown areas by crop for the year; (2) enumeration of tasks assigned for fulfillment, with an indication of their volume; (3) agrotechnical time limits of the agricultural tasks by crops; (4) data on the number of machines on the farm and information on their condition.

The problem consists in the following. Having the initial data, the required number of machines (tractors, combines, trucks) must be determined, as well as their optimal distribution by type of work, in order to complete the entire planned volume of production with a minimum expenditure of labor and means.

For all crops grown on the farm, technological maps were drawn with a consecutive enumeration of all tasks within the best agrotechnical time limits. For each type of task, tractors of the required make were foreseen in combination with agricultural machines and other means of production.

In planning the ratios of various machines for similar operations one must keep in mind the experience of leading state farms and also the achievements of the scientific agricultural research institutions. The use of narrow-row seeding of grains must be sharply increased and the sowing of grains must be accompanied by the simultaneous application of mineral fertilizers. Crop-harvesting tasks, carried out by varied means must constitute 60 percent of the total volume of work and by direct combining, 40 percent. This combination makes it possible to harvest the grains with the number of harvest machines available.

The use of more refined work methods and types of machines in field operations (postharvest gathering of straw, introduction of new vehicle chassis, tractors, agricultural implements, loading and unloading installations and transport) will make it possible to greatly decrease labor and other expenditures. This cannot, however, be foreseen for all crops in view of the absence of appropriate machines and implements for the complex mechanization of production. Often manual labor and horse transport are used.

The compilation of technological maps is preceded by the calculation of technico-economic aggregate data and the selection from these data of the most effective for all types of agricultural work. Aggregates totaling 140 were selected and for each there was calculated: the production per hour of seeding time and per working day, consumption of fuel and lubricating material, expenditures of labor in man-hours per unit of area, and direct production expenses per unit of work in monetary terms. All initial data for the computer were prepared in the form of a table. The large size of the table did not allow us to include it completely in this article. As an example, only a fragment is given (table 1).

Mathematically the problem was formulated in the following manner. Given are n types of agricultural tasks that are prepared within the assigned agrotechnical time limits and have the following capacities:

$$Ps\,(Ps \geqslant 0; s = 1, 2, 3, \ldots, n).$$

Work is performed by m various tractors, the productivity of which is a_{is} assigned by the matrix

$$\{a_{is}\}(i = 1,2,3, \ldots, m); s = 1,2,3, \ldots, n; a_{is} \geqslant 0)$$

The direct production expenditures c_{is} related to the performance by a tractor of a given type is determined by the matrix

$$\{c_{is}\}\;(c_{is} \geqslant 0, s = 1,2,3, \ldots, n; i = 1,2,3, \ldots, m).$$

Still to be determined is $X_{isk} \geqslant 0$ $(i=1,2,3, \ldots, m; s=1, 2,3, \ldots, n; k = 1,2,3, \ldots, t)$; the intensity of utilization of the machine for the sth task in a kth period, that is, the number of machines of the same type that can be used for the work of each type within a definite period (for example, $X_{isk} = 30$; this means that for the period of five days, six machines are needed (30:5), and $y_i \geqslant 0$ $(i=1,2,3, \ldots, m)$ is the intensity of use of all the machines of the first type. The unknowns are determined by transforming to a minimum of the linear form:

$$\sum_{s=1}^{n} \sum_{k=o}^{t} \sum_{i=1}^{m} C_{isk} X_{isk} + a \sum_{i=1}^{m} C_i Y_i \to \min.,$$

where a is the coefficient of effectiveness of investments comprising $1 / 20$; c_i is the cost of a given tractor type.

Limitations are the following:

$$\sum_{i=1}^{m} a_{isk} X_{isk} = \Gamma_{ks} \qquad \begin{array}{l} s = 1,2,3, \ldots, n \\ k = 1,2,3, \ldots, t \end{array} \qquad (1)$$

TABLE 1

PRODUCTIVITY OF MECHANISMS IN HECTARES (NUMERATOR) AND
DIRECT EXPENDITURES IN RUBLES (DENOMINATOR)
IN VARIOUS TYPES OF AGRICULTURAL WORK
(Early Spring Harrowing and Other Work: Total = 44)

	CROPS						
	Spring Wheat	Barley	Oats	Buckwheat	Millet	Peas	Fodder Root Crops
Types of Tractors							
DT-75 DT-74	$\frac{158}{0.29}$	$\frac{158}{0.29}$	$\frac{158}{0.29}$	$\frac{158}{0.29}$	$\frac{158}{0.29}$		$\frac{158}{0.29}$
DT-54 DT-54A	$\frac{121}{0.24}$	$\frac{121}{0.24}$	$\frac{121}{0.24}$	$\frac{121}{0.24}$	$\frac{121}{0.24}$		$\frac{121}{0.24}$
MT3-50	$\frac{66}{0.30}$	$\frac{66}{0.30}$	$\frac{66}{0.30}$	$\frac{66}{0.30}$	$\frac{66}{0.30}$		$\frac{66}{0.30}$
Agrotechnical time limits of work		15-20 April	20-25 April	20-25 April	19-20 April	26 April	26 April
Work capacities (in hectares)	11,700	700	3,490	110	50		55

where Γ_{ks} is the volume of work that must be performed during each kth calculation period for each sth type of work.

$$\sum_{s=1}^{n} X_{isk} - Y_i \leqslant 0 \qquad \begin{array}{l} i = 1,2,3, \ldots, m \\ s = 1,2,3, \ldots, n \\ k = 1,2,3, \ldots, t \end{array} \qquad (2)$$

The first limitation guarantees the fulfillment of the entire volume of work within definite calendar time limits (nt). The second establishes that the sum of intensities of machine use of one type of machine on all tasks must be less or equal to the intensity of utilization of the total number of these machines in the economy (mt). If the work is performed in somewhat less time than the duration of the short period, then the utilization intensity of machines must be multiplied by the coefficient $p\,/\,r$ where p is the number of days in the period and r the number of days in the course of which the work is fulfilled.

Sixteen kinds of agricultural crops and 44 types of agricultural tasks were examined; to fulfill this work 140 aggregates were selected. The entire period of work is divided into T ($T=10$) large periods equal to the t of small periods ($t=3-6$). Each small period consists of 5 days. The entire period of agricultural work from 5 April to 14 October is divided into 38 small periods.

In solving the problem for each large period the number of equations and unknowns did not exceed the number provided by the parameters of the program. The calculations were conducted with a standard program taking into account the data of the preceding periods. In connection with this the limitations are divided into two types.

For the first large period:

$$\sum_{s=1}^{n} X_{isk} - Y_i \leq 0 \qquad \begin{array}{l} i = 1,2,3, \ldots, m \\ s = 1,2,3, \ldots, n \\ k = 1,2,3, \ldots, t \end{array}$$

where Y_i is the utilization intensity of the total number of machines of t type needed during the first large period. The following large periods (2) are expressed thus:

$$\sum_{s=1}^{n} X_{isk} = Z_i = d_i; \quad \sum_{s=1}^{n} X_{isk} = Z_i < d_i.$$

Here the utilization intensity of the total number of machines of i type for the large period being calculated is $Y_i = Z_i + d_i$, where d_i is the greatest utilization intensity of all machines of i type of the preceding large periods obtained in calculation; Z_i is the additional utilization intensity of machines of i type in the large period being calculated.

The parameters of this program make it possible to solve problems with no more than 80 equations and a practically unlimited number of unknowns. Therefore, each large period is composed of such a number of smaller periods that the number of equations for its calculation would not exceed $nt + mt \leq 80$.

As a result of the calculations there were obtained: a plan for performing mechanized work with minimum direct production expenditures and capital investments, a graph for loading machines for the whole period of agricultural work, and an optimal technical requirement (see table 2).

The Ust'Ordynskiy State Farm has 50,892 hectares of land, of which 28,979 hectares is agricultural land that includes 22,199 hec-

TABLE 2

OPTIMAL REQUIREMENT OF POWER MACHINES
BY THE UST-ORDYNSKIY STATE FARM

Machine Types	Number of Machines	Machine Types	Number of Machines
C-100	13	MTZ-5 LC	10
DT-75	20	T-28	6
DT-54	17	DT-20	13
T-40	9	T-16	1
MTZ-2	4	CK-3	13
MT3-2	–	GAZ-51	16
MTZ-50	6	CK-4	17
		ZIL-585	16

tares of plowed land. In addition, the state farm has its long-term use of 985 hectares of arable plow land from the state land fund. The entire sown area in 1964 was 22,167 hectares and was distributed by crops in the following manner.

Total grain legumes including	16,650	hectares
wheat	11,700	hectares
potatoes	150	hectares
vegetables	50	hectares
fodder	5,317	hectares
corn for silage	2,550	hectares
sugar beets	140	hectares

Grain crops in the structure of sown land occupy 75.1 percent of the land.

From table 3 it is clear that the state farm needs to supplement the tractor park at the expense of high power machines of the five-ton type, C-100. There are only three of them and according to the calculations 13 are needed. The number of three-ton tractors (DT-75) must be reduced by 17. The number of tractors of type DT-54 must also be considerably reduced.

With regard to wheel tractors the Ust'Ordnynskiy State Farm at present uses mainly wheel tractors of the 1.5-ton class. There are 17 MT 3-50 tractors and 12 more are on order, although by selecting

TABLE 3

COMPARISON OF CALCULATED NUMBER OF POWER MACHINES
IN THE ECONOMY WITH THOSE AVAILABLE

TYPES OF MACHINES	NUMBER OF MACHINES	CALCULATED NUMBER OF MACHINES	COST IN THOUSANDS OF RUBLES		
			Of One Machine	Of All Machines	Of Calculated Number of Machines
C-100	3	13	3,270	9,810	42,510
DT-75	37	20	2,460	91,020	49,200
DT-54	55	17	2,250	12,370	38,250
T-40	2	9	2,100	4,200	18,900
T-38	1	4	1,700	1,700	6,800
MTZ-50	29	6	2,200	63,800	13,200
MTZ-5 LC	11	10	2,100	23,100	21,000
T-28	2	6	1,800	3,600	10,800
DT-20	4	13	1,200	4,800	15,600
T-16	1	1	1,150	11,500	1,150
RAZ-51	22	22	1,390	30,580	22,240
ZIL-585	7	7	2,000	14,000	31,000
Total	186	131	.....	371,500	271,650

agricultural crops and the volume of work in this farm, the maximum number of tractors of this class and type need not exceed six. The number of 0.9-ton and 0.6-ton wheel tractors (that is, T-40, T-28, and DT-20) must be increased since they are the most acceptable for cultivated crops and have greater maneuverability; they can also be used for various transport and loading work. The machines of brand MT 3-50 are the most expensive of all wheel tractors and do not operate at full capacity where it is more profitable to use tractors T-40, T-28, DT-20. Calculations show that at present the farm must have 28 tractors of these instead of eight. Thus, a radical change is needed in the structure of the tractor park including a reduction of approximately 1.2 times the number of power machines.

In connection with such a change in the structure and number of machines in the machine tractor park, costs will be greatly reduced. Thus, the cost of caterpillar machines, obtained through calculation (without including production expenditures) is 41 percent lower than the cost of the total number of these tractors on the farm. The same is true for the wheel tractors. As a whole, the machine-tractor fleet receipts according to calculations will be 99,860 thousand rubles, or 38 percent less expensive than the present situation, which will considerably reduce the labor and other expenditures for the mechanized work.

The simplex method of linear programming used in solving this problem is the simplest, most universal, and best developed. It makes it possible to solve a problem in parts and to compose a program with comparatively few parameters, which are limited to 30–40 equations and 150–200 unknowns. Only the operative memory of the computer is used, and this considerably reduces the time for solving one matrix with maximum permissible dimensions. At the same time the method of linear programming does not permit the solution of complex problems, for example, the calculation of the total required machine-tractor parts, including agricultural machines, and the division of the problem into parts leads to some inaccuracies. For complex problems special methods are needed that permit significant reduction in the expenditure of machine time and time to prepare the initial data.

The formulation and solution of this problem once more proves the complexity and labor-consuming work in planning agricultural production but also the necessity of broader application of mathematical methods and computer techniques.

SECTION V

Industrial Geography

Introductory Note

Industrial geography is a well defined subfield of Soviet economic geography. At the same time, however, there exists a high degree of interrelationship with concepts and problems treated in other subfields of economic geography; for example, the territorial production complex and questions of economic regionalization, problems of labor resources and migration as dealt with in population geography, and industrial nodes and other problems of urban geography.

The great interest in industrial geography and theoretical and practical problems of industrial location has existed since the early days of the Soviet regime. "As early as April, 1918, in his 'Draft of a Plan of Scientific Technical Work,' Lenin called on the Academy of Sciences to work on the problem of rational industrial location."[1] The leading role assigned to industrialization in Soviet economic planning and development, the numerous location decisions required by the enormous programs of planned industrial development and postwar reconstruction, and the scale and essential irreversibility of the investments involved placed heavy demands on planning agencies and design institutes that have often turned to geographers and economists in academic institutions and institutes for assistance. In response, a very considerable body of literature has been produced, much of it apparently unnoticed, or at least rarely cited, by Western geographers and economists.[2]

Included in the substantial literature on industrial location are many works that deal directly or indirectly with the development of an explicitly Soviet, socialist framework of theory and principles of the distribution of industry. In the course of developing their theories Soviet geographers and economists were, of course, aware of classical statements by Western scholars such as those by Weber, Engländer, and Hoover. Weber's *Theory of Industrial Location*, translated into Russian in 1926 with an introduction by Baranskiy, is the Western work best known to Soviet writers. A number of economists and geographers initially accepted or employed Weber's principles and approach. Later his theories were severely criticized, primarily because they are based on the principle of least cost from the standpoint of the individual plant and imply a short-term outlook, and ultimately rejected as incompatible with the theory and objectives of Marxism-Leninism.[3]

In its place has evolved a theory based on "the objective laws of socialism." This theory has as its goal that of maximizing the productivity of social labor and is derived from statements by Marx, Engels, and Lenin, as well as from resolutions and official pronouncements of the Communist party. In the postwar period these elements have been formalized into a set of locational planning principles. These principles or guidelines have been described as follows:

1. Move industrial enterprises closer to sources of raw materials and to final consumers in order to reduce freight costs.

2. Plan the distribution of plants among economic regions in such a way that they can develop special industries utilizing available natural resources most efficiently. This will facilitate the territorial division of labour. On the other hand, each region should strive to become economically self-sufficient.

3. Distribute industrial production evenly throughout the country in order to utilize all human and natural resources in all regions.

4. Abolish the contradiction between cities and rural areas which is based on difference between industrial and agricultural production.

5. Secure the industrialization and cultural development of all regions inhabited by national minorities.

6. Strengthen the defense capacity of the country.

7. Facilitate the international division of labor among the countries of the socialist bloc.[4]

These principles are those used for determining general, regional location; specific location within a region is then generally determined by comparing production costs at alternate locations.[5]

The general locational principles would appear overtly conflicting and thus difficult to apply. It can be argued that the contradictions are more apparent than real and that in practice they are frequently resolved by Party resolutions that set policy priorities.[6] Nevertheless Kolosovskiy's frank comments of 1938 still have a ring of timeliness: "The absence of a detailed location theory has led in several cases to utilitarianism and errors in actual plant location. A planning engineer should be able to translate all his findings into the language of mathematics, precise numerical indicators. He cannot be guided by general principles in selecting sites for a single plant or a combination of plants. Our economists restrict themselves to criticism of bourgeois methods of plant location (the Weber and Engländer schools) and do not provide our socialist industry with a soundly based methodology for solving location and industrial-combination problems."[7]

Though hampered by the obvious problem of incorporating the required social welfare considerations and the difficulties of working within the context of the Soviet price system, Soviet economic geographers have continued to work toward more quantitative analytical solutions to their location problems. The work of Isard that has been translated into Russian and the mathematical and spatial modeling approaches of the regional scientists have elicited sympathetic interest for methodological, though not theoretical, reasons.[8] There is every reason to anticipate that Soviet work in industrial location analysis also will soon be characterized by a mathematical, model-building approach.

The readings included in this section include the theoretical-philosophical, the substantive-interpretative, and tentative steps toward quantitative and modeling approaches. As do articles in other sections they reflect the current concern with economic efficiency and a fuller utilization of labor resources.

The first article in this section, "Certain Questions on the General Theory of Distribution of Industry," is a general statement by Probst of the Soviet view on location theory and the principles or laws to be followed in the distribution of industry. A distinction is drawn between general economic laws and specific laws derived from them. Although the distribution of industry under capitalism is seen as responsive to particular laws under the general law of surplus value, it is in contradiction with the law of social development— the law of raising the productivity of social labor—that exists only under socialism. Various specific laws in the realm of distribution are then elaborated, including the law of the distribution of extractive industry, the law of distribution of manufacturing industry, the law of concentration of production, laws of specialization, cooperation, and combination of industrial areas, the law of comprehensive harmonious development of the productive forces of regions, and the law of proportionality.

The second selection of this section, by Korneyev, "The Development of Industrial Complexes of Economic Regions," is largely a substantive presentation that includes considerable information in tabular form. Marked regional differences in industrial development, specialization, and labor resource availability are shown to exist. By use of diverse indices it is shown that the level of industrial development of the individual economic regions still varies considerably; this is attributed largely to differences in stages of development prior to the Soviet era. The industrial specialization of economic regions is then taken up, and the difficulty of determining such specialization by a single set of indicators is discussed. Specialization, Korneyev states, is in any case not a goal in itself but is secondary to achieving a higher social productivity; some existing irrationalities of specialization are mentioned. The article concludes with a discussion of labor resources. Solution to the problem of achieving fullest utilization of labor resources depends on a greater involvement of female labor and seasonally available agricultural workers; a rational distribution of appropriate industries is needed to solve these problems and to assist in further equalization of levels of economic development.

The third article of this section, "Toward a Quantitative Evaluation of the Economic-Geographic Location of an Industrial Enterprise," is an attempt to analyze the influence of certain aspects of location on the net costs of production. Net costs of production are shown to have great regional variation. A partial explanation lies in the regional variation in raw-material costs; because of the system of "market demand zones," enterprises do not necessarily receive lower cost local raw materials. Labor expenditures are also shown to vary greatly spatially due to differences in wage levels and productivity; although the author considers the principle of regional wage supplements justified by regional differences in expenses, "unjustifiable differences in magnitude exist." The author considers vital the development of more complete quantitative and qualitative measures to determine rational industrial distribution especially "under Soviet conditions where the effect of the law of price is limited."

In the final article of this section, "The Kursk Magnetic Anomaly," Novikov uses a problem of industrial location and raw materials supplies of the iron and steel industry to demonstrate the application of the concepts of a territorial model of the national economy and a sectional/interregional model of the Soviet iron and steel industry. He begins by arguing that the goal of achieving maximum growth of the aggregate social product and national income cannot be solved simply by minimizing the operating costs of production and transportation but will necessitate the aid of a territorial model of the national economy achieved by the method of successive approximation. The requirements of such a model are explored and a conceptual sub-component, a sectional/interregional model of the Soviet iron and steel industry, is introduced. The circulation of goods in the iron and steel industry is examined in this framework, with particular reference to the potential use of the Kursk iron ore deposits. The optimal use of these deposits Novikov argues is as a source for iron and steel industry in the Urals rather than in Central Russia, a conclusion that contradicts conventional economic solutions.

1. Yu. G. Saushkin, "A History of Soviet Economic Geography," *Soviet Geography: Review and Translation* 7, no. 8 (October 1966):4–5.

2. See, however, the excellent commentary and review by I. S. Koropecky, "The Development of Soviet Location Theory Before the Second World War," part 1, *Soviet Studies* 19, no. 1 (July 1967):1–28; ibid., part 2, *Soviet Studies* 19, no. 2 (October 1967):232–44.

Extensive citation of Soviet industrial location literature can be found in Saushkin, "A History of Soviet Economic Geography."

3. See Koropecky, "The Development of Soviet Location Theory," parts 1 and 2, for a detailed examination of Soviet reaction to Weber's theories.

4. Koropecky, "The Development of Soviet Location Theory," part 1, pp. 23-24. See also R. E. Lonsdale, "Industrial Location Planning in the Soviet Union," *The Professional Geographer* 13, no. 6 (November 1961):11-15, and V. S. Klupt, "Geography of Industry," in *Soviet Geography: Accomplishments and Tasks*, ed. C. D. Harris, American Geographical Society Occasional Paper, no. 1 (New York, 1962).

5. Lonsdale, "Industrial Location Planning in the Soviet Union." p. 13.

6. Koropecky, "The Development of Soviet Location Theory," parts 1 and 2.

7. As quoted in Saushkin, "A History of Soviet Economic Geography," p. 34.

8. See T. A. Reiner, "Introduction to the Russian Edition of Isard's *Methods of Regional Analysis* with a Commentary by Thomas A. Reiner," Regional Science Research Institute Discussion Paper Series, no. 18 (1967). See also Yu. G. Saushkin, "The Construction of Economic Models of Regional and Local Territorial Production Complexes," *Soviet Geography: Review and Translation* 2, no. 4 (April 1961):63.

A. YE. PROBST

Certain Questions on the General Theory Of the Location of Industry

THE ECONOMIC LAWS OF SOCIALISM AND THE LAWS OF DISTRIBUTION

The geographic distribution of productive forces is determined by the means of production. For this reason, to every socioeconomic structure there corresponds a historically unique type of distribution of productive forces and a characteristic system for the territorial organization of production. Fundamental economic law and the system of economic laws connected with it determine the nature of every social means of production, all its chief aspects, and the principles for its development. The fundamental economic law, along with its system of related economic laws, determines also all the particular laws operating in the separate sectors and areas of production.

Economic laws are divided into universal (or general) ones, inherent in all social structures, and historical ones, valid only within a historically defined social structure; that is, they are divided into general historical and specific historical laws. Moreover, the specific historical economic laws, valid in general for any given historical type of production, are altered in accordance with the specific features of each separate sector and area of production. The general economic laws for any given historical type of economy are manifested in different ways in each economic area or sector in the form of particular, or specific laws, peculiar only to that

Translated and abridged from A. Ye. Probst, *Razmeshcheniye sotsialisticheskoy promyshlennosti* [The Location of Socialist Industry] (Moscow, 1962), pp. 7–27.

given economic area. Friedrich Engels wrote, in a letter to Schmidt: "Wherever there exists a division of labor on a societal scale, the various labor processes become independent of one another," and they follow "their own course" and "their own laws." Each such sector "acquires its own development and . . . has its distinctive laws and stages determined by its own nature."[1]

In all such particular, or specific, economic laws acting in the different sectors and areas of production, the same traits are inherent that are possessed by the more general economic laws that regulate the economy as a whole. They, too, are historical and objective laws, like the economic laws for which they constitute the concrete manifestation. The particular economic laws at work in the different sectors of production constitute the object of study of corresponding economic subdisciplines—industrial economics, agricultural economics, transportation economics, and so forth.

The laws of territorial distribution of production, investigated by economic geography,[2] and also in part by economic subdisciplines, are particular, or specific, economic laws. Those economic laws that hold in general for any given historical type of social reproduction become modified into particular laws for the location of production or its territorial organization under the influence: first, of complex and diverse interactions between production and the geographic environment; second, of the varying circumstances brought about by the spatial combination of various production areas or, on the contrary, by the spatial dismemberment of a particular type of production; and third, of specific characteristics in the economy of different sectors.

Thus, the economic law of socialism, the law of a steady rise in the productivity of social labor, is modified in given specific areas of the economy into a number of particular, or specific, laws for the distribution of production: the law of economizing on the social expenditures for transportation by bringing industry to the sources of raw materials and fuel or to regions where the finished product is consumed (depending on the economic characteristics of the different sectors of industry) is the specific manifestation in a given economic sphere of the law of economizing labor, which holds in general for all sectors and spheres of production.

The laws of socialization of labor and production appear in the form of the particular laws of specialization of production, occupational specialization of labor, and productive specialization of regions. In particular, the law of social division of labor receives its specific spatial manifestation in the form of the law of territorial division of labor. The law of concentration of production has its particular manifestation in the form of the territorial organization of production.[3] The law of systematic, proportional development of the economy is transformed in any given concrete activity: it is modified in the particular, or specific, law of comprehensive and harmonious development of the productive forces of regions.

In investigating and discovering the different economic laws of socialism, one must investigate the actual forms in which each such law is manifestated in the various branches and spheres of production and also how they apply to the territorial organization of socialist production. In addition, it is necessary in studying the specific laws of distribution of socialist production to indicate which general economic law is being manifested in the form of each such specific law.

There exists an intimate organic bond and an interaction between the economic laws that hold in general for the total socialist economy and the specific laws peculiar to each different sector and area of production, since in their totality they form a dialectical unit. Dialectical materialism gives great significance to determining the relations between the concrete and the abstract. Proper understanding of every concrete and particular phenomenon is possible only from the viewpoint, and on the basis, of the general, the abstract. Likewise, consideration of concrete phenomena in all their diversity is the means to understanding the abstract.

A proper understanding of the relations between the specific laws of location of socialist production and the general economic laws of socialism will enable us to better explain and reveal the particularities in each law and the mechanism by which it actually works. This is all the more important since the economic laws of socialism, as well as the specific economic laws that constitute the concrete form in which they are manifestated in the various branches and spheres, alter in the different stages of development of socialist society.[4] Moreover, the developmental dialectic is often disregarded among us, both for specific economic laws and for those that hold in general for the whole economy.

LAWS OF LOCATION OF SOCIALIST PRODUCTION

Socialism has its own, historically distinctive type of distribution of production, different in principle from the distribution of production in other economic systems. The geographic location of socialist production is determined by specific laws, which are manifestations of the general economic laws at work for the whole socialist economy. The economic laws of socialism differ in content, direction, and mechanism of action from the economic laws of capitalism, and in this respect it is inevitable that there be a cardinal difference between the distribution of socialist and capitalist industry.

The economic laws of location, both under socialism and under capitalism, are objective laws, and it is this fact that determines the objective circumstances under which the location of production takes form, in both socialist and other economic structures. The objective economic laws under socialism become known to men and are consciously applied (utilized) in planning and organizing the economy. In this way they differ essentially from the economic laws of capitalism, which act in a spontaneous fashion outside man's consciousness as some sort of blind force. However, the capacity for economic laws becoming known under socialism in no way reflects on their objective nature: they remain independent of the will and desires of people and can be neither reversed nor revised. The planning of the location of socialist production must be fully founded on objective economic laws. The objectivity and, in consequence, the rationality of our plans for the distribution of socialist industry are defined by their correspondence to the objective specific laws of the socialist location of production.[5]

The progressiveness of the socialist economic system, as compared to the capitalist and to every previous economic system, lies in the more rapid growth of productive forces and in the possibility of attaining a higher level of productivity of social labor—goals unattainable with any other means of production. In the historical competition between the two economic systems, "capitalism may be decisively defeated and will be decisively defeated through socialism's creation of a new, much higher level of labor productivity."[6] For this reason, it is through the building of socialism that "there moves into the foreground the essential task of creating a higher social structure than capitalism; that is, raising the productivity of labor, and in connection with this (and for this purpose) highly organizing it."[7] Labor is the foundation of the existence of man and of society. It is only in the process of work that an exchange of matter takes place between man and nature and that the utility values necessary to the existence of human society are created. The level of productivity of social labor determines the degree of its subjugation, and, hence, the degree of progress in social development. Social development, progress in human society, proceeds on the basis of raising the productivity of social labor. It is for this very reason that Marx and Engels formulated, in the capacity of the "universal economic law," the "law of growing labor productivity."

Raising the productivity of social labor, even though it is a universal economic law in accord with the basic tenets of Marxism, proceeds and finds realization in varying ways in different economic structures. Under capitalism this law falls into conflict with the law of surplus value that is peculiar to this means of production. Striving for maximum profit does not coincide with ensuring the greatest productivity of social labor; such a contradiction the apologists of capitalism endeavor to conceal in every way possible. Under capitalism, the universal economic law comes into an antagonistic conflict with other, specifically capitalist laws, and so it is inevitable that there arise constant deviations from this universal law and that the latter is expressed under capitalism only as a constantly violated tendency.

Under socialism, the law of rising productivity of social labor acts directly and spontaneously as a known, unconditional law of social development. Marx pointed out that "economy of time, in the same way as the systematic distribution of the workers' time in the various branches of production, *remains the first, basic economic law* based on collective production."[8] Lenin underscored the prime importance of this law, and at the same time, its direct and spontaneous action under socialism. "The productivity of labor, this, in the last analysis, is the chief, the most important thing for the triumph of the new social structure."[9] He further pointed out that socialism appears as "a systematic organization of the social-production process for ensuring the welfare and all-round development of all members of society."[10]

The location of production, like the geographical division of labor in general, is one aspect in the organization of production: its territorial organization. And this aspect of the organization of socialist production, as well as socialist production as a whole, is subordinate to and regulated by the economic laws of socialism. The law of continual growth in the productivity of social labor determines the unbroken expansion and improvement of production on the basis of higher technical levels, as well as the location of production and ensures ever greater rationality in this location.

The criterion for development and improvement of production under socialism in accord with its economic laws is the raising of the productivity of social labor. This is the criterion, too, for determining the rationalization and improvement in the territorial organization of production under socialism and the economic efficiency of alternatives in the location of socialist production. In the new Program of the Communist Party of the Soviet Union particular stress is laid on the fact that a rational location of industry must insure economy in social labor. In the realm of location the law of steady rise in the productivity of social labor is modified in the form of certain specific laws.

First, it appears under the form of the *law of location of the extractive industry*, which stipulates for any given mineral the industrial utilization of only those deposits, and in such sequence, quantities, and ratios of extraction among the various deposits, so that the sum total of social expenditures for extraction (including initial processing and preparation) and the transportation of these minerals will be minimal. A specific trait of the extractive industry is that it may be located only where the resources of proper minerals exist. However, not all deposits of minerals may be economically exploited. On the basis of this specific law it is determined which of the existing deposits ought to be industrially exploited, and in what order, what quantities, and with which interrelationships for each given period. This in turn depends on the possible location of consumption of the product extracted. This law determines both the rational distribution of the extractive industry and, in conjunction with it, the rational regionalization of the consumption of its products.

Second, one of its forms of manifestation is the *law of location of the manufacturing industry*, which requires that it be moved to the sources of raw material, fuel, and energy or to the regions where products are consumed (depending on the specific characteristics of the economics of the different sectors of industry) in order to lower total social expenditures for production and transport of products. The law of moving industry to sources of raw material, fuel, and electrical energy and to areas where products are consumed was, in essence, formulated by Lenin as early as 1918, and its import has been underscored on more than one occasion in resolutions of the Party congresses, the Central Committee of the CPSU, and the Soviet government.

The territorial distribution of natural resources, the distribution of labor resources, and the geography of product consumption do not coincide one with another. Production under socialism must be so organized as to ensure maximum economy of expenditures in overcoming the consequences of this lack of spatial coincidence for the three conditions of production cited. By properly utilizing the specific laws of location for socialist production, it is possible to find the most efficient means for resolving this complex task. The forms of social organization of production are determined under socialism by the corresponding economic laws of concentration, specialization, cooperation, and combination. Each of these laws is manifested in its particular form—as a specific law for the location of production.

The concrete forms in which the law of concentration of production manifests itself and the optimal number of industrial establishments is modified in space, since depending on the geographic distribution of industry and on the regionalization of finished product consumption, as well as on a number of other territorial considerations, there are changes in the forms and boundaries of concentration of industry and in the optimal number of industrial establishments. In particular, increasing the distance over which products are transported may limit the degree of concentration and the efficiency of consolidating industrial enterprises.

The laws of specialization, cooperation, and combination of industrial-production areas undergo an analogous modification under the influence of territorial conditions. The economic efficiency and rational limits of specialization, as well as the forms of cooperation and combination, depend closely on a number of territorial conditions: the regionalization of product consumption, the location of sources of raw material and fuel and their spatial combinations, and so forth. Interaction between "intra-production" patterns and territorial ones results in a new synthesis of patterns and in the special forms of their manifestation. In particular, the law of collectivization of labor and production and their forms of social organization appear in the specific law of the rational combination in space of different industrial establishments and types of production into single territorial-production complexes.

Raising the productivity of social labor, as a result of the concomitant spatial combination of various types of production and productive establishments; the rational forms of such combination, as separate territorial-production complexes within the boundaries of a geographic center ("nodes" or "points") or economic regions (that is, in the form of regional production complexes); the interdependences among them; the limits and direction of their development; all this is determined by specific laws that take the form of the economic laws of socialization of production, holding in general for the whole of the socialist economy. The economic law of social division of labor also determines specialization in sectors of production, and the territorial division of labor. In its specific form, this law determines as well the productive specialization of territorial-production complexes, that is, of different economic regions and economic centers (or "nodes").

The law of systematic and proportional economic development has its own specific manifestations in relation to the territorial organization of social production. This economic law, which is a general one for the whole economy, determines not only the systematic and proportional development of the various branches of social production but the proportional development of the different economic regions, as well as the proportional development of all types of production within each economic region. One form of this economic law is the specific law of comprehensive, harmonious development of the productive forces of regions. As a result of the action of this law, such a combination is achieved in the development of economic sectors in each region, as well as proportions among them such that there is attained a maximum and diversified development of productive forces in the region and the greatest possible increase in the productivity of social labor. Because of this law, the one-sided development of productive forces in different regions and their extraordinarily narrow specialization are eliminated under socialism, whereas under capitalism they lead to negative and even unnatural consequences.

A second form in which the stated law appears is the specific *law of proportionality in rates of development of the various economic regions*. The action, under the conditions of socialism, of these specific laws ensures systematic and proportional development of the whole economy and an increase in productivity of labor in production as a whole, and at the same time resolves a number of particular problems. Through the workings of these laws preponderant development of the productive forces of economically backward regions is achieved, that is, the consequences of their irregular development under capitalism are overcome as well as the consequences of national inequalities and the oppression of nationalities under the bourgeois system and the consequences of an irregular distribution of industry that itself was the result of the antagonistic, contradictory development of different regions under capitalism.

Liquidation of the disproportion among sectors that came into existence under capitalism is realized under conditions of socialism on the basis and within the limits of the law of systematic, proportional development of the economy. The rate of liquidation is subject to a certain regularity. More systematic, harmonious geographical distribution of socialist industry and, in this connection, more rational territorial organization of the whole of socialist production are the usual result, a synthesis of the action of specific laws of location of socialist production as definite manifestations of the law of systematic, proportional development of the economy.

The political economy of socialism is still a young science, as the historical period is relatively short during which the socialist system of economy has existed and developed. For the moment there have not yet been revealed in their entirety all the economic laws of socialism, and many of them still require additional study in depth. Moreover, as the socialist economy develops and enters a period of expanded building of communism, alteration occurs in the economic laws of socialism themselves and their mechanism of action. And this circumstance necessitates their special study.

In contrast with the general economic laws for the whole of the socialist economy, there has been considerably less investigation of their concrete manifestations in the various branches and realms of production, that is, of specific laws, especially in the area of location. The further development of the science concerned with territorial organization of socialist production demands more profound study of all specific laws for the location of socialist production, as well as a full systematization of such laws.

The rational location of socialist industry must be based on the complete and joint utilization of all the potentialities for a maximal increase in the productivity of social labor created by both natural and social conditions. The different ways (investigated below) for rationalizing the location of socialist industry complement one another and interact intimately, since, in the last analysis, they all result in a rise in the productivity of social labor, in the creation of a new geography of industrial production.

In accordance with the actual characteristics of each branch of industry, there must be distinguished the most efficient method for its location, as well as the most efficient combination in such location of interrelated branches of industry. Proper solution of this problem calls for painstaking economic analysis and a detailed calculation of the economic efficiency of various alternatives in the location of each branch of industry. All these investigations must be grounded on the Marxist-Leninist methodology of determining economic effectiveness.

The criterion of economic efficiency for location alternatives, both of industry as a whole and of its various branches, is a rise in the productivity of social labor; that is, the degree to which, in any given variant of distribution, total social expenditures for the output of products and for their transportation to the consumer may be lowered under any given social stock of capital expenditures. Consequently, the indices of comparative efficiency for location alternatives in a given sector of industry must, in theory, be the total value of all products at the place where they are consumed and the total value of all corresponding capital expenditures.

However, in practice, unfortunately, we are denied the opportunity for the time being of operating with the indices of product value. In spite of the express instructions of the founders of Marxism on the importance and necessity of direct computation of value under socialism, we have been unable to this day to compute product value. In practice the sole admissible index for measuring the productivity of social labor at present is the cost price of the product. But cost price coincides neither in qualitative nor in quantitative terms with value. Change in cost price only indirectly reflects change in the value of a product. Therefore, the problem consists in utilizing cost-price indices for products in the different alternatives of location in their maximally comparable form and in introducing corrections in such a way as to ensure that their ratio most fully expresses the ratio of product cost. It is not possible here to discuss this problem in greater detail, and besides it has been specifically dealt with in the literature on the subject.[11]

The raising of labor productivity is usually achieved by equipping labor with the means of production. The alternative location of industry that secures the lowest cost for a product may at the same time require a greater capital investment. For social reproduction in general it is completely immaterial through which supplementary capital expenditures the rise in labor productivity is achieved in each sector of production and in each locational alternative. Since the fund of social accumulation is always limited by social reproduction conditions, for society in general it is important to secure the maximal rise in productivity of social labor with a given fund of accumulation; that is, with a given stock of capital expenditures. The proper solution to this problem may be obtained only by approaching it from the standpoint of the economy as a whole. For this purpose it is necessary to determine the ratio between supplementary capital expenditures and the decrease in unit cost of the product in each alternative of production location. This leads to the necessity of utilizing the formula established for determining the economic efficiency of a new technique.

The economic efficiency of different alternatives in industrial location may properly be determined through a comparison, not of production outlays right at the plants, but of the cost of products at the place where they are consumed. It is only by this means that one may properly determine the economic efficiency of various alternatives in the location of production.

This significantly complicates the methodology and calculations of economic efficiency, since it is necessary to determine for the alternatives being compared not only the immediate production outlays at the plant but also transportation expenditures and all related expenditures for storage, loading, and unloading if such are also required. In this connection it is extremely important to get a proper notion of actual social expenditures for transportation. For this purpose it is impossible to utilize transportation rate scales, since, in the first place, these do not correspond to the factual transportation expenditures for the different types of freight; and, in the second place, rail transportation rates have been set up on the average for the whole railroad network of the USSR, without regard to the actual shipping conditions for the different routes and the direction of shipping (empty and loaded).

In order to determine the comparative efficiency of the different alternatives in industrial location it is necessary that one's efforts be grounded on actual or projected freight expenditures along specific routes to definite localities of consumption, taking into account the dependence of the expenses on volume of freight turnover for the routes, on the technical equipment, and on the possibilities of using empty rolling stock, and so forth.

In determining the unit cost (value)[12] of comparable products at the place of their consumption, much depends on the correctness of initial transportation indices and the method applied in calculating them, and also on the analysis of their dependence on any number of factors. This is all the more important, as there frequently is an intimate relation between the distribution of large industrial establishments and the transportation indices themselves: the location of industrial establishments is to a considerable extent determined by the level of transportation indices. But the level of the indices is in turn dependent on the given locational variant, in as much as the latter may considerably affect the volume of freight turnover and hence, the type of traction and a number of technical arrangements.

Unit cost at places of consumption varies with an increase in the distance from the place of production. Its value attains a minimum in points of consumption that are located in the immediate vicinity of the production point. On the other hand, at these places the unit cost of products imported from other districts achieves a maximum. As areas of consumption get further away from one point of production and approach another such point, the difference in the unit cost of products should become gradually lower until a point of equality is reached. For this reason in calculating the economic effect of a production-location alternative, it is incorrect to base the calculations on the comparative unit cost of its products and imported products at the nearest points of consumption, where the difference in unit cost of local and imported products is at a maximum. In such cases the proper calculation of economic effect should be based on the value (unit cost) as an *average* for the whole region where the product is consumed.

In order to calculate the average product value for the whole region where it is consumed, the following primary data are needed: first, the cost of making the product; second, the cost of transporting the product to the different points of consumption, which will permit the determination of product value at those various points of consumption; third, the rational limits of the region where the product is consumed, that is, the delimitation of regions of consumption; fourth, the scale of total consumption of the given product within the rationally bounded region where it is consumed and the ratio of the several localities in overall consumption (balance) of the product. Data on the scale and geography of consumption are necessary in order to calculate the average value of the product throughout the region, since only the latter valuation can serve as the basis for determining the economic efficiency of comparable production location alternatives.

Computation of such average product value over the region of its consumption often requires rather complex calculations, but carrying them out may be considerably facilitated through application of modern mathematical methods, in particular linear programming. There has long been an attempt to use mathematical methods to solve a number of problems in the location of production. Considerable experience in such applications has already been built up. However, it necessitates profound critical analysis and, in the main, a combination of modern mathematical methods (which can play only a subordinate role) with general Marxist theory on the location of industry. Applying methods of mathematical analysis can be effective only on the condition that the economic content (quality) of the categories and the theoretical foundation for quantitative connections between them are determined beforehand. The fault of many authors who apply mathematics to economics consists in attempting to determine some sort of economic regularities by moving from mathematics to economics instead of offering quantitative expression to theoretically based patterns in economics.

Computation of the corresponding indices of total (including transportation expenditures) capital outlays per unit output at the point where they are utilized is fully analogous to computation of the average value of products over a region.

1. Karl Marx and Friedrich Engels, *Sochineniya* [Works], 28:256.

2. Sun Ching-chi, *Ekonomicheskaya geografiya kak nauka* [Economic Geography as a Science] (Moscow, 1959), p. 41.

3. R. M. Kabo, *Ob'ektivnye svyazi mezhdu razvitiyem narodnogo khozyaystva razmeshcheniem strasley proizvodstva i protsessem formirovaniya rayonov* [Objective Connections between the Development of the Economy, the Distribution of Industries, and the Formation of Regions], in *Izvestiya Vsesoyuznogo Geograficheskogo Obshchestva* [Proceedings of the All Union Geographical Society], no. 1 (1956), p. 66.

4. Economic laws arising in other social systems also change according to the development of the latter and changes in the conditions in which they become manifested. Marx pointed out, for example, the "*specific* development that the law undergoes with the rise of capital" (*Teorii pribavochnoy stoimosti* [Theories of Surplus Value] [Moscow, 1954], 1:55). The same was noted by Engels: "These economic laws of commodity production become altered at different stages in the development of this form of production" (Marx and Engels, *Sochineniya*, vol. 16, part 1, p. 150).

5. In our literature there have been attempts to attach special significance and mootness to the question of the problem of differences between the patterns and the principles of the location of socialist production. To us such a way of posing the problem seems puerile. For the planning principles of the socialist economy are nothing but a certain kind of *formulation* of knowable economic laws and their action. Realization of the action of the laws cannot fail to correspond to the principles of their application. It is another thing if actual economic policy and the *practice* of applying these laws significantly de-

viates from the laws of socialist location, that is, from the principles or theoretical fundamentals for economic planning (economic policy). These deviations may be the result of
subjective errors, especially of voluntarism. But one cannot allow that the theoretical
principles, that is, the theoretical fundamentals (these terms are synonymous), do not coincide with the laws, for the aim of every theory is to discover the contents of laws.

Ya. G. Feygin has attempted to give other grounds for the differences between principles
and laws, which it is impossible to agree with. "The principles of location," he writes, "are
the result of the knowledge of objective regularities" ("On the Study of Contemporary
Questions in the Distribution of Productive Forces," *Voprosy ekonomiki* [Problems of
Economics], no. 1 (1960), p. 62). Through such a determination, only if the unknowableness
(complete or partial) of objective laws is acknowledged one may assert the existence of
differences between the laws and the principles of location.

6.   V. I. Lenin, *Sochineniya*, 29:394.

7.   Ibid., 27:227.

8.   *Arkhiv Marksa i Engel'sa* [Archives of Marx and Engels] (1935), 4:119.

9.   Lenin, *Sochineniya*, 29:394.

10.   Ibid., 24:430.

11.   In particular, these questions are dealt with in A. Probst, *Ekonomicheskaya effektivnost' novoi tekhniki. Metodologiya opredeleniya* [The Economic Efficiency of New Machinery: The Methology of Determination] (Moscow, 1960).

12.   For the present the practical possibility exists of calculating only the unit cost and
not the value of a product. However, in solving location problems the unit cost should be
viewed as the only index, even though indirect, available to us for value (change in unit cost
should reflect in value). Therefore, in spite of the deviation of unit cost from value, from
the general methodological standpoint of the solution to this question, the same requirements
should be made of unit cost indices as of value indices.

A. M. KORNEYEV

The Development of Industrial

Complexes in Economic Regions

THE INDUSTRIAL COMPLEXES OF ECONOMIC REGIONS

In each economic region of the country a singular economic complex is developing with a unique combination of industry, agriculture, and other economic branches and a distinctive role in the territorial division of labor. This distinctiveness is due to the particular features of the natural environment and the particular economic and historical conditions contributing to the economic development of each region. These same features also determine the uniqueness of the industrial complex of each economic region.

The development of sectors of the economy is closely related to the general development of the national economy of the country and of its economic regions. The mutual relationships are due to a number of factors: resources, especially minerals; the rational use of labor resources; the rational utilization of industrial by-products and wastes; the creation of public transportation and other communications; and the expediency of organizing industrial complexes connected by co-operative suppliers.

The diversity of conditions in the various regions determines the expediency of specializing the individual economies of the economic

Translated and abridged from *Promyshlennost' v khozyaystvennom komplekse ekonomiches-kikh rayonov SSSR* [Industry in the Economic Complex of Economic Regions of the USSR] ed. A. M. Korneyev (Moscow, 1964), pp. 5–18, 23–33.

regions of the country and of a territorial division of labor. The utilization of favorable conditions in determining the economic specialization of each region and for the production of various types of products is a important factor in raising the productivity of labor.

The branches of specialization have a great and sometimes a determining influence on the formation of the whole economic complex of the region. As a rule, they cause the development of associated branches. For example, the development of ferrous metallurgy evokes a parallel development of the iron ore and coal industry. By utilizing the by-products of coke production, which usually occurs in metallurgical plants, coke chemistry develops; by utilizing slag cinder, cement plants and other industries arise. The needs of ferrous metallurgy demand a corresponding development of energy supplies. Some associated branches can become supplementary branches of the industrial specialization of the economic region. By exerting an influence on the creation of new industrial centers and on the growth of urban population, various branches of specialization thereby influence the development of other branches that serve the needs of the local population and economy: the food industry, production of building materials, and others. Therefore, economic specialization within economic regions requires the development not only of the branches of specialization but also an associated complex of mutually related branches. Consequently, specialization and the comprehensive development of the economy of the regions define the unique process of economic development of economic regions, a process that is directed toward a fuller utilization of favorable conditions for the development of a suitable complex that will help to raise the productivity of social labor.[1]

The specialization and comprehensive development of the economy of economic regions means the fullest rational utilization of natural and labor resources, as well as other conditions available in each economic region, for the creation of a specific structure of the economy with that proportionate development of individual sectors most profitable for the national economy as a whole.

Industry occupies an important place in the economic complex of each economic region. The structure of the economy and the role of industry in the economic complexes were formed under the influence of the general industrial development of the country. The place of industry in the national economy of the country in 1962 can be characterized by the indices shown in table 1. Industry produced about two-thirds of the gross social product and created more than one-half of the national income. It is true that the number of workers and employees occupied in industry and construction as a percentage of the total number of workers engaged in production was only 33.3

TABLE 1

THE ROLE OF INDUSTRY IN THE NATIONAL ECONOMY
(As Percentage of the Total)

	Industry	Construction	Agriculture	Transport and Communications	Other Branches
Number of population occupied in national economy (without students and servicemen). . . .	34		35	7	24
Number of workers, employees, collective farmers.	27.5	5.8	31.9	8.6	26.2
Basic production funds at end of 1962	48.3	3.3	22.9	20.8	4.7
Gross social product.	63.2	9.9	16.8	4.0	6.1
National income	52.2	9.4	22.3	5.2	10.9

SOURCE: Narodnoye khozyaystvo SSSR v 1962 godu [National Economy of the USSR in 1962] (Moscow, 1963), pp. 53, 64, 368, 451, 453-54, 482.

percent, that is, one-third. However, the proportion of the population occupied in industry and construction in the national economy was somewhat higher, and this proportion has had a tendency to rise rapidly. Thus, in 1913, 9 percent of the work force of the national economy was occupied in industry and construction; in 1940 it was 23 percent and in 1962, 34 percent, that is, almost the same number that was occupied in agriculture, in which in 1913 was 75 percent but was only 35 percent in 1962. The proportion of workers and employees in industry and construction in the total number of people occupied in social production already exceeds the proportion occupied in agriculture. Moreover, agriculture has begun to produce a considerable part of its production for the needs of industry. In 1959, for example, from the production of agriculture only 33 percent went directly for consumption and 39.6 percent for industrial processing, 27.4 percent of the production was used for reproduction in agriculture itself (seeds, feed) and for other needs.[2]

The levels of industrial development of the individual economic regions differ considerably. At present there is no single index that can describe exactly the level of economic development of each of the regions, relative to one another. In order to define the level, a number of indices, can be used (table 2).

To determine the level of industrial development of economic regions of the country, one can use, together with other indices, the index of per capita production of the gross product. But the sectoral structure of industry in the region exerts an especially great influence on this index. For example, the regions of Western and Eastern Siberia are similar in the number of industrial-production personnel as a percentage of the All-Union total, and also in the proportion of industrial-production personnel to the population of these regions (10.5 and 10.2 percent respectively); these figures are close to the average of the Soviet Union (10.4 percent). Yet the per capita production of gross product in Western Siberia exceeds the average level in the USSR; this is due to the predominance of the processing industry in the sectoral structure. In Eastern Siberia where the role of the extractive industry in the sectoral structure is great, the gross production per capita is considerably lower than the average in the USSR. On the other hand, the amount of basic production funds per capita in Eastern Siberia exceeds the Union average index; but in the region of Western Siberia this index is below the Union average index. In Eastern Siberia the proportion of urban population in the total population is higher than in the Western region. This may serve as an additional indirect index for a comparative evaluation of the levels of industrial development of the regions.

TABLE 2

INDICES OF THE LEVEL OF INDUSTRIAL DEVELOPMENT OF ECONOMIC REGIONS IN 1960

MAJOR ECONOMIC REGIONS	REGION'S SHARE OF TOTAL POPULATION OF USSR (In Percent)	PROPORTION OF URBAN POPULATION IN TOTAL POPULATION OF THE REGION (In Percent)	PROPORTION OF WORKERS AND EMPLOYEES IN THE TOTAL POPULATION OF THE REGION (In Percent)	NUMBER OF INDUSTRIAL-PRODUCTION PERSONNEL AS PERCENTAGE OF ALL UNION	PROPORTION OF INDUSTRIAL-PRODUCTION PERSONNEL IN TOTAL POPULATION OF THE REGION (In Percent)	INDUSTRIAL PRODUCTION BASIC FUNDS PER CAPITA, AS A PERCENTAGE OF ALL-UNION AVERAGE
Northwestern	5.4	67.5	40.5	8.6	16.4	151.3
Central.	11.8	63.7	39.9	19.2	17.2	105.3
Volga-Vyatka	3.9	42.3	26.7	4.6	12.3	79.5
Central-Chernozem. . .	4.1	29.5	22.3	2.6	6.5	53.8
Volga.	5.9	51.2	30.5	6.3	11.0	123.1
Northern Caucasus. . .	5.7	43.6	26.9	4.3	7.8	84.6
Urals.	8.9	58.6	32.4	11.6	13.6	156.4
Western Siberia. . . .	4.9	55.3	32.1	5.1	11.0	107.7
Eastern Siberia. . . .	3.4	54.9	32.5	3.3	10.2	110.3
Far Eastern.	2.0	72.1	36.9	2.4	12.0	179.5
Donets-Dnieper . . .	7.9	68.1	31.7	10.9	14.2	174.4
Southwestern	9.6	31.2	18.5	5.4	5.9	41.0
Southern	2.4	51.0	27.8	1.8	7.7	41.8
Western.	2.9	49.7	30.2	2.9	10.5	76.9

TABLE 2--Continued

MAJOR ECONOMIC REGIONS	REGION'S SHARE OF TOTAL POPULATION OF USSR (In Percent)	PROPORTION OF URBAN POPULATION IN TOTAL POPULATION OF THE REGION (In Percent)	PROPORTION OF WORKERS AND EMPLOYEES IN THE TOTAL POPULATION OF THE REGION (In Percent)	NUMBER OF INDUSTRIAL-PRODUCTION PERSONNEL AS PERCENTAGE OF ALL UNION	PROPORTION OF INDUSTRIAL-PRODUCTION PERSONNEL IN TOTAL POPULATION OF THE REGION (In Percent)	INDUSTRIAL PRODUCTION BASIC FUNDS PER CAPITA, AS A PERCENTAGE OF ALL-UNION AVERAGE
Transcaucasia	4.6	47.2	21.3	2.8	6.4	112.8
Central Asia	6.7	36.4	18.1	2.7	4.3	43.6
Kazakhstan	4.7	44.5	29.4	2.5	5.6	74.4
Belorussia SSR	3.8	33.8	23.0	2.5	6.8	41.0
Moldavia SSR	1.4	23.9	14.6	0.5	4.1	28.2
Total in the USSR .	100.0	50.1	29.0	100.0	10.4	100.0

In their combination, the indices shown in table 2 make it possible to single out regions having a higher level of industrial development. Here belong, first of all, the Center, Northwest and also the Donets-Dnieper and the Ural regions. These four regions have a higher proportion of industrial-production personnel in the total population (14–17 percent as opposed to 10.4 percent on the average for the USSR) and other relatively higher indices (for example, in the per capita processing of gross production of industry). Regions with a relatively low level of industrial development include: Central Asia, Kazakhstan, Transcaucasia, Central-Chernozem, and Southwest regions and also the Belorussian and Moldavian SSRs. The proportion of industrial-production personnel to population in these regions does not reach even 7 percent; gross industrial production per capita is 30 to 50 percent lower than the average level in the USSR.

The differences in the levels of industrial development of the various economic regions of the country are caused by a number of factors, above all by the great backwardness in the development of the economy in some regions at the beginning of social construction. Thus, in spite of the rapid rate of the development of industry in the Kazakh Republic, the level of industrial development, which from 1913 to 1962 exceeded the average Union rate by 36 percent, still lags considerably behind the average Union level. The rate of development of industry in the regions of the Center and Northwest has been considerably lower than the average All-Union rate, but, because these regions had the most developed industry in the prerevolutionary period, they remain the most developed at the present time. It must, however, be noted that as a result of the higher rate of industrial development in the formerly backward regions, especially in the East, and the relatively lower rate of development of industry in the most developed industrial regions of prerevolutionary Russia, the difference in the levels of economic development among the economic regions of the country has decreased considerably. As a result of applying socialistic principles to the location of production forces, there is occuring a process of equalization of the levels of regional economic development.

The rate of industrial development and, consequently, the attained level of industrial development in the various regions of the country have been influenced by a number of factors. The most important of them are the existence of hydroelectric power and other resources previously investigated and accessible to industrial utilization and the existence of cadres of qualified workers. The time factor was also significant.

Each economic region has special conditions for the development of industry. The specific combination of factors influencing its development also determines the character of formation of industrial complexes. The character of industrial complexes of the economic regions can be defined by analyzing the data concerning the sectoral structure of industry in the regions of the country, the data about the interregional, interbranch connections, and the data on interregional economic connections of industry. The industrial structure of the economic regions is described by data on the share of the various branches in the gross production of the entire industrial region (see table 3) and by data on the share of the branches in the total number of industrial-production personnel of the region (see table 4).

The branch structure gives an idea of the essential differences in the industrial complexes of the economic regions of the country and of the relative role of the industrial branches in the industrial complex of each region. The largest branches of the processing industry—machine-building, light industry, and food industry—have a considerably higher share in the industrial complexes of a number of economic regions than in the economy of the country as a whole (in gross production as well as in number of personnel). These regions include the Central, Western, Southern, Central-Chernozem, and Central Asian regions; the Moldavian SSR has the highest share. Not all the above-mentioned branches are developed to the same degree in each region. For example, in the Southern and Central-Chernozem regions and also in Moldavia the food industry has the highest share; in the Central Asian region, the light industry—more precisely cotton-refining and food; in the Central region, light industry and machine-building. A high share of the food industry in the industrial complex characterizes, as a rule, a relatively high level of development of agriculture in the region. The exception is the Far East where the fishing industry provides about half of the gross production and employs about two-thirds of the personnel in the food industry of the region.

In the industrial complexes of the remaining regions the share of of the extracting industry and the directly related metallurgy, electrical energy, and building-material industries have a relatively high proportion that considerably exceeds the average All-Union proportion. Among such regions are the Urals, Eastern Siberia, Donets-Dnieper, Kazakhstan, and Western Siberia.

In the industrial complexes of the remaining regions the share of the branches mentioned is close to the average All-Union share or deviates from it only in one of the indices under consideration: either in gross production or in the number of industrial-production personnel.

The above-cited data concerning the branch structure of industry by regions augment each other and reflect the character of the index accepted for the description of the branch structure. The structure of industry according to gross production serves as a broader base for showing the character of the industrial specialization of the economy of the region than does the structure according to the number of workers. However, an evaluation of the role of the sectors in the industrial complex, from the point of view of utilizing the major productive forces of the country, can only be made on the basis of the structure of industry according to the number of workers employed in these branches. Branches of specialization must be, first of all, those sectors of industrial production that are either impossible in other regions in the amounts needed by the country or that demand greater expenditures of social labor. The establishment of a group of branches in which it is expedient to specialize the economy of a region is a problem of determining the proper direction for the formation of the economic complex of the region.

The first step in revealing the sectors in which the economies of regions actually specialize must be an examination of the sectoral structure of industry in each economic region in comparison with the sectoral structure of industry in the country as a whole. For example, the high share of industrial output of machine construction in the Volga-Vyatka region or of light industry in the Central region gives one reason to think that the economies of these regions are specializing in the industrial production of the respective branches of industry. However, in order to make the sectors of specialization apparent it is necessary to take into account the general level of industrial development and the level of development of the individual sectors in the economic regions. To a certain degree this level is reflected in comparing the gross industrial production per capita with the average per capita production in the Soviet Union.

Thus, in the industry of Transcaucasia the proportion of light industry is greater than that of the Union as a whole (23.9 percent as against 20.6 percent). But the per capita production of light industry is 23.1 percent of the average for the Soviet Union. In Central Asia the proportion of light industry in the gross industrial output is very high (46 percent); the production of light industry per capita exceeds the average national per capita production by 9.6 percent and in the textile industry by 51.2 percent. And yet, on the basis of these data one cannot draw the conclusion that light industry as a whole is the sector of industrial specialization in the economy of Central Asia.

In Central Asia the norms of per capita production of the majority of the major types of articles of the light industry lag considerably behind the average national norms. Forty-five percent of the gross

TABLE 3

THE STRUCTURE OF INDUSTRY OF ECONOMIC REGIONS IN 1960

(In Percent)

(By Gross Production)

Major Economic Regions	All Fuel Industries	Coal Only	Production of Electric and Thermal Energy	Machine-building and Metallurgy	Lumber, Paper, Wood-working Industry	Building Materials Only	Other Sectors of Heavy Industry	Light Industry	Textile Industry Only	Food Industry	Fishing Industry Only
Northwestern	2.5	1.6	1.6	27.4	13.5	3.1	8.4	19.4	8.7	20.5	6.2
Central	1.6	0.7	1.5	23.7	3.4	2.7	9.8	40.4	29.6	13.3	0.2
Volga-Vyatka	1.6	...	2.3	35.7	13.4	2.2	9.7	18.3	6.6	15.0	...
Central Chernozem	0.1	...	1.7	25.2	2.7	4.1	14.0	13.9	6.8	36.8	0.1
Volga	8.1	5.3	4.4	30.8	4.8	4.5	6.6	16.8	7.0	21.3	1.4
Northern Caucasus	9.6	1.9	1.9	18.0	4.9	4.5	5.5	16.4	6.6	34.8	1.4
Urals	6.4	...	3.7	24.5	8.5	3.6	30.1	7.9	1.5	13.4	0.4
Western Siberia	12.0	10.4	3.3	26.3	5.8	3.7	13.7	13.5	4.9	19.2	0.3
Eastern Siberia	5.8	3.9	5.4	15.0	18.4	5.5	7.4	10.5	3.2	21.1	0.7
Far Eastern	7.9	6.4	4.4	20.0	13.1	5.7	5.2	6.4	0.1	36.2	17.8
Donets-Dnieper	12.1	11.8	2.1	23.1	2.0	4.0	28.3	8.5	2.3	17.9	0.3
Southwestern	2.3	1.3	1.9	16.7	9.7	5.1	4.0	19.2	5.2	37.7	0.2
Southern	0.7	...	2.4	23.5	3.2	5.2	3.8	15.5	4.9	42.4	4.7

TABLE 3--Continued

Major Economic Regions	All Fuel Industries	Coal Only	Production of Electric and Thermal Energy	Machine-building and Metallurgy	Lumber, Paper, Wood-working Industry	Building Materials Only	Other Sectors of Heavy Industry	Light Industry	Textile Industry Only	Food Industry	Fishing Industry Only
Western	2.5	..	1.9	16.4	8.8	3.9	2.2	28.3	14.9	32.7	7.4
Transcaucasia . . .	9.9	0.9	2.6	12.3	3.4	4.4	10.4	23.9	14.5	30.6	0.6
Central Asia. . . .	3.4	1.2	1.7	12.0	2.6	4.4	3.5	46.0	35.0	25.0	0.5
Kazakhstan. . . .	6.8	5.9	3.9	12.7	4.8	7.2	20.2	17.3	7.3	25.9	1.1
Belorussia SSR. . .	2.6	..	2.2	22.5	9.6	4.1	2.2	26.6	12.8	27.5	0.1
Moldavia SSR. . . .	..	..	1.4	7.4	4.8	5.0	0.8	19.9	4.5	60.0	0.2

TABLE 4

STRUCTURE OF INDUSTRY OF ECONOMIC REGIONS IN 1960
(In Percent)
(By Average Number of Industrial Production Personnel)

Major Economic Regions	All Fuel Industries	Coal Only	Production of Electric and Thermal Energy	Machine-building and Metallurgy	Lumber, Paper, Wood-working Industry	Building Materials Only	Other Sectors of Heavy Industry	Light Industry	Textile Industry Only	Food Industry	Fishing Industry Only
Northwestern	3.8	2.1	1.2	32.1	23.9	4.6	5.9	14.8	6.2	8.8	3.4
Central	3.8	2.0	1.1	35.8	7.1	4.8	6.1	29.6	20.4	5.6	0.1
Volga-Vyatka.	2.7	..	1.4	41.4	23.0	3.6	7.0	12.2	4.0	6.1	..
Central Chernozem . .	0.5	..	2.9	38.3	5.9	8.8	9.4	13.8	5.4	18.1	0.2
Volga	2.5	..	2.1	46.7	9.1	6.9	5.4	13.1	4.5	10.4	2.3
Northern Caucasus . .	10.8	9.0	2.2	28.6	10.0	9.4	3.9	14.1	3.2	16.4	2.3
Urals	4.9	3.1	1.7	32.9	17.0	5.4	22.0	7.5	1.4	6.1	0.7
Western Siberia . . .	13.3	12.8	2.5	35.4	12.4	6.1	10.0	11.0	3.4	7.8	0.4
Eastern Siberia . . .	5.3	4.0	3.2	19.2	29.4	6.4	10.0	8.6	2.0	8.1	0.8
Far Eastern	8.1	7.4	3.1	24.8	19.5	6.4	7.6	7.1	..	21.8	14.6
Donets-Dnieper. . . .	22.3	22.0	1.3	30.5	3.3	6.6	18.7	7.8	1.7	7.8	0.3
Southwestern.	3.7	1.7	2.2	23.7	16.8	10.8	2.8	17.4	4.7	17.6	0.3
Southern.	0.3	..	2.5	36.6	5.8	9.0	3.7	18.4	6.7	18.8	4.2

TABLE 4--Continued

Major Economic Regions	All Fuel Industries	Coal Only	Production of Electric and Thermal Energy	Machine-building and Metallurgy	Lumber, Paper, Wood-working Industry	Building Materials Only	Other Sectors of Heavy Industry	Light Industry	Textile Industry Only	Food Industry	Fishing Industry Only
Western.	4.8	. .	1.6	23.2	15.4	6.7	1.7	25.5	13.2	15.9	5.3
Transcaucasia. . .	8.1	1.6	2.1	20.2	6.5	8.8	10.3	25.2	13.7	14.2	0.9
Central Asia . .	4.2	2.7	3.3	22.8	5.1	9.3	5.8	32.4	18.9	13.7	0.9
Kazakhstan . .	8.5	7.8	5.5	20.8	7.5	11.5	17.1	14.7	4.0	12.7	2.5
Belorussia SSR . .	10.3	. .	2.4	25.4	17.0	7.9	2.0	19.5	6.9	11.5	0.3
Moldavia SSR . ．	. .	. .	2.7	16.1	11.1	15.1	0.6	25.8	9.2	27.7	0.6

TABLE 5

PRODUCTION OF MAJOR ECONOMIC REGIONS AS A PROPORTION OF THE ALL-UNION PRODUCTION IN 1960
(In Percent)
(In Their Natural Units)

	Northwestern	Central	Volga-Vyatka	Central-Chernozem	Volga	Northern Caucasus	Urals	Western Siberia	Eastern Siberia	Far East	Donets-Dnieper	Southwestern	Southern	Western	Transcaucasia	Central Asia	Kazakhstan	Belorussia SSR	Moldavia SSR
Iron ore.	2.0	0.6	...	3.6	...	...	26.1	4.0	1.6	...	51.4	...	4.0	...	1.3	...	5.4	...	...
Cast iron	2.7	1.6	...	2.3	...	...	32.5	7.1	...	...	51.7	...	...	...	1.5	...	0.6	...	...
Steel	3.9	2.5	2.2	0.6	2.9	1.8	33.6	7.0	0.8	0.7	39.7	0.1	0.2	0.2	2.6	0.5	0.5	0.2	...
Rolled steel.	4.2	1.5	2.4	0.6	2.2	1.6	33.0	8.0	0.8	0.5	41.0	...	0.4	0.3	2.4	0.4	0.6	0.1	...
Coal.	3.5	8.3	...	...	...	6.3	12.1	16.4	7.2	4.2	32.8	0.8	...	...	0.6	1.5	6.3	...	...
Gas	3.4	1.8	...	...	16.0	29.0	3.9	...	...	0.7		30.3		0.9	12.4	1.5	0.1	...	...
Electric energy	6.1	10.4	2.8	1.8	9.3	3.3	18.7	7.6	5.6	1.7		18.5		1.6	4.6	3.0	3.6	1.2	0.2
Automotive:																			
trucks.	...	34.3	48.5	...	0.8	...	8.2	...	...	...	1.3	...	...	...	1.7	...	...	5.2	...
passenger cars.	...	45.3	36.1	...	18.2	...	...	...	...	...	0.4	...	...	...	...	...	...	...	...
Electric motors (alternating current): from 0.25 to 100 kilowatts (3 phase) .	0.3	19.2	2.1	...	...	...	15.8	13.2	...	...	21.3	...	1.6	8.4	12.6	2.7	...	2.6	0.2
above 100 kilowatts . .	17.4	37.5	...	...	...	...	13.3	...	...	...	10.3	...	...	9.7	11.8	...	...	...	...

TABLE 5—Continued

	Northwestern	Central	Volga-Vyatka	Central Chernozem	Volga	Northern Caucasus	Urals	Western Siberia	Eastern Siberia	Far East	Donets-Dnieper	Southwestern	Southern	Western	Transcaucasia	Central Asia	Kazakhstan	Belorussia SSR	Moldavia SSR
Chemical fibers	6.0	43.5	...	0.1	0.3	15.9	1.9	6.0	9.8	...	...	6.7	...	...	...	0.4	2.4	7.0	...
Mineral fertilizer	4.2	14.9	2.4	1.2	...	...	24.7	4.2	0.1	...	20.7	5.5	1.6	5.7	3.3	8.1	3.4	...	...
Shipments of processed:																			
lumber.	26.6	7.5	9.9	0.4	1.5	1.2	20.4	4.9	14.2	5.0	0.2	3.7	...	1.8	0.3	...	0.5	1.9	...
paper.	32.2	4.7	10.8	0.3	1.4	0.9	21.2	0.1	0.4	8.1	0.9	5.2	0.7	9.1	1.4	0.4	...	2.9	...
cotton cloth.	3.8	77.4	1.5	0.1	0.7	0.2	0.2	1.5	1.3	0.1	0.2	0.6	1.0	3.0	3.4	4.9	0.3	0.1	...
wool cloth.	4.7	56.9	0.5	3.6	7.7	3.6	1.6	0.6	0.4	...	3.2	1.4	0.5	5.5	3.2	0.4	1.2	4.4	0.01
granulated sugar. . . .	...	0.7	0.1	16.9	1.3	5.2	0.3	0.6	...	0.6	13.5	46.9	2.5	5.1	1.1	2.1	1.9	0.5	2.8
fish	38.4	0.3	0.1	0.1	7.7	3.9	1.6	0.5	0.6	24.5	0.9	0.2	...	11.7	1.9	1.5	3.4	0.2	...
vegetable oil	1.0	0.8	0.9	6.3	5.7	18.2	1.1	0.7	...	3.2	17.4	6.1	4.8	0.7	2.1	21.5	3.2	0.9	5.4

production of light industry and 20.7 percent of the entire industry of the region are contributed by the cotton-refining industry, which involves the primary processing of raw cotton. The most important branch of specialization of the economy of the region is cotton growing. It is this branch that basically determines the profile of the economic complex of Central Asia. Cotton growing causes the development of a number of branches of industry: chemical fertilizers and machine-building specialized in the production of machines and equipment for the cotton industry; it causes also the development of the cotton-refining industry. Since raw cotton is not a transportable product, cotton-refining plants are situated directly in the microregions of cotton growing. In the prewar years sowing of cotton was expanded into the so-called new regions of cotton growing in the European part of the country. Cotton-refining plants were erected for the primary processing of raw cotton. When, for reasons of expediency the sowing of cotton was discontinued in these regions, the cotton-refining plants were liquidated. Although raw cotton therefore does not participate in the interregional exchange of Central Asia, cotton growing is nevertheless an important branch of specialization of the region. The cotton-refining industry, however, serves as an adjunct to cotton growing; it is a branch accompanying cotton growing and it cannot independently be a branch of economic specialization of the region, although it is its product, cotton fiber, that enters the interregional exchange.

The data on shares of branches in the gross production of the total industry of economic regions and on the output of this product on a per capita average give a most general idea of the character of industrial specialization of the economy of the regions. One can determine the specialized production of a region only by examining the data on production and interregional connections between closely related branches of industry, and even between individual types of production. Thus, the comparative data on the share of production of economic regions in the output of specific types of industrial production of the Union in its natural expression (see table 5) and on the per capita output of this production (see table 6) make it possible to approach this question tangibly: In what types of production does the industry of the various regions specialize? It becomes apparent here that despite a low proportion of production of a complex sector in the gross production of industry, and a generally low level of industrial development, some regions nevertheless specialize in the production of individual articles of a given sector. For example, in the Northern Caucasus and in Eastern Siberia the general level of development of the chemical industry is considerably lower than the average Union level. But these regions occupy second and third place in the Union in the production of chemical fibers.

TABLE 6

PRODUCTION OF THE MOST IMPORTANT PRODUCTS IN THEIR NATURAL UNITS
IN THE ECONOMIC REGIONS IN 1960
(As Percentage of the Average Union Production)

	Northwestern	Central	Volga-Vyatka	Central Chernozem	Volga	Northern Caucasus
Iron ore	38.0	6.0	...	86.0	...	...
Cast iron.	49.7	14.2	...	57.3	...	...
Steel.	71.5	21.4	57.0	15.6	49.0	32.4
Rolled steel	76.7	12.9	62.2	14.6	36.6	27.9
Coal	64.7	71.6	...	...	...	109.9
Gas.	63.6	13.6	...	...	268.2	504.5
Electric energy.	112.1	89.5	73.0	43.2	156.8	57.5
Automotive:						
trucks	...	294.1	1,247.0	...	11.8	...
passenger cars	...	387.7	930.8	...	304.6	...
Electric motors						
(alternating current):						
from 0.25 to 100						
kilowatts (3 phase). .	6.9	164.9	55.7	...	...	1.5
above 100 kilowatts. . .	322.2	322.2	...	...	...	...
Chemical fibers.	110.2	375.5	...	...	...	279.6
Mineral fertilizer	76.4	128.0	62.1	29.7	...	...
Shipments of processed:						
lumber	500.0	66.7	258.3	8.3	25.0	25.0
paper.	597.3	40.2	283.0	8.0	23.2	16.1
cotton cloth	69.6	669.2	38.2	1.7	12.5	4.4
wool cloth	87.5	487.5	12.5	87.5	131.2	62.5
granulated sugar	...	6.1	...	412.6	21.1	91.2
fish	706.7	3.0	1.5	2.2	130.4	68.9
vegetable oil.	17.8	6.8	21.9	153.4	95.9	317.8

TABLE 6--Continued

	Urals	Western Siberia	Eastern Siberia	Far East	Donets-Dnieper	Southwestern	Southern	Western	Transcaucasia	Central Asia	Kazakhstan	Belorussia SSR	Moldavia SSR
Iron ore	292.0	82.0	46.0	···	644.0	···	166.0	···	26.0	···	114.0	···	···
Cast iron.	365.3	145.2	···	···	651.5	···	···	···	33.3	···	12.4	···	···
Steel	378.9	143.6	22.6	32.6	501.1	1.1	7.2	5.2	57.2	6.8	9.9	4.8	···
Rolled steel	371.9	164.4	25.2	23.4	510.4	···	17.4	9.9	52.6	5.6	12.7	2.4	···
Coal	136.3	336.4	214.2	207.8	412.9	8.0	···	···	12.0	22.6	133.6	···	···
Gas.	45.5	···	···	36.4	···	150.0	···	31.8	268.2	22.7	1.8	···	0.3
Electric energy. . . .	210.6	155.7	167.3	83.3	···	92.3	···	56.7	96.3	44.7	76.0	32.5	16.5
Automotive: trucks . .	94.1	···	···	···	11.8	···	···	···	35.3	···	···	135.3	···
passenger cars	···	···	···	···	1.5	···	···	···	···	···	···	···	···
Electric motors (alternating current): from 0.25 to 100 kilowatts (3 phase).	177.9	271.0	···	···	267.9	···	66.4	293.1	271.8	39.7	···	68.7	13.7
above 100 kilowatts. .	155.5	···	···	···	133.3	···	···	344.4	255.5	···	···	183.7	···
Chemical fibers. . . .	21.4	124.5	292.8	···	···	70.4	···	···	···	6.1	50.0	···	···
Mineral fertilizer . .	278.0	86.4	4.5	···	261.4	57.3	64.1	198.9	71.4	120.6	73.1	···	···

TABLE 6--Continued

Shipments of processed:	Urals	Western Siberia	Eastern Siberia	Far East	Donets-Dnieper	Southwestern	Southern	Western	Transcaucasia	Central Asia	Kazakhstan	Belorussia SSR	Moldavia SSR
lumber.	233.3	100.0	433.3	250.0	2.5	41.7	1.7	66.7	75.0	0.2	8.3	50.0	2.5
paper	241.1	1.0	123.2	400.9	11.6	54.5	...	320.5	30.4	6.2	...	75.9	...
cotton cloth.	2.7	31.8	38.2	5.4	2.7	6.1	29.1	107.8	74.3	73.3	6.7	2.0	1.7
wool cloth.	18.7	12.5	12.5	...	37.5	12.5	37.5	193.7	68.7	6.2	25.0	118.7	6.2
granulated sugar. . .	3.4	12.6	0.7	29.9	170.4	490.8	21.4	177.9	23.8	31.6	40.1	13.3	202.0
fish.	17.8	11.1	17.8	1,206.7	11.9	1.5	9.6	410.4	41.5	22.2	71.9	4.4	1.5
vegetable oil	12.3	15.1	...	158.9	220.5	64.4	198.6	26.0	45.2	319.2	67.1	24.7	386.3

The output of these fibers per capita in these regions exceed the average Union per-capita output by almost three times.

Not only can the branch that finishes the product, delivers the product, and participates in the interregional exchange be classified as a branch of industrial specialization of the economy of a region, but the allied branches connected with the end product can be included as well.

The development in the Ural region of machine-building utilizing a considerable part of locally produced metals may serve as an example of the development of *allied branches of specialization*, in this case of ferrous metallurgy and machine-building. Even if all the metal produced in the Ural region were fully utilized by the machine-building of this region, even then metallurgy would not lose its significance as a branch of industrial specialization in the Ural region. It is the metallurgical industry that has been exercising the major influence on the formation of the character of the industrial complex of the Ural region. The structure of machine-building in this region has developed under the influence of metallurgy and the large-scale production of metal in the region. And if the iron and steel industry in the Urals produces one-third of the cast iron, steel, and rolled metal in the country, and if the production of these types of products is 3.5 to 4 times as large per capita in the region as in the country, the significance of ferrous metallurgy as the most important branch of industrial specialization cannot change depending on what region uses what quantity of metal produced in the Ural region.

Much attention is paid in the literature to questions concerning the specialization of economic regions and to indices that characterize the branches of specialization. However, the most important question of great practical significance is ignored: How economically effective is the specialization of the economy of a given region in the production of particular types of products?

The branches of specialization must come about under the influence of natural-climatic, economic, and other conditions favorable for their development. The development of these sectors in individual regions must contribute to economical expenditures of social labor while solving the problem of satisfying the needs of the national economy. Therefore, the most important criterion in defining branches of economic specialization for a region must be the data that characterize the comparative expenditures of labor for production output in these branches of the various regions. The specialization of the economy in a definite product is not a goal in itself, but only one of the means of the economy of social labor, whereby the specialization of the region guarantees a higher productivity of labor only if this region has relatively more favorable conditions for the development of suitable sectors of the economy or industry.

The development of branches of specialization increases the territorial division of labor. The participation of the region in interregional direct exchange with a part of its production, or indirectly with a component part of the production of allied branches, expresses the degree of specialization[4] of the region in products of the cited branches. The larger the part of the total production that is sent out, the higher is the degree of specialization of the region in the given type of product.

The existing interregional exchange must be examined and evaluated from the point of view of whether it contributed to the solution of All-Union problems with minimum expenditures of social labor. If it is stipulated by a specialization contradicting these principles, then the specialization of the pertinent regions must be reexamined. For example, the Far East ships out a number of articles of the machine-building and metals-processing branches. Moreover, the production for export comprises a large part of its production. In 1960, 68 percent of its hoisting transport equipment and 78 percent of its compressors were sent to the European part of the country. Of the total equipment no more than 10 to 12 percent was locally utilized, and of forge-press equipment only 2 to 3 percent. A considerable amount of equipment produced in the Far East is sent to the Ukraine, the West, and the Moldavian SSR. Even in Magadan Oblast half of the diesel motor fuel machinery produced there is sent to western regions of the country. The metal for the production of metal articles is brought in from regions thousands of kilometers away, and the articles made from this metal are sent in the opposite direction. The cost of some types of production is three to four times higher in the Far East than that of a corresponding product in other regions, and as a result of transport expenditures is even higher for the consumer.

Similar economically irrational specialization exists in other regions; this confirms the importance of and the necessity for a practical evaluation of historically accumulated specializations in economic regions, taking into account all factors that determine the rational specialization of the economy of the economic regions.

An analysis of the structure of industry in the economic regions according to the number of industrial-production personnel occupied in the various branches makes it possible to find out the following: the place in the industrial complex of branches in which either men or women predominantly are employed; the role of seasonal branches of industry in the economic complex of the region and the possibility of their combination. A suitable combination of branches of industry in the economic regions will have great significance for a fuller utilization of the labor resources of the country and a reduction of the strain on the labor balance in some regions.

LABOR RESOURCES

Although the significance of natural wealth is very great it characterizes only the potential possibilities of a region. To mobilize this wealth and to put it into the service of society is only possible through the labor of man. Therefore, labor resources are one of the most important conditions for the development of industry; they influence the formation of pertinent industrial complexes. At the same time, the development of productive forces, above all industry, can guarantee the fullest utilization of labor resources in production, provided there are suitable combinations of branches in the economic complex of the various economic regions.

The resources of the labor force will be expanded by: a natural increase in the population reaching working age; the release of part of the labor force occupied in agriculture, which will be the result of the comprehensive mechanization of agricultural work and other measures resulting from a continual technological progress; and the involvement of the labor force occupied in the domestic and secondary establishments in production. The utilization of the last source is at the present directly related to the solution of the problem of the rational distribution of industry and the comprehensive development of the economy of the economic regions. This is the main objective of our investigation.

The utilization of labor resources and their possible reserves by economic regions can be seen in table 7. The census of 15 January 1959 best shows the state of the labor resources utilization and possible reserves in the regions of the country. The total labor resources are comprised of the population of working age, 16 to 60 years for males and up to 55 years for females. A small part of the population in this age group are invalids. Moreover, a part of the population above 16 years of age is studying and cannot participate in production. These two categories of persons in the census totaled 7.6 percent. In the future the proportion of these categories will change somewhat. The changes occurring in the age structure of the population of the postwar period and the introduction of compulsory secondary education will raise the proportion of students over 16 years of age. On the other hand, there will be a decrease in the number and proportion of invalids of working age since a considerable number of them, the war invalids, will belong to the group of population beyond working age.

Of the total population occupied in branches of the national economy and industry, 4.2 percent are people of 60 years and above and 4.1 percent of ages 55 to 59.[6] These same groups comprise the supplementary labor resources used in the national economy. Basically they compensate for the number of students of working age who do not participate in production.

People occupied in private domestic agriculture (more than 5 million) and the group of working-age people out of school who are dependents serve as supplementary reserve cadres of production. This group comprises almost 13 million, of which 9.5 million live in the cities. The total labor reserves of these two groups are 18 million, of which 91 percent are women. Consequently, the problem of maximum utilization of existing labor resources of the country is, first of all, a problem of involving female labor in production.

The proportion of the population composed of labor reserves of these groups varies considerably by economic regions. The proportion of the number of people employed in private, domestic agriculture, and nonstudying dependents of working age out of the total number of population of working age is highest in Kazakhstan (21.2 percent), in Transcaucasia (19.7 percent), and in the Far East (19.2 percent). Moreover, in Kazakhstan and the Far East there is a labor shortage. In Eastern and Western Siberia, where there are also difficulties in supplying the industrial enterprises with working cadres, the proportion of the number in these groups of the population is above average, 16.5 percent and 17.8 percent respectively. At the same time the proportion of these groups in the population in the Central region is 10.4 percent as against 14.9 percent of the average in the Union. Of the population in these groups women comprise 94.3 percent in the Far East, 93.6 percent in Western Siberia; in Kazakhstan, 93 percent, in Eastern Siberia, 93 percent, that is, the proportion of women occupied in personal and secondary agriculture and non-studying dependents of working age in these regions is higher than the Union average. It must be noted, however, that in the Soviet Union more than half, and in the region of the Far East two-thirds, of the number of nonstudying female dependents have children under 14 years of age. A part of the women, due to a lack of a network of preschool children's institutions and public dining facilities, must stop working and devote their time to tending the children and household.

With the development of the food industry and public dining facilities, with the creation of a suitable network of children's institutions, and with the establishment of other services, conditions will be favorable for freeing women from household work. Consequently, the actual labor reserves made up of nonstudent female dependents occupied at the present time in household work will grow. The utiliza-

TABLE 7

LABOR RESOURCES IN THE ECONOMIC REGIONS OF THE USSR

MAJOR ECONOMIC REGIONS	POPULATION OF WORKING AGE 15 JANUARY 1959		NUMBER EMPLOYED		EMPLOYED IN PRIVATE, DOMESTIC AGRICULTURE (In Thousands)		
	Thousands	As Percentage of Population	Thousands	Percentage of Working Age Population	Total	Urban Population	Village Population
Northwestern	7,010.4	61.1	5,669.3	80.9	100.0	14.1	85.9
Central	14,966.2	60.4	12,118.9	81.0	263.6	40.8	222.8
Volga-Vyatka	4,672.2	56.6	3,793.4	81.2	199.0	20.4	178.6
Central Chernozem . .	4,921.3	56.6	3,844.2	78.1	330.4	18.3	312.1
Volga	7,152.7	57.4	5,474.4	76.5	328.2	26.4	301.8
Northern Caucasus . .	6,830.5	58.0	5,004.2	73.3	423.3	42.1	381.2
Urals	10,642.7	57.2	8,223.8	77.3	392.1	62.2	329.9
Western Siberia . . .	5,809.2	57.2	4,313.3	74.2	404.4	84.2	320.2
Eastern Siberia . . .	3,995.6	57.4	3,028.5	75.8	196.2	33.2	163.0
Far East	2,608.9	60.0	1,919.4	73.6	76.3	25.6	50.7
Donets-Dnieper . . .	10,217.4	61.7	7,593.8	74.3	347.3	128.8	218.5
Southwestern	11,623.9	57.4	9,409.6	80.9	615.0	85.9	529.1

TABLE 7—Continued

MAJOR ECONOMIC REGIONS	POPULATION OF WORKING AGE 15 JANUARY 1959		NUMBER EMPLOYED		EMPLOYED IN PRIVATE, DOMESTIC AGRICULTURE (In Thousands)		
	Thousands	As Percentage of Population	Thousands	Percentage of Working Age Population	Total	Urban Population	Village Population
Southern.	3,041.1	60.0	2,358.0	77.5	100.0	16.8	83.2
Western	3,443.1	57.4	2,660.8	77.3	127.9	14.4	113.5
Transcaucasia . . .	5,092.0	53.6	3,647.1	71.6	247.9	38.5	209.4
Central	6,774.3	49.6	5,093.6	75.2	304.0	38.0	266.0
Asia	4,979.7	53.6	3,522.5	70.7	343.5	46.8	296.7
Kazakhstan. . . .	4,446.5	55.2	3,635.1	81.8	195.3	29.5	165.8
Belorussia SSR. . . .	1,593.9	55.3	1,385.8	86.9	41.1	5.1	36.0
Moldavian SSR . . .							
Total in the USSR. .	118,921.6	57.4	92,695.7	77.4	5,035.5	771.1	4,264.4

TABLE 7—Continued

MAJOR ECONOMIC REGIONS	DEPENDENTS OF INDIVIDUALS, NONSTUDENTS (In Thousands)			INCLUDING EMPLOYED IN PRIVATE, DOMESTIC AGRICULTURE AND DEPENDENTS OF INDIVIDUALS, NONSTUDENTS					
				In Thousands			As Percentage of Population Working Age		
	Total	Urban Population	Village Population	Total	Men	Women	Total	Men	Women
Northwestern.	678.0	511.8	166.2	778.0	57.6	720.4	11.1	0.8	10.3
Central	1,287.5	940.1	347.4	1,551.0	167.5	1,383.5	10.4	1.2	9.2
Volga-Vyatka. . . .	346.8	224.1	122.7	545.7	58.0	487.7	11.7	1.3	10.4
Central Chernozem . . .	382.0	233.1	148.9	712.5	75.2	637.3	14.5	1.6	12.9
Volga	747.6	551.4	196.2	1,075.8	95.6	980.2	15.0	1.3	13.7
Northern Caucasus . . .	855.6	596.2	259.4	1,278.9	104.7	1,174.2	18.7	1.5	17.2
Urals	1,179.3	892.7	286.6	1,571.5	121.4	1,450.1	14.8	1.2	13.6
Western Siberia . . .	632.4	492.2	140.2	1,036.8	66.1	970.7	17.8	1.1	16.7
Eastern Siberia . . .	469.9	327.0	137.9	661.1	46.8	614.3	16.5	1.1	15.4
Far East.	425.4	326.4	99.0	501.7	28.2	473.5	19.2	1.1	18.1
Donets-Dnieper. . . .	1,441.7	1,305.0	136.7	1,789.0	119.7	1,669.3	17.5	1.2	16.3
Southerwestern. . . .	912.8	613.3	299.5	1,527.7	189.0	1,338.7	13.1	1.6	11.5
Southern.	347.5	273.7	73.8	447.5	34.4	413.1	14.7	1.1	13.6
Western	399.9	270.4	129.5	527.8	62.7	465.1	15.3	1.8	13.5

TABLE 7--Continued

MAJOR ECONOMIC REGIONS	DEPENDENTS OF INDIVIDUALS NONSTUDENTS (In Thousands)			INCLUDING EMPLOYED IN PRIVATE, DOMESTIC AGRICULTURE AND DEPENDENTS OF INDIVIDUALS, NONSTUDENTS					
				In Thousands			As Percentage of Population Working Age		
	Total	Urban Population	Village Population	Total	Men	Women	Total	Men	Women
Transcaucasia.	753.7	616.7	137.0	1,001.6	105.6	896.0	19.7	2.1	17.6
Central.	850.9	589.3	261.6	1,155.0	123.5	1,031.5	17.0	1.7	15.2
Asia.	712.5	468.3	244.2	1,056.0	73.6	982.4	21.2	1.5	19.7
Kazakhstan	338.7	228.1	110.6	534.0	64.5	469.5	12.0	1.4	10.6
Belorussia SSR	103.1	71.8	31.3	144.2	14.1	130.1	9.0	0.8	8.2
Moldavian SSR.									
Total in the USSR .	12,860.8	9,531.6	3,328.7	17,895.8	1,608.2	16,287.6	14.9	1.3	13.6

SOURCE: Itogi Vsesoyuznoy perepisi naseleniya 1959 g., SSSR[Results of All-Union Census 1959, USSR] (Moscow, 1962), pp. 98-101. Itogi Vsesoyuznoy perepisi naseleniya 1959 g., RSFSR [Results of All-Union Census 1959, RSFSR] (Moscow, 1963), pp. 158-175. Itogi Vsesoyuznoy perepisi naseleniya 1959 g., Ukrainskaya SSR [Results of All-Union Census 1959, Ukrainian SSR] (Moscow, 1963), pp. 74-79.

tion of these labor reserves will depend considerably on the development of pertinent economic complexes in these regions of the country. Working conditions in some branches of industry limit the possibility of using female labor, and as a result in some regions there remain considerable reserves of female labor force outside the sphere of production. The involvement of these reserves in production is of great social and economic significance. It can also be a great influence on the formation and structure of industrial complexes in the economic regions.

The participation of women in production is continually growing. The proportion of employed women in the total number of workers and employees has risen from 27 percent in 1929 to 48 percent in 1962.[7] The proportion of female workers in the total number of industrial workers has also risen. In 1913, it was 24.5 percent; in 1940, 42.9 percent; in 1959, 44.2 percent; and for January 1962, 45.6 percent. Whereas in 1913, 63.1 percent of all female workers in large industries worked in the textile and sewing branches of industry, and only 2.4 percent in machine-building and metal-processing, in January 1959 the proportion of women workers in machine-building and metal-processing rose to 28.1 percent. The proportion of women working in the textile and sewing industry fell to 22.1 percent.[8]

In essence, the labor conditions in the various branches of industry determine the distribution of male and female labor among the branches of industry. This distribution is described by the data on the proportion of men and women in the total number of industrial production personnel and workers in the industrial branches. The highest proportion of women in the number of industrial production personnel and in the number of workers is characteristic for the branches of light industry, especially the knitted fabric and sewing industry. Women comprise 83.6 percent and 82 percent respectively of the number of workers in these branches as against a 45.6 percent average in all industry. Of the branches of heavy industry, only in the chemical branch do women somewhat outnumber men. But in other branches of heavy industry male labor is used predominantly.

It must be noted, however, that in the last several years there has been a tendency toward a rise in the proportion of men in the number of workers and industrial production personnel in a number of branches of heavy industry. Thus, from 1 January 1959 to 1 January 1962 this proportion has risen in ferrous metallurgy from 69.9 to 70.6 percent; in machine-building and metal-processing from 60.7 to 61.1 percent; in the cement industry, from 63 to 63.7 percent. But in the textile industry the proportion of female labor has risen from 72.3 to 72.9 percent and in the food industry from 53.1 to 54.4 percent.

The ratio of men to women among workers in the various branches of industry is also reflected in the composition of the working force by economic regions. This ratio can be seen in table 8. Depending on the structure of industry in the economic regions, variations can be observed in the proportion of men and women in the total number of industrial production personnel and industrial workers (see table 9). The proportion of women in the number of industrial production personnel in the various economic regions varies from 33.3 percent in the Far East to 54.2 percent in the Center. In the various oblasts this variation is still greater. Thus, in Primorskiy Kray it is 27.7 percent, and in the Ivanovo Oblast, 61.2 percent.

The differences in the structure of industrial production personnel in the regions is determined mainly by the branch structure of the industry. In the Far East the proportion of branches, in which mainly men are occupied—forestry, fishing, and nonferrous metallurgy, branches of machine-building (power, forge-press equipment, and metal-processing)—is high. For branches of light industry employing predominantly female labor, the proportion is only 7.1 percent in contrast to the average of 16.6 percent in the Union.

In the Center a high proportion of the industry branches employ predominantly women—light industry, machine building branches (instruments and electrical machines), and branches of the chemical industry (chemical-pharmaceutical, and manufacturing of synthetic fibers). As a result of such a structure of the branches of machine-building and the chemical industry, the proportion of women employed in these branches in the Center is higher than the average Union proportion. The ratio of male to female labor in the Center is due to a relatively weak development of ferrous and nonferrous metallurgy, fuel, and other branches of the extracting industry, in which predominantly male labor is used. All this determines the high proportion of women in the industrial production personnel of the Center.

Thus, the proportion of women in industrial production personnel is, as a rule, lower in regions where branches of heavy industry, especially the extracting industry, predominate. On the other hand, in regions where there is a high proportion of branches of industry, especially light industry, employing predominantly women, the proportion of female labor is higher. A comparison of data on the proportion of women in industry with the proportion of women in the population in the regions reveal definite relationships. In the Center the highest proportion of women in the number of industrial production personnel (54.2 percent) corresponds to the highest proportion of women in the population (57.1 percent); and in the Far East the lowest proportion of women in the number of industrial personnel (33.3 percent) corresponds to the lowest proportion of

TABLE 8

PROPORTION OF MEN AND WOMEN IN THE TOTAL NUMBER OF INDUSTRIAL-
PRODUCTION PERSONNEL AND WORKERS, 1 JANUARY 1962

	INDUSTRY PRODUCTION PERSONNEL		WORKERS	
	Men	Women	Men	Women
All industry	54.5	45.5	54.4	45.6
Ferrous metallurgy	68.1	31.9	70.6	29.4
which includes mining ores of ferrous metals	76.8	23.2	79.2	20.8
Nonferrous metallurgy.	69.0	31.0	70.5	29.5
which includes mining ores of nonferrous metals	75.1	24.9	76.4	23.6
Fuel industry and production of products from coal, oil and shale.	78.2	21.8	79.4	20.6
Coal	81.6	18.4	82.6	17.4
Oil extraction	75.1	24.9	77.0	23.0
Production of electrical energy and thermal energy	70.8	29.2	71.2	28.8
Machine-building and metal-processing, which includes:	60.5	39.5	61.1	38.9
production of forge-pressed equipment. . . .	70.6	29.4	74.0	26.0
appliances	53.6	46.4	53.6	46.4
repair work.	77.0	23.0	79.2	20.8
Chemical industry (without forest chemistry and the hydrolysis of wood).	49.2	50.8	49.4	50.6
mining chemical (not including salt chemical).	72.4	27.6	74.4	25.6

TABLE 8--Continued

	INDUSTRY PRODUCTION PERSONNEL		WORKERS	
	Men	Women	Men	Women
basic chemistry.	60.3	39.7	62.6	37.4
chemical-pharmaceutical.	32.3	67.7	30.4	69.6
Forest, paper, and wood-working industry (including forest chemistry and hydrolysis of wood)	64.4	35.6	64.8	35.2
which includes:				
forest exploitation.	74.2	25.8	75.6	24.4
wood working (including matches) . . .	54.9	45.1	54.2	45.8
Building-materials industry.	59.9	40.1	59.4	40.6
Light industry (including production of rubber footwear and production of tanning materials).	26.8	73.2	25.8	74.2
which includes:				
cotton	28.3	71.7	27.6	72.4
sewing	19.4	80.6	18.0	82.0
Food industry.	46.7	53.3	45.6	54.4
including fishing (without production of fish canning).	71.5	28.5	69.7	30.3

TABLE 9

PROPORTION OF WOMEN IN THE POPULATION AND IN INDUSTRIAL-PRODUCTION PERSONNEL 1 JANUARY 1962

Major Economic Region	Population 15 January 1969	All Industry	Ferrous Metallurgy	Nonferrous Metallurgy	Fuel Industry	Production of Thermo-Electric Energy
Northwestern.	55.9	52.5	30.1	37.7	25.3	35.3
Central	57.1	54.2	35.9	44.5	25.5	33.2
Volga-Vyatka.	56.5	48.5	35.9	31.9	39.4	33.4
Central Chernozem . .	56.6	46.8	28.4	...	36.7	26.0
Volga	55.8	48.4	36.2	32.0	31.4	29.4
Northern Caucasus . .	55.2	43.7	30.9	32.1	23.3	26.9
Urals	54.8	44.2	34.6	34.7	25.7	33.2
Western Siberia . . .	53.9	43.5	27.9	25.6	18.9	28.6
Eastern Siberia . . .	52.0	40.4	25.3	27.4	26.5	32.1
Far East.	51.1	33.3	29.6	21.4	19.3	26.9
Donets-Dnieper. . . .	55.1	36.7	30.8	32.0	18.2	29.2
Southwestern.	56.1	40.7	32.1	18.1	14.8	18.9
Southern.	55.3	44.3	32.0	50.0	37.7	26.6
Western	55.2	46.2	34.2	...	29.6	19.1
Transcaucasia	53.0	41.0	20.9	16.9	28.8	22.8
Central Asia.	52.0	44.6	20.0	26.0	23.3	24.8
Kazakhstan.	52.5	40.9	29.4	25.8	17.4	30.5
Belorussia SSR. . . .	55.5	47.6	28.5	....	27.4	23.4
Moldavia SSR.	53.7	44.0	...	....	...	16.3
USSR	55.0	45.5	31.9	31.0	21.8	29.2

TABLE 9--Continued

Major Economic Region	Machine Building and Metallurgy Processing	Chemical Industry	Forest, Paper and Woodwork Industry	Building Materials Industry	Light Industry	Food Industry
Northwestern.	39.8	56.8	39.2	43.9	80.0	49.6
Central	42.8	55.3	39.4	44.6	74.1	64.8
Volga-Vyatka. . . .	44.1	48.3	41.0	48.6	72.6	63.8
Central Chernozem . .	38.0	51.0	34.4	41.8	73.7	52.4
Volga	42.7	51.6	41.4	42.9	74.5	56.9
Northern Caucasus . .	35.9	47.8	35.4	37.9	70.8	53.7
Urals	43.6	48.9	36.9	46.5	78.9	63.7
Western Siberia . .	43.6	53.2	32.0	44.3	78.3	58.9
Eastern Siberia . .	35.5	59.4	28.7	41.9	77.9	61.2
Far East.	29.6	54.8	27.2	35.3	84.7	39.1
Donets-Dnieper. . .	37.0	47.1	41.4	42.9	73.1	53.8
Southwestern. . . .	30.0	47.7	25.7	31.0	68.6	43.6
Southern.	28.5	46.7	42.4	33.1	72.2	55.2
Western	35.4	51.5	30.4	34.4	68.7	47.5
Transcaucasia . . .	27.4	38.2	24.2	18.7	68.5	47.0
Central Asia. . . .	31.1	37.1	29.5	30.3	63.5	46.3
Kazakhstan.	28.9	38.1	36.5	37.9	72.8	54.5
Belorussia SSR. . .	35.9	57.6	35.5	43.8	71.6	55.4
Moldavia SSR. . . .	26.8	54.6	33.3	26.4	63.7	41.9
USSR	39.5	50.8	35.6	40.1	73.2	53.3

women in the total population of a region (51.1 percent) in the Union.[9]
The revelation of the nature of this interdependence is of great
significance for determining the direction of development of the
economic complex in the various regions of the country and for
organizing and planning the migration of the population.

One more tendency must be emphasized: the lower the proportion
of women employed in the total industry of the region, the higher
it is in certain individual branches as compared with the Union
average. Thus, in the Far East, the proportion of women in the
industrial production personnel of light industry is 84.7 percent in
contrast to a 73.2 percent average for the USSR. The differences
in the structure of the industry of the Far East could have resulted
in the somewhat higher proportion of women in the light-industry
personnel. But this tendency is apparent also when comparing the
data for separate branches. In the sewing industry in the Far East,
the proportion is 90.2 percent as opposed to an 80.6 percent average
in the USSR. The same tendency is apparent when the data for the
furniture industry is compared. In the Far East these data are 46.5
percent as opposed to the Union average of 42.9 percent. No less
indicative are the data on the chemical industry. In the chemical
industry of the Far East the proportion of women in total personnel
is 54.8 percent in contrast to 50.8 percent in the Union.

Therefore, a lower proportion of women in the population of a region does not exercise an influence on the proportion in the personnel of separate branches of industry, and, consequently, also not
on its proportion in the personnel of the total industry of the region.
The higher proportion of women employed in individual branches of
Far East industry as compared with the average Union indices in
the same branches reveals the more limited possibilities for employing female labor in that region. This even causes migration of female
labor force from the region.

The significance of the structure of industry for the solution of
the problem of a fuller utilization of labor reserves of a region
can be shown in the example of one of the largest industrial centers
of the country, Leningrad. In Leningrad after the war, the proportion
of women in the population was considerably higher than that of
men. According to the census of 1959, it was 57.8 percent in the
Leningrad City Soviet. On 1 January 1960 women made up 56.4 percent
of the number of workers and employees in the city, which is close
to the proportion of women in the population. In considerable measure
this was achieved by a rational relationship of branches with a varied
composition of male and female labor force.

Machine-building and metal-processing have first place in the
industry of the city. But there are combined in them such branches
as power machine-building, instruments, and others with a varied

proportion of male and female labor. The number of women in the chemical industry and also in branches of the light industry, which together with the food industry utilize the labor of 26.4 percent of the industrial production personnel of the city, is relatively large (in the USSR as a whole these branches utilize 26.6 percent of all industrial personnel).

As the experience of Leningrad industry has shown, the solution of the problem of the fullest utilization of labor resources of the country by involving female labor in production must be based on a suitable combination of branches employing predominantly male labor with branches employing predominantly female labor.

Agriculture can also provide considerable reserves of manpower for work in industry. The collective farmers of various economic regions do not work to the same degree. The sown area per collective farmer in 1961 was on the average: in Transcaucasia, 1.73 hectares; in the Belorussian SSR, 2.93 hectares; in Western Siberia, 14.52 hectares; and in Kazakhstan, 12.45 hectares as compared to the Union average of 5.44 hectares. Differences in the structure of agriculture in these regions cause some differences in the work load per collective farmer. But in all regions, even with the existing level of technology, these figures can increase considerably. In regions where the possibilities for expansion of areas are limited, considerable labor resources can be freed for work in industry. A further development of technology and the comprehensive mechanization of agricultural work will increase the labor reserves still more. The Southwestern, Transcaucasia, and Central-Chernozem regions, and the Belorussian and Moldavian SSRs will have especially large labor reserves when freed from agricultural work. The labor reserves of some regions will serve in part as a supplementary source of labor reserves for the eastern regions where the increased need in labor reserves will exceed the possible natural increase.

By a rational combination in the development of seasonal branches of industry, and by utilizing in industry the free time of agricultural workers between seasons, it will be possible to increase the degree of utilization of labor resources. For example, in 1960 the difference in the number of workers of the collective and state farms from May to October and from December to March varied from 6.5 million to 14.8 million people. The deviation from the average yearly number was –28.1 percent in January and +24.8 percent in June. In 1962 the monthly deviations in the number of workers in agriculture remained approximately the same as in 1960.

In the future these variations can decrease as a result of suitable combinations of branches of agriculture and under the influence of technical progress in agricultural production. But the production seasons will remain; therefore, an important national economic problem is to find possibilities for utilizing the free work time of agricultural workers. The utilization of free work time reserves in agriculture can proceed by combining the work time in agriculture with some seasonal branches of industry. To such branches may belong, for example, the sugar industry and in part the canning industry, in which the work season does not coincide with peak periods of work in agriculture. Other industries can be created that are associated with satisfying local needs in the certain product.

We have examined two very important conditions for the development of industrial complexes in economic regions—the natural resources and labor resources—and the conditions of their rational utilization. Other factors also have an influence on the development of industrial complexes in the regions. Significant among them is the achieved level of productive forces in the regions of the country. The economic complex of an economic region is a continually developing organism. The development of individual branches of industry and the appearance of new branches does not occur in isolation. It can augment and change the direction of specialization. But the achieved level of economic development, its structure and specialization, is always the starting point for further economic development. For example, the Ural region produces 26.1 percent of the iron ore, 32.5 percent of the pig iron, 33.6 percent of the steel, and 33 percent of the rolled steel of the country. This predetermines the development of the economic region and its specialization in the future. The Volga-Vyatka region produces 48.5 percent of the trucks and 36.1 percent of the passenger cars of the USSR. In this region the pertinent cadres of workers and technical-engineering personnel have been created. And, of course, automotive manufacturing together with the forest and paper branches will determine the industrial specialization of the region in the future.

A comparison of the proportion of the total production of the most important types of industrial articles and the average per capita production of these types of articles in the region with the Union average norms will show the role of these regions in the production of these articles in the Union and to some degree show the industrial specialization of their economy. The data about the production level of the most important items will show the starting points for determining the direction of the production of these articles in the various regions of the country. If, for example, the sewing and confectionary industries and a number of others are to be brought close to the consumer, then these branches of industry must be developed first of all in regions less provided with these articles.

The achieved levels of industrial development of the economic regions vary greatly, as has been shown. There are regions with highly developed industry (Center, Northwest, Donets-Dnieper) and regions with less developed industry (Central-Chernozem, Central Asia, and the Moldavian SSR). The level of industrial development also determines to a considerable degree the general level of economic development of the region, although it is not the only indicator. Therefore, the further development of industry in the regions of the country must take into account the achieved levels of industrial development in the economic regions. More than that, taking into account these levels and other conditions associated with them is of significance for the correct approach to the rational distribution of industrial branches among the regions of the country. It is the rational distribution of industry, as was pointed out in the Program of the CPSU [Communist Party of the Soviet Union] that must assist in the further equalization of the levels of economic development in the regions of the country. Consequently, the achieved level of industrial development of the economic regions is an important initial condition for solving the problem of further development of their economic complexes.

1. Some bourgeois economists, trying to slander the principles of distribution of productive forces in the socialist countries, interpret the task of comprehensive development of the economy as an undertaking to create a closed economy. For example, in the book *The Political Economy of Communism*, by P. Wiles, the author writes: "The communist policy is rather irrational in distribution. It uses the following critera: (1) regional autarchy, (2) regional specialization. These two criteria, directly contradicting each other, are often cited together in theoretical works without any hints or recognition that in reality they are incompatible. P. Wiles, *The Political Economy of Communism* (Oxford, 1962), p. 150.

In connection with this one must be reminded of the special indication of the necessity to "struggle decisively against an outdated understanding of the complex economy as a closed economy" as found and approved by the Twenty-first Congress of the CPSU concerning control figures of the development of the national economy of the USSR for 1959–65 (*Kontrol'nyye tsifry razvitiya narodnogo khozyaystva SSSR na 1959–1965 gody* [Control figures of the Development of the National Economy of the USSR for 1959–1965] (Moscow, 1959), p. 64.

2. *Narodnoye khozyaystvo SSSR v 1960 godu* [The National Economy of the USSR in 1960] (Moscow, 1961), p. 144.

3. The Central Statistical Administration includes in this branch the preparation of wood pulp, the production of firewood cross ties, and soft resin.

4. Some authors call this index, specialization in depth.

5. The labor consumption of the branches of industry is characterized by the total need for manpower, which is determined not only by the expenditure of labor for the production of a unit of output, which is of great significance in evaluating the labor consumption of the production of mutually exchanged objects, but also by the amount of their production.

6. *Itogi vsesvouznoy perepisi naseleniya 1959 goda* [Summary of the All-Union Population Census in 1959 in the USSR] summary volume (Moscow, 1962), p. 117.

7. *Narodnoye khozyaystvo SSSR v 1960 godu* [The National Economy of the USSR in 1960], p. 642; *Narodnoye khozyaystvo SSSR v 1962 godu* [The National Economy of the USSR in 1962] (Moscow, 1963), p. 459.

8. *Zhenshchina v SSSR* [Woman in the USSR] (Moscow, 1960), pp. 37–39.

9. The proportion of the women in the population is given according to data for 15 January 1959.

YE. YE. LEYZEROVICH

Toward a Quantitative Evaluation of The Economic–Geographic Location Of an Industrial Enterprise

Economic geography is interested in the industrial enterprise as the simplest component of the geographic division of labor and views it in mutual association with other elements of the territorial-production complex to which the enterprise belongs and its interrelations with similar enterprises in the composition of other territorial-production complexes, that is, from both the regional and branch viewpoint.[1] The interconnections and interrelations of enterprises of interest to economic geography are dictated by the characteristics of the enterprise itself and its economic-geographic location. Among the types of economic-geographic location that have the greatest influence on the distribution of productive enterprises and their development in a given place, the following should be noted.

1. Location with respect to economic resources (centers of the extractive and processing industries, agricultural regions, logging regions, and so forth, on which depends the possibility of obtaining cheap raw materials, fuel, electric energy, semifinished products, building materials, and equipment for the enterprise and also a sufficient amount of merchandise for the personnel of the enterprise.

Translated from *Kolichestvenniye metody issledovaniya v ekonomicheskoy geografii* [Quantitative Methods of Investigation in Economic Geography] (Moscow, 1964), pp. 62–89.

2. Location with respect to market demands, which determines the additional expenditures for transporting the finished product to the consumer. The possibility of effective calculation of demand also depends on this location.

3. Location with respect to centers of population, which affects the possibility of providing the enterprise with manpower in general and qualified manpower in particular, and which affects the possibility of utilizing the latest achievements in industry and scientific-technological thought.

4. Location with respect to transport lines, which permits the obtaining of raw materials and semifinished products and the shipping out of finished products, and which facilitates production and marketing cooperation of the enterprise. Location directly on transport lines enables the enterprise to specialize in servicing transportation or to develop by the partial utilization of freight flows passing along the transport lines.

5. Location with respect to industrial enterprises with which the given enterprise has economic ties and also with respect to industrial enterprises that manufacture products similar to the product of this enterprise. This type of location affects the specialization of the enterprise, the assortment of articles, the expedient scale of production, and also the character of the industrial cooperation of the enterprise.

6. Location with respect to the country's borders, which affects the expedient scale, profile, and specialization of the enterprise. Enterprises situated closer to the borders can more actively participate in foreign trade operations and in servicing the army but must not, for strategic considerations, have a large proportion of national production or be the only producer of a given product.

The effect of economic-geographic location on the distribution and development of enterprises has been noted more than once. It is very important to find and study the more concrete forms of this influence. Above all, such study must involve the elaboration of methods of quantitative evaluation of the economic-geographic location of an industrial enterprise. One of the basic methods for such an evaluation is the evaluation of the influence of the economic-geographic location on the index of comparative effectiveness of investments, which is the most comprehensive quantitative index of the effectiveness of the location and development of an industrial enterprise in any geographic locality.

The calculations of comparative effectiveness of investments are made by comparing the net cost of the product and the specific investments and the time needed for the return of the investments. This time is determined by comparing the additional investments` (that is, their difference under various alternatives; in economic-geographic investigations, the different alternatives of location) and economy in net cost. The time of return of additional investments is: $T = K_1 - K_2/C_2 - C_1$, where K_1 and K_2 are the investments in the comparable alternatives, and C_1 and C_2 are the net cost of yearly production for the same alternatives. Under present conditions, this time is considered acceptable for industrial branches if it does not exceed seven years.[2] The index of comparative effectiveness of investments is determined, as was noted, by comparing investments (initial expenditures) and the net costs (current expenditures). In this article the task is not to evaluate the influence of the economic-geographic location for the time of return as a whole but only to analyze the influence of the economic-geographic location on the net costs of production—one of the most important economic indices of the productive activity of an industrial enterprise.

The net cost of production of an industrial enterprise is influenced by a number of technical and economic factors. Enterprises manufacturing homogeneous products, distributed in various geographical localities, other conditions being equal, do not have the same net cost of production. For example, for 77 types of products manufactured by various Sovnarkhoz of the RSFSR in 1957, judging from the calculations made from data of the Central Statistical Bureau of the republic, the coefficient of variation of the net cost of a homogeneous product was $\pm 17\%$, and for some types of products the maximum net cost exceeded the minimum by a factor of two or more. One of the most important causes for such a phenomenon is the differences in the economic-geographic location of the enterprises. The economic-geographic location influences the net cost of the product of an industrial enterprise and affects the individual types of expenditures of which this net cost is composed. First of all it influences material expenditures (for raw material and other materials, fuel, energy, and amortization) and labor expenditures (wages and deductions for social insurance). The structure of expenditures by percentage of total costs of production for various sectors of industry for 1961 are shown in table 1.[3]

THE EFFECT OF THE ECONOMIC-GEOGRAPHIC LOCATION OF AN ENTERPRISE ON MATERIAL COSTS AS A COMPONENT PART OF THE NET COST OF PRODUCTION

Due to the different economic-geographic locations of enterprises they must use raw material, fuel, and electric energy coming from various places, that is, having different conditions of production and, correspondingly, a different net cost at the place of production. Moreover,

TABLE 1

SECTORS	TOTAL EXPENDITURES	RAW MATERIALS	FUEL AND ENERGY	AMORTIZATION	WAGES AND SOCIAL INSURANCE	OTHER
All industry.	100	69.2	5.0	3.6	18.9	3.3
Ferrous metallurgy.	100	56.5	14.2	6.1	20.0	3.1
Coal.	100	32.3	3.7	6.4	50.4	7.2
Mining.	100	17.0	4.1	7.7	62.5	8.7
Oil extraction.	100	7.6	10.5	46.8	23.0	12.1
Gas	100	35.6	15.8	20.5	20.1	8.0
Peat.	100	18.5	7.7	14.4	50.9	8.5
Production of electrical and thermal energy.	100	4.7	55.0	20.4	15.2	4.7
Machine-building and metal-processing.	100	57.0	3.8	3.8	31.3	4.1
Chemical.	100	69.2	7.6	4.2	16.1	2.9
rubber-asbestos	100	83.8	2.8	1.3	10.9	1.2
Timber, paper, and woodworking.	100	45.9	5.5	4.1	36.1	8.4
logging	100	24.3	5.2	5.2	50.0	15.3
paper	100	56.8	14.5	4.9	20.5	3.3
Building materials industry .	100	42.0	12.3	6.6	29.9	9.2
cement.	100	25.7	33.9	10.7	20.6	9.1
Light industry.	100	88.5	0.9	0.8	8.9	0.9
textiles.	100	89.8	1.0	0.9	7.6	0.7
cotton refining	100	96.3	0.6	0.4	1.7	1.0

the enterprises supply the necessary raw material, fuel, electric energy, and the like, from varying distances and by different types of transport, that is, with different transport expenditures. As a result, depending on the economic-geographic location of the enterprises, great differences occur in the cost of the raw materials, fuel, and electric energy consumed, that is, in the elements that comprise the material expenditures of an enterprise.

The net cost of pig iron in two large metallurgical plants of the USSR in 1956 varied from 169 to 354 rubles (on the old price scale). The cost of raw material and fuel (metal charge, fusing agents, coke), that is, the magnitude of material expenditures, varied for the same plants within limits of from 159 to 317 rubles per ton of pig iron, or 158 rubles. The cost of delivery to the metallurgical plant of raw material and fuel, necessary for the production of one ton of cast iron, varied for these two plants, according to approximate calculations, from 15 to 75 rubles.[4]

The reason for using raw material, fuel, and electric energy, which differ greatly in the conditions of their production and net cost, is the fact that in large-scale consumption of individual types of raw material, fuel and also electric energy, we cannot limit utilization to the richest sources of raw materials, fuel, and electric energy but are compelled to put into circulation the poorer sources as well. Almost every type of raw material, fuel, and electric energy is produced in a number of places in our country. Therefore, for the majority of types of raw material, fuel, and also electric energy, our country is divided into a number of market demand zones, each of which is served by a definite source of raw material, fuel, or electric energy. The borders of the zones pass along so-called points of equal net costs of production to the consuming enterprise. Enterprises using raw materials, fuel, and electric energy are in practice assigned to definite zones. For these enterprises, there economic-geographic location is determined (1) by assignment to one zone or another, and (2) by the transport situation in this zone with respect to the source of raw material, fuel, and electric energy, which influences the transport expenditures in delivering the raw material, fuel, and electric energy to the enterprise. The assignment to one zone or another and the location within it determine in the final analysis the net cost of the raw material, fuel, and electric energy used by the enterprise.

As can be seen from the example of the metallurgical plants, location with respect to raw materials and fuel often acquires a decisive significance in the determination of the net cost of the product of an enterprise. The main factor here is, as a rule, not the transport location with respect to the source of raw material, fuel, or electric energy, but the location in the market demand zone of one source or another of raw materials, fuel and electric energy, since the differences in net cost of these items to the producing enterprise are considerably greater than the differences in transport expenditures for the enterprises, located differently [within a demand zone] with respect to their sources of raw material, fuel, and electric energy.

The work on the regionalization of coal consumption in Eastern Siberia[5] shows the advantage of dividing the region from the Sokur-Yurga-Abakan border with the Mongolian People's Republic on the west to the Neva in the east, into nine market demand zones attached to the Itat, Nazarov, Chernogor, Uglugkhem, Irsha-Borodin, Cheremkhov, Gusinoozer, and Chernov coal beds. The net cost of the coal where it is mined varies from 1.1 rubles to 3.5 rubles per ton. The differences in cost of transport from the place of mining to the consumers varies from 0.0 to 1.8 rubles per ton. As a result the net cost of coal for the consumer varies from 1.1 to 4 rubles per ton. Thus, for different consumers the net cost of coal, which is a considerable part of the material costs of coal-using enterprises, varies under the influence of location in one market demand zone or another as well as under the influence of location within this zone. The location by zone is more important than the location within the zone.

In the work cited, these indices were calculated for an idealized scheme. However, up to the present time, due to the fact that less coal is mined than used in the European part of the country and in the Urals, 12 million tons of Kuznetsk coal are shipped annually to the regions of the Volga and Center, and 22 to 24 million tons to the Urals. Therefore, there is not enough Kuznetsk coal to supply all the demands of Western Siberia, and Eastern Siberian coal is shipped in, mainly Cheremkhov coal (in 1957, 6.3 million tons). At the same time Eastern Siberia receives about one million tons of coal from the Far East. Thus, all the zoning of market demand for coal is shifted to the West. East of the Kuznetsk Basin consumers do not fall into the zone of the coal beds closest to them but into a zone of deposits situated further east, which provides a more expensive coal. Thus, the importance of location by zone as compared with the importance of distance to the nearest deposits, proves to be even greater in practice than in the calculation schemes.

Yet there are groups of enterprises for which location in a zone does not exist, and the influence of the economic-geographic location on the material component of costs is determined solely by transport expenditures; for example, groups of enterprises using a unique raw material (unique not in potential reserves, but in actual output). In 1955, 83 percent of all phosphorous raw material used at superphosphate plants consisted of apatite concentrate produced in Murmansk Oblast.[6] Therefore, almost all enterprises of the superphosphate industry were in fact in a single zone of influence. Their economic-geographic location, relative to the raw material, was determined exclusively by distance to the deposits and was expressed by comparing the transport costs for the delivery of the raw material. A similar situation existed in our country with respect to oil when all oil supplies came from the Caucasus before it was extracted in the Volga region. At that time the situation of enterprises relative to the resources of oil was determined exclusively by their transport situation with respect to the oil regions of the North Caucasus and Transcaucasia. And so we see that economic-geographic location has an important influence on the material component of net cost; moreover, this influence has a dual character: a zonal character and a linear, transport character.

The economic-geographic location influences most strongly the material component of net cost for material-consuming, fuel-consuming, and energy-consuming enterprises, which includes the majority of enterprises of ferrous metallurgy, basic chemistry, plywood, matches, paper, cement, glass, meat-milk, sugar and alcohol industry, regional electric stations, enterprises of heavy- and transportation-machinery production, and enterprises of a number of branches of the building materials industry.

The enterprises of these branches usually have a high proportion of material costs included in the net cost of the product (more than 50 percent) and, therefore, must have an economic-geographic location guaranteeing them sources of basic economic resources (fuel, raw material or energy, depending on the profile of the enterprises), usually within their own economic region.[7]

Location relative to resources is of little influence for industrial enterprises using raw material or semifinished products, the production of which demands great expenditures per ton of finished product: economically transportable raw materials and semifinished products (stainless steel, nonferrous metals, synthetic fiber, cotton fiber, yarn, and so forth). Such enterprises, although the material component may form a high proportion of their net cost, as, for example, in light industry, depend little in their distribution on proximity to the raw material base, but more on other factors, which we will consider below.

Location relative to resources is also of little significance for enterprises of some branches of the extractive industries (iron ore, coal mining, oil extraction, shale, chemical, logging, brick, and porcelain pottery) for which the raw material resources enter into the very composition of the enterprise and exist within it and for which the specific consumption of other materials, fuel, and electric energy is not great. For enterprises of these branches of industry, a low proportion of material component in the net cost is characteristic, usually less than 40 percent.

THE INFLUENCE OF ECONOMIC-GEOGRAPHIC LOCATION ON LABOR
EXPENDITURES AS A COMPONENT PART OF THE NET COST
OF THE PRODUCT

Labor expenditures (wages together with supplementary costs) are the second important component part of net cost. The proportion of labor expenditures in the net cost of the product is determined by the expenditure of wages per unit of production, and the latter depends on the level of wages (hourly wages) and on labor productivity (expenditure of working hours per unit of production). The level of wages and labor productivity differ considerably for the separate economic regions of our country. One can judge the levels of wages by comparing the level of average monthly wages of a worker in the Far East with the level of the average monthly wages in the other economic regions and republics.[8] [The average monthly wage in the Far East is set at 100.00.]

Far East	100.0	Latvian SSR	64.3
Eastern Siberia	74.5	Georgian SSR	63.8
Karelian ASSR	72.0	Northern Caucasus	63.5
Northwestern (including			
Leningrad)	71.5	Turkmenian SSR	63.3
Urals	69.7	Central (including Moscow)	61.7
Kazakh SSR	69.4	Azerbaydzhan SSR	61.7
Estonian SSR	69.3	Uzbek SSR	60.7
Northern	69.1	Volga Region	60.2
Ukrainian SSR	67.7	Armenian SSR	59.2
Western Siberia	66.2	Lithuanian SSR	56.3
Kirgiz SSR	65.7	Belorussian SSR	50.7
Tadzhik SSR	65.2	Moldavian SSR	49.2

In this distribution of the average monthly wage levels by economic regions, it is difficult to see at once any kind of regularity. However, definite tendencies do show, namely: (1) a higher level of wages in the distant, sparsely populated regions and a lower one in the central and southern, more densely populated regions; (2) a higher level of wages in regions where heavy industry predominates (Urals, Estonian SSR, Ukrainian SSR) and a lower one in regions of prevailing light industry.

We are interested in the first tendency, which is the result of various kinds of regional additions to the wages that in some places are quite considerable.[9] The question of wage supplements has not yet been regulated in our labor laws. There are unjustifiable differences in the magnitude of such supplements for some parts of the country. However, by itself the existence of regional wage bonuses is justified: (1) by the differences in natural conditions, in particular those which influence the quantity and composition of commodities necessary for the inhabitants of these regions; (2) by the differences in the conditions of supplying the population with food and industrial merchandise. The latter issue depends to a certain extent on the differences in the economic-geographic location of the regions. For large economic regions it is a question of differences in the macrolocation with respect to the main agricultural and industrial regions of the country. For small regions, meso- and micro-location with respect to local agricultural regions acquire significance.

The macrolocation of some economic regions with respect to the main agricultural and industrial regions of the country is calculated by a system of zonal prices, which regulate the market. The location of small regions with respect to local agricultural regions finds an expression in the degree to which the population uses the kolkhoz [collective farm] market, where the prices are higher than the government retail prices. The existence of a system of zonal prices and differences in the degree of utilizing the kolkhoz market, caused by dissimilar conditions of supplying the population with food and industrial merchandise, explain why the per capita expenses of the population of the different regions are also dissimilar. These differences must be taken into account and are at the present time partially taken into account by the regional supplements to the wages.

A correctly established system of regional wage supplements must, as an ideal, guarantee an equal distribution of labor resources in the country. However, in reality this does not occur, and for two reasons. The first is subjective and temporary: it is the absence of a sufficiently differentiated system of regional wage supplements, which must be developed as soon as possible. The second reason is objective: it is impossible for the flow of labor resources to keep up with the changes occurring in the demand for labor in the various parts of the country. Therefore, there always will be in some regions discrepancies between supply and demand in labor resources; they will vary in degree and duration.

As a result of differences in the provision of some regions with labor resources, there are differences in the level of cultural and domestic services provided to the population and differences in the productivity of labor. The differences in the level of cultural and domestic service occur between regions of our country as well as within regions. As one moves away from the central regions of our country, and within regions as one moves away from their centers, there occurs, as a rule, a deterioration of living conditions. In regions and localities situated far from the central regions and regional centers, housing, medical services, the network of children's establishments, and secondary and higher education services are poorer.[10] This is, first of all the result of insufficient numbers of medical, youth, and communal workers who

are needed in the peripheral regions, especially in the sparsely populated regions, where it is more difficult to organize the proper cultural-
domestic services. Poor cultural-domestic conditions are compensated by higher wages, which attract the labor force.

Differences in labor resources are also one of the reasons for the regional differences in the productivity of labor. Where there is a deficit
in labor force, by using the system of hourly wages (and this system
is prevalent), wages are paid for less work than in regions provided
with adequate labor force. This occurs in the following manner: in
staffing individual establishments all positions are usually filled. But
where there is not enough qualified labor force and there is no other
possibility, the positions are filled with a less qualified labor force. As
a result, a machinist or lathe operator in Moscow with earnings of
100 rubles per month produces more on an hourly pay basis than a
machinist or lathe operator with the same pay in the Far East, where
the qualification of this worker and, consequently, his labor productivity are usually lower.

Piece-wages partially equalize the disparity, but nevertheless differences in productivity do exist and lead to an increase in labor expenditures and their share in the net cost of products in the eastern regions of the country.

For example, according to data by G. A. Zakharov[11] (Institute of
Economics of the Siberian section of the USSR Academy of Sciences) the net cost of an over-haul of a Zil-150 automobile at the auto
repair plant of the Magadan Sovnarkhoz, is 4.5 times higher than in
the Moscow auto repair plant no. 5, and the expenditure of wages is
6.4 times as high, that is, the share of labor expenditures in the net
cost of the product is 40 percent higher.

The total effect of all these factors, which are influenced by the economic-geographic location, on the increase of the proportion of labor
costs can be illustrated by an example of machine-building. In republics and oblasts, the proportion of expenditure in this branch, according to statistical data published in different years, varies in the following manner (labor expenditures as a percentage of the net cost of the
product).

USSR	31.3	Saratov Oblast	37.0
Rostov Oblast	27.7	Chelyabinsk Oblast (1960)	31.3
Azerbaydzhan SSR	27.9	Kirov Oblast (1959)	31.6
Belor. SSR (1955)	27.9	Moldavian SSR (1960)	32.0
Mordvinian ASSR (1958)	28.5	Bryansk Oblast (1959)	32.5
Volgograd Oblast (1961)	28.9	Checheno-Ingush ASSR (1956)	32.6
Orel Oblast (1959)	30.7	Lithuanian SSR (1958)	33.5
Ul'yanovsk Oblast (1960)	31.1	Kirgiz SSR (1959)	37.5
Uzbek SSR (1957)	34.7	Irkutsk Oblast (1957)	38.3
Novosibirsk Oblast (1960)	36.0	Amur Oblast (1956)	44.9
Dagestan ASSR (1957)	36.0		

As can be seen from these indices, there are no eastern oblasts among the six republics and oblasts having the smallest proportion of labor expenditures, and there are three eastern oblasts among the six republics and oblasts having the highest proportion of labor expenditures. In our opinion, such a phenomenon does not occur by chance: it is explained by the fact that a good location of a machine-building enterprise, with respect to densely populated regions where there are large labor resources and with respect to regions of mass production of food and industrial commodities, affects the lowering of wage expenditure per unit of production and, correspondingly, the lowering of the proportion of labor expenditures in the net cost of the product.

As a more concrete example one can point out that each of the sub-regions within the territory of the Krasnoyarsk Kray (Southern Yenisey, Achinsk-Krasnoyarsk and Lower Angara), possess very large sources of hydroenergy that can serve as a basis for the development of homogeneous industrial complexes. The lightly populated Lower-Angara subregion will specialize exclusively in energy-consuming industries having a low labor component. An electrical machinery plant with a typical average labor capacity of 1,000 workers, placed not in the Southern Yenisey but in the Lower Angara subregion, would require that an additional 2,000 people be settled and provided with housing. This would cause an additional expenditure of 6 million rubles and at a ten-year return would increase the amortization from 4 to 6 percent. The additional charge for remoteness would increase the portion of labor expenditures 20 percent, that is, from 33 to 40 percent. As a result, the net cost of the product would increase 10 percent.

The effect of the economic-geographic location on labor expenditures is most evident; therefore, it deserves attention and consideration in labor-consuming branches where the utilization of human labor as compared with machine labor is most developed. In these branches the expenditures for wages and social insurance exceed by far the expenditures for amortization of equipment. Labor expenditures especially exceed expenditures for equipment amortization (more than eight to ten times) in coal mining, shale, rubber-asbestos, machine-building, logging, plywood, matches, glass, porcelain-pottery, textile, sewing, and leather-footwear industries. For the enterprises of these branches, great significance can be attached to the location with respect to population, cultural, and economic centers of the country, that is, to places of concentration of major labor resources.[12]

The effect of the economic-geographic location on labor expenditures varies in the branches of the processing and extractive industry. The expenditures for wages in each branch depends on the technological pattern that determines expenditures in work hours per unit of production, and on the wage scale for labor. In the processing industry the technological pattern depends very little on the natural conditions of the locale of the enterprise. Therefore, similar enterprises of the processing industry have a different proportion of labor expenditures due mainly to different economic-geographical locations. The effect of the economic-geographic location is more directly apparent.

It is different in the extractive industries where the technological pattern differs considerably from enterprise to enterprise depending on the mining conditions. Here it is often thought that the differences in the net cost of the product are determined exclusively by the differences in the mining conditions. This, in fact, is not quite the case. The main element in the net cost of the product in an extractive industry is labor expenditures, and these expenditures depend not only on the mining conditions but also on wage conditions. The latter, as we noted, depend on the economic-geographic location of the deposits. For deposits situated in remote regions, the expenditures for the labor force, which is attracted from other regions and by the incentive of additional pay, greatly increase the net cost of the product.

In the Karaganda open-pit basin in 1960 the net cost of mining one ton of coal was 1 ruble, 81 kopeks; 75 kopeks were for labor expenditures, all the other expenses came to 1 ruble, 6 kopeks. The average monthly productivity of labor was 233 tons per worker. The Karakichinsk coal bed surveyed in the mountains of Tyan 'Shan' has mining conditions approximating in complexity the Karaganda conditions. The planned labor productivity here will be 189 tons per month per worker, that is, only 19 percent lower than in Karaganda. However, the net cost of coal will be 3.4 rubles per ton, since the labor expenditures will be 2.5 rubles per ton, while all other expenditures will be only 0.9 rubles per ton, or less than in Karaganda. The high labor expenditures for the recovery of Karakichinsk coal (recalculated per ton they are 2.7 times as high as in Karaganda) sharply increase the net cost of mining and make exploitation of the bed inexpedient; this is due to the necessity of increased wages associated with an economic-geographic location of the deposit far removed from populated localities.

THE EFFECT OF ECONOMIC GEOGRAPHIC LOCATION
ON THE TRANSPORT COMPONENT OF NET COST

We examined above the influence of economic-geographic location on the net cost of the product of an enterprise—on its production cost. In the final analysis, however, of major significance for the national economy is not the net cost of the product at the enterprise, but its net cost to the consumer. The net cost of the product to the consumer is composed of the full net cost of the product at the enterprise and the transport cost for delivery of the finished product to the consumer. For some branches of industry the absolute magnitude and the proportion of these costs are very high. In the final price of fuel at the place of consumption, the proportion representing transport costs is more than 50 percent; in the final price of timber, 40 percent; metals, more than 20 percent; fuel oil, 19 percent; kerosene and cement, 15 percent.[13] The index is large for chemical and some other products. Therefore, in resolving the problem concerning the location or development of an enterprise, the location of the existing or future enterprise with respect to the consumer acquires great significance.

According to their dependence on location with respect to the consumer, all types of industrial production can be divided into strongly dependent, weakly dependent, or of an intermediate position. Of course, such a division will be tentative. The following considerations can serve as its basis. In our country distances of up to 200 kilometers are considered short for rail shipments; trucks compete with the railroads.[14] The average distance in our country can be considered 200 to 800 kilometers, since the average distance of a rail shipment is 800 kilometers. A distance above 800 kilometers is considered long distance. The average proportion of expenditures for railroad shipments in the USSR is about 4 to 4.5 percent[15] of the total sum of expenditures of the national economy.

Those types of industrial production whose transport expenditures for the finished products exceed 4 to 4.5 percent when shipped by rail for short distances (up to 200 kilometers) can be considered strongly dependent on location with respect to the consumer. When the transport expenditures for average distances (from 200 to 800 kilometers) exceed 4 to 4.5 percent, production can be considered intermediately dependent on location with respect to the consumer. In those cases in which transportation over a long distance (more than 800 kilometers) does not exceed 4 to 4.5 percent, production can be considered weakly dependent on the location with respect to the consumer.

In adopting the average net cost of transfers as 0.3 kopeks per ton-kilometer, one can consider the production of enterprises that have a net cost of less than 15 rubles to be greatly dependent on location with respect to the consumer, the production of enterprises with a net cost of 15 to 60 rubles per ton to be intermediately dependent, and the production of enterprises with a net cost above 60 rubles per ton to be weakly dependent. Products with a net cost above 150 to 200 rubles per ton are nearly independent of location with respect to the consumer. By asserting that some enterprise is independent of location with respect to consumer we have in mind full material independence, but only in the sense that the location with respect to the consumer is hardly reflected in the net cost of the product to the consumer. In choosing the locality for such an enterprise the question of location of future consumers is usually not examined in detail. For example, one ton of aluminum is very expensive. The expenditures for the transportation of the finished product, even for 3,000 kilometers, do not exceed 4 percent of net cost. Therefore, in choosing a place for an aluminum plant no attention is paid to the question of future consumers. This question is overshadowed by the more important ones: the position with respect to raw materials, electric energy, and so forth. After the plant is built, the distribution of its products will be determined by the conditions of the national aluminum market, that is, the product will not be sent just anywhere but in conformity with the consumption balance of aluminum as it develops in the various parts of the country.

An additional reservation must be made. We spoke about net cost of transfers. However, usually in economic practice, in contrast to projected conditions, one has to deal with tariffs for railroad transfer. For the products of [tariff] group A, these tariffs are close to the net cost of transfers, with the exception of nonferrous metals, machines, and chemical products. For the products of [tariff] group B they are, as a rule, from 2 to 4 times above net cost. Taking this into account, the final cost criteria of dependence on the location with respect to the consumer will be the following net cost per ton of product.

Industrial Production	[Tariff] Group A (Rubles)	[Tariff] Group B (Rubles)
Strong dependence on location with respect to consumer	to 15–20	to 30–50
Average dependence on location with respect to consumer	15–20 to 60–70	30–50 to 150–300
Little dependence on location with respect to consumer	60–70 to 150	150–300 to 300–600
Practically independent of location with respect to consumer	above 150–200	above 300–600

The cost of transfer depends not only on the freight weight but also on the manner of loading the car.[16] However, this is taken into account by the tariffs in such a way that it has no special effect on the cited indices. Moreover, the fact that additional expenditures—that of transporting the freight to the railroad and delivering it from the railroad station to the warehouse of the customer—have not been included in the net cost has no essential effect on these indices.

On the basis of the data shown above one can refer the following basic types of production to the category of those strongly dependent on the location with respect to the consumer: extraction of coal, limestone, peat, bauxite, procurement of wood and hay,[17] production of kerosene, diesel fuel, fuel oil, sulfuric acid, pulp, cement, mineral building materials, reinforced-concrete, and sugar beets. To the category of average dependence on location with respect to the consumer belong: production of cast iron, ordinary steel and rolled steel, rails, refractory articles, copper and zinc concentrates, benzines, metallurgical equipment, cast iron ware, caustic soda, salt, and plywood. Weakly dependent on location with respect to the consumer are the following: production of sheet and quality steel, pipes, aluminum oxide, cast iron, iron ware, simple but heavy machines and machinery, cellulose, paper, grain, flour, and soap. Finally, the production activities hardly influenced by location with respect to the consumer, are: the production of tin, fast-cutting stainless steel, nonferrous and rare metals and their rolled products, complex machines, machinery and instruments, dyes, synthetic fiber, yarn, footwear, meat products, canned food, and tea.

This classification was based on data about the net cost of individual types of industrial production at the largest enterprises of the RSFSR in 1957 and is subject to change over time, since the selection criterion chosen—that of average percent of transport expenditures in the national economy, and the average net cost of the transfer indices differ for various time periods.

The basis of the classification is cost and not natural physical indices since the criterion of distribution of enterprises is the reduction of social expenditures and not reduction of the volume of transport work. This must be noted, since many existing classifications of distribution of branches with respect to consumers are based exclusively on weight indices.

This article does not examine the effect of the economic-geographic location on such important economic indices as capital outlay and time of return of investments, since this question must be the object of a special investigation. However, we think that the examples cited of the effects of the economic-geographic location on the net cost of production sufficiently attest to the possibility of finding ways for a quantitative evaluation of the effect of economic-geographic location on the industrial enterprise. Further work in this direction can attract an ever greater amount of factual material. In the final analysis this will make it possible to find some ways to proceed to the construction of an orderly method for the evaluation of the role of economic-geographic location in the distribution and development of enterprises of the various branches of industry situated in different parts of our country. This, in turn, will make it possible to give more concrete practical recommendations for the distribution of industry with respect to economic-geographic location and also to work out a series of norms to use in the construction of mathematical models of industrial location.

With these considerations in mind, it must be remembered that only a complete calculation of quantitative and qualitative factors will give the desired criteria for the rational location of industrial enterprises. This is especially important to consider under Soviet conditions, where the effect of the law of price is limited.

1. This refers also to agricultural enterprises.

2. *Tipovaya metodika opredeleniya ekonomicheskoy effekitvnosti kapital'nykh vlozheniy i novoy tekhniki v narodnom khozyaystve SSSR* [A Standard Method of Determining the Economic Effectiveness of Investments and New Machinery in the National Economy of the USSR] (Moscow, 1960), pp. 9–11.

3. *Narodnoye khozyaystvo SSSR* [National Economy of the USSR] (Moscow, 1962), p. 195.

4. R. S. Livshits, *Razmeshcheniye chernoy metallurgii SSSR* [Distribution of Ferrous Metallurgy in the USSR] (Moscow, 1958), pp. 226–27.

5. Yu. A. Sokolov, "Regionalization of Coal Consumption in Eastern Siberia," in *Razvitiye proizvoditel'nykh sil Vostochnoy Sibiri. Toplivo i toplivnaya promyshlennost'* [Development of Productive Forces in Eastern Siberia, Fuel and the Fuel Industry] (Moscow, 1960).

6. *Osobennosti i faktory razmeshcheniya otrasley narodnogo khozyaystva SSSR* [Distinctive Features and Factors in the Location of Branches of the USSR National Economy] (Moscow, 1960), p. 264.

7. A quantitative evaluation does not rid us from the necessity of also considering qualitative factors. For example, the meat industry is so distributed that two of our four largest meat combines are in the raw material zones (Semipalatinsk and Ulan-Ude) and two in consumption centers (Moscow and Leningrad).

8. B. Bukhanevich and M. Sonin, "On Interregional Regulation of Wages in the USSR," *Voprosy ekonomiki* [Problems of Economics], no. 1 (1957), p. 21.

9. For example, in the administrative unit "Turkmenneft," which extracts oil in southwestern Turkmenia, the additional pay associated with regional bonuses comprises 25 to 32 percent of the total wages. A detailed description of regional bonuses is found in the book by A. G. Aganbegyan and V. F. Mayer, *Zarabotnaya plata v SSSR* [Wages in the USSR] (1959), chap. 7, Sec. 2.

10. This is convincingly shown in an interesting work by D. D. Moskvin, "The Influence of Various Factors Determining the Standard of Living on the Distribution of Labor Resources and Population Stability," in *Problemy razmeshcheniya proizvoditel'nykh sil* [Problems of the Distribution of Productive Forces] (Moscow, 1960).

11. G. Prudenskiy, "Problems of the Use of Labor Resources," *Voprosy ekonomiki,* no. 1 (1962), p. 57.

12. Some of the industrial branches listed were mentioned as branches in which the economic-geographic location affects the material component of net cost. The others belong to the branches in which location with respect to consumers is of importance. However, this involves no contradiction. In contrast to location with respect to raw materials or location with respect to consumers, location with respect to populated centers does not play a decisive role in distribution. It is an additional factor that plays an active role only when location with respect to resources or location with respect to consumers does not give any advantages to the alternative distributions of the enterprise under consideration.

13. S. F. Kuchurin, *Tarify zheleznykh dorog SSSR* [Railroad Tariffs in the USSR] (Moscow, 1957), p. 55.

14. *Sfery primeneniya zheleznodorozhnogo i avtomobil'nogo transporta pri perevozke gruzovna korotkiye rasstoyaniya* [Spheres of Application of Railroad and Automobile Transport in Transporting Freight Over Short Distances] (Moscow, 1961), p. 5.

15.   We calculated it as the ratio of railroad freight transport expenditures to the sum of expenditures in industry and agriculture in the USSR. The value of expenditures on railroad transport we took from the work *Transportnyye izderzhki v narodnom khozyaystve SSSR* [Transportation Expenses in the National Economy of the USSR] (Moscow, 1959), pp. 63, 68, and 69, and the sum of expenditures in industry and agriculture was considered as 80 percent of the gross production of these branches of the national economy.

16.   According to data of A. Ye. Probst cited in *Voprosy razmeshcheniya proizvodstva i ekonomicheskogo rayonirovaniya* [Problems of the Location of Production and Economic Regionalization], p. 174, 100 percent of the freight capacity of cars is used in transporting coal, ore, cement, salt, sugar, chemical, raw materials, and mineral fertilizers, 60 to 75 percent in transporting cotton, leather raw material, metal, meat, lumber, sugar beets, and paper; 28 to 30 percent in transporting clothing, footwear, agricultural machinery, and raw cotton, and less than 20 percent in transporting automobiles, tractors, and combines.

17.   At first glance it may seem strange to assert that mining of coal, iron ore, peat, bauxite, and the procurement of wood and hay are forms of production dependent on location with respect to the consumer. However, for the production of coal, iron ore, peat, bauxite, wood and hay it is not correct to speak of significance of location with respect to the raw materials, for there is none. For these branches, the raw material base is part of the production itself. The degree of utilization of the raw material base and the scale of production for these branches is no doubt strongly dependent on their location with respect to the consumer.

V. P. NOVIKOV

The Kursk Magnetic Anomaly:

A Promising Iron-Ore Base

For the Iron and Steel Industry

Of the Urals

There are many varieties of economic models and many forms and techniques of mathematical interpretation, construction, and solution of such models. While acknowledging the contributions made in this area by the mathematical economists, we must point out that the formulation of a model that would offer concrete proposals for improving the policy of location and territorial organization of the productive forces of the nation as a whole was first advanced by economic geographers in the form of the territorial model of the national economy.[1]

Many miscalculations in the economic geography of the Soviet economy can be attributed to the lack of a long-range territorial model. If we are to achieve an optimal territorial structure of productive forces on the basis of quantitative relationships between various alternative distributional patterns and rates of economic growth, we will require the organization of such a territorial model of the national economy of the USSR, a key instrument for an economically and mathematically programmed territorial management of the economy. If we are to raise the rate of economic growth of the USSR, we will have to review thoroughly the long-range directions of development

Reprinted with modifications from *Soviet Geography: Review and Translation* 10, no. 2 (February 1969):43–86 by permission of the publisher. The article originally appeared in *Geografiya i khozyaystvo* [Geography and Economics] (1968), pp. 28–62.

and changes in the structure of capital investment, employment, production, and the relations between the West, the Urals, and the East of the Soviet Union.

WHAT IS A TERRITORIAL MODEL OF THE NATIONAL ECONOMY?

By territorial model of the Soviet economy we mean a schematic, mathematically exact reproduction of the principal regional relations, the economy between the branches of material production, the nonproductive [service] sector, and transportation within all regions and the country as a whole.

Preplanning prediction of economic development requires not simply the establishment of the aggregate of functional relationships in the economy in their areal aspect but also the construction of these areal relationships for the future. The construction of a territorial model of the economy involves the construction of a whole system of sectional, regional, and industrial models that are combined by a general interregional model of the country. It is the construction of this interregional model that is the most important task. It involves knowledge about the basic chain reactions, effects, and areal interrelationships because the future efficiency of the economy will depend on proper consideration, changes, and formation of these aspects.

Under the conditions of the economic reform and the changeover to planning on an industry-wide basis, a territorial model of the Soviet economy, down to industrial nodes, would represent an essential tool for raising the level of management of the economy of the entire country. The construction of such a model would make use of the results of particular design and planning projects, geological prospecting, and other research organized for the purposes of model construction.

All preplanning measures and the research-and-development policy in the field of location and investment must be subordinated to the task of developing the most efficient relationships (chain reactions) that can be established as a result of computations of a high degree of accuracy and strict coordination of research in all economic sectors, regions, and in the country as a whole. If we want to assure the purposefulness and effectiveness of all research on the location of productive forces of the Soviet Union, we will need the art of prediction and the proper, purposeful orientation of all institutes, organizations, and other participants. This will require, as soon as possible, a formulation of the basic elements of the territorial model that would stem from the basic contradiction in the territorial economy of the USSR and the further development of the division of labor between the West, the East, and the zone of contact between them, and would have to be tested experimentally in the territorial model over time.

The territorial model should reflect the aggregate of the basic decisions whose implementation would insure optimal economic results stemming directly or indirectly from the regional combination and territorial make-up of productive forces, from geographical shifts in the distribution of production and population, and in the use of natural resources and conditions; the model should further reflect the advantages of developing the productive capacity and improving the structure and territorial organization of production and the circulation of goods, and better proportionality of development of the productive forces as a whole, of industries and regions, both in time and in space. The territorial model should contain instructions stating by which course, under which conditions, and through the use of which reserves, optimal economic results can be achieved in each of the sectors of the economy and in the economy as a whole, and when.

The formulations of the aims and concepts of the territorial model should be based on an assessment of present and future aggregate costs and the results of the operation of the entire economy, on an effective realization of reserves, and on the constructive results of the programmed economic regulation of development of mutually exclusive industries in various parts of the country, such as can be achieved by territorial economic models. The decisive starting point in the construction of a territorial model of the Soviet economy should be a comprehensive and economically complete assessment of the spatial factor. Such an assessment should include quantitative relationships between the distribution of the aggregate of economic activities and the rates of development of all social production. It should also reflect the magnitude of economic results compared not only with production costs but with the costs required to overcome space.

Instead of working out a set of variants of the structure and magnitudes of production in West and East, we should first work out variants of the conditions and factors that determine the feasibility and acceptability of a plan. Then, by gradually selecting only the better decisions, we could find a plan that would be optimal in terms of a certain criterion. In theory, we can find a plan that is optimal in all respects by building a dynamic model of the Soviet economy both with interindustry and interregional relationships. But, for a number of reasons, such a model cannot yet be built. For the time being, therefore, we have to work on a territorial model of the Soviet economy by breaking down the over-all problem into separate interrelated industrial and local problems in which the industrial and local optima would not differ from the global optima.[2] But this requires that all the variants of conditions and factors of development of a particular industry that would have been included in the overall problem also be included in the particular problem. The constraints of the industry problem (or the regional problem) and their criteria will not contradict the constraints and the criteria of the overall problem of the national economy. The very least that should be achieved is that the particular problem reflect conditions and factors that do not significantly conflict with the territorial model of the Soviet economy.

The essential elements of the territorial model should be the distribution throughout the national territory of individual industries and types of products, and the development and location of the productive forces of the country's regions as components of the national economy. The territorial model, however, cannot be a simple sum of these elements. It should be a higher form of generalization and areal synthesis that would dictate not only the direction of corrections but also the magnitude of change in the industrial and areal structure, in the economic geography of the Soviet economy, on which the rate of growth would directly depend.

The elaboration of sectional/territorial models for individual industries as components of the territorial model of the Soviet economy as a whole would help speed the development of productive forces of the USSR at a given level of capital investment. The distribution of resources by industries and the constraints on consumption will have to be established in the course of the solution of the total problem (the territorial model of the entire economy), but the upper and lower limits can be obtained indirectly because the development of the economy in any particular plan period is predetermined to a large extent by the conditions of economic reproduction in the preceding period.

If we want to eliminate contradictions between the criteria of the particular (industry or regional) problems and the total economic problem of the territorial model of the Soviet economy, we will have to express the indicators of these criteria in terms of the indicators of differentiated costs, taking into consideration both direct relationships and feedback, that is, we will have to compute individual cost items in the industry (or regional) problems on the basis of estimates of the optimal plan (the territorial model of the economy as a whole).

Certain approximations of the estimates of the optimal plan can be obtained on the basis of previously optimized schemes of development and location of fuel and energy industries and other extractive and raw-material industries and the determination of differentiated indicators of transport costs and labor costs.[3] By their mathematical character, the economic-mathematical estimates of the optimal plan are limiting values that characterize the effectiveness of each factor and each condition in the problem of increasing costs as one proceeds from the use of the limited better resources to the use of the less limited worse resources, that is, the influence of the given factor on the ultimate function.[4]

Only on the basis of strict quantitative accounting of objectively formed patterns of location of production is it possible to work out the systematic course of the process of optimal adjustment of the regional structure of the economy, investment, and interregional relations (proportions, links, interactions) to the interest of development of the country as a whole. A correct evaluation of the complex set of existing regional conditions and the peculiarities of interregional relations is important for the elaboration of a general strategy of capital construction on a nationwide basis because excessive commitment and scattering of funds among many small projects would affect the rates of completion of productive capacities and other projects.

Increases in the effectiveness of existing fixed capital and of potential productive capacity (resulting from investment in the production sector) of individual regions as well as of individual industries as a result of shifts in location should be the basic criterion for correct planning of investment and employment on an areal basis.

The elaboration of an optimal structure of employment of labor resources, capital investment, production, and relationships between individual parts of the country requires the consecutive and repeated solution of a system of models of various kinds and different economic content. The determination of the most effective make-up of territorial-production complexes of appropriate types and sizes, and their specialization within the framework of the country as a whole or of a territory of a certain order, is an important element in the programming of the directions of the process of development and improvement of the economic geography of the Soviet economy and the territorial organization of productive forces. But only a territorial model makes it possible to work out the course of this process with the detail and the establishment of specific proportions that are needed for decision-making.

The computations that would ultimately insure a solution of the problem of searching for an optimal make-up of territorial-production complexes can be simplified by carrying them out by stages and separately for different groups of industries and types of problems as well as for generalized, consolidated zones: the European part of the USSR (without the Volga), the Volga-Urals-Ob' zone, and the East, with subsequent identification of regional specializations within these three zones (a division of the country into two zones is no longer adequate at this stage of research). Earlier, in the first stage of research, when we sought to elaborate methods of determining the direction of regional specialization and the magnitude of development of the basic group of industries in the various regions, we assumed the existence of only two zones (West and East) and two groups of industries—labor-intensive and energy-intensive.

The determination of zonal specialization in terms of extractive industries (structural links I and II [the author here is referring to the model depicted graphically in figure 1 in N. M. Budtolayev, V. P. Novikov, and Yu. G. Saushkin, "On Methods of Drawing Up a Territorial (Spatial) Model of the National Economy of the USSR," which is presented in Section 2 above]) and key raw-material and energy industries (structural links III and IV of the territorial model) means to identify the country's most economical resources subject to inclusion in the economy, and transformation into raw-material products and energy within each of the economic zones on the basis of differences in the availability of resources and the benefits of exchanging resources or their products between zones. Elaboration of the structural model of the Soviet economy by Yu. G. Saushkin[5] makes it possible to formulate a new problem, which in the final analysis involves the construction of a model of location of production and of links between the whole aggregate of industries that extract the primary raw material and fuel resources and transform these raw-material resources and fuels into primary fabricated materials and energy, and, on the basis of that model, to determine the proportion of investment and employment in three groups of industries: extractive industries, those transforming raw materials and fuels, and the other branches of manufacturing industries in the three zones of the country.

Determination of the specialization of zones and regions of extractive industry mainly means identifying the most economical directions of processing of regional resources and the optimal relationships between the branches of production of primary fabricated materials and energy (structural links IV and III of the territorial model), on the one hand, and the other branches of manufacturing industry, on the other hand (links V and VI of the model).

The actual effectiveness of the development of link IV in various production cycles of a particular region will, first, depend on the extent to which links I and II are economical within the given region (including the conditions of imports of resources for links I and II). The effectiveness will also depend on a reduction or increase of the cost of developing alternative regions and mutually exclusive industries in the country as a whole, and of organizing the transportation of other bulk resources except raw materials, that is, fuel and finished products. Shifts in the location of each cycle of industries will depend on the relative requirements of labor, capital, and fuels of the given cycle of industries or part of it. Such requirements are based, first, on the presence or absence of the possibility of raising the effectiveness of the productive capacity of the branches of the cycle linked by the principle of an integrated vertical combine and areally adjacent industries of particular regions, and, second, on the orientation, length of haul, and magnitude of interregional freight flows that are produced or eliminated by the locational shifts, and their relative effectiveness under various locational variants.

The forced development of an effective specialization of zones and regions by extractive industries, those transforming resources into raw-material products and energy, and other manufacturing industries should raise the effectiveness of the nation's productive capacity and rationalize the structure of employment by sectors of the economy and by industries. It should also intensify the movement of manpower among regions, changing the share of each part of the country in population and production and speeding the settlement of the regions east of the Urals.

No single cycle of industries predetermines the magnitude and structure of the growth of regional production as much as the iron and steel cycle—machine-building, which accounts for two-fifths of the industrial employment in the Soviet Union. We should therefore give priority to establishing how the development of the initial, intermediate, and final stages of this cycle of industries can be combined in various parts of the country and what measures should be taken to concentrate increments in iron-mining, iron and steel, machine-building, and steel fabrication in the optimal regions and thus obtain the maximum effect on a nationwide basis.

The proposed magnitude of development of the national economy, especially capital construction and development of machine-building, will require an increase in the production of iron and steel, with steel, in particular, rising from 91 million metric tons in 1965 to 250 million metric tons according to present long-term plans. Under these circumstances, what should be the relationships and structure of capital investment in the consecutive links of this key production cycle: the extraction and beneficiation of iron ores and coking coals (I and II), the production of iron and steel (IV), machine-building and steel fabrication (V and VI) between the regions of the West, the Volga-Urals zone, and the East of the Soviet Union? Analysis has shown that the rates of growth of machine-building and steel fabrication, which play such a decisive role in eliminating manual labor in agriculture and elsewhere and in raising the productivity of labor, depend directly on the relationship between the location of increments of production in these industries between the West and East of the USSR.

We know that a substantial jump in the rise of effectiveness of the productive capacity of machine-building and steel-fabricating industries and others is observed after the most difficult stages associated with the initial mechanization and the raising of the level of mechanization to 50 percent in all subdivisions. According to V. Ya. Feodoritov, the costs per released worker are cut to one-half when the base level of mechanization is raised above 25 percent, and to one-fourth when the base level of mechanization is raised above 75 percent, and the costs then become lower than the present value of plant and equipment per worker in these industries.[6]

Thus, after the elimination of manual labor, which is characterized by the lowest level of mechanization, capital investment in higher machine-building and metal-fabricating industries is more advantageous than the construction of new enterprises. It releases a large number of workers, who should then be given new jobs as a result of expanding employment in existing machine-building centers. Employment in machine-building and metal fabrication already exceeds one-third of the nation's total industrial employment and continues to rise at a more rapid rate than total employment in all sectors of the economy and in industry.

In seeking to construct a model of the development and location of productive forces in the European part of the USSR, the Volga-Urals-Ob' zone, and the East, we have to start from the main factor that determines the effectiveness of all costs. This principal factor is the varying labor requirements of industry. It operates as the basic cause of differences in total capital costs because the labor requirements of an industry will determine the magnitude of costs related to the creation of the nonproductive services that must be provided to varying degrees when industry is established in existing urban and rural communities in the West and when new cities and towns are built from scratch in the East. The total value of nonproductive fixed capital is almost 50 percent higher than the value of productive assets in the USSR as a whole, two to three times higher in machine-building centers, and five times higher in textile centers.

The development of machine-building and metal fabrication is taking place under different conditions in the West and the East. In the East, wages and construction costs are higher, there are fewer opportunities for expanding or reconstructing existing enterprises, and less favorable conditions in terms of the supply of needed types of rolled steel, the availability of markets, and the concentration and specialization of production. As a result, the return obtained from investment in productive capacity, services, and transportation needed to produce a certain increment in the output of the machine-building industry will be two to three times lower in the East than in the West, and the products of the eastern machinery plants will be of lower quality and will cost the consumer more.

These facts alone would be sufficient reason to alter the existing locational pattern in the machine-building industry in favor of the West. But, in addition, such a course is dictated by the need for concentrating all available manpower and all economic potentialities in the East and in the Urals on the forced development of extractive and primary processing industries as soon as additional resources have been made available by the cancellation of the proposed construction of new enterprises of labor-intensive industries, especially consumer goods, machine-building and metal fabrications, in the East and in the Volga-Urals zone.

A decision not to press the development in the East of machine-building and metal-fabricating industries, in accordance with the territorial model of the Soviet economy, would make available tremendous manpower and economic resources for sharply increasing the volume and rate of development of other industries that would be more beneficial for the country, and for carrying out fundamental structural shifts in the economy of the East and of the country, dictated by the interests of the entire Soviet Union.

The transfer of planned increments in the productive capacity of the machine-building and metal-fabricating industries from the East to the West and of energy-intensive industries from the West to the East would, in addition to all the other benefits, help to reduce the shortage of fuels and electric power in the West and, first, make available resources needed for an increase in the energy supply per worker, which would be less costly than an increase in the amount of fixed capital per worker. Second, the transfer would reduce transport costs of fuel from the East to the West, as well as the transportation of a large part of the machinery production. At the same time, raw material for the energy-intensive industries and the products of western machine-building and metal-fabricating industries could be shipped eastward at almost no extra cost (availability of empty eastbound freight-carrying capacity) over huge distances, with the machines being shipped mainly in assembled form (except to Central Asia and the Soviet Far East, where part of the machines can be sent more efficiently in dismantled form).

The concentration of capital investment in machine-building and metal fabrication in the West, which will greatly enhance the effectiveness of productive capacity and rates of growth in this group of industries, will in turn stimulate additional demand for ferrous and nonferrous metals and new fabrication materials, which is even now not being fully satisfied.

A SECTIONAL INTERREGIONAL MODEL OF THE SOVIET
IRON AND STEEL INDUSTRY AS A COMPONENT OF THE
TERRITORIAL MODEL OF THE ECONOMY

There are many aspects to the problem of constructing a territorial economic model of an industry. In addition to trends of technological progress, the development of the territorial division of labor, and changes in the regional effectiveness of development of various branches of an industry, a territorial model should include a model of resource use: available reserves of mineral resources and an estimate and comparison of the time periods over which the recoverable reserves of various regions will last at a given magnitude of resource use. A territorial-economic model in general includes a model of the optimal combination of the future structure of regional economies and the most effective system of relations between them. A territorial/sectional model of an industry should provide a quantitative picture of how the entire chain of elements of the particular industry should be distributed over the territory of the country, and how goods related in one way or another to that industry should move from one place to another.

In its spatial aspect, a model should also include a comparison of the effectiveness of the future links between consecutive elements of the industry. Such a comparison should not limit itself to tracing the geography and the economics of present and proposed links but should involve the design of a system of regional ties between branches of the industry based on a comparison of the sum total of possible alternatives of constructing the interregional links between all structural elements: extractive activities (I) and primary processing (II) with the iron and steel industry (IV), and the iron and steel industry with the structural elements that turn the steel into finished products in the form of producer goods (V), consumer goods (VI), and elements of productive, nonproductive, and transport assets (VII).

The establishment of an optimal system of such links does not necessarily require a minimization of transport costs, although it is based on computations that include the most reliable determination of changes in the transport component. The design of an optimal geography of relationships is based on a comparison of alternative systems of interregional and intraregional relationships; in other words, on a comparison of parallel calculations of the aggregate costs and the results of each of the alternative regional systems of location of branches of the particular industry and the related circulation schemes of the total production of all branches of the industry in the country as a whole.

The territorial structure and organization of production of the entire iron and steel industry should insure maximum increments in the production of pig iron, crude steel, and rolled products. The increments are predetermined by the most effective combination of the following distributional elements: (1) the geography of resources, extraction, and transportation of iron ore and of coking and steam coals based on the different cost levels in various regions, including transport costs; (2) the geography of the present productive capacity of the iron and steel industry and of the potential capacity (capital investment), taking into account the possibility of expansion of existing capacity and the limiting factors of water supplies and nonproductive services in each of the iron and steel centers; (3) the geography of consumption by regions and interregional exchanges of various types of iron and steel products; and (4) the geography of production and consumption by regions and interregional flows of various kinds of steel products and machines. The economic benefit derived from the greater availability of iron and steel products in the economy will be enhanced if the increased supply of such products will be associated with an improvement in the location and territorial organization of production of machines, metal products, other fabricated materials, and, of course, iron and steel themselves.

The regional geography of production and consumption of iron and steel depends not only on their costs of production and transportation but also on the costs of production and transportation of iron ore, coking and steam coals, and of steel-consuming products. In turn, the relationships between these costs and the investment needed to expand the existing raw-material and fuel base and the productive capacity of the iron and steel industry or to create entirely new iron and steel centers in certain regions will determine the most economical interregional proportions in the development of the industry. Consequently, the choice of regional investment and resource use must take into account the extent to which national and regional production increases in the iron and steel industry and its derivative industries and the production and transport costs of iron ore, coal, iron and steel, and fabricated machines will change in accordance with the adoption of a particular alternative distributional pattern.

The present discrepancy between the principal regions of production and consumption of iron and steel and the existence of heavy interregional freight flows make it possible to formulate the direction and magnitude of effective shifts in the geography of the iron and steel industry even before locational models of the industry and of iron and steel consumption by regions can be worked out in accordance with the territorial model of the national economy.

The elaboration of a model of the iron and steel industry requires an analysis of the locations where the productive capacity of the iron and steel industry can be most economically increased, of the most desirable assortment of production by regions, the locations of the largest, most accessible, and highest-grade resources of iron ore, and the economics of iron-ore extraction. The analysis should also cover the sources and amounts of fuel needed by the iron and steel industry, the resulting interregional flows, and economic indicators. One has to determine to what extent changes in the regional volumes of production of iron ore and iron and steel will affect the magnitude and orientation of future flows of iron and steel, ores, and fuels and how the resulting transport costs will compare with the costs of nonproductive services needed to develop the industry.

A preliminary comparison of alternative future locational patterns of the iron and steel industry suggests that the future distribution of the iron-ore and iron and steel industries should be different from the one indicated simply on the basis of a minimization of costs if only because of differences in the expected periods of existence of ore mines and differences in costs for the provision of nonproductive services and transportation. But, so far, no detailed locational study has been made even of the iron-ore industry. The time has now come to carry out Lenin's instructions about the need for developing the resources of the Kursk Magnetic Anomaly, that great series of iron-ore deposits with unparalleled reserves and a uniquely high grade of ores.

The largest iron-ore reserves in the USSR and in the world are concentrated in the Kursk Magnetic Anomaly, whose total reserves amount to many trillions of tons of iron, and in the highest-grade ores alone to more than 30 billion tons of iron. For purposes of comparison, the USSR extracted in 1962 a total of 184 million metric tons of crude iron ore containing 80 million tons of iron; the crude ore yielded 128 million tons of usable ore (70 million tons of iron).

Half of the present iron-ore output of the Soviet Union stems from Krivoy Rog, whose reserves (17.6 billion tons with an average metal content of 39.8 percent) constitute 37 percent of the recoverable reserves (40 billion tons with an average metal content of 48 percent, or 19.2 billion tons of iron) of the Kursk Magnetic Anomaly, which now yields 6 to 7 percent of the total iron-ore production of the USSR. The long-range reserves of the Kursk Magnetic Anomaly are estimated at more than 40 billion tons of high-grade ore and 10,000 billion tons of iron quartzites.

According to preliminary data, the cost per ton of iron in usable magnetite of the KMA mines (4.3 to 4.5 rubles) and the investment required per ton of production (25 to 30 rubles) are much lower than in the Urals without Magnitogorsk (production costs 9.5 to 10 rubles, investment 35 to 65 rubles per ton of iron), in Kazakhstan (costs 5.5 to 10 rubles per ton and investment 50 to 60 rubles per ton), or in Siberia (costs 12 to 15 rubles and investment 50 to 63 rubles per ton). If we also consider the investment in transportation needed to deliver ore along the high-density traffic routes from Kazakhstan to the Urals and from the mines of Siberia and Kazakhstan to the iron and steel mills of the eastern regions, the total investment required per ton of iron in the usable ores of Siberia would amount to 85–150 rubles, which is more than the investment per ton of pig iron needed in the KMA and the Urals. Further development of the iron and steel industry will depend to a large extent on the cost of ore, which tends to account for a growing share of the total cost of pig iron (68.3 percent of the pig iron cost at the Kuznetsk mill and 46 to 50 percent at the mills of the Urals without Magnitogorsk, compared with 18 to 36 percent of the mills that use Krivoy Rog ore).

In contrast to the geological reserves of nonferrous and rare metals, which are concentrated in the eastern regions and the Urals, the most effective and virtually inexhaustible iron-ore resources of the country are found in the central European part of the USSR, of which the northern half has the greatest demand for iron and steel. The demand of the regions of the northern half of the European part of the USSR (after out-shipments of certain types of iron and steel products) was 17.6 million tons in 1961 and is continuing to increase. (In-shipments of iron and steel into these regions amounted to 17.6 million tons in 1961 and out-shipments 3 million tons, making a net deficit of 14.6 million tons.)

The iron and steel deficit is also expected to be large in the countries of the Council of Economic Mutual Assistance (Comecon), which will be receiving an increasing flow of iron ore and coking coal. It would be more economical, instead, to supply these countries with pig iron from the southern European part of the USSR, where the consumption of iron and steel is also likely to increase sharply.

The iron-ore deposits of the KMA are situated more favorably with respect to the principal iron and steel consumer than are the mills of the Urals and the South, with their raw-material base, not to speak of the eastern mills. The length of haul of ore from the Yakovlevo deposit of the KMA to an iron and steel plant proposed for the Gor'kiy area would be 917 kilometers and to Cherepovets, 1,377 kilometers (to Moscow, 750 kilometers, and to Leningrad, 1,363 kilometers), while the distances to Moscow and Leningrad from the present major iron and steel centers are as follows: from Dnepropetrovsk—1,046, 1,613 kilometers; from Donetsk—1,111, 1,733 kilometers; from Chusovoy —1,538, 1,868 kilometers; from Nizhniy Tagil—1,769, 2,100 kilometers; from Chelyabinsk—1,953, 2,362 kilometers; from Magnitogorsk—2,317, 2,726 kilometers; from Temirtau—3,036, 3,445 kilometers; from Novokuznetsk—3,065, 4,074 kilometers. (Imports of iron and steel into the European part of the USSR from the Urals and the East amounted to 10.8 million tons in 1961, of which 8.15 million tons moved into the northern half of the European part. The areas of origin of the iron and steel flow to the European part were: Urals-Volga region 9.4 million tons, Kazakhstan 0.3, Siberia 1, and Far East 0.03 million tons.)

As a result of the existing geography of production of iron ore and iron and steel, millions of tons of rolled products, crude steel, and pig iron are being hauled over distances of thousands of kilometers despite the fact that the largest, richest, and most economical reserves of iron ore are near the consuming areas of iron and steel. The average length of haul of rolled products in the USSR is in excess of 1,200 kilometers and continues to increase because the planned shifts in the location and specialization of iron and steel mills tends to increase the areal discrepancy between producing and consuming areas. At the same time there is a growing discrepancy between the geography of reserves and the geography of extraction of iron ore, whose principal deposits are concentrated in the KMA and are several times closer to the main consumers than the deposits being mined, and all the more closer than the ore sources now proposed for exploitation. For example, the iron and steel industry of the Urals, which developed originally on local iron ore and was greatly expanded under Soviet rule as a result of the long haul of coals for coking, is now entering a new stage of development (relying to an ever-increasing degree on long-haul raw materials).

In 1963 the Urals produced 16.5 million tons of pig iron (29 percent of the Soviet total compared with 37.5 percent in 1950) and 26 million tons of crude steel (32.8 percent of the Soviet total compared with 39.2 percent in 1950). The shuttle system that played such an important role in the days of the Urals-Kuznetsk combine in reducing the cost of coking-coal hauls over a distance of 2,000 kilometers has long been completely disrupted. But because of the relatively low cost of transportation and of the ore, the cost of pig iron and steel produced in the Urals is among the lowest in the Soviet Union.

In connection with the depletion of the most economical iron ore of Magnitogorsk and other high-grade deposits, the Urals has long ceased to supply iron ore to the Kuznetsk mill, and since 1960 has in fact been importing increasing amounts of ore from Kazakhstan. These ore hauls coincide with the westward hauls of coking coal, and the two raw-material flows add up to 2.5 to 3 tons per ton of pig iron. For this reason, long-range plans call for limitations on the production in the Urals not only of pig iron but also of crude steel and rolled products (according to present plans of the Council for the Study of Productive Forces [SOPS], the share of the Urals in the Soviet production of pig iron is to decline to 25.5 percent by 1970). The iron-ore base of the Nizhniy Tagil steel complex in the middle Urals is represented by the low-grade Kachkanar ore (16 percent iron) whose economical exploitation depends to a large extent on the multipurpose use of the ore and the extraction of vanadium as a by-product; in other words, on the demand for vanadium in the Soviet Union.

But, according to data of Gipromez (Iron and Steel Design Institute), even with a reduced rate of growth of the iron and steel industry of the Urals, the depletion not only of the ore deposits but also of the tailings at Magnitogorsk by 1970 will make the iron and steel mills of the southern Urals (Magnitogorsk, Chelyabinsk, Novotroitsk) 50 percent dependent on the ore (8 million tons in terms of iron content) of the Rudnyy complex of Kazakhstan, whose designed capacity will be 9.4 million tons in terms of iron content. By 1970 it is expected that ore with an iron content of 17.1 million tons will have to be imported into the southern Urals out of total needs of 21.1 million tons, or 81 percent.

To cover the needs of the southern Urals mills and of the Karaganda iron and steel plant, there are plans to increase ore production in

Kazakhstan to 12.5 million tons by 1970 and 22.6 million tons by 1975 (in terms of iron content). Magnetite ores will be adequate to meet requirements only until 1982, before mining of this type of ore is expected to decline. At the same time existing Siberian iron and steel mills will require, in addition to expansion of present mines, an additional ore capacity of 1.3 million tons (in terms of iron) by 1970 and 2.8 million tons by 1975, and, in case of construction of a new Siberian plant, an additional 2 million tons by 1975.

In connection with the extraction of the needed amounts of raw material east of the Urals, plans call for the construction of a dozen mines and concentrators as well as cities and towns and the required communications. As will be shown below, such a solution of the problem of supplying the iron and steel mills of the Urals and the East would mean a tremendous and totally unjustified waste of social labor. What is needed here is a computation of economic costs and expectable results for the entire iron-mining industry, for every alternative locational pattern of iron mining and the iron and steel industry as a whole, and not just for individual deposits, groups of plants, and regions. That is the whole purpose of elaborating an interregional model of the iron and steel industry as part of a territorial model of the Soviet economy.

The integrated computational approach that we are trying to adopt involves the following: a computation for each alternative locational pattern of the costs and results in the immediate sphere of extraction of the raw material and its primary processing; a comparison of the total production costs with the cost of transporting the aggregate of goods of the particular industry and the required amount of fuel within the country as a whole; a computation of the effectiveness of a particular locational pattern of raw materials in terms of the magnitude of consumption and quality in various regions, including the effectiveness at subsequent production stages of the industry; and a selection of the locational pattern that would minimize aggregate costs and maximize results for the industry as a whole in all regions, including any feedback that may arise in the process of development of interregional exchanges.

The specific economic formulation of the sectional model of the iron and steel industry should include the concept of development of the industry that was elaborated in the course of work on the territorial model of the national economy, and further refined in the course of its detailed elaboration. This formulation should satisfy the requirements of the general economic formulation of the problem, for the solution of which each industrial model is elaborated as a system of information processing within the framework of the territorial model of the economy as a whole; that is, a system that would create optimal conditions for the development of other industries and would insure a locational pattern that would yield a maximum increase of production in the shortest possible time and with minimal costs to the consumer and minimal investment in the economy as a whole.

By relying on integrated computations of effectiveness of various alternative shifts in the location of the entire industry, the sectional model of the iron and steel industry is a key element of the territorial model of the Soviet economy. Such a sectional model should be the first to be elaborated, and it may help find the optimal locational pattern of the iron and steel industry, which would allow correction to be introduced into existing plans.

A PLAN OF THE CIRCULATION OF GOODS IN THE SOLUTION OF A SECTIONAL MODEL OF THE IRON AND STEEL INDUSTRY

Both the territorial model of the Soviet economy as a whole and its component, the model of the iron and steel industry, should represent the most effective organic combination of a model of location of production and a model of territorial circulation of goods. A circulation model has decisive significance for the design of the future locational pattern of industries that involve the processing of large amounts of raw material and fuel. A model of the principal freight flows and car flows makes it possible to investigate the interplay of regions in the course of development and the effect of freight flows in stimulating the formation of economic regions. A model thus helps to uncover reserves and additional possibilities for the effective use of the potential of these regions.

A major defect in the system of relationships between the western regions, in which are concentrated the greater part of the population, productive capacity, communications net, and urban places, and the eastern regions, with their particularly effective natural resources and conditions of production, is the great excess of westward freight traffic over eastbound traffic. This causes a tremendously large movement of empty cars from the western border of the USSR to Lake Baykal.

The predominant development of labor-intensive industries, which produce mainly finished products, in the heavily populated western regions through the expansion and reconstruction of existing productive capacity and accumulated assets will require a great increase in the flow of raw material and fuel from the less populated eastern regions. Extractive industries (categories I and II) and other primary and fuel industries (III and IV), as well as the construction industry, are the principal users of transportation, whose cost comes close to being half of the cost of production. Therefore, the question of organizing the rational use of communications cannot be reduced simply to the determination of the demand of various industries for freight transportation or to the establishment of certain average lengths of haul, and penalties for exceeding them. The method of evaluating the transportation component in establishing optimal interregional proportions must be completely changed. The existing territorial division of labor, involving as it does the magnitude, length of haul, orientation, and combination of basic interregional freight flows, the available facilities of transport communications, their provision with equipment, and the degree of use of traffic capacity, all these determine the magnitude of operating and capital costs produced by the formation and changes of freight flows (car flows) between regions, which in turn affect regional specialization and the basic territorial proportions.

Changes over time in the empty eastbound car flows are of particular interest in this connection. On the basis of data assembled by N. N. Kolosovskiy and now at our disposal, the total capacity of empty eastbound dry-goods cars at the western boundary of the Urals was 6 million tons in 1945, and the capacity of empty westbound tank cars 1 million tons. In 1961, according to computed data checked against actual statistics of the Railroad Ministry, the eastbound flow of empty dry-goods capacity through the western boundary of the Volga region and the Udmurt ASSR was 38 million tons (not counting empty waterway tonnage), and through the eastern boundary of the Urals, 55 million tons. According to forecasts of the Institute of Complex Transport Problems, the excess of westbound over eastbound dry-goods traffic across the Volga River is expected to be 88 million tons by 1970, and the flow of empty cars from the western frontier is expected to be 56 million tons. The freight capacity of empty dry-goods cars passing through the Volga meridian thus rose almost seven times from 1945 to 1961, and was expected to increase by 15 times, from 1945 to 1970.

According to the Institute of Complex Transport Problems, an empty dry-goods capacity of 24 million tons can be expected to move eastward by 1970 in the RSFSR across the meridian between Pskov and Bryansk, including more than 20 million tons in four-axle half-cars. In the Ukraine, the eastbound flow of empties (not counting tank cars) is expected to be 44 million tons north of Poltava and 86 million tons south of Poltava. And the total eastbound movement of empty dry-goods cars between Pskov and Dzhankoy is expected to be 155 million tons by 1970, and across the meridian between Gor'kiy and Volgograd 80 million tons (114 million tons if the Pechora Railroad is included).

The growth of empty dry-goods capacity will be accompanied by a growth of oil and gas flows. At the western boundary of the Volga region, for example, the total empty capacity of eastbound tank cars was expected to rise from 38 million tons in 1960 to 80 million tons in 1970; at the western boundary of the Urals from 60 to 103 million tons; at the eastern boundary of the Urals from 48 to 83 million tons; east of the Irtysh River from 50 to 68 million tons; east of the Ob' River from 60 to 100 million tons; and east of the Yenisey River from 10 to 12 million tons. According to analyses of data, not only by the Institute of Complex Transport Problems but also by the State Institute of Technical-Economic Surveys and Designs of Rail Transportation, the Council of Economic Mutual Assistance, and other organizations, transcontinental runs of empty cars will continue to increase.

The present carrying capacity of empty rail cars in the USSR exceeds the nation's total rail traffic of 1950, and transport assets engaged in empty runs (their percentage is increasing and amounted to more than 40 percent of the loaded runs in 1964) are equivalent at least to one-half of the total assets of the iron and steel industry.

Until lately, locational patterns of production were considered optimal if they minimized production costs in the given industry. The transport component was evaluated either on the basis of freight rates or on the basis of costs according to the shortest-distance principle. The magnitude of the transport costs calculated by this method differs substantially from the actual cost increments involved in the formation of particular interregional links, depending on whether they tend to increase or to reduce empty car runs. We estimate the actual transport costs in terms of economic distances expressed as the cost of moving a loaded train (over a corresponding distance in kilometers).

Table 1, which shows the difference between the shortest rail distance (in numerator) and the economic distance (in denominator) between centers of regions, was displayed in May 1964 at the symposium on economic geography of the USSR at the Fourth National Congress of the Geographical Society of the USSR. The denominators of the table make up a table of economic distances, which represents a matrix of costs involved in forming new interregional freight flows. A comparison between numerators and denominators shows the tremendous differences between economic distances and shortest rail distances. If we take the cost involved in forming freight flows from the northern half to the southern half of the European USSR as 100 percent, then the costs involved in forming other interregional flows can be simply read off in percentages, by dividing each denominator by 10. According to the data in table 1, the cost of a newly formed flow of bulk commodities from Moscow (northern half of European USSR) to Krasnoyarsk (Siberia)—equivalent to the cost of moving a loaded train over a distance of 850 kilometers—is almost 12 times lower than the cost of moving goods from Khabarovsk (Far East) to Moscow, and 8.5 times lower than the cost of moving goods from Krasnoyarsk to Moscow.

The method of computation of the table of interregional economic

TABLE 1

THE SHORTEST AND THE ECONOMIC DISTANCES
BETWEEN REGIONS OF THE USSR

EXPORTING REGIONS	IMPORTING REGIONS					
	Northern Half of European USSR	Southern Half of European USSR	Volga-Urals	Kazakhstan and Central Asia	Siberia	Far East
Northern half of European USSR	650	1111 / 1000	1400 / 350	3500 / 600	4150 / 850	8500 / 7025
Southern half of European USSR	1111 / 1800	850	1950 / 450	3700 / 750	4800 / 900	9100 / 7160
Volga and Urals	1400 / 3500	1950 / 3700	1100	2400 / 350	2900 / 550	7200 / 7000
Kazakhstan and Central Asia	3500 / 5400	3700 / 6500	2400 / 3500	1500	3600 / 500	7800 / 6750
Siberia	4150 / 7200	4800 / 8500	2900 / 4800	3600 / 5000	1800	4250 / 6995
Far East	8500 / 9800	9100 / 10900	7200 / 7500	7800 / 7700	4250 / 2625	2000

NOTE: Numerator--shortest rail distance between regional centers; denominator--economic distance, calculated for dry goods. Distances are calculated in kilometers.

distances takes into account the existing and expected future imbalance in bulk freight flows between the West and the East, and the
system of empty car runs from the West to the East, as well as the car
flows generated by major industrial centers and regions and attracted
by other regions that absorb the empty cars. The computational
methods also take into account the possibility of transforming economic distances into value indicators related to the particular type of
freight and the loading of various types of cars. Existing railroad
freight rates do not take these important circumstances into account,
nor does the territorial differentiation of wholesale prices, which is related to freight rates.

So far, we lack any kind of economic lever such as, for example,
extremely reduced freight rates for certain categories of enterprises.

Until recently, industrial indicators unsuitable for the solution of locational problems have been used in planning and operational calculations. Only according to the latest indicators of the Institute of Complex Transport Problems (1965) does the cost of moving dry goods from west to east in the direction of the empty car runs (0.02 to 0.025 kopecks per ton-kilometer) turn out to be six to seven times lower than in the heavily traveled westbound direction (0.12 to 0.17 kopecks per ton-kilometer). This is in complete agreement with the results obtained back in 1962 by the present author, together with N. M. Budtolayev.[7]

The predominance of raw-material and fuel shipments from the East occurs despite the relative similarities of industrial economies in the West and the East. This is due to the fact that the western regions export large amounts of raw material and fuel abroad, and the productivity of labor in many extractive industries in the East is higher (for a smaller amount of capital assets) than in the West. As we have shown, however, not all raw-material industries are necessarily more economical in the East than in the West, especially if we take into consideration the pattern of goods circulation and empty-car runs in the country, the locational pattern of the productive capacity of the manufacturing branches of these industries, and their provision with raw material and fuel. This is particularly true of the economical iron-ore resources concentrated in the West and the resources of oil concentrated in the Volga-Urals-Ob' zone.

The low cost and small capital requirements of the iron-ore deposits of the Kursk Magnetic Anomaly combine with an advantageous economic-geographical situation of the ore basin and an effective interplay between the new flows of ore, fuel, and steel resulting from its development and the existing pattern of goods circulation in the Soviet Union. That interplay of circulation would help utilize existing reserves in the national economy and eliminate or reduce disproportions.

The structure, magnitude, and changes of Soviet exports and the predominance of bulk commodities in exports are well known. Together with the flow of export products through western frontier stations and ports, we have also been witnessing a rapid growth of the flow of raw material and fuel into the densely populated western European regions.

A comparison of computational data about car runs with the actual data of the Railroad Ministry about the transfer of empty cars from one railroad to another shows the greatest discrepancy between computed and actual empty-car runs in the southern half of the European USSR. This is due to the fact that timber shipments from the North (in 1965 the Ukraine shipped in 18 million tons, and the 1970 figure is expected to be 17 million tons plus exports), machinery shipments from the North, and so forth, make only two-thirds as effective a use of car capacity as the coal shipments from the southern half of the European USSR. As a result, some of the cars from the Center, the Northwest, Belorussia, and the Baltic republics, instead of returning empty directly to the East over the northern trunk railroads, are needed to carry northern timber and machinery to the southwestern and southern European regions of the USSR, and only then join the eastward movement of empty-car capacity through the Left-Bank Ukraine, the Central Chernozem region, and the Balashov trunk line.

Table 2 shows the operating expenses, capital investment in rolling stock and fixed installations, and the derived cost of delivering a ton of raw material along those railroad lines in existence or near completion in the present five-year plan (as of 1970). It can be concluded from table 2 that the transportation of a ton of iron ore to the iron and steel mills of the Urals is more economical from the Lebedi concentrator or mine of the Kursk Magnetic Anomaly than from Rudnyy in northwest Kazakhstan. In our calculations we used indicators for the proposed Yakovlevo iron mine as being representative of the average indicators of the Kursk Magnetic Anomaly to demonstrate our point that the production capacity of the Kursk Magnetic Anomaly should be expanded to cover the ore deficit at the Urals mills. Actually, ore for the Urals would be shipped from those mining districts of the KMA that are the nearest the Urals.

The distances between these mining districts—Staryy Oskol and Novyy Oskol—and the Urals were measured along proposed rail lines, but the calculations were also made for existing lines supplemented by potential trunk routes that were inaugurated in the seven-year plan of 1959-65 (for example, the line Yuryuzan'-Beloretsk-Magnitogorsk, which already appears on maps).

To accommodate the growth of eastward traffic, there are plans for straightening and double-tracking present rail lines that would further reduce the length of haul from the KMA to the Urals by 150 to 250 kilometers or more. If ore from the Lebedi mine and later from the Lebedi concentrator were to be transported to the Urals, only the section between Staryy Oskol and Kasortnoye (65 kilometers) would be overloaded, according to the Institute of Complex Transport Problems. Even if the eastbound empty-car traffic generated by the Moscow region were to be used for the shipment of KMA ore to the Urals, the additional cost would be 30 to 50 kopecks per ton of ore.

Empty cars now move from the middle Urals to Karaganda and Ekibastuz by north-south lines via Chelyabinsk and Kartaly. Since empty cars would become available in the southern Urals after discharging their KMA ore loads, the present north-south runs of empty cars in

TABLE 2

OPERATING COSTS, CAPITAL INVESTMENT IN ROLLING STOCK AND FIXED INSTALLATIONS, AND THE DERIVED COST OF TRANSPORTING TO THE IRON AND STEEL MILLS OF THE URALS ONE TON OF IRON ORE FROM THE KURSK MAGNETIC ANOMALY AND FROM KAZAKHSTAN
(Without Initial and Final Terminal Operations)
(In Rubles Per Ton)

IRON AND STEEL MILLS OF THE URALS	ORE SHIPMENTS FROM RUDNYY (KAZAKHSTAN)			ORE SHIPMENTS FROM LEBEDI AND STOYLA (STARYY OSKOL DISTRICT) OF KURSK MAGNETIC ANOMALY		
	Operating Expenses	Capital Investment	Derived Cost (*)	Operating Expenses	Capital Investment	Derived Cost (*)
Magnitogorsk.	0.36	0.553 0.989 0.436	0.56	0.46	0.360	0.53
Chelyabinsk	0.43	0.717 1.499 0.782	0.73	0.53	0.360	0.60
Novotroitsk	0.42	0.736 1.310 0.574	0.80	0.52	0.360	0.59
Nizhniy Tagil	1.0	0.884 2.863 1.979	1.40	0.75	0.360	0.82
Chusovoy.	0.96	1.345 3.619 2.274	1.68	0.65	0.360	0.72

TABLE 2--Continued

IRON AND STEEL MILLS OF THE URALS	ORE SHIPMENTS FROM POGROMETS AND CHERNYANKA (NOVYY OSKOL DISTRICT) OF KURSK MAGNETIC ANOMALY			DIFFERENCES BETWEEN THE COST OF TRANSPORTING ONE TON OF IRON ORE FROM THE KURSK MAGNETIC ANOMALY AND FROM RUDNYY (KAZAKHSTAN)		
	Operating Expenses	Capital Investment	Derived Cost (*)	Excess (+) In Operating Expenses	Savings (-) In Capital Investment	Savings (-) In Derived Cost (*)
Magnitogorsk.	0.30	0.25	0.34	+0.10 to -0.06	-0.629 to -0.739	-0.03 to -0.22
Chelyabinsk	0.36	0.25	0.41	+0.10 to -0.07	-1.140 to -1.250	-0.13 to -0.32
Novotroitsk	0.35	0.25	0.40	+0.10 to -0.07	-0.950 to -1.060	-0.21 to -0.40
Nizhniy Tagil	0.58	0.25	0.63	-0.25 to -0.42	-2.503 to -2.613	-0.58 to -0.77
Chusovoy.	0.48	0.25	0.53	-0.31 to -0.48	-3.259 to -3.369	-0.86 to -1.15

NOTE: (*) The derived cost is the sum of the operating expenses and 20 percent of the capital investment.

the Urals could be eliminated and be redirected toward the east. This would help reduce empty-car runs roughly by the same distance that would be added by the additional movement of empty cars from Moscow to the KMA deposits because the distance from the KMA to Karaganda is only slightly greater than the distance from Moscow to Karaganda.

We can conclude on the basis of the foregoing that the present flow of empty cars from the European part of the USSR could be used to carry KMA ore to the Urals almost without an increase in the overall distance traveled, namely, to the extent of 20 million tons capacity. If the total distance traveled by the empty cars in the European part of the USSR and to the Urals were to be increased by just 200 to 300 kilometers, the capacity of the empty rolling stock that could be used to carry KMA ore to the East would amount to at least 50 million tons after 1970.

Information about future freight flows and empty-car runs is of great significance in deciding the future ore and coal orientation of the iron and steel industry in various regions and the rates and locations of future development. The indicators worked out by the Institute of Complex Transport Problems according to the principle of taking advantage of existing empty-car runs to transport bulk commodities within the available empty-car capacity seem completely reliable. There is no doubt that the difference between the cost of generating freight traffic in the direction of present empty-car runs as opposed to the direction of heavy freight traffic lines will continue to increase.

1.   N. M. Budtolayev, V. P. Novikov, and Yu. G. Saushkin, "On the Methods of Constructing a Territorial (Spatial) Model of the National Economy of the USSR," *Vestnik Moskovskogo Universitete,* seriya geografiya [Herald of Moscow University, Geography series], no. 6 (1964); translated in *Soviet Geography* (November 1965).

2.   *Osnovnyee metodicheskiye polozheniya po optimal'nomu otraslevomu planirovaniyu* [Basic Methodological Principles of Optimal Industry Planning] (Novosbirsk, 1966).

3.   L. A. Kozlov, "Primeneniye matematicheskikh metodov v optimal'nom planirovanii razvitiya i razmeshcheniya otrasley proizvodstva" [The Application of Mathematical Methods in Optimal Planning of the Development and Distribution of Branches of Production], *Nauchniy trudy Novosibirskogo Universiteta* [Scientific Works of Novosibirsk University], no. 3 (1965).

4.   A. G. Abanbegyan, "Primeneniye ekonomiko-matematicheskogo modelirovaniya pri reshenii zadach optimal'nogo razmeshcheniya proizvoditel'nykh sil" [The Application of Economic-Mathematical Modeling for the Solution of Problems of Optimal Distribution], *Izveztiya Sibirsk Otdel, Akademii Nauk SSSR* [Proceedings of the Siberian Section of the Academy of Sciences of the USSR], No. 6 (1967), pp. 6–14.

5.   Budtolayev, Novikov, and Suashkin, "Methods of Constructing a Territorial Model."

6.   V. Ya. Feodoritov, *Ekonomicheskaya effektivnost' proizvodstvennogo apparata* [Economic Effectivenesss of Production Capacity] (Moscow, 1965).

7.   N. V. Ovchininskiy, A. V. Turkin, A. N. Korobov, *Voprosy razvitiya chernoy metallurgii v tsentral'nykh rayonakh SSSR* [Problems of Development of the Iron and Steel Industry in the Central Regions of the USSR] (Moscow, 1961).

SECTION VI

Transportation Geography

Introductory Note

Until recently, transportation geography has occupied an unexpectedly minor role in Soviet economic geography.[1] The immense size of the USSR, the dispersed patterns of resources, production, and consumption, the crucial role of transportation in development planning and in regional shifts in development, the rapid rise in transportation use (even more rapid than the rise in output) all should have induced an extraordinary interest in transportation problems on the part of Soviet geographers. As with Western geography, the delayed growth of interest in transportation can perhaps be partly traced to the more traditional concern of economic geographers with aspects of production rather than questions of infrastructure and consumption. In the Soviet context this may have been reinforced by other factors: the past attitude toward transportation as a nonproductive or only *indirectly productive* form of economic activity; the conscious endeavor to minimize capital investments in transportation in order to divert funds to industrial development; and the concomitant effort to achieve transportation adequacy instead through intensification of use and efficiency of operation.

The growing volume of literature in transportation geography, as most writings in Soviet economic geography, reflects a concern with practical problems of national planning and also a high degree of interrelationship with problems and themes of other subfields of Soviet economic geography.[2] As indicated in Nikol'skiy's review in the initial article of this section, the field has developed, though belatedly, in response to the demands of planning; a considerable portion of the research in the field is carried on by specialized research institutions or under contract to planning agencies, and much of it therefore remains unpublished. Geographic studies of transportation are often directed to the interrelationships of transportation and such problems as resource development, industrialization, development of industrial nodes, economic regionalization and questions of interregional balances of trade, and regional specialization.[3] Conversely, a considerable concern with transportation questions is frequently encountered in works at least nominally within other subfields of economic geography.

In his review article Nikol'skiy categorizes the major topics studied as of the early 1960s: surveys of the geography of transportation of the USSR, theory, transportation networks, transportation relations to other economic activity, and regional transportation studies. In more recent years, as illustrated in some of the articles of this section, there has been increasing emphasis on questions of long-term projection of transportation needs, on passenger movements as well as freight movements, and on a more rigorous economic evaluation of transportation alternatives.

The second article in this section, "The Transport Factor and the Development of Economic Regions," by Popova, is a traditional examination of transportation in relation to regional economic development. Essentially informational, it presents characteristics of the Soviet transportation system and levels of development by economic regions. The author discusses such transport indicators as traffic density, network density, and network configuration. Empirically derived formulas showing relationships of network density, traffic, and economic development are briefly presented. Economic regions are classified by level of transportation development with the purpose of discovering the relationship between transport development and specialized and diversified development. The paper concludes with a discussion of the varying regional role of individual types of transportation, interregional ties, and regional transport balances.

In the third article of this section Kibal'chich presents an approach to the projection of interregional passenger movements. He states that methods used to calculate transport needs have not yielded adequate estimates and have resulted in past investment miscalculations. Kibal'chich believes that it is possible to develop useable projections for periods up to twenty years by beginning with large economic regions and working down to individual passenger generating points. Some of the required assumptions regarding changes in national and regional mobility, population distribution, and so forth, are discussed in some detail. For example, national mobility is expected to increase in proportion to per capita income whereas past regional differences in mobility are expected to lessen. The actual calculation is carried out by constructing a matrix of total interregional passenger movements, followed by a division between rail and air passengers, and finally by assignment of traffic along major routes.

In the final article of this section, "Long Distance Transport of Energy," Kuznetsov and Chuvilkin evaluate alternative methods of supplying the deficit areas of the European Center and Urals with fuel from Siberia and Central Asia. For selected energy source areas, alternative transportation methods are evaluated by employing a form of benefit-cost analysis. The authors recognize, however, that considerations other than transport cost alone must enter into the final decision.

1. This is not to say that the topic has been completely neglected in either prerevolutionary or Soviet economic geography; for example, the technique of mapping commodity flows was apparently first worked out and used in nineteenth-century Russia. See A. I. Preobrazhenskiy, "Economic Cartography," *Soviet Geography: Review and Translation* 7, no. 9, (November 1966): 41. However, the emergence of transportation geography as a distinct subdiscipline is a recent phenomenon.

2. The reader is referred to *Soviet Geography: Review and Translation* especially the issue of September 1964 (vol. 4, no. 7).

3. Because of these direct concerns Soviet transportation geography may be yet to develop a methodology more applicable to planning in developing countries than its sister discipline in the West. There is the further consideration that the Soviet historical experience of the relation of transportation and development and the conscious policy of allowing transport to be a lagged factor is probably more relevant than is the Western experience of transportation as a leading factor. A recent examination casts doubts on the "Western concept of infrastructure as a necessary precursor of economic expansion." See Holland Hunter, *Soviet Transport Experience: Its Lessons for Other Countries,* Brookings Institution (1968); also Robert N. Taaffe, *Rail Transportation and the Economic Development of Soviet Central Asia,* University of Chicago, Department of Geography, Research Paper no. 64 (1960).

I. V. NIKOL'SKIY

Research in the Geography of

Transportation in the USSR

The geography of transportation as a special branch of economic geography has developed rather recently in the USSR. In prerevolutionary Russia there were no works on the geography of transportation other than some dealing with questions of economics and the history of development of the transportation network in its geographical aspects.[1]

The development of Soviet geography of transportation was connected with socialist development. In the twenties the ideas of the GOELRO [State Commission on Electrification of Russia] plan raised major economic problems concerning transportation that caused a number of investigations of nationwide railroads, their classification, analysis of freight flows, and questions concerning the influence of transportation on economic development of the country.[2] Since the twenties special courses on the geography of transport have been given in transportation and economic institutes (Moscow and Leningrad institutes of transport engineers, at the Leningrad M. I. Kalinin Polytechnichal Institute, and the Moscow I. V. Plekhanov Institute of the National Economy) and also in institutions offering a specialization in geography (Moscow and Leningrad universities and the V. I.

Translated from *Vestnik Moskovskogo Universiteta,* seriya geografiya [Herald of Moscow University, Geography series], no. 3 (1963), pp. 3–11.

Lenin Pedagogical Institute). At the present time, courses in the geography of transport or the geography of communications are given in many universities of the country, in economic and transportation institutes, and in technical institutes.

In 1925 for the first time in the USSR, a short book on the geography of transportation was published by G. G. Sitnikov, then in 1930 a larger work appeared by S. V. Bernshteyn-Kogan in which the first chapter, "General Principles of the Geography of Transportation," is devoted to methodological questions.[3]

In creating cadres of economic-geographers specializing in the field of transportation, a large role was played by special expeditions on transportation economics and by comprehensive geographic expeditions, which have been conducted for many years by the Faculty of Geography of Moscow University in various regions of the USSR. The transport-economic investigations of Moscow University were conducted under assignment from design institutes, and their work was very seldom published. Both graduate and undergraduate students of Moscow State University took part in the expeditions of the geography faculty.[4]

The university geographers participated in drawing up methodological directions and instructions for comprehensive economic road research that was subsequently reflected in the character of the economic-geographic considerations.[5] In studying the problems of transportation, the economic-geographers did not go far in working out theoretical problems in the geography of transportation. Their main efforts were devoted to fulfilling the practical tasks connected with the development and distribution of transportation, to improving transportation communications, and to regionalizing freight transportation.

The problem of the development and distribution of transportation interests a wide circle of specialists; not only economic geographers, but to an even greater extent, economic and technical specialists participate in its solution. Economic factors occupy a decisive place in the geography of transportation; the major principles in the development and distribution of transport are associated with them, and they have an important role in regional differences in transportation. As a result of this it is difficult to draw a line between economic-geographical and purely economic works. Many works on the economics of transport of the USSR also contain geographical characteristics.[6]

At the present time investigations on the geography of transport are conducted at the Institute of Comprehensive Transport Problems of the USSR Gosplan, the Institute of Geography of the USSR Academy of Sciences, by the Faculty of Geography at the Moscow State University, at the transportation-economics research institutes, at the Transport Economics Institute, the Institute of the Geography of Siberia and the Far East of the Siberian section of the USSR Academy of Sciences, several departments at universities and other higher educational institutions, and also some branches of the USSR Academy of Sciences.

The geography of transportation in the USSR is studied under the following major topics: (1) General survey works and textbooks on the geography of transportation in the USSR; (2) Problematical and theoretical questions; (3) Geography of transportation networks; (4) Transport-economic relations and the geography of transportation; (5) Investigations of transportation in individual regions.

In spite of the large number of composite works on particular problems of the geography and economics of transport, there are few directly concerned with the geography of transportation of the USSR. Books on this subject have been published at wide intervals, and only in recent years have several works[7] been published; among them is a book published in 1960 by the author of this survey. In it are given the principles of development of the geography of transportation in the USSR, and a geographic survey of types of transportation, and economic transportation regions are described.

Some investigators assume that the geography of transportation only involves its distribution, which leads to the creation of purely descriptive works deprived of economic-geographical analysis. One subject of study under the geography of transportation must be its study as a link in the industrial-territorial complex. In connection with this there arises a group of geographical problems on transportation that are associated not only with the distribution of transport but also with the analysis of its distributed elements, which exist under specific economic and natural conditions.

Problems of the geography of transportation have been considered in articles by I. I. Belousov and I. V. Nikol'sky.[8] Although many theoretical and practical works have been devoted to questions of zoning and the regionalization of transportation, insufficient attention has been given to comprehensive economic transportation regionalization. Some investigators assume that transportation, as an important link of the industrial-territorial complex, is closely connected with the whole national economy and therefore no transportation regions should be distinguished. However, due to its specific characteristics, transportation differs from other branches of production. It has regional differences that provide reasons for conducting a special regionalization on the basis of transportation economics. The study of questions concerning the interdependence of transportation and industry, the influence of branches of industry on transportation and also the investigation of properties of transportation in economic regions were of great significance to the geography of transportation.

The works of I. I. Belousov most clearly express the role of economic regionalization in the study of transportation. He thinks that between the development of economic regions and transport there exist interconnections: transportation of freight, the distribution and exchange of commodities, the specialization of regions, the geographical division of labor, and the development of economic regions.[9] The regional balances of production and consumption, according to Belousov, are the initial material for planning freight turnover and freight flows, and consequently are also the basis for establishing a rational single transportation network. The ratios between production and transportation are decided within the economic regions.

Geographers have not given enough attention to the study of transportation nodes. This field is more thoroughly represented in the technical literature, especially that on railroad junctions. In the book *Principles for Developing Transportation Junctions*,[10] nodes are examined in combination with all types of transportation, and their types are analyzed according to diagrams of network outline: transportation nodes at the terminal points of trunk lines; radial transportation nodes; transportation nodes attenuated in length; radial-circle transportation nodes, radial-semicircle transportation nodes. The location of transportation nodes, the combination in them of various types of transportation, the properties of network outline, the variety of functions they fulfill, their different roles in the national economy, and other economic-geographical aspects have complicated the classification and study of transportation nodes.

The geography of transportation networks has always been one of the major objects of study in the geography of transportation. Some investigators tried to find mathematical formulas that would determine the service area requirements for the transportation network. N. N. Kolosovskiy derived an empirical formula according to which "the required length of railroad network in the USSR is proportional to the energy supply of the country, and proportional to the length of the anticipated haul, or, put in another way, proportional to the square root of the territory and inversely proportional to the planned freight intensity of freight traffic."[11] This formula has not been used, because it does not take into account the specific properties of the various regions. V. I. Panferov[12] proposed a method for the evaluation of a local communications network and derived a formula determining the rational concentration of the local communications network. The norms of a rational concentration obtained from these calculations do not make it possible to establish the directions of communication lines.

Proceeding from the role of individual railroads in the system of economic regions, Kolosovskiy worked out an original classification of the railroad network. Proceeding from principles of regionalization of the country, he grouped the whole network of railroads into three categories, and then in each category he distinguished classes of roads depending on their freight intensity and the role that they play in the transportation network. Many works have been devoted to the problems of the development of a network of separate types of transportation.[13] Now attention is being given to the problem of creating a unified transportation network in the USSR.[14]

The problems of transporting individual types of freight, their zoning and regionalization, interregional communications and the formation of freight flows have attracted the attention of both economic geographers and economists. In the 1950s at the Institute of Comprehensive Transport Problems of Gosplan under the direction of Belousov, a large project was undertaken on interregional communications and the establishment of the trend of the major freight flows. Belousov[15] established a connection between economic regions and market zones or zones of transportation of individual products. The first are distinguished mainly by the combination of productive forces and industries; the second, on the basis of distribution and exchange of commodities. The economic regions, in his opinion, are basic, and the transport zones are their derivatives. Rational transportation zones can be established on the basis of his network of economic regions by working out regional balances of production, the technical structure of enterprises and the planning of marketing and shipping. Defects in this coordination lead to the formation of irrational transportation.

Great attention has been given to the analysis of transportation of individual types of freight.[16] Factors and principles in the geography of transportation were revealed that are closely connected with the distribution and properties of production of individual bulk types of products.

Many works have been devoted to the study of transportation of individual regions of the country. The majority of these works have dealt with the problems of developing transportation networks, hauling, interregional communications, the improvement of navigation on major waterways, problems of transportation development in new regions, and also the study of the work of transportation nodes and the development of seaports. Many works have not been published and are in the archives of scientific and project institutions. Among works describing the transportation of a region as a whole the following have been published: G. I. Granik on Yakutia and Magadan Oblast; Ye. P. Gutsev on Belorussia; V. N. Kuznetsov on Bashkiria; A. V. Miroshnikova and N. K. Razdobud'ko on Latvia; I. V. Nikol'skiy on Kazakhstan; S. M. Khodzhayev on Uzbekistan; and P. N. Chernov on Uzbekistan and Turkmenia.[17]

The rapid development of productive forces in the eastern regions of the country, gave rise to a large number of investigations on the transportation of these regions. Major investigations were conducted on Siberia and especially its southern part.[18] The transportation development in the northeastern regions of the USSR with respect to economic geography was the subject of study of B. V. Belinskiy, G. I. Granik, F. V. D'yakonov, Ye. D. Rodin, S.˙V. Slavin, and S. S. Tsenin.[19] Many major investigations were devoted to the development of a northern sea route, the physical-geographical conditions of navigation, and the history of development of navigation. Recently works have been published elucidating the questions concerned with the geography of transportation in connection with the northern sea route.[20] Major investigations on the economics and geography of transportation have been conducted in regions of Kazakhstan.[21] Comparatively few works have been published that describe transportation of regions of the European USSR or Transcaucasia in its economic-geographical aspects.

1. A. I. Voyeykov, *Budet li Tikhiy Okean glavnym torgovym putem zemnogo shara?* [Will the Pacific Ocean be the Main Trade Route of the Earth?] (St. Peterburg, 1911); A. D. Bilimovich, *Tovarnoye dvizheniye na russkikh zheleznykh dorogakhs* [Freight Traffic on Russian Railroads] (Kiev, 1902); Bliokh, *Vliyaniye zheleznykh dorog na ekonomicheskoye sostoyaniye Rossii* [Influence of Railroads on the Economic State of Russia] (St. Petersburg, 1880); A. I. Chuprov, *Zheleznodorozhnoye Khozyaystvo* [Railroad Economy], vol. 2 (Moscow, 1910). (A number of works are devoted to the history of the development of the transportation network and to the description of ports in Russia.)

2. O. A. Kibal'chich, "The Influence of Lenin's GOELRO plan on Soviet Transportation Geography," *Geografiya i khozyaystvo* [Geography and Economy], no. 8 (Moscow, 1960). With regard to the geography of transportation, see also G. I. Petrusevich, *Istoriya razvitiya zheleznodorozhnykh gruzovykh potokov s 80-kh godov* [History of the Development of Railroad Freight Flows Since the Eighties] (Moscow, 1930); I. A. Poplavskiy, *Transport Sovetskogo Soyuza s narodnokozyay-stvennoy tochki zreniya* [Transportation of the Soviet Union from the Point of View of the National Economy] (Moscow, 1925); and A. Rybinikov, "The Influence of Railroad Transportation on the Formation of Agricultural Regions" in *Saratov-Millerovo* (Saratov, 1928).

3. S. V. Bernshteyn-Kogan, *Ocherki geografii transporta* [Essays on the Geography of Transportation] (Moscow-Leningrad, 1930); G. G. Sitnikov, *Geografiya transporta* [The Geography of Transportation] (Moscow-Leningrad, 1925).

4. B. F. Kosov, "The Geographic Expeditions of Moscow University," *Geografiya v Moskovskom Universitete za 200 let* [Geography at Moscow University Over 200 Years] (Moscow, 1955); A. L. Lutskiy, "Economic Geography at Moscow University," *Ucheniye zapiski Moskovskogo Universiteta* [Teaching Notes of Moscow University], no. 4, Geography (Moscow, 1940); I. V. Nikol'skiy and Ye. E. Tsedeler (with the participation of Ya. F. Antoshko), "The Faculty of Geography After the Great October Socialist Revolution," *Geografiya v Moskovskom Universitete za 200 let.*

5. *Instruktsiya i metodicheskiye ukazaniya k proizvodstvu kompleksynykh dorozhno-ekonomicheskikh izyskaniy po avtoguzhevym dorogam* [Instruction and Methodological Directions for Comprehensive Economic Investigation of Auto and Cart Roads] (Moscow, 1939); N. F. Krapivin and I. V. Nikol'skiy, *Vremennaya instruktsiya po titul'num dorozhno-eko-*

nomicheskim izyskaniyam avtoguzhevykh dorog respublikanskogo i mestnogo znacheniya [Provisional Instruction on the Subject of Economic Investigations of Auto and Cart Roads of Republic and Local Significance] (Moscow, 1948).

6. V. G. Bakeyev, *Morskoy transport SSSR za 40 let* [Forty Years of Maritime Transport in the USSR] (Moscow, 1957); *Voprosy razvitiya transporta i svyazi v SSSR* [Problems of the Development of Transportation and Communications in the USSR] (Moscow, 1948); B. N. Gladtsinov, *Voprosy razvitiya truboprovodnogo transporta SSSR* [Problems of the Development of Pipeline Transportation in the USSR] (Moscow, 1958); Yu. I. Koldomasov, *Kompleksnoye razvitiya transporta SSSR* [Comprehensive Development of Transportation in the USSR] (Moscow, 1961); S. F. Koryakin, I. L. Bernshteyn, and Yu. F. Ellinskiy, *Ekonomika morskogo transporta* [Economics of Maritime Transportation] (Moscow, 1959): A. A. Mitaishvili, *Vnutrenniy vodnyy transport SSSR i puti povysheniya yego eknomichnosti* [Internal Water Transport in the USSR and Ways to Increase Its Efficiency] (Moscow, 1957); V. S. Protasov and P. P. Sidorov, *Ekonomika rechnogo transporta* [The Economics of River Transport] (Moscow, 1958); *Transport SSSR* [Transportation of the USSR], part 1 (Moscow, 1960), part 2 (Moscow, 1961); T. S. Khachaturov, *Ekonomika transporta* [The Economics of Transportation] (Moscow, 1959); *Ekonomika transporta* (Moscow, 1957).

7. S. V. Bernshteyn-Kogan, "Advances Over 30 Years in the Geography of Railroad and Water Transport," *Voprosy geografii* [Problems of Geography], no. 6 (1957); I. V. Nikol'skiy, *Geografiya transporta SSSR. Uchebno-metodicheskoe ukazaniye i kratkiy konspekt otdel'nykh razdelov kursa* [Geography of Transportation of the USSR. Textbook and Methodological Instruction and a Brief Abstract of Individual Sections of the Course] (Moscow, 1958); also his *Geografiya transporta SSSR* [Geography of Transportation of the USSR] (Moscow, 1960); also his "Development and Distribution of Transportation in the Seven-Year Plan," in *Razvitiye i razmeshcheniye promyshlennosti i transporta v semiletke* [Development and Distribution of Industry and Transportation in the Seven-Year Plan], ed. A. T. Kurshchev and I. V. Nikol'skiy (Moscow, 1960); also his "The Main Trend of Development and Distribution of Means of Communication" *Geografiya v shkole* [Geography in School], no. 5 (1962); P. L. Sarantsev, *Geografiya putey soobshcheniya* [Geography of the Means of Communication] 2d ed. revised (Moscow, 1962); T. S. Khachaturov, *Razmesche-niye transporta v kapitalisticheskikh stranakh i SSSR* [Distribution of Transportation in Capitalist Countries and in the USSR] (Moscow, 1963).

8. I. I. Belousov, "Problems of the Geography of Transportation," *Sovetskaya geografiya* [Soviet Geography] (Moscow, 1960); I. V. Nikol'skiy, "Some Problems on the Geography of Transportation in the USSR," *Nauchnyye doklady vysshey shkoly, Geologo-geograficheskiye nauki* [Scientific Papers of the Higher Schools, Geologic-Geographic Science], no. 4 (1958).

9. I. I. Belousov, "On the Role of Transportation in Economic Regionalization," in *Trudy MTIPP*; N. G. Bochkarev, "On the Question of the Correlation between Volume of Production and Volume of Transport in the USSR," *Vroposy ekonomiki zheleznodorozhnogo transporta* [Problems of the Economics of Railroad Transportation] (Moscow, 1948); N. N. Kazanskiy, "Problems of the Investigation of Transportation in Economic Regions," *Izvestiya Akademii Nauk SSSR*, seriya geografiya [Proceedings of the Academy of Sciences of the USSR, Geography series], no. 6 (1961): M. S. Minakov, "Conditions for the Economic Expediency of Creating Local Metallurgical Bases and the Role of Transportation. Problems of the Economics and Planning of Transportation," *Trudy konferentsii molodykh spetisial-istov* [Works of the Conference of Young Specialists], part 1 (Moscow, 1960); A. A. Nikol'skiy, *Razmeshcheniye pishchevoy promyshlennosti i transportnyy faktor* [Distribution of the Food Industry and the Transportation Factor], part 1 (Moscow, 1960): E. I. Popova, "Major Characteristics and Factors in the Distribution of Transportation," in *Osobennosti i faktory razmeshcheniya transporta* [Characteristics and Factors of Distribution of Transport] (Moscow, 1960); Ye. A. Probst, "Transportation and the Distribution of Industry," in *Voprosy razmeshcheniya proizvodstva i ekonomicheskogo rayonirovaniya* [Problems of the Distribution of Industry and Economic Regionalization] (Moscow, 1960); N. Ye. Razdina, "The Influence of Transportation on the Distribution of Individual Branches of Industry," *Trudy konferentsii molodykh spetsialistov*, part 1; Ye. D. Khanukov, *Transport i raz-meshcheniye proizvodstva* [Transportation and the Distribution of Industry] (Moscow, 1956): G. A. Tsaritsyna, *Rol' zheleznykh dorog v razvitii sel'skogo khozyaystva SSSR* [The Role of Railrods in the Development of Agriculture in the USSR] (Moscow, 1958).

10. S. V. Zemblinov, V. A. Burakov, A. H. Obermeyster, A. A. Polyakov, V. A. Persianov, K. K. Tal', and V. P. Khodatayev, *Osnovy postroyeniya transportnykh uzlov* [Principles of the Development of Transportation Junctions] (Moscow, 1959).

11. N. N. Kolosovskiy, *Osnovy ekonomicheskogo rayonirovaniya* [Principles of Economic Regionalization] (Moscow, 1958).

12. V. I. Panferov, *Mestnaya set' putey soobshcheniya i usloviya eye formirovaniya* [The Local Network of Transportation Networks and Conditions for Its Formation] (Moscow, 1961); see also his "Grounds for the Development of a Local Network of Communications," *Geografiya i khozyaystvo*, no. 11 (Moscow, 1961).

13. I. G. Aleksandrov, *Postroyeniye plana razvitiya zheleznykh dorog* [Drawing Up a Plan for the Development of the Railroads] in *Materialy transportnoy komissii. Akademii Nauk SSSR.* [Materials of the Transport Commission. Academy of Sciences of the USSR] (Moscow, 1934); N. G. Bochkarev, "New Railroad Lines and Their Significance in the Development of Transport Ties," in *Voprosy razvitiya zheleznodorozhnogo transporta* [Problems of the Development of Railroad Transportation] (Moscow, 1957); G. M. Matlin, "Prospects for the Development of Future Inland Waterways on the Basis of Comprehensive Hydrotechnical Development," *Rechnoy transport* [River Transport], no. 8 (1958); V. I. Petrov, *Voprosy razvitiya set zheleznykh dorog* [Problems of the Development of a Railroad Network] (Moscow, 1957); A. I. Preobrazhenskiy, "New Railroads in the USSR," *Geografiya v shkole*, no. 4 (1957); T. S. Khachaturov, *Transportnaya set' SSSR* [The Transportation Network of the USSR] (Moscow, 1960).

14. S. P. Blank, "On the Problem of the Proportional Development of Various Types of Transport in the USSR," *Rechnoy Transport*, no. 8 (1958); N. N. Kazanskiy, "Economic Regionalization and the Problem of A Unified Transport Network in the Soviet Union," in *Problemy ekonomicheskikh svyazey i transporta* [Problems of Economic Ties and Transportation] (Moscow, 1962); V. N. Obraztsov, "Major Principles of the Development of a Transportation Network in the USSR," *Izvestiya Akademii Nauk SSSR,* seriya geografiya, no. 10 (1940); also his "On the Question of a Comprehensive Theory of Transportation," *Izvestiya Akademii Nauk SSSR,* Department of Technical Sciences, nos. 10–11 (1945); V. I. Petrov, "A Single Transportation Network in the USSR," *Zheleznodorozhnyy Transport* [Railroad Transport], no. 11 (1961); T. S. Khachaturov, "Complex Problems of Transportation," *Vestnik Akademii Nauk SSSR* [Herald of the Academy of Sciences of the USSR], no. 2 (1957); see also his "Comprehensive Development of a Transport System in the USSR," *Voprosy ekonomiki* [Problems of Economics], no. 9 (1962).

15. I. I. Belousov, *Osnovy mezhrayonnykh svyazey i perevozok* [The Bases of Interregional Ties and Transport], author's abstract (Moscow, 1958).

16. I. I. Belousov, *Ratsionalizatsiya perevozok khlebnykh gruzov i razmeshcheniye mukomol'noy promyshlennosti i elevatorno-skladskogo khozyaystva* [The Rationalization of Grain Transport and the Distribution of the Flour Mill Industry and Elevator-Granary Economy] (Moscow, 1957); see also his "On Nonrational Transport," *Geografiya i khozyaystvo* (Moscow, 1958); also his *Mezhrayonnyye svyazy i perevozki khlebnykh gruzov* [Interregional Ties and Grain Transport] (Moscow, 1958); I. I. Belousov, N. N. Kazanskiy, and V. S. Varlamov, "The Long-Run Development of Interregional Ties and Freight Flows," *Voprosy geografii*, no. 7 (1962); *Voprosy ratsionalizatsii perevozok gruzov* [Problems of the Rationalization of Freight Transport] (Moscow, 1957); L. Kats, "Problems of Fuel Balance and Interregional Ties," *Voprosy ekonomiki*, no. 6 (1957); see also his *Perevozka uglya po zheleznum dorogam* [Transport of Coal by Rail] (Moscow, 1959); O. A. Kibal'chich, "Experience in Formulating Hypotheses of Prospective Interregional Passenger Flows," *Voprosy geografii*, no. 57 (1962); also his "Problems in the Geographical Study of Passenger Transport," in *Problemy ekonomicheskikh svyazey i transporta* [Problems of Economic Communication and Transport] (Moscow, 1962); A. S. Kviitsinskiy, "The Rationalization of Timber Transport," *Zheleznodorozhnyy transport*, no. 6 (1960); also his "Economic Indices of Effectiveness of the Rationalization of Transportation Ties," in *Trudy konferentsii molodykh spetisialistov*, part 1; also his *Ratsionalizatsiya perevozok lesnykh gruzov i metodika opredeleniya eye effektivnosti* [Rationalization of Timber Transport and a Method for Determining Its Effectiveness], author's abstract (Moscow, 1961); Yu. Koldomasov, "Economic Regionalization and Improvement in Planning," *Planovoye khozyaystvo* [Planned Economy], no. 1 (1962); I. Langunov and K. Sheyman, "Distribution of Productive Forces and the Rationalization of Freight Transport," *Planovoye khoz-

vaystvo, no. 11 (1958); M. S. Minakov, "Problems of the Rationalization of Interregional Production Ties," *Voprosy ekonomiki*, no. 3 (1961); B. I. Shifirkin, "Distribution of Productive Forces and the Rationalization of Transport," in *Razvitiye zheleznodorozhnogo transporta v semiletkii* [The Development of Railroad Transportation in the Seven-Year Plan] (Moscow, 1960).

17. G. I. Granik, "Major Problems of the Development of Transport," in *Problemy razvitiya promyshlennosti i transporta Yakutskoy ASSR* [Problems of the Development of Industry and Transportation in the Yakut ASSR] (Moscow, 1958); also his *Transport Magadanskoy oblasti* [Transportation in the Magadan Oblast]; Ye. P. Gutsev, "Transport of the Belorussian SSR and Prospects for Its Development," *Voprosy ekonomiki Belorussii* [Problems of Economics in Belorussia], no. 1 (1960); V. N. Kuznetsov, *Transport Bashkirii* [The Transportation of Bashkiria] (Ufa, 1960); also his *Geografiya transporta Bashkirskoy ASSR* [Geography of Transportation of the Bashkir ASSR], author's abstract (Ufa, 1960); A. V. Miroshnikova and N. K. Razdobud'ko, "The Transportation of the Latvian SSR," in *Voprosy ekonomiki transporta* [Problems of the Economics of Transport] (Riga, 1961); I. V. Nikol'skiy, "The Geography of Transportation of Kazakhstan," *Geografiya i khozyaystvo*, nos. 3–4 (1958); S. M. Khodzhayev, *Transport Uzbekistana* [The Transport of Uzbekistan] (Tashkent, 1961); P. N. Chernov, "Transport of Uzbekistan," *Voprosy ekonomiki transporta*, no. 16 (Tashkent, 1960); also his "Transport of Turkmenistan," in *Trudy Instituta ekonomiki AN TSSR* [Works of the Institute of Economics of the Academy of Sciences of the Turkmen SSSR], vol. 3 (Ashkhabad, 1961).

18. *Voprosy razvitiya transportnoy seti Sibiri* [Problems of the Development of the Transport Network of Siberia], in *Trudy Transportno-energeticheskogo instituta* [Works of the Transport Energy Institute] (Irkutsk, 1960); B. B. Gorizontov, "Development of Transport in the Yenisey Basin," *Rechnoy transport*, no. 7 (1958); Ye. V. Yevreyskov, "The Increase of Transport Ties in the Kuznetsk Basin," in *Izvestiya Vostochnykh Filialov AN SSSR* [Proceedings of the Eastern Affiliate of the Academy of Sciences of the USSR], no. 1 (1957); N. N. Kazanskiy, "Economic Development of Southeastern Siberia and the Problem of Transport," in *Sibirskiy geograficheskiy sbornik* [Siberian Geographical Collection] (1962); G. T. Krasheninnikov, "Inland Water Transportation of the Eastern Regions of the Soviet Union," in *Isvestiya Vsesoyuznogo geograficheskogo obshchestva* [Proceedings of the All Union Geographical Society], vol. 93, no. 4 (1961); R. I. Nekrasova, "The Development of Transport Ties of the Buryat-Mongolian ASSR," *Geografiya i khozyaystvo*, no. 2 (1958); also her "The Interregional Transport-Economic Ties of the Eastern Siberian Economic Region and Their Rationalization," *Geografiya i khozyaystvo*, no. 9 (1961); also her "Problems of the Development of Transport of the Baykal Region in Relation to Its Industrial Development," in *Trudy konferentsii molodykh spetsialistov*, vol. 3; V. I. Petrov and S. S. Tsenin, "The Development of the Economy of the Eastern Regions and Their Transportation Development," *Zheleznodorozhnyy transport*, no. 10 (1956); *Razvitiye proizvoditel'nykh sil Vostochnoy Sibiri. Transport* [Development of the Productive Forces in Eastern Siberia: Transport] (Moscow, 1960); T. S. Khachaturov, "Prospects for the Development of Transport in Eastern Siberia," *Izvestiya Sibirskogo otdeleniya Akademii nauk SSSR* [Proceedings of the Siberian Section of the Academy of Science of the USSR], no. 12 (1958).

19. B. B. Belinskiy, "The Transport Significance of the River Aldan in the Development of the Northeastern Yakutia," *Voprosy geografii Yakutii* [Problems of Geography of Yakutia], no. 2 (Yakutsk, 1962); I. I. Granik, "River Transport in the Basins of the Rivers Yana, Indigirka, and Kolyma," *Rechnoy transport*, no. 3 (1958); also his *Transport Severo-Vostoka Yakutskoy ASSR* [The Transport of the Northeast Yakut ASSR], *Problemy Severa* [Problems of the North], no. 3 (1959); also his "Transportation," in *Problemy razvitiya proizvoditel'nykh sil Magadanskoy oblasti* [Problems of the Development of the Productive Forces of Magadan Oblast] (Moscow, 1961); also his "Chronicle of Navigation on the Rivers of Yakutia," *Letopis' Severa* [Chronicle of the North], vol. 3 (Moscow, 1962); F. V. D'yakonov, "Problems of the Geography of Trunk-Line Transportation of the Northeast of the USSR," *Trudy Vostochno-Sibirskogo filiala Sibirskogo otdeleniya, AN SSSR* [Works of the Eastern Siberian Section, the Academy of Sciences of the USSR], no. 32 (1960); Ye. D. Rodin, "Ways of Developing the Small Rivers of Western Yakutia," *Rechnoy transport*, no. 6 (1958); S. V. Slavin, *Promyshlennoye i transportnoye osvoyeniye Severa SSSR* [The Industrial and Transport Development of the North of the USSR] (Mos-

cow, 1961); S. S. Tsensin, "Transport Development of the Basin of the River Lena," *Rechnoy transport*, no. 8 (1958); also his "Problems of the Development of the Economy and Transport Ties of the Northeast of the USSR (Yakut ASSR and Magadan Oblast) and the Role of the Northern Sea Route," *Problemy Severa*, no. 3 (1959).

20. I. A. Fel'dman, "Problems of Export of Siberian Timber Through Northern Seaports," *Razvitiye proizvoditel'nykh sil Vostochnoy Sibiri. Transport*; I. L. Freydin and T. I. Shlykova et al., *Sovremennyyee vneshniye transportno-ekonomicheskiye svyazi rayonov Kraynego Severa, raspolozhennykh k vostoku ot Yeniseya* [Contemporary External Transport-Economic Ties of Regions of the Extreme North Situated to the East of the Yenisey] (Leningrad, 1959).

21. A. Bekkulov and K. Mizambekov, *Stal'nyye magistrali Kazakhstana* [Steel Trunklines of Kazakhstan] (Alma-Ata, 1960); D. R. Bogorad, "Problems of Rationalization of Transport Ties of Kazakhstan," *Zheleznodorozhnyy transport*, no. 1 (1959); I. V. Lavrova, "Transport-Economic Ties of Kazakhstan," *Trudy Instituta ekonomiki AN Kazakhskoy SSR* [Works of the Institute of Economics, Academy of Sciences, Kazakh SSR], no. 3 (1959); I. V. Lavrova, A. Bakkulov, et al., "Problems of Improving the Organization of Freight Work of the Major Types of Transport in the Transportation Junction of Kazakhstan Based on the Example of the Alma-Ata Junctions," *Kazakhskoy SSR* [Kazakh SSR], no. 6 (1962); I. V. Nikol'skiy, "Problems of Developing a Transport Network in Eastern Kazakhstan," in *Voprosy ekonomicheskoy geografii Vostochnogo Kazakhstana* [Problems of the Economic Geography of Eastern Kazakhstan] (Moscow, 1961); V. A. Osorgin, I. V. Lavrova, et al., *Voprosy kompleksnogo razvitiya transporta v rayonakh osvoyeniya tselinnykh i zalezhnykh zemel' Severnogo Kazakhstana* [Problems of the Comprehensive Development of Transportation in Regions of Developing Virgin and Long-Term Fallow Lands in Northern Kazakhstan] (Alma-Ata, 1957); *Promyshlennost' transport Zapadnogo Kazakhstana* [Industry and Transport of Western Kazakhstan] (Alma-Ata, 1955).

22. S. V. Bernshteyn-Kogan, "Major Features of the Historical Geography of the Moscow Water-Transport Junction," *Voprosy geografii*, no. 27 (1951); V. S. Varlamov, "Transport-Economic Ties of the Orenburg Industrial Junction," in *Trudy konferentsii molodykh spetsialistov*, part 2; P. G. Ivantsov, "Novorossiysk Port," in *Trudy Gosudarstvennogo po proyektirovaniyu morskikh i sudoremontnykh predpriyatiy* [Works of the State on Projected Seaprts and Maintenance Enterprises], no. 4 (1957); N. F. Izotov, "Transport and Its Role in the Formulation of Regions of the Far East," in *Sbornik statey kafedry Marksizma-Leninizma Khabarovskogo instituta inzhenernogo transporta* [Collection of Articles of the Department of Marxism-Leninism of the Khabarovsk Institute of Engineering Transport], no. 1 (1960); R. I. Kverenchkhiladze, "Transport and Interregional Ties of Lower Imeretia," in *Trudy Instituta Geografii AN Gruzinskoy SSR* [Works of the Institute of Geography of the Academy of Science of the Georgian SSR], no. 13 (1960) (in Georgian); I. M. Korobkov, "Economic Significance of Transport Development of Small Rivers in the Perm Economic Administrative Region," *Geografiya i khozyaystvo*, no. 8 (1960); B. A. Makovskiy and M. M. Glaskov, *Volzhskiy basseyn i yego rechnoy transport* [The Volga Basin and Its River Transport] (Moscow, 1958); N. A. Molva, "The Port Odessa," in *Trudy Gosudarstvennogo Instituta po proyektirovaniyu morskikh portov i sudoremontnykh predpriyatiy*, no. 4 (1957); V. F. Pavlenko, "Interregional Transport-Economic Ties (Based on the Examples of Central Asia)," *Planovoye khozyaystvo*, no. 9 (1962); S. I. Prokhorov, *Promyshlennost' i transport Gor'kovskoy oblasti* [The Industry and Transport of Gor'kiy Oblast] (Gor'kiy, 1958); also his *Transport* [Transport] (Gor'kiy, 1960); N. K. Razdobud'ko, "Some Problems of the Transport-Economic Ties of the Baltic Republics and the Lowering of Unit Cost," in *Voprosy spetsializatsii promyshlennosti i kompleksnogo razvitiya narodnogo khozyaystva Pribaltiyskikh sovetskikh respublik* [Problems of the Specialization of Industry and the Comprehensive Development of the Economy of the Soviet Baltic Republics] (Riga, 1959); N. A. Retyunskiy, "The Port Nakhodka," in *Trudy Gosudarstvennogo Instituta po proyektirovaniyu morskikh portov i sudoremontnykh predpriyatiy*, no. 4 (1957); G. Stepanyan and A. Agababov, "Railroad and Truck Transport in the Republic," *Narodnoye khozyaystvo Armenii* [National Economy of Armenia], no. 6 (1960); A. A. Taltyn', "The Seaports of Soviet Latvia and the Prospects for Their Development," in *Razvitiye narodnogo khozyaystva Latviyskoy SSR* [The Development of the Economy of the Latvian SSR] (Riga, 1961).

YE. I. POPOVA

The Transport Factor and the

Development of Economic

Regions

THE DEVELOPMENT OF TRANSPORT IN THE ECONOMIC REGIONS

In order to make apparent the role of transport in the development of the territorial division of labor, the level of development of transport in the economic regions must be determined; this level must correspond to the level of development of production of the economic region. Formerly there was a certain lag in the development of transport behind the needs of the rapidly developing economies of the regions, especially the eastern regions. This shows up in the inadequate transport facilties, the difficulties in transporting bulk freight (in particular, seasonal shipments of agriculture), the insufficient development of mixed communications with use of various types of transport, the great deal of time needed to deliver freight to the consumer, the problems of spoilage in transit, and, finally, in the large national economic expenditures for delivery of freight into remote regions. In regions of new development the development of transport as an imperative condition preceding in time the develop-

Translated and abridged from the article in *Zakonomernosti i faktory razvitiya ekonomiche-skikh rayonov* [Laws and Factors of Development of Economic Regions of the USSR], ed. Ya. G. Feygin, M. A. Vilenskiy, and D. D. Moskvin (Moscow, 1965), pp. 222–32, 236–44.

ment of production has not always been observed, and this has prolonged construction and raised costs due to transport expenditures.

The problem of determining the most rational levels of development
of transport in economic regions is one of the least studied and clarified problems in the economic literature. We shall take up only the
system of indices of the level of transport development, the description of the regions according to these indices and the grouping of regions according to the level of transport developed.

The level of transport development in the economic regions is described by the following basic indices:

1. freight volume (in billions of ton-kilometers) and shipments (in
millions of tons) on all types of freight transport of the regions, and
the relative significance of the regions according to these indices;

2. the length of railways, auto roads, navigable-river routes, pipelines (in kilometers) and their density per 1,000 square kilometers of
territory;

3. the degree of utilization of the individual types of the transport
networks or their density of freight traffic (in ton-kilometers per kilometer);

4. economic indices of work for the individual types of transport
(cost of transfer, relative capital investments).

The indices of the volumes of freight turnover and transfers are
the basic ones in determining the level of transport development in the
region. As can be seen from table 1, the economic regions of the
USSR differ in relative significance of freight turnover and transfers.
In the large industrial and agricultural regions, where there are
branches of the economy that result in bulk freight, freight turnover
and transfer are the most significant. About 65 percent of freight
turnover and more than 60 percent of freight transfers occur in the
Central, Northwestern, Volga, Urals, Donets-Dnieper, Western Siberian, and Kazakh regions. The volumes of freight turnover and
transfers are considerably less in regions where branches resulting
in bulk freight are absent or developed on a small scale. The Volga-
Vyatka, Central-Chernozem, Southern, Baltic, Central Asian and Far
Eastern regions have only 15 percent of freight turnover and about
17 percent of transfers.

The difference in relative significance of a region in freight turnover
and transfers is due to the different economic distance of transfers,
which depends on the areal extent of the regions, the location of production and population, and the composition and extension of the
transport network. It is precisely the considerable distance of freight
transfers resulting from the dimensions of the eastern regions, the re-

TABLE 1

FREIGHT VOLUME AND FREIGHT SHIPMENTS IN THE ECONOMIC REGIONS IN 1962
(As Percentage of USSR Total)

ECONOMIC REGIONS	REGIONAL SHARE OF TOTAL	
	Freight Volume*	Shipments
Northwest.	6.3	6.8
Center	9.2	9.5
Volga-Vyatka	3.1	2.2
Central Chernozem.	3.9	2.7
Volga.	12.7	8.7
Northern Caucasus.	4.8	5.5
Ural	10.9	9.6
Western Siberia.	9.0	6.3
Eastern Siberia.	6.8	5.1
Far East	3.3	2.9
Donets-Dnieper	8.3	13.2
Southwest.	4.7	4.7
Southern	1.3	2.3
Baltic	1.6	2.8
Transcaucasia.	1.8	3.2
Central Asia	2.2	4.1
Kazakhstan	8.0	7.8
Belorussia	1.8	1.9
Moldavian SSR.	0.3	0.7

NOTE: *The volume of freight was calculated by the author. In
freight volume and shipments, shipments going abroad by maritime
transport and some types of shipments by truck were not taken into
account; hence, total freight volume and shipments in the USSR are
somewhat smaller than in the published statistical yearbooks of the
Central Statistical Department of the USSR.

moteness of industrial centers and centers of consumption from
each other, and the great extension of the railroad network mainly
along a latitudinal direction, that cause the volumes of freight turnover
in the eastern regions and their relative significance in the freight
turnover of the Union to exceed the corresponding indices of transfers
to other regions. In the majority of regions of the European part of the
country, where the transfer distance is less, the relative significance of
freight turnover is below that of transfers. This is most characteristic
for the Donets-Dnieper, Southern, Transcaucasia, and Baltic regions.

At the present time there are no statistical data on the distance of
transfers by rail and water transport within an economic region. Only
for automotive transport is it possible to calculate the transfer distance in the economic regions with sufficient accuracy. But automotive
transport is not indicative in this respect, since the main part of its

transfers is within cities and within primary administrative regions. Therefore, the uniqueness of an economic region lies in the composition and dimension of its territory, the level of industrial development, the distribution of production and consumption, the composition and configuration of the trunk lines in the transport network; the load has little effect on trucking distances. These shipments in a kray, oblast, and republic differ little in average distance. Thus, the average trucking distance in the Yakut ASSR in 1962 was 19.8 kilometers; in the Moldavian SSR, 14 kilometers; in the Lithuanian SSR, 11.1 kilometers; and in Magadan oblast, 42.3 kilometers.

The distance of freight shipments in each region by all types of transport (excluding automotive transport) was determined by calculation. In a number of regions with a large freight transit this distance proved to have been overstated (in particular in the Central-Chernozem and Volga-Vyatka regions). The greatest distance of shipments is in the eastern regions of the country—in Eastern and Western Siberia and in Kazakhstan (up to 1,000 kilometers). A distance of from 500 to 700 kilometers is characteristic for the Northwestern, Volga-Vyatka, Volga, Urals, and Far Eastern regions. In the Central, Northern-Caucasus, Donets-Dnieper, Southwestern, Baltic, Central Asian, and Transcaucasian regions the distance varies from 300 to 500 kilometers.

Changes in freight distances will occur in the economic regions as a result of increased integration in the economy, improved distribution of production and consumption in the regions, and a higher level of transport development. However, even in the future, freight distances in the eastern regions will certainly be greater than in the western regions. However, because of the increase in transport service and a more rational distribution of freight flows between types of transport, freight distances will be lowered.

The next important index of the level of development of transport in the economic regions is the length of the transportation network and the density of the network. Usually the latter index is considered basic in describing the level of transport development, since it is most susceptible to analysis. Without denying the great significance of this index in the present analysis and in the determination of the future level of transport development, its limitations must be noted. The density of a network does not give an idea of the degree of its utilization, and, moreover, it depends on the size of the territory of the region. Therefore, in determining the density of a transport network, one must take into account the developed territory that provides a definite load to transport, rather than the entire area (including the unpopulated and sparsely populated regions of the tundra or desert).

The densest network of railroads (with a density of over 20 kilometers per 1,000 square kilometers) is characteristic for the old, developed regions of the European part of the Union: Central, Central-Chernozem, Baltic, all regions of the Ukrainian SSR, Belorussia, and the Moldavian SSR. They have more than 40 percent of the railroads in the Union. The average density of the network of roads (from 10 to 20 kilometers per 1,000 square kilometers) is in such regions as the Volga-Vyatka, Volga, Northern-Caucasus, and Transcaucasia, where this network was created later than in the western regions of the country. These regions have about 16 percent of the road network whereas the regions of the western half of the USSR (without the Urals) carry more than 65 percent. The Northwestern, Ural, Western Siberian, Eastern Siberian, Far Eastern, Kazakh, and Central Asian regions, where large-scale industrial development began only in the years of Soviet rule, have 44 percent of the network with a density of from 0.8 to 6.7 kilometers per 1,000 square kilometers of territory. Even if one omits the Far North, the density of the transport network in these regions is low.

Comparison of the regions by the network density of hard-surface roads shows approximately the same correlation as that of railroads. In the regions of the European part of the Union (Donets-Dnieper, Southwest, Southern, Baltic, Central, Transcaucasia, Belorussian and Moldavian SSRs), the density of hard-surface roads is considerable (from 40 to 120 kilometers per 1,000 square kilometers); more than 50 percent of the length of the network is concentrated in them. But in the eastern regions the density of hard-surface roads varies within the limits of 1 to 20 kilometers per 1,000 square kilometers; these regions have 19 percent of the network. In automotive transport, the difference in the distribution of density of the network of paved roads among the regions is more apparent than in railroad transport.

An analysis of the density of a transport network must be augmented by an analysis of the configuration of this network that develops under the influence of the distribution of production and population. A definite type of distribution of transport network corresponds to each type of economy in a region and to each type of distribution of production. A favorable configuration of the network, even at a comparatively low density, makes it possible to service the economy of the region with the least transport expenditures. An unfavorable configuration of the network at a low density increases the difficulties in transport service to the region. The higher the density of a network, the more possibilities there are for improving its configuration and, consequently, of facilitating the processes of freight transfers from the point of production to the trunk-line network and back.

The comparison of the character of distribution of the railroad network in regions of different types, in particular the regions of the Baltic and Eastern Siberia, is interesting. The railroad network density in the Baltic is 42 times as high as in Eastern Siberia (and the density of hard-surface highways is 100 times as high). For the Baltic the combination of the network of trunk-line latitudinal roads to ports and border points, and meridional roads of inter- and intra-regional significance is characteristic: they form a complex honeycombed configuration. Almost all administrative regions of the Baltic republics are served by railroads (including narrow-gauge railroads). The absence of large beds of minerals, a high population density, a large number of populated localities with a diverse processing industry, the presence of large ports, and a developed, agricultural production over the whole territory of the region, all these features of distribution of the production of the region, have put an imprint on the configuration and composition of the network. In turn, the considerable density of the transport network and its favorable configuration facilitates the process of transporting freight from the place of production to the place of consumption and contributes to the establishment of direct transport connections between all industrial centers and the majority of populated localities in the region.

In Eastern Siberia there is one large double-track railroad line in a latitudinal direction; adjoining it are roads serving regions situated to the south and north of the trunk line. Vast areas remain outside the railroad belt; for them the main type of transport are the systems of the rivers Yenisey, Lena, Selenga and of Lake Baykal. The distribution of production is concentrated in the railroad belt or along navigable rivers connected by transhipment points with the railroads. Most branches produce heavy flows for transport along the trunk line.

The configuration of the transport network of Eastern Siberia is such that the newly developing industrial centers are distributed mainly along the trunk line, thereby increasing its load still more. This cannot be accepted as correct, neither from the point of view of the general development of the economy of the region, nor from the point of view of exploitation of the railroad line. An increase in density of the transport network and a modification of its configuration will contribute to a higher degree of comprehensive development of the economy in Eastern Siberia.

The index of density of freight traffic, or traffic density, is a very important complementary index to the index of network density. There exists a definite correlation between network density and its load: the lower the network density, the higher the load per kilometer of road and vice versa. Many facets of transport activity, especially its quality indices, depend on the degree of density of freight traffic. The higher the density of freight traffic of the railroad lines, the greater their technical equipment and the more favorable are the indices of labor productivity and net cost of transfers. In regions with a high density of freight traffic the net cost of transfers is lower than the average Union net cost.[1] In regions with a high network density, for example, in the

European part of the country and in Transcaucasia, the density of freight traffic is below the Union average (with the exception of the Donets-Dnieper region), and the net cost is higher (with the exception of the Chernozem-Center, which has a large volume of highly economical transit shipments, and the Donets-Dnieper region). For regions with an average network density of railroads, a higher density of regions is on a par with the average Union level. This group (Volga-Vyatka, Volga, Northern Caucasus, and Ural regions) is transitional between regions with a dense network and a comparatively small load and the group of regions with a sparse network and a large load.

The regions of the eastern part of the Union and the Northwest have an exceedingly sparse network of railroads, but the load of this network is considerably above the average Union load (in Western Siberia, about 3 times the average). Only in some regions of the Northwest and also in the Far East and in Central Asia is the load of the network comparatively small. In the eastern regions, with their heavy load, the net cost of shipments is lowest as compared with other regions.

The development of increasing transfers in the eastern regions occurred as the result of new construction as well as of the technical reequipment of trunk lines for the purpose of increasing their traffic capacity and capacities in general. The concentration of freight flows immediately improved the technical-economic indices of transport work. However, the concentration of shipments on active lines is possible only to a certain limit after which the lowering of net cost is insignificant. From the national economic point of view, the concentration of transfers is not always beneficial, since it sometimes leads to a high territorial concentration of industry along the loaded trunk line. A concentration of industry along a transport trunk line inevitably causes the development of local lines and approach roads on which the transfer net cost is usually several times as high as on the trunk line. As a result of this the total expenditures for shipments in such a region increases considerably, and the effect of the transfer concentration on the trunk line is diminished. At the same time the location of industry in regions with a small network load leads to an increase of the load of this network and to the improvement of its technical-economic indices.

An orientation in location of industry only to trunk lines with the lowest net cost of transfer (other conditions being equal) is not always actually economical for the national economy. Only when the sum of expenditures (operating and capital) for production and transport of the product to the consumer is minimal does the location have an actual economic effect. The lowering of net cost and relative capital expenditures for transport is thus important, but it is not the major and only problem in a rational location of industry.

For truck transport the dependence of technical-economic indices on the density of the network is characteristic; the greater the proportion of hard-surface highways, the more favorable are the technical-economic indices of truck transport. In regions where the proportion of hard-surface roads exceeds the average Union index (23.6 percent), the net cost of shipments is below the average Union net cost. For example, in the Baltic, where the proportion of hard-surface roads reaches 34.6 percent, the net cost is 89 percent of the average Union net cost. In Eastern Siberia improved roads comprise only 15 percent of the network, and the net cost is 121 percent of average Union net cost.

The technical-economic indices of river transport are influenced first of all by the quality of the waterways and the magnitude of freight flow. The more favorable the conditions for exploitation of river routes and the greater their load, the lower the net cost of shipments. Most heavily loaded are the river routes of the Volga, Volga-Vyatka, and the Central regions. Associated with them is the Ural region, since the basin of the river Kama is a most important component of the Volga-Kama river route, the largest in the country in freight turnover of interregional significance. These regions have more than 60 percent of the freight turnover and about 20 percent of the navigable network. In this basin the average density of shipments is about 3 million ton-kilometers per kilometer of waterway; this is 300 percent above the Union index. As a result the net cost of shipments of Volga shipping routes is from 40 to 60 percent of the net cost of shipments by river transport in the RSFSR.

In the regions of the Northwest, Western Siberia, Eastern Siberia, and the Far East the very extended river network is lightly loaded, and the net cost of shipments for some shipping lines is 350 percent of the average in the RSFSR. An increase in the load of these main lines, in particular the inclusion of the Northwest into a single deep-water river system of the European part, and a related increase in the volume of long-distance interregional shipments will have a favorable influence on the work of river transport and will improve its technical-economic indices.

The length and load of the transport network in economic regions has its rational limits. Too dense a network leads to an incomplete use of traffic capacity and to an increase in the net cost of transfers, although it contributes to the fullest serving of the national economy. An insufficient density of the network impedes the development of the economy of the region and increases the distance of freight shipments to the trunk lines and back, which in total also has an influence on the increase in net cost of transfer (although, as has already been mentioned, the concentration of shipments on the trunk lines leads to a lowering of net cost).

It is very complicated to derive the optimum proportions between the length of the network and its technical equipment on the one hand, and the amount of transfers and freight turnover, on the other. A generally accepted method used to establish the optimum density of transport network, depending on the level of industrial development, has not yet been worked out. Various authors have considered different factors that influence the density of the network. Thus N. N. Kolosovskiy[2] hypothesized that the length of a railroad network must be directly proportional to the total energy supply of the country and the estimated length of haul and inversely proportional to the planned intensity of freight traffic. This, however, has not found application in computations.

M. M. Protod'yakonov determined the relationship between the level of industrial development as a whole and the level of transport development in economic regions by the following formula:

$$G_{pr} = K \cdot a^{0.85} \cdot H^{0.5},$$

where G_{pr} is the average density of freight traffic of the railroad network (in thousands of ton-kilometers per kilometer); a is the side of the square of the tentative design (in kilometers) based on the density of the network; H is the average population density (population per square kilometer); K is a coefficient based on the degree of industrial development of the country, which was obtained as the result of processing statistical data on the level of economic development of 32 countries.[3] The deficiency of this formula lies in the fact that the author did not give an exact method of determination of all values used in the formula, but only derived their approximate value empirically.

The relationship between the density of the network of transport trunk lines and the volume of shipments is given in the same author's calculations.[4] In a general form the formula of rational density of a railroad network is thus:

$$G = K \sqrt{\frac{q \, C_n}{C_m}}$$

where K is a coefficient dependent on the time of occupancy adopted in the calculations; q is the density of the freight mass (in tons per square kilometer); C_n is the full net cost of transporting freight on access roads (in rubles per ton), the average for the region; C_m is the cost of constructing one kilometer of trunk line (in rubles).

At any rate, the formula proposed by V. I. Panferov can be applied to a comparatively small territory with a simple configuration of its transport network. The author tested his principle using data on transport networks in a number of oblasts and autonomous republics, where his calculated indices (estimated maximum distance of hauling freights to the railroads) differ from the actual within limits of 7 to 15 percent. The use of a proposed relationship applicable to the areas of oblasts and krays is in itself of great practical significance. The main position of the author, that the freight mass as an expression of a definite level of economic development of a region demands a corresponding transport density, is indisputable. The determination of the average net cost of hauling freights to trunk lines is very complex since in practice this hauling is accomplished by railroad, river, and especially automotive transport. Accepting some type of transport as a basis of net cost for hauling distorts the actual requirement of network density.

An optimum correspondence between the development of the regions, density, and load of network has as its criterion the minimal total outlay for the production and transport of the product within the region under the determining influence of production. The specific features of regional economic complexes and the resulting differences in level of economic development, even in the future, will be a source of differences from region to region in the levels of development of transport.

An analysis of common indices of the level of development of transport in economic regions makes it possible to group the regions according to the present level of transport development with the purpose of finding the influence of the level of transport development on the specialization and comprehensive development of regions during the period when the material and technical basis of communism is being created (see figure 1).

The first group of regions includes the western and southern regions of the European part of the Union that have a dense network of railroad and hard-surface roads. The degree to which this territory is supplied with these types of transport is much higher than the average for the Union. River and pipeline types of transport are insufficiently developed. The degree of loading of this network is less than the average for the Union or is at the level of Union indices. The qualitative indices of transport work are less favorable (that is, net cost is above that of the average for the Union). This group includes the following regions: the Baltic, Belorussia, Southwest, South, and Transcaucasia. The Donets-Dnieper region, the Center and Chernozem Center belong to the same group, but their railroad density of freight traffic is high, consequently, the technical-economic indices are more favorable. For the first group of regions, the raising of the level of transport development in association with planned production development will occur in connection with the utilization of internal reserves, in particular further development of diesel and electric traction, and the use of heavy trucks and specialized machines, the expansion of the network of railroad and highways in itself is not the main means of providing for growing transfers; it will serve the same

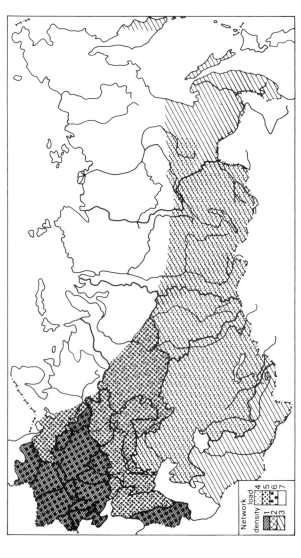

Fig.1. The Level of Transport Development of
the Economic Regions of the USSR

1—high, 2—average, 3—low 4—high, 5—above average, 6—average, 7—low

goal as providing transport access to newly discovered beds of useful minerals and to large new industrial construction and populated areas, the straightening and removal of circuitous routes, and the improvement of the configuration of the transport network in some administrative regions where the network is not sufficiently developed. As a whole, the rates of growth of the network of railroads and highways in this group will be below that of the average in the Union. An increase in the load of the active network has a favorable influence on the technical-economic indices. The construction of a direct waterway from the Black Sea to the Baltic Sea (uniting the Dnieper and Neman rivers) will increase the role of river transport in the total freight turnover and transfers in this group of regions.

Great tasks are to be performed in developing pipeline transport, which will help to rationalize the transportation of crude oil from the regions of extraction to petroleum-processing plants and the transfer of petroleum products from refineries to the large consumers.

The second group of regions consists of the Northwest, Volga-Vyatka, Volga, Northern Caucasus and the Ural regions. Its territory is not as well provided with a transport network and has a large volume of freight turnover and a comparatively high load of transport lines exceeding the average Union load. The qualitative indices of the railroad and river network are quite favorable. Characteristic for this group of regions is a dense network of pipelines with a heavy load. In the future, the development of transfers will proceed first of all by a further expansion of the road network, and in particular by the creation of trunk lines of interregional significance (for example, a second outlet from the Northern Caucasus to the regions of the Center, and the outlet from the Southern Siberian trunk line to the Urals and Volga), and also by the growth of a network of improved roads, by the connection of regions by means of national highways, and by the growth of a road network of intraregional significance. The necessity for further growth in the network of improved highways in the Volga and Ural regions must be especially emphasized. The network of pipelines will be further developed. The reserves for the growth of traffic capacity make it possible to develop the growing transport on the active trunk lines, first of all, on the railroads and pipelines. River transport, as before, will have an important place; a great deal of construction is going on and will continue (the Volga-Baltic and the Kama-Pechora connection).

Regions of the third group include Western Siberia, Eastern Siberia and Kazakhstan. This group has a low density transport network; however, the freight turnover and density of freight traffic of this network are very large and have a tendency toward further growth. The qualitative indices of railroads are most favorable precisely in these regions with a great density of traffic on technically first-class trunk lines. As concerns river transport and truck transport, the net cost of transfers by these means is higher than the Union average due to the small use of rivers and a great deficiency of improved highways.

The regions of the third group are characterized by huge, poorly developed territories. Thus, the regions of the Far North (the national okrugs [ethnically based administrative units]) comprise more than 53 percent of the territory of Western Siberia and more than 64 percent of the territory of Eastern Siberia (including the Yakut ASSR). The transport network of the regions of the Far North is very sparse; it consists of the northern sea route, river transport on the Ob', Yenisey, and Lena rivers, road transport, and air transport. One can hardly expect the creation of a dense transport network in these regions. The selective development of natural resources will not require the creation here of major trunk lines with immense investments and a comparatively small freight turnover. Therefore, in speaking of further ways of developing transport, we mean the developed part of the territory that has a trunk line of transport. For this group of regions (Middle Siberian, Northern Siberian, Southern Baykal, and Ural-Pechora), the main problem in developing a freight turnover growing at a very high rate, is a further growth of the network and the creation of trunk lines of Union and interregional significance, as well as lines of intraregional significance, for the development of beds of mineral deposits and forests, virgin lands, and the oil-bearing regions of Tyumen' Oblast.

The necessity for a faster rate of development of automotive roads of intraregional significance must be especially emphasized; it is the least developed in the regions of Siberia. Pipeline transport is acquiring great significance, especially in Western Siberia.

The construction of great hydroelectric stations on the rivers of Siberia will change the working conditions of river transport. The formation of large water reservoirs and the utilization of new types of river vessels make it possible to increase the volume of transported freight and to improve service to the territories orientated to river ways.

Of course, for the third group of regions too, the further utilization of reserves of the existing network is important. But due to the heavy load of the main railroad trunk lines the reserve traffic capacity is not as great as in other regions. The technological reconstruction being conducted at present concerns transport of the immediate future. In the future it will be expedient to increase the length of the railroad network by about 100 to 150 percent, which will raise the total extension of the transport network of the Union approximately 20 percent over the level of 1960.

The regions of the fourth group—the Far East and Central Asia—are characterized by few transport lines as well as by a comparatively small load and by a reserve freight traffic capacity. The development of transport in the future will consist in the growth of the network of trunk line and local significance, and also in the increase of traffic density on the existing lines. The network growth of roads of intraregional significance is especially necessary in the Far East where the territory has very few highways. Of great significance for Central Asia will be the construction of a pipeline network, including interregional gas pipelines, from very rich gas deposits to fuel-deficient regions.

The levels of transport development in the economic regions will continue to differ inasmuch as in the future there will continue to be some differences in the level of development of the economy of the regions and in their structure. However, the marked differences observed at the present time will be liquidated as the result of new transport construction and further technical reconstruction.

THE TRANSPORT COMPLEX OF ECONOMIC REGIONS

A successful development of regional economic complexes demands a suitable development of the transport complex in which all types of transport in the region are connected and their coordination and technological unity provided for. The development and improvement of transport complexes in regions, as component parts of a single transport system of the country, will become an important task in the development of transport in the USSR during the period of creation of the material and technical base of communism.

The comprehensive development of transport in the economic regions means: (1) a rational distribution of freight flows between the individual types of transport in conformity with the economic effectiveness of their utilization in a given region; (2) a proportional development of the individual types of transport (this refers also to the development of trunk and local transportation lines); (3) the development of individual links within each type of transport and the elimination of bottlenecks impeding the general development of a given type of transport; (4) a technological unity of the transport process from the place of origin of the freight flow to the place of its destination.

The criterion of economic effectiveness of a comprehensive development of transport in a region is the minimum total expenditures for freight transfer from the point of production to the point of consumption and the reduction in time for the process of freight transfer. An up-to-date transport complex of economic regions depends on the general distribution of the individual types of transport in the country and the regions as well as on the conditions of development of the regions. The natural conditions of the region and the historical aspects of its development are of great significance.[5]

The combination of railroad and automotive types of transport is common to the transport complexes of all regions; the first provides for the major part of freight volume in ton-kilometers, the second for the major part of shipments in tons. The other types of freight transport are developed in the various regions depending on their natural conditions (rivers, seas), on the presence of oil and gas extraction, petroleum-processing, and the volume of consumption of petroleum products and gas (pipelines).

The development of transport leads to an ever greater elaboration of the transport complex and to a better coordination of the work of the separate links within the economic region. Although at present there are, as a rule, three types of transport in the regions: rail, automotive, and water, in the future there will be an additional development in each region of pipeline transport and electric-power transport of interregional significance.

As can be seen from table 2, the Volga, Northern Caucasus, Far East, and Central Asia have the most complete transport complex.

TABLE 2

PARTICIPATION OF THE INDIVIDUAL TYPES OF FREIGHT TRANSPORT
IN THE TOTAL FREIGHT VOLUME OF THE ECONOMIC REGIONS IN 1962
(Total Freight Volume of a Region = 100)

ECONOMIC REGIONS	RAILROAD	RIVER	COASTAL	AUTOMOTIVE	PIPELINE
USSR.	84.0	5.7	2.2	4.4	3.7
Northwest	85.2	8.3	2.6	3.9	. . .
Center.	85.0	7.7	. . .	5.2	2.1
Volga-Vyatka.	76.2	13.5	. . .	3.3	7.0
Central Chernozem . .	95.7	0.5	. . .	2.9	0.9
Volga	71.4	12.3	0.9	2.7	12.7
Northern Caucasus . .	82.6	5.4	3.7	5.8	2.5
Urals	86.0	8.7	. . .	2.9	2.4
Western Siberia . . .	83.1	4.2	. . .	2.9	9.8
Eastern Siberia . . .	88.5	4.5	0.0	3.4	3.6
Far East.	66.7	5.5	21.9	4.7	1.2
Donets-Dnieper. . . .	92.2	2.0	0.8	4.7	0.3
Southwest	94.6	. . .	. . .	5.4	. . .
Southern.	68.8	2.8	19.1	9.3	. . .
Baltic.	83.0	3.2	4.4	9.4	. . .
Transcaucasia	69.6	. . .	20.3	8.1	2.0
Central Asia.	77.0	0.7	10.5	9.7	2.1
Kazakhstan.	92.4	0.7	0.2	6.3	0.4
Belorussia.	90.1	3.1	. . .	6.8	. . .
Moldavian SSR	88.0	. . .	. . .	11.1	. . .

NOTE: The freight volume of sea transport was calculated without shipments abroad, automotive transport without special shipments; pipeline transport, without the pipeline "Druzhba."

Expecially favorable is the combination of different transport types in the Volga region, where river and pipeline transport provide for 25 percent of the freight turnover.

In Kazakhstan and the Donets-Dnieper region, all types of freight transport are represented. The dominating position is occupied by railroad transport: in Kazakhstan this is due to the weak development of all other types of transport; in the Donets-Dnieper region this is a result of the huge volume of freight turnover in railroad transport of the coal-metallurgical regions of the Ukrainian SSR, and consequently the considerable volume of freight turnover by sea (almost twice as much as in Transcaucasia) and river transport is only about 3 percent.

In the remainder of the regions not all types of freight transport are represented; this can be explained (1) by the geographic position of the region (remoteness from the sea) and (2) the insufficient development of pipeline transport. The shifts in distribution of the pipeline network, in particular the construction of pipelines in the Northwest and Baltic, will make it possible to raise the degree of integration of transport in this group of regions. In the future one should expect a greater development of maritime transport in the North and Baltic regions of the country, which will increase the portion of sea transport in the freight turnover of the Northwest and Baltic.

In this group of regions one must point out the considerable proportion of shipments by river transport in the Center, Volga-Vyatka, Ural, and Northwestern regions, the pipeline transport in Western Siberia and the Volga-Vyatka region, automotive transport in the Baltic and the Southern, and sea transport in the South. This has an effect on a certain lowering of the portion of railroads in the total freight turnover in these regions.

The Central region, in addition to railroads and automotive transport, is also provided with a ramified system of river lines whose freight turnover is significant. At the present time the pipeline transport there is developed on a large scale; this increases the complexity of the transport network.

The Chernozem Center, in contrast to the Central region, does not have good conditions for the development of river transport; automotive transport is insufficiently developed, therefore railroads carry more than 95 percent of the freight volume. The same situation exists in the Southwestern region. The creation of pipeline transport will secure a high degree of integrated transport in these regions.

The general tendency for all regions in the future will be a certain lowering of the proportion of railroad transport in freight volume and an increase in the proportion of all other types of transport. This process will go on independently in each region, with consideration of the characteristic features of the development and distribution of separate types of transport in the country as well as for the demands of the regions themselves. The participation of river transport in the regions of the European part of the country is increasing due to the creation of a single deep waterway and the improvement in the utilization of active canals and also the construction of new ones. An increase in the freight volume of river transport by the development of intraregional transport will take place in Eastern and Western Siberia and the Far East. For the other regions the prospects for an increase of the proportion of river transport in total freight volume are relatively low.

The huge growth of the network of pipelines and improved highways in the country as a whole will lead to an increase in the proportion of these types of transport in almost all regions. The increase in the role of pipeline transport will be greatest in such regions of petroleum-processing and concentrated consumption as: the Center, Chernozem Center, Volga-Vyatka and Donets-Dnieper, Baltic, Western Siberia, Kazakhstan, and Eastern Siberia. The transport complex will thus become more complete, and the improvement of the organization of mixed communications will make it possible to more effectively utilize all types of transport.

One of the most important functions of the transport complex of regions is the provision of interregional connections, realized by the trunk-line types of transport. Therefore, the more trunk lines there are in the transport of a region, the larger the volume of interregional communication. Table 3 shows the interregional connections of regions and their intensity, reflecting the degree of development of trunk-line types of transport and also the approximate volumes of interregional connections by regions (minus pipeline transport). From the point of view of provision with trunk lines, the Volga, Northern Caucasus, Donets-Dnieper, Southern, Central Asian, and Kazakh regions are in the most favorable situation; their interregional connections are realized by railroad, sea, river, and pipeline types of transport. The least favorable in this respect is the position of the regions of Siberia where only two types of transport—railroad and pipeline—carry the freight flow of interregional exchange (in Eastern Siberia the northern sea route must be added; however, the volume of interregional sea transport in this region is very small). All other regions are supplied with three types of trunk-line transport; railroad, pipeline, and either river (Center, Volga-Vyatka, Chernozem Center, Urals, and Southwest), or sea (Blatic and Transcaucasia).

The participation in interregional exchange of pipeline transport especially increases the role and significance of interregional exchange of the Volga region, and to a lesser degree of the Urals and Northern Caucasus, since pipeline transport in these regions will also increase.

TABLE 3

INTERREGIONAL CONNECTIONS OF THE ECONOMIC REGIONS IN 1962

REGION OF SHIPMENT ORIGIN*	NUMBER OF REGIONS THAT HAVE RAILROAD CONNECTIONS		NUMBER OF REGIONS THAT HAVE COASTAL CONNECTIONS		NUMBER OF REGIONS THAT HAVE RIVER CONNECTIONS	
	Volume		Volume		Volume	
	More than 1 Million Tons	More Than 100,000 Tons**	More than 500,000 Tons	Less Than 500,000 Tons	More Than 100,000 Tons	Less Than 100,000 Tons
Northwest.	10	8	..	3	3	4
Center	13	5	..	..	4	1
Volga–Vyatka . . .	8	10	..	..	4	1
Chernozem Center .	8	9(1)	..	..	1	1
Volga.	16	2	1	3	5	1
Northern Caucasus.	10	8	2	6	3	3
Urals.	17	1	..	..	5	1
Western Siberia. .	9	9	..	..	..	3
Eastern Siberia. .	6	7(5)	..	2	..	2
Far East	2	7(9)	..	4	..	1
Donets–Dnieper . .	14	4	3	..	5	4
Southwest.	6	11(1)	..	..	..	..
Southern	3	12(3)	1	5	2	2
Baltic	3	11(4)	..	3	1	1
Transcaucasia. . .	3	14(1)	5	3	..	..
Central Asia . . .	2	13(3)	2	2	1	..
Kazakhstan	8	9(1)	..	4	2	..
Belorussia	1	15(2)	..	..	1	2
Moldavian SSR. . .	..	9(9)	..	..	..	1

NOTE: *means including the Moldavian SSR; **specifies, in parenthesis, less than 100,000 tons.

Table 4 shows the participation of the various types of transport in realizing interregional connections. The main interregional freight flow is carried by railroads that connect all regions. The other types of transport realize interregional exchange, (1) in considerable but smaller dimensions, or (2) between a limited number of regions. The combination of types of transport in hauling out freight is most favorable in the Volga region, Northern Caucasus, and South where pipeline and water types of transport are of great significance in shipping out freight. The role of sea transport is especially great in shipping freight from Transcaucasia, the South, and Central Asia (sea transport in the Black Sea and Caspian basins).

Much freight is shipped by river transport in the Volga-Vyatka, the Volga, Urals, and Donets-Dnieper regions. Finally a heavy interregional flow of oil and petroleum products is accomplished by pipeline transport from the Volga region and a lesser one from the Northern Caucasus.

TABLE 4

THE ROLE OF TRUNK-LINE TYPES OF TRANSPORT IN SHIPPING OUT FREIGHT
FROM THE ECONOMIC REGIONS IN 1962
(The Sum Shipped Out = 100 for Each Region)

Economic Regions	Railroad	River	Coastal	Pipeline
USSR.	86.8	5.1	3.1	5.0
Northwest	98.2	1.2	0.6	. . .
Center.	93.2	6.8	. . .	. . .
Volga–Vyatka.	85.6	14.4	. . .	. . .
Chernozem Center. . .	99.6	0.4	. . .	. . .
Volga	61.0	10.8	1.0	27.2
Northern Caucacus . .	86.0	5.4	6.2	2.4
Urals	90.4	9.6	. . .	. . .
Western Siberia . . .	97.4	. . .	. . .	2.6
Eastern Siberia . . .	100.0	. . .	. . .	. . .
Far East.	91.5	3.4	5.1	. . .
Donets–Dnieper. . . .	94.2	4.0	1.8	. . .
Southwest	100.0	. . .	. . .	. . .
Southern.	85.7	2.2	12.1	. . .
Baltic.	91.6	3.7	4.7	. . .
Transcaucasia	42.9	. . .	57.1	. . .
Central Asia.	69.2	0.7	30.1	. . .
Kazakhstan.	98.6	1.4	. . .	. . .
Belorussia.	96.7	3.3	. . .	. . .
Moldavian SSR	100.0	. . .	. . .	. . .

The largest freight flows come from the Donets-Dnieper and the Ural regions, which together account for about one-third of the total freight, and from the Volga and Western Siberia regions, which provide almost half of the interregional exchange. The proportion of shipments from the regions of the Northwest, Center, North Caucasus, and Kazakhstan is more than 25 percent. The share of all the other regions is comparatively small.

A large concentration of freight flow on a comparatively few transport trunk lines is characteristic of interregional connections. In this category belong the railroad trunk lines connecting the western and eastern regions including the Siberian trunk line, the latitudinal roads of the Urals region, and the roads connecting Moscow with the Urals. In the regions of the European part of the country they are the trunk lines between the Donets-Dnieper and Northern Caucasus regions and the Center and Northwest, and also between the Volga and Donets-Dnieper regions and the Northern Caucasus region. On these railroads the traffic density is several times that of the average Union traffic density.

In river transport a heavy interregional flow is characteristic for the entire Volga-Kama system, including the Volga-Don Canal. This main line connects the regions of the Center, Volga-Vyatka, Urals, Volga, and Northern Caucasus. A lesser flow presently connects these regions and the Northwest. The river main line second in significance, accommodating interregional shipments of the Donets-Dnieper, South, and Belorussian regions, is the Dnieper. As yet the interregional connections on the Irtysh between Kazakhstan and Western Siberia are insufficiently developed.

Considerable interregional connections in sea transport exist in the two basins of the country, the Black and Caspian seas. In the first, large-scale transfers connect the Northern Caucasus, Transcaucasia, Donets-Dnieper, and Southern regions; in the second, the regions of the Transcaucasia, Central Asia, Volga, and Kazakhstan.

One cannot but note the large concentration of freight flow by interregional pipeline, which acquires a special significance in the interregional connections of the Volga region with the regions of the European part of the country and Siberia. The gas pipelines connecting the regions of the Northern Caucasus, Volga, Donets-Dnieper, and Southwest with the Center, West, Northwest, and also Central Asia with the Urals, are becoming an important factor of interregional exchange.

The degree of development of the transport complex can be seen from the transport balance of the regions. The fuller and higher the level of development of the transport complex, the more intensive its interregional connections in shipments and in production. The indices of the shipping out and the shipping in of freight by specific types of transport make it possible to judge the level of development of the transport system of a region, its capacity of transport, and its technological equipment.

The transport balance of a region reflects the general balance of production and consumption; it is that portion which is associated with the transport of production. That part of the production which is used at the place of production does not enter into the composition of the transport balance, yet for many types of production this part is considerable. On the other hand, repeated shipments of freight by various types of transport increase the production reflected in the transport balance. The transportability coefficient of some products is more than

1 (as a result of repeated transfers), and for others it is less (when a certain part of the production is used locally). Thus, lower transportability coefficients are characteristic for lumber, coke, ferrous metals, and grain. At the same time the transportability coefficient of oil, salt, and sugar exceeds 1. As a result of this, the transport balance of any freight can differ considerably from the balance of production and consumption of the same type of product (in comparable weight units of measurement). Nevertheless, the transport balance of individual types and subtypes of freight reflects the specialization of a region and defects in its economic complex.

There are serious difficulties in working out the transport balance of a region, because the necessary statistical data by regions of the country are absent. Interregional freight communication is worked out only for railroad, river, and sea types of transport. At present it is not possible to consider the place of the automotive and pipeline types of transport in interregional transfers.

However, since interregional trucking is still insignificant in volume and is carried out between neighboring oblasts and krays of the various economic regions, the error will not be great if they are put into the category of intraregional shipments and not included in the transport balance.

Pipeline transport is occupying an ever more noticeable place in petroleum transfer. In 1962 its proportion in the total transfers of petroleum freight by three types of transport[6] was more then 43 percent. In regions of oil extraction—the Volga, Northern Caucasus, and Transcaucasia—pipeline transport noticeably changes the transport balance. At the same time oil and petroleum products supplied by pipeline to other regions (Volga-Vyatka, and Center) increase the negative balance of these regions. Therefore, in working out the transport balance of regions, the pipeline transport must be taken into account. We have attempted to determine the interregional transfers of freight by pipeline transport in 1962.

A serious defect of transport statistics, creating difficulties in determining a transport balance, is the duplicate count of shipments which seems impossible to calculate. The duplicate count is especially great for lumber and grain freight and occurs mainly in regions where mixed shipments are developed and where cross hauls occur in sending freight abroad. Great difficulties are encountered in analyzing the shipments of the four types of transport and in the incomparability of the majority of items of freight. The most detailed itemizing of interregional freight exchange is applied in rail transport, but for water transport there is a different nomenclature.

In the calculations of regional transport balances, shipments by railroad, sea (coastal), river, and pipeline types of transport were taken into account. The above-mentioned defects in transport statistics could not be completely avoided in these calculations. Therefore, the results of the analysis are only tentative and can serve only as an illustration of the basic tendencies of special regional features. Table 5 shows the proportion of incoming and outgoing freights (the sum of four types of transport).

TABLE 5

THE REGIONAL SHARE IN THE SHIPPING IN AND
SHIPPING OUT OF FREIGHT IN 1962
(As Percentage of USSR Total)

Economic Regions	Shipped In	Shipped Out
Northwest.	4.7	5.5
Center	12.7	6.2
Volga-Vyatka	4.9	3.7
Chernozem Center	4.8	2.8
Volga.	6.8	16.9
Northern Caucasus.	5.9	6.7
Urals.	11.4	10.5
Western Siberia.	5.1	8.4
Eastern Siberia.	2.9	3.3
Far East	1.4	0.6
Donets-Dnieper	7.7	16.8
Southwest.	8.3	3.7
Southern	4.9	2.9
Baltic	4.4	1.2
Transcaucasia.	2.1	2.2
Central Asia	2.5	1.5
Kazakhstan	4.9	6.0
Belorussia	3.7	0.7
Moldavian SSR.	0.9	0.4

Outgoing freight considerably exceeds incoming freight in the Donets-Dnieper, Northern Caucasus, Volga, Western Siberian, Eastern Siberian, and Kazakh regions. These regions ship out bulk freight: coal, petroleum, ferrous metals, lumber materials, and grain. They supply the regions that have an unfavorable balance in these items. Yet the regions that ship out these cargoes themselves have a deficit in some of these items: the Volga region in coal; the Donets-Dnieper, Northern Caucasus, and Eastern Siberia regions, in petroleum; the Donets-Dnieper, North Caucasus, and Kazakh regions, in lumber; and Eastern Siberia, in grain. Heavy freight flows going beyond the regions mentioned basically determine the volume and direction of interregional exchange of the country. At the same time the presence of such flows also predetermine the character of the transport network of these regions, its technical equipment and traffic capacity. Indeed, all these regions have a technically advanced and powerful transport network providing for large internal as well as external flows. Here it must be noted that the bulk freight usually has a localized origin; for example, Eastern Siberia sends coal in great quantities from three stations and lumber from 21 stations; Western Siberia sends coal from 13 stations, the Donets-Dnieper and Northern Caucasus region send coal from 35 stations, and ferrous metals from 12. The outgoing flows thus are concentrated in comparatively few directions with technically advanced equipment.

The Center, Central-Chernozem, Urals, Volga-Vyatka, Baltic, Southwestern, Southern, and Central Asian regions have a negative transport balance. The predominant development of the processing industry and the comparatively limited resources of raw materials and fuel have made these regions ones that receive large bulk freight shipments. Some of these regions having a negative transport balance ship out some types of bulk freight (the Volga-Vyatka region ships out lumber in considerable amounts; Central Asia, petroleum; the Chernozem-Center, grain; the Ural region, ferrous metals and lumber).

Since the incoming bulk freight is usually dispersed to many receiving stations and the outgoing freight is represented by nonbulk freight, the concentration of flows in the directions of regions with a negative transport balance is considerably less than in regions with a positive balance. Therefore, the demands for technical equipment and traffic capacity of the trunk lines in these regions are not as great. The only exceptions are those regions through which a large transit flow passes. Transit flows of bulk freights are characteristic for a number of transport main lines of the Ural, Volga-Vyatka, Central-Chernozem and Center regions, where, in addition to everything else, there is a concentration of incoming flows for Moscow and Moscow Oblast.

The specific structure of production and the border position of the regions of the Far East and Transcaucasia put them in a special position when characterizing the transport balance. In the Far East it is negative and in Transcaucasia positive. In the Far East, due to its remoteness from industrially developed regions of the country and the presence of local sources for the production of bulk types of products, the deficit in transport balance is due to shipped-in petroleum and ferrous metals. Transcaucasia has a positive transport balance due to the shipping out of petroleum and manganese ore; all other bulk freight is shipped in.

An analysis of the transport balance of regions based on the balance of production and consumption and an analysis of shifts in freight flows, as a result of the economic development of the regions, are the basis for showing the need for development and technological reconstruction of transport routes of the regions. The comparison of existing transport installations, their technical equipment and traffic capacity, and the increasing volume of transfers in specific directions of the transport system make it possible to show the capacities lacking and capital investments necessary in the development of these capacities.

The solution of the problem of integrated development of transport in economic regions as part of a single transport system of the Union demands an accounting of the economically valid spheres of action of the various types of transport and also an accounting of the economic and natural characteristics of transport development under the specific conditions of each region. A comparison of alternatives of development of freight flows by various types of transport in specific directions of the network of economic regions makes it possible to solve this problem with the maximum economic effect.

1. Exceptions are associated with the specific features of the work of railroads under the given regional conditions (the relation of steam, electric, and diesel traction; the road's profile; and the structure of freight volume).

2. N. N. Kolosovskiy, *Osnovy ekonomicheskogo rayonirovaniya*[Principles of Economic Regionalization] (Moscow, 1958), p. 122.

3. V. I. Panferov, *Mestnaya set' putey soobshcheniya i usloviya eye formirovaniya* [The Local Transport Network and Conditions of Its Formation] (Moscow, 1961), p. 22.

4. Ibid., p. 68.

5. *Osobennosti i faktory razmeshcheniya otrasley narodnogo khozyaystva SSSR* [Characteristics and Factors of the Distribution of Branches of the National Economy in the USSR] (Moscow, 1960), chap. 14.

6. Minus sea and truck transport.

O. A. KIBAL'CHICH

An Attempt to Develop a Projection of

Future Interregional Passenger Flows

The new Program of the Communist Party of the Soviet Union stresses that one of the most important tasks in the field of transport is to fully satisfy the various needs of our population for transportation over short and long distances. It is expected that the total volume of nonlocal passenger transportation in the USSR (excluding automobile trips) will increase at least fourfold by 1980. The mobility of the population of the USSR measured by the number of trips will increase even more for the same period.

The most important factors in the growth of passenger communications in our country in the next two decades (1961–1980) will be: (1) increased economic development in all economic regions of the USSR especially in the East; (2) increase in the total size of population and an increase in its migration rate; (3) improvement of material living conditions and a rise in the cultural level of all strata of population in the USSR; a decrease in the length of the working day and an increase in available free time; (4) expansion of the transport network as a result of new rail and auto road construction, the opening of new airlines; a considerable increase in the number of transport carriers, in particular, automobiles, buses, planes; technical progress in transport (increase

Translated from *Voprosy geografii* [Problems of Geography], no. 57 (1962), pp. 180–93.

in traffic speeds and so forth); and (5) lower passenger tariffs, especially in air and automobile transport (bus); improvement of the quality of passenger traffic and the organization of mixed passenger transportation (rail-air, automobile-rail, air-automobile, and others). All these factors, by exerting a decisive influence on the growth of national passenger traffic, will determine the distribution of passenger movement along the network of transport lines in the USSR and change the actual geography of passenger flow.

For the purpose of long-term national economic planning, the forecast of total volume of All-Union passenger traffic and its distribution according to types of transport as well as the forecast of the dimensions of passenger traffic along the main routes of the network of communications in the USSR are equally important. The projection of future passenger flows along major routes is necessary first of all as the basis for the program of development of the network of trunk lines of communications itself, and also for determining requirements in transportation equipment (locomotives, passenger cars, planes, buses, sea and river vessels) and in capital investments for the construction of railway stations, airports, bus stations, garages, hotels, and other facilities. Previous miscalculations in planning capital investments in the development of passenger transportation were due to the fact that, in the elaboration of long-term national economic plans, there was no determination and scientific substantiation of impending growth in the volume of passenger flows.

Studies (conducted in the Institute of Comprehensive Transport Problems of the Council of the National Economy of the USSR, with the participation of the author) on the prediction of passenger traffic development for all types of USSR transport in the coming two decades, have shown that the methods of transportation economics associated with the usual technical-economic calculations for the determination of future passenger flows are insufficient. Geographical methods of analysis are required for this purpose in order to reveal the relationship of passenger trips to the location of productive forces, population settlement, and characteristics of occupation and everyday activity.[1]

Results of our studies have shown that the prediction of passenger flows along the main paths of the transport system of the USSR can be projected over a period as long as twenty years. It has proven adequate to consider the principles of development and distribution of passenger transport in connection with the growth and distribution of population in such comparatively large territorial units as the major economic regions of the USSR. One must keep in mind, however, that the study of trends in the development of passenger communications for each separate passenger generating point—a city, workers' settlement, state farm, and so forth—will have to be done at the next stage of national economic planning when, on the basis of a general twenty-year plan, the more specific seven-year, five-year, and one-year plans of passenger transport development can be worked out.

In the prediction of the geography (that is, paths and dimensions) of interregional passenger flows for the next twenty years it is necessary to distinguish the following stages: (1) determination of the total future volumes of passenger transportation by all types of transport in the USSR, distinguishing between long-distance departures by rail or air; (2) calculation of the distribution of passenger departures by various economic regions of the country; (3) establishment of the future territorial structure of interregional passenger communications of each economic region; (4) determination of the dimensions and directions of future interregional communications; (5) generalization of the volume of communications in each instance mentioned above in order to estimate future passenger flows by the main routes of the transport network of the USSR; (6) distribution of the estimated patterns of passenger transport according to the relative technical-economic advantages of each.

It must be noted that the role of geographical methods of analysis is most important at the second and third stage of projecting interregional passenger flows; at the first and sixth stage the analytical methods of transport economics are decisive.

The influence of each of the above-mentioned groups of national economic factors (geographical, economic, technical, and so forth) on the mobility of the population has not yet been sufficiently studied. At the same time, studies of transport economics by Ye. V. Mikhal'tsev and B. M. Parakhonskiy have revealed the presence of a correlation between the rates of growth of mobility (total for all types of passenger transport in extra-urban communication) and growth rates of the national per capita income. This relationship shows up strongly during periods of peaceful development of the country (see table 1).

TABLE 1

	National Income	Mobility of Population
Prewar Period (1937 as percentage of 1928)	354	327
Postwar Period (1958 as percentage of (1950)	197	171

From these data[2] it can be seen that in the prewar as well as in the postwar period, population mobility in the USSR increased proportionately to national per capita income, lagging somewhat behind the growth rate of the latter (by 10 to 15 percent). A similar correlation exists between the growth rates of the mobility of the population and the real incomes of workers. For the period from 1913 to 1958, the real income of Soviet workers increased approximately five times and that of peasants almost six times.[3] During the same period the mobility of the population of the USSR increased by 420 percent. A parallel growth in population mobility and per capita norms of national income is explained by the fact that the increase in national wealth and the associated intensification of economic life and stimulation of sociopolitical life are accompanied by an increasing mobility of the population for work and social purposes and also by the fact that the increased real incomes of the workers related to the growth of the national income creates conditions for an increased number of trips for personal purposes (visiting of relatives and acquaintances, rest, excursion-tourist trips) that are financed, as a rule, by the personal income of the workers.

It is assumed that the growth rate of mobility of the population and the per capita norm of national income will remain parallel. The correctness of this assumption is corroborated by the fact that the seven-year plan of development of the national economy in the USSR foresees an increase in per capita national income of 40 to 45 percent and of real incomes of the working people of nearly 40 percent with a growth of mobility in extra-urban communications of no less than 50 percent for the same period (see table 2).

One of the important principle questions in working out a projection of interregional passenger flows for a general plan is the selection of a network of major economic regions for the analysis of interregional passenger communications in the past and for their prediction for the future.

The fact is that the former thirteen-unit and the present seventeen-unit network used for national-economic planning is not very suitable for a construction of interregional passenger flows, since it includes such exceedingly large economic regions as the European North, Volga, Kazakhstan, Western Siberia, and Eastern Siberia. In particular, it does not permit separating the passenger flows that differ in their composition, for example, those from the Industrial Center to the Volga and back (one flow is from Moscow to Gor'kiy and Kazan', the other to the Middle Volga, the third to the Lower Volga), or determining the exact Siberian destination (Western Siberia or Eastern Siberia) of the passenger flows going from Siberia via Kazakhstan to Central Asia and back.

TABLE 2

PASSENGER TRANSPORT AND VOLUME IN INTERCITY CONVEYANCE
BY VARIOUS TYPES OF TRANSPORT IN THE USSR

TYPES OF TRANSPORT	1958		1965	
	Passengers In Millions	Passenger-Kilometers	Passengers In Millions	Passenger-Kilometers
All transport in the USSR	...	190	...	312
Automobile transport, public and private	...	7	...	20
General purpose transport (excluding public automobiles)	2,368	183	4,005	292
Long-distance conveyance, by rail and air transport	259	128	325	189

On the other hand, it would be very complicated to employ a more detailed system of economic regionalization in the analysis of interregional passenger travel, such as the system of N. N. Kolosovskiy[4] or that of Moscow State University.[5] An attempt to combine the systems of economic regionalization of the USSR with the system of administrative division employed in the calculations of the Ministry of Communications of the USSR would prove nearly impossible to perform. An escape from the difficulty was found by creating an auxiliary system of major territorial units (economic regions),[6] based on the system of economic regions of the USSR by N. N. Kolosovskiy and Moscow State University. As a result of special recalculations we succeeded in combining within these new boundaries all passenger departures by rail and air transport in 1950 and 1958 as well as the population numbers for the same period. The forecast of population distribution by economic regions of the country was exceptionally useful for calculating future mobility of the population by economic regions of the USSR.

The decisive factors in the population redistribution by economic regions since the establishment of the Soviet government were: (1) geographic differences in reproduction (natural increase) of the population; and (2) migratory displacements of population related to the settlement and economic development of regions little developed in prerevolutionary time and to the redistribution of the population between villages and urban centers in the course of industrialization of the USSR.[7] In the forecast of future population distribution by economic regions, the influence of the natural increase of the population and normal migration on the population size and its redistribution between economic regions have been kept uppermost in mind.

The determination of the population numbers by regions with regard to natural increase was conducted on the basis of the method proposed by the Central Statistical Administration of the USSR for the forecast of All-Union population growth.[8] The geographical differences in the natural increase of the population were also taken into account. The latter are related first of all to the differences in the age-sex structure and also to differences in city and village populations. Local social and nationality characteristics (for example, in the republics of Central Asia) have a great influence on the amount of natural increase of population, particularly in villages.

The eastern regions (Central Asia, Kazakhstan, Central and Eastern Siberia) and some European parts of the USSR (Transcaucasia, the North, and the Urals) will have the highest natural increase of the economic regions of the country. A low natural increase will continue in such regions as the Baltic, Northwest, Central, Eastern and Central Ukraine.

Based on the projected regional balance of labor resources in the next twenty years, a redistribution of about three million to four million able-bodied men and women will be needed, mainly from the European part of the country to the eastern regions. In addition, tens of millions of people will have to be resettled during this period from rural areas to cities and workers' settlements.

In former years the following economic regions served as the main source of personnel for the eastern regions and for new developments in the country: the Northwest (Novgorod and Pskov oblasts), the Industrial Center (Kalinin, Smolensk, and other oblasts), the Chernozem Center, the Central Ukraine, the European North (Vologda oblast). In the future the Western Ukraine, Belorussia, Moldavia, Volga-Vyatka region and Northern Caucasus must be considered the most important regions of population emigration. Regions of new settlement in the future will be, first of all, Central and Eastern Siberia and Central Kazakhstan. As can be expected, the normal migration to the Far East and Yakutia will be less significant. In the European part of the country the influx of population from neighboring regions will take place in the Industrial Center and Baltic republics (in particular Estonia). Thus, in the next twenty years considerable shifts will occur in the regional distribution of population. A decrease in the proportion of total national population will be characteristic for such economic regions as the Chernozem Center, Western (Belorussia-Baltic), Central, Volga-Vyatka, and also for the economic regions of the Ukraine. A considerable increase in population will occur in Central Asia, Kazakhstan, and Central and Eastern Siberia, in the Far East, Urals, and Transcaucasia.

A comparison of materials on the change of passenger flows for 1950–58 was conducted for two types of transport—rail and air—that play a decisive role in interregional passenger communications. Passenger transport by automobile and water were not analyzed in detail because bus and car transport were mostly intraregional rather than interregional, and the passenger flows following the main river lines (Volga, Dnieper, Yenisey, and others) are comparatively small.

An analysis by economic regions of the USSR of the relation between departure of passengers by rail or air transport and the size of population of the regions (this association is better called passenger *intensity* of economic regions) shows that in former years there were great differences in the levels of passenger intensity in the various economic regions, in intraregional as well as interregional passenger communications. In 1958, for example, the largest departures of passengers in intraregional and interregional communications per capita were in the Northwest and Industrial Center, which showed the role of Moscow and Leningrad—the largest economic-organizational and cultural centers of the country. Above the national average in mobility was passenger intensity of the distant and border regions of the country that have railway communications: the southern part of the Far East, Eastern Siberia, the European North and the Union republics of the West. On the other hand, the remote economic regions without railroads (Yakutia and the Northeast RSFSR) had a passenger intensity of 67 to 75 percent less than the central regions of the country. Regions with a predominance of rural population: Central Asia, Transcaucasia, Central Chernozem region, Volga and Western Siberia also had a low passenger intensity.

In earlier years, too, there were differences in the level of passenger intensity in the various economic regions of the USSR. In 1950, for example, the southern part of the Far East, the Northwest RSFSR, the European North, the Union republics of the West and Transcaucasia had a comparatively high level of passenger intensity. A low passenger intensity was characteristic for Central Asia, the Chernozem Center, Volga, Kazakhstan, and Western Siberia. It is significant that the highest rate of passenger intensity from 1950 to 1958 occurred in precisely those economic regions where the urban population increased considerably during these years: in Central and Eastern Siberia, the Urals, Kazakhstan, the Northwest RSFSR and the Industrial Center.

Even greater differences were characteristic for the interregional passenger intensity of the different economic regions of the USSR. In 1950, for example, the Northwest RSFSR exceeded Central Asia in interregional passenger intensity by more than 600 percent. In 1958 these differences were somewhat lower, but, as before, the interregional passenger intensity of economic regions of the European part of the country, better equipped with a transport network, was two to three times higher than the interregional passenger intensity of the eastern regions of the USSR.

There is reason to assume that in the next twenty years, in association with further development of the productive forces in all economic regions of the country, a considerable increase in mobility of the rural population, cheaper passenger transport, development of rapid transit means, in particular aviation, and under the influence of a number of other factors, the differences in the level of passenger intensity between the economic regions of the USSR will lessen somewhat but will, nevertheless, still be significant, because of the geographic differences in territorial structure of the economy, forms of settlement of the people, their way of life, transportation facilities, and territorial structure of transportation. Therefore in calculating the volumes of future regional passenger departures all economic regions of the USSR were divided into three groups according to the level of their passenger intensity: (1) a comparatively high level of passenger intensity; (2) average level of passenger intensity; (3) comparatively low level of passenger intensity. The differences of the economic regions in interregional passenger intensity were also considered.

An analysis of shifts in interregional passenger communications for

1950–58 showed that in contrast to the prewar period—when the highest rates of growth in passenger transport were characteristic mainly for the central and southern regions of the European part of the USSR—an increase occurred during postwar years in passenger communications between regions of the European part of the country: the Left Bank of the Ukraine (Donets Basin and Crimea), Northern Caucasus, Transcaucasia, Northwest RSFSR and European North on the one hand, and Western Siberia, Kazakhstan, Central Asia, and Central Siberia, on the other. These changes in passenger communications reflected the shifts in the distribution of industry and population that occurred in the postwar period in our country (the exploitation of forest and mineral wealth of the European North, the development of virgin land in Kazakhstan and the Altay region, and new industrial development in Western and Central Siberia), and they also reflected the growth in the material prosperity of workers and the increased number of trips to Crimean and Caucasian resorts and to regions of mass tourism (Ukraine, Moscow, Leningrad, etc.) that were associated with material prosperity.

It is certain that in the next twenty years, as economic development in the East continues to expand, a further increase in passenger communications will occur between the western and eastern regions of the country; however, the future shifts in interregional passenger communications eastward must not be overestimated. An analysis of the rates of growth of interregional passenger communications for 1950–58 showed that a considerable part (more than 60 percent) of the increase of rail and air transport of passengers was communication among the European parts of the USSR. The highest rates of growth of transport were shown for passenger communications of the following pairs of economic regions (in billions of passenger kilometers):

> Industrial Center–Northern Caucasus.......................3.8
> Right-Bank Ukraine–Left-Bank Ukraine..................2.9
> Industrial Center–Right-Bank Ukraine.....................2.8
> Industrial Center–Left-Bank Ukraine 2.6
> Industrial Center–Transcaucasia 2.3
> Northwest–Northern Caucasus...............................2.1
> Industrial Center–Urals...2.1
> Industrial Center–Volga 2.1

Passenger transportation between economic regions of the Asian part of the USSR increased from 1950 to 1958 only by 6.7 billion passenger-kilometers (8.4 percent of the All-Union increase of interregional passenger communications).

In the twenty-year forecast of the territorial structure of passenger communications, in addition to taking into account the shift in passenger communications to the East, attention was given to another stable tendency in the development of passenger transport—the increase of communications with the southern resort regions of the country (the Northern Caucasus, Transcaucasia, Crimea, Moldavia, and others).

Prospective interregional flows were calculated in the following manner. (1) On the basis of the calculated volume of passenger departures to economic regions of the country and the forecast of the territorial structure of passenger transportation, a summary table (matrix) of interregional passenger exchange was constructed. (2) Spheres of utilization of rail and air transport in interregional communications of passengers by various regions were also revealed by economic methods of analysis. We then calculated the distribution between these two types of transport and derived two auxiliary tables of interregional passenger exchange. (3) By generalizing the passenger communications of each of these tables according to the common routes, future passenger flows were revealed for each type of transport along the main routes of the transport networks of the USSR.

What shifts will occur in the next twenty years in the geography of interregional passenger flows?

In 1958 the decisive role in interregional passenger flows belonged to railway transport. The most significant passenger transport was from the Industrial Center to the South—to Khar'kov and the Donets Basin with branches to the Crimea and Northern Caucasus. In the summer, passenger trains on round trips carried daily 35,000 to 40,000 persons between Moscow and Khar'kov. Second in size was the passenger flow in an eastern direction from Moscow to the Urals, Siberia, and back. On two railway trunk lines (from Moscow to Buy, Kirov, Perm', and Sverdlovsk, and from Moscow to Gor'kiy, Kazan', and Sverdlovsk) about 25,000 passengers were carried daily in both directions by more than 25 pairs of long-distance trains. The passenger flows in northeastern (Moscow to Leningrad and back), western (from Moscow to Minsk, Vil'nyus, Riga and back), southwestern (from Moscow to Kiev and Odessa and back), and trans-Ukrainian (from L'vov to Kiev, Dnepropetrovsk, and Donetsk and back) directions were also quite significant in 1958.

The passenger flows of air transport in 1958 could not compare with the passenger flows of railway transport. In 1958, the largest air transport routes were the eastern (from Moscow to Kazan', Sverdlovsk, Omsk, Novosibirsk, Irkutsk, and Khabarovsk and back) and southern (from Moscow to Voronezh and Rostov with branches to Adler and Tbilisi and also back from these airports to Moscow). Planes carried thirty to fifty times fewer passengers than the railroads.

It is expected that in the period from 1975 to 1980 aviation will take first place in volume of passenger transport in interregional communications of the USSR. The largest air passenger flow will be that from the central regions of European Russia to the southern, especially to the resort regions of the USSR (Crimea and Northern Caucasus) and back.

In absolute quantities this flow will equal the parallel passenger flow by rail transport. The total volume of interregional transport of passengers in the southern direction will in the future exceed the existing passenger flow by at least two to three times. The passenger flow in an eastern direction by air will be quite considerable (exceeding by 100 to 150 percent the passenger flow on the latitudinal railway trunk lines). In 1980 on the eastern air lanes, aviation will carry daily in each direction several thousands of passengers. This will demand numerous daily flights of such planes as the TU-104 and IL-18. Passenger air transport between Moscow and Leningrad, between regions of European Russia, Kazakhstan and Central Asia, and between Western, Central and Eastern Ukraine will be of considerable magnitude.

In railway transport, aside from the southern directions, passenger transport will remain large in volume between: Moscow and Leningrad, the Industrial Center and the Central Ukraine, and the Industrial Center, Belorussia, and the Baltic republics. The trans-Ukrainian passenger flow will exceed by 50 percent the existing passenger flow; the extent of passenger transport by rail in an eastern direction will be 33 to 50 percent less than at present.

In conclusion, it must be emphasized that the largest growth in interregional passenger flows is expected in 1980 along routes connecting the largest economic regions of the country: the Industrial Center, Eastern Ukraine, Northwest RSFSR, Urals, and Central Siberia. This means that, as in freight transport, passenger transport will retain the tendency, already planned in former years, toward a concentration of flows on the main trunk line routes that are the axes of the country's economic life.

1. It must be noted that in Soviet economic geography, in particular the geography of transportation and the geography of population, the study of passenger communications, especially of long-distance transportation, is absent. The types of economic-geographic literature widely circulated in our country are, as a rule, regional monographs, which do not contain an analysis of the geography of passenger communications of the economic regions of the USSR. It is characteristic that the existing economic-geographic expeditionary projects do not even orient the investigators to the study of the mechanisms of formation and growth of passenger communications. See, for example, V. I. Lavrov and V. V. Pokshishevskiy, "Some Methodological Questions of the Conduct of Field Work in Economic Geography and the Compilation of Partial Regional Outlines," *Izvestiya Vsesoyuznogo Geograficheskogo Obshchestva* [Proceedings of the All Union Geographical Society], no. 3 (1947); Yu G. Saushkin, "Economic-Geographic Investigations. Investigation of the Geography of Population," in *Spravochnik Puteshetvenika i Krayeveda* [Handbook of the Traveler and Regional Ethnographer], vol. 2 (Moscow, 1950).

In our opinion, the creation of a method of studying the geography of passenger transportation in the economic regions of the USSR and by smaller geographic areas for the purpose of long-range planning of this transportation is an urgent problem considering the great prospective development of passenger transport in the next twenty years.

2. *Narodnoye khozyaystve SSSR v 1958 godu* [The National Economy of the USSR in 1958], pp. 98–99.

3. Ibid.

4. See N. N. Kolosovskiy, *Osnovy ekonomicheskogo rayonirovaniya* [Principles of Economic Regionalization] (Moscow, 1958), pp. 120–21.

5. See Yu. G. Saushkin and T. M. Kalashnikova, "Basic Economic Regions of the USSR," *Voprosy geografii* [Problems of Geography], no. 47 (1959), pp. 42–73.

6. The following 19 territorial units were distinguished in the USSR: (1) Northwest RSFSR, (2) European North, (3) Industrial Center, (4) Chernozem Center, (5) Union Republics of the West (including Kaliningrad Oblast), (6) Left-Bank Ukraine, (7) Right-Bank Ukraine, (8) Northern Caucasus, (9) Union Republics of Transcaucasia, (10) Volga, (11) Urals, (12) Kazakhstan, (13) Union Republic of Central Asia, (14) Western Siberia, (15) Central Siberia, (16) Eastern Siberia, (17) South of Far East, (18) Yakutia, (19) extreme Northeast RSFSR.

7. World War II also had a certain effect on the change in the geography of population of the USSR because of the large losses suffered by the civilian population in regions under occupation and because of the evacuation of millions of people from fertile regions to the East, where part of the evacuated population took up permanent residence. See S. A. Kovalev, "Shifts in the Geography of Population of the USSR over 40 Years," *Geografiya v shkole* [Geography in School], no. 5 (1957), pp. 11–12.

8. V. N. Starovskiy, "On a Method of Forecasting the Growth of Population in the Soviet Union," *Vestnik Akademii Nauk SSSR* [Herald of the Academy of Sciences of the USSR], no. 2 (1960), pp. 44–45.

V. B. KUZNETSOV
O. D. CHUVILKIN

Long-Distance Transport of Energy

The large increase in the production of electric energy demands enormous resources of fuel and hydroelectric energy. The geography of fuel and hydroelectric reserves is in sharp inbalance with the existing regions of energy consumption: the European part of the USSR (including the Urals) has 9 percent of the total energy resources but consumes about 80 percent of the total energy. In the future this will change; the share of the Asian part will increase, but fifteen years from now with the absolute growth of electric consumption the share of the European part will nevertheless be large. The natural energy resources of the European USSR will not be able to fully cover the demand and as a result the fuel-energy balance of this territory is deficient.[1]

This deficit can be wiped out in three different ways: (1) by using new sources of energy not yet utilized or utilized on an insignificant scale; (2) by applying new methods of energy conversion providing significant economy of primary fuel resources (the efficiency of common steam-turbine power plants is only 33 percent and can hardly be raised in the future); (3) the transport of energy over long distances (1,000 to 3,000 kilometers) from the surplus regions of Siberia, Kazakhstan,

Translated from *Vestnik Moskovskogo Universiteta*, seriya geografiya [Herald of Moscow University, Geography series], July-August 1965, pp. 80–83.

and Central Asia to the center of the European part of the USSR and the Urals region.

Undoubtedly our energy production will follow all approaches in various degrees. But technical and economic difficulties still stand in the way of utilizing new sources of energy. The same is true of the new methods of converting energy that are still at the experimental stage. Therefore, even though in the more distant future all energy production will shift to a new basis (owing to the low effectiveness of traditional means of obtaining electric energy), the problem of eliminating the deficit in the energy balance of the European part of the USSR in the period 1965–80 can be solved only by means of the long-distance transmission of energy from the Asian part.[2]

The additional amount of energy needed by the central regions of the European part of the USSR can be obtained on the basis of gas from Kazakhstan (the pre-Caspian lowland, the Aral-Sor deposit) and Central Asia (Turkmen SSR, Darvaza deposit). The economy of the Ural region will utilize central Asian gas (Uzbek SSR, Gazli deposit), Ekibastuz and Siberian coal, and the hydroenergy of Siberian rivers.

In transferring large quantities of energy for long distances, an *economically justified choice of transport type* is one of the most important preliminary stages. The quantities of transferred energy are determined, on the one hand, by the needs of the energy system of the Center and the Ural regions, and on the other, by the fuel resources at hand; but energy can be transmitted either by placing the power plants where fuel is being extracted and by building transmission lines, or by transporting fuel by rail or gas pipeline and placing the power plants near the consumer. Only the problem of transporting electric energy from hydroelectric stations and from coal-mining regions where the coal is not transportable due to its technical characteristics (for example Kansk-Achinsk) can be simply solved.

An economic comparison of the various types of energy transport has been conducted by individual authors and by organizations in our country and abroad.[3] The results of these works are contradictory and cannot be compared due to the differences in methods and initial data. Moreover, calculations were often made without considering the economic-geographic characteristics of the project, real quantities of transmitted energy and the time required for putting the project into operation.

The main aspects of the method accepted by the authors are as follows:

1. Determination of actual quantities of necessary energy transport derived from the yearly requirements of the recipient energy systems;

2. Guarantee of the same energy effect, that is, receipt by the consumer of the same quantity of energy under any transport alternative.

For this reason, a correction factor, taking into account the value of losses in electrical-transmission lines was added to the capacity of power plants situated in the fuel-extraction region.

3.   Calculation of the varied time of investments by assigning them to the year of development of the projected alternative power service.[4] This was least accounted for in economic calculations, although clearly other conditions being equal, the more profitable transport alternative is that in which the major investment takes place at a later time;

4.   Consideration of the basic expenditures and outlays in related branches, in this case in fuel extraction;

5.   Calculation of the cost of railroad transport in terms of the marginal expenditures and that of pipe line and electronic transport in terms of the total expenditures. In calculating the marginal expenditures, the existing transport structures and their freight flows are taken into consideration; the freight charges for the required amounts of freight are expressed only in terms of the expansion of traffic capacity of the old transport system. In calculating the total expenditures the total cost of developing the transport route (or its proportionate share if other freight is carried) is applied to the freight flow under consideration. Of course, with such a method of calculation, the transport cost is higher than in the first case. We calculated the transport cost of coal by rail from the marginal expenditures taking into account the actual features of the various roads.

The case is different with gas pipe lines and electrical-transmission lines, since there are no all encompassing networks that, with comparatively small changes, can be used to transmit large amounts of gas and electricity. In each case, special new lines must be built.

The final determination of the comparative effectiveness of the alternatives was clearly defined by the period of return for additional investments, and the economies of annual operating costs from the formula worked out by the USSR Academy of Sciences and adopted in economic practice:[5]

$$\frac{K_1 - K_2}{C_2 - C_1} = T \gtreqless T_n$$

where K_1, K_2 are the total investments of the first and second alternatives; C_1, C_2 are the annual operating costs of the first and second alternatives; T is the period of return of additional investments; and T_n is the standard period of return (over eight years).

We also applied the method for determining total expenditures required for the implementation of each of the alternatives for a period of fifteen to twenty years.

In all, five connections of Siberia and Central Asia with the European Center and the Ural regions were examined:

1. Turkmen SSR: gas of the Darvaza deposit (Center)
2. Kazakh SSR: gas of the Aral-Sor deposit in the Caspian lowland (Center)
3. Kazakh SSR: Ekibastuz coal (Urals)
4. Siberia: Kuznetsk coal (Urals)
5. Uzbek SSR: Gazli gas deposit (Ural).

The Donets Basin-Center connection (6), which serves for purposes of energy redistribution within the European part of the USSR, was also examined.

As a result of the calculation of all major types of expenditures (fuel extraction, power plants, trunk lines, local gas pipelines, collecting and distributing networks), the following data have been obtained (see table 1).[6] An analysis of the table with the help of the formula of economic effectiveness of the compared alternatives leads to the following conclusions: The transfer of electric energy from Turkmenia (Darvaza region) to the Center pays off in 14 years, that is, a period greater than the standard. However, it still can become efficient if (because of the intersystem effect) a lowering of total established power of the unified energy systems by approximately 700,000 kilowatts is permissible when the electrical transmission line is built. The transmission of electric energy from the Kazakh SSR (Aral-Sor region) to the Center is higher in investment expenditures than the transport of gas; however, the operating costs for both alternatives are approximately equal. Therefore, building of a line of electric-transmission is not efficient. For the Kazakh SSR (Ekibastuz)—Urals interconnections—the larger investment expenditures in electrical transmission line pay off, because of lower annual operating costs, in less time (6.3 years) than the standard, and the transport of electric energy is more profitable than the railroad transportation of the fuel. Such a situation is due to the small reserve traffic capacity of the local railroads and the low caloric value of the Ekibastuz coal. It is more rational to consume the coal in the mining region. The transport of electric energy from the Kuznetsk Basin to the Urals from the Donets Basin to the Center is not efficient, since the railroads in these directions have a large reserve of traffic capacity. From the Uzbek SSR (region Gazli) to the Urals region, it is more profitable to transmit the gas by pipe line, since the building of lines for electricity transmission, which is about equal in annual expenditures, is considerably higher in investments.

For the final choice of a type of transport many other facets of this question must be examined; among these should be the reliability of energy supply, the possibility of obtaining an intersystem effect, the

TABLE 1

INTERCONNECTION		ALTERNATIVE	CHARACTER OF TRANSPORTATION TRUNK LINE	LENGTH OF TRANSPORT TRUNK LINE (Kilometers)
1. Turkmen SSR-Center	1a	Gas Transport	Two lines of gas pipes	2,600
	1	Transport of Electrical Energy	Two lines of electrical transmission, direct current	2,300
2. Kazakh SSR-Center	2a	Gas Transport	Two lines of gas pipes	1,100
	2	Transport of Electrical Energy	Two lines of electrical transmission, direct current	800
3. Kazakh SSR-Urals	3a	Transport of Coal	Railroad	1,350
	3	Transport of Electrical Energy	One line of electrical transmission, alternate current	1,150
4. Siberia-Urals	4a	Transport of Coal	Railroad	1,900
	4	Transport of Electrical Energy	Two lines of electrical transmission, direct current	1,700
5. Uzbek SSR-Urals	5a	Transport of Gas	One gas pipe line	2,000
	5	Transport of Electrical Energy	One line of electrical transmission, direct current	1,900
6. Donets Basin-Center	6a	Transport of Coal	Railroad	1,040
	6	Transport of Electrical Energy	Two lines of electrical transmission, direct current	750
	6b	Transport of Electrical Energy	Four lines of electrical transmission, alternating current	840

TABLE 1--Continued

INTERCONNECTION	ALTER-NATIVE*	INVESTMENTS (Millions of Rubles)	OPERATING COSTS (Millions of Rubles)	TOTAL COSTS (Millions of Rubles)	FIRST-YEAR OPERATING COSTS OF THE PROJECTED ALTERNATIVE (Millions of Rubles)
1. Turkmen SSR-Center	1a	1,938	1,413	3,351	143.5
	1	2,182	1,215	3,398	117.4
2. Kazakh SSR-Center	2a	1,695	1,130	2,825	110.1
	2	1,983	1,147	3,130	111.5
3. Kazakh SSR-Urals	3a	389	308	697	39.6
	3	401	298	694	37.7
4. Siberia-Urals	4a	1,802	2,504	4,306	230.0
	4	2,595	2,760	5,355	241.0
5. Uzbek SSR-Urals	5a	744	523	1,267	61.7
	5	908	552	1,460	63.7
6. Donets Basin-Center	6a	1,538	2,007	3,544	213.0
	6	1,927	2,159	4,086	228.0
	6b	1,938	2,201	4,145	232.0

*See page 1 of table for definitions and characteristics of alternatives.

universality of various types of transport, the economic development of territories adjacent to the transport trunk lines and so forth.

1. Ya. A. Mazover, A. S. Nekrasov, and V. K. Svayel'ev, "The Future Geography of the Fuel and Energy Economy of the USSR," *Voprosy geografii* [Problems of Geography], no. 57 (1962).

2. This position has been developed in the works of many authors. See, for example, L. A. Malent'ev and Ye. O. Shteynkhaus, *Ekonomika energetiki SSSR* [The Economics of Energy in the USSR] (Moscow-Leningrad, 1963).

3. M. M. Albegov, "Comparative Savings in the Transmission of Electrical Energy and the Transport of Fuel," *Elektricheskiye stansii*, no. 9, (1960); M. A. Vilenskiy, *Elektrifikatsiya SSSR i razmeshcheniye proizvoditel'nykh sil* [Electrification of the USSR and the Distribution of Productive Forces] (Moscow, 1963); L. A. Malent'ev and Ye. O. Shteynkhaus, *Ekonomika energetiki SSSR*; Ye. A. Probst, *Razmeshcheniye sotsialisticheskoy promyshlennosti* [Location of Socialist Industry] (Moscow, 1962); G. Falomo, "The Economics of Long-Distance Transport of Fuel and Electric Energy Transmission," in *Voprosy razvitiya zarubezhnoy elektroenergii* [Problems of the Development of Foreign Electric Energy] (Moscow-Leningrad).

4. The period of construction and order of investment were determined by *Normy prodolzhitel'nosti stroitel'stva predpriyatiy, puskovykh kompleksov, tsekhov, zhaniy i sooruzhenty* [Norms for the Duration of Building Enterprises, Complexes Set to Open, Shops, Buildings and Equipment], Building Norms 164–61 (Moscow, 1961).

5. *Tipovaya metodika opredeleniya ekonomicheskoy effektivnosti kapitalovlozheniy i novoy tekhniki v narodnom khozyaystve SSSR* [Standard Method for the Determination of Economic Effectiveness of Investments and of New Machinery in the National Economy of the USSR] (Moscow, 1960).

6. All expenditures on the table are assigned to the year of development of the projected alternative power source.

SECTION VII

Population Geography

Introductory Note

The geographic study of population in the Soviet Union has been one of the most rapidly advancing areas of economic geography. Serious attention was first given to the study of population shortly before World War II by such geographers as N. N. Baranskiy and R. M. Kabo.[1] In the postwar period population geography, and particularly the study of urban population, was greatly accelerated and encouraged by such geographers as Saushkin, Pokshishevskiy, Konstantinov, Kovalev, and Mayergoyz.[2] By the decade of the 1960s the importance of population geography is attested to by the series of conferences and symposia organized around this theme by geographers. The First Interdisciplinary Conference on Population Geography[3] was held in Moscow in 1962, followed by a series of regional and special-purposes conferences such as the All-Union Conference on Demographic Problems of Central Asia (Tashkent, 1965), the All-Union Conference on Problems of Settlement and Labor Resources in the Far North (Magadan, 1965), and the Symposium on Problems of Settlement and Populated Places in the USSR (Moscow, 1966). Additionally, the rapidly growing number of studies in population geography is ample evidence of the place of this field in Soviet economic geography.[4]

In the system of Soviet geographical sciences, population geography is considered a special economic geographic discipline. It is distinguished as being different from the systematic branches of economic geography (transport geography, manufacturing geography) inasmuch as "population is the object of any economic activity" and the "characterization of its distribution acquires a synthetic, general economic-geographic significance."[5] Although there was some controversy earlier over whether population geography should occupy an independent position equivalent to economic and physical geography, there is now general agreement among Soviet geographers that population cannot be studied as an isolated phenomenon but must be considered within the framework of the prevailing social and political organization and as a vital link in the production process.[6] It is this same argument, of course, that is used to criticize the *human* or *social* geographies commonly distinguished in the West.

In the Soviet context population geography is a broad field encompassing a number of problem areas. The field was defined in a resolution of the First Interdisciplinary Conference that stated that "population geography is a social-geographic discipline concerned with the study, over time, of the facts and patterns of the distribution of population (territorial systems of populated places), its growth, structure, migration, and other characteristics as part of the process of social reproduction. This requires the study of population in its territorial aspect, as a productive force and as a consumer of material and spiritual goods, as well as its own reproduction."[7] A later definition by Pokshishevskiy, found in the initial article of this section, differs slightly in that the field is expressly labeled a branch of economic geography and emphasis is placed on the dynamics of population change and the search for laws of a spatial nature.[8] In general then, population geography has four major areas of concern: (1) general questions of population ranging up the hierarchy of scales from the regional to the national level; (2) questions of urban geography (in this collection urban geography is treated separately, see Section 8); (3) questions of rural population distribution; and (4) questions concerning the historical geography of population. In more specific, scientific terms, the major objective of study is the discovery and application of laws governing the distribution of population, migration, the formation of a system of populated places, and the development of the material elements of the various-sized populated places.[9]

Another evaluation of the needs and role of population geography in the Soviet Union took place at the Second Interdisciplinary Conference on Population Geography[10] in 1967 in Moscow. The conference highlighted the growth of the field especially in terms of the recent emphasis on urban and rural population problems, man-environment relations including aspects of medical geography, migration, interdisciplinary contacts, and the application of mathematical methods. The deficiency of the discipline most noted was the dearth of research in the theoretical area. It is also interesting to note that the conference proposed the creation of an institute of demography within the Academy of Sciences and called for the creation of departments of population geography in the universities.

In summary, the field of population geography has evolved very rapidly in the last ten years. There has been, and remains, a great emphasis on urban problems, although other directions of research are emerging. The most recent and interesting of these new areas include the geography of services, the geography of consumption, and the geography of living conditions.[11] These new emphases reflect the recent trends discernable in Soviet internal policies in general and should result in a series of interesting studies useful to Western scholars for their intrinsic value as well as for comparative purposes.

The four selections included in this section are designed to reflect a cross section of Soviet research on problems of population geography (exclusive of urban research), starting with a philosophical-methodological statement and continuing to the problem of variations in natural growth and the examination of internal migration.

In the lead article Pokshishevskiy emphasizes that the study of population by economic geographers focuses on the role of man as a producer and consumer in the economic cycle. He stresses that population geography is not concerned with population per se but rather with the spatial aspects of population in its production-consumption role. The danger of separating population geography from economic geography is argued, and the roles of related fields such as ethnography, demographic statistics, and labor economics are discussed. The development of geographic studies of population are traced through the prerevolutionary and Soviet periods, and the main, immediate tasks for the field are enumerated.

Nevel'shteyn's paper on territorial differences in natural growth is one of the few Soviet geographical studies to examine crude fertility and mortality figures at less than a national scale. In the introduction, a set of population growth categories is outlined that is very similar to the basic divisions usually associated with demographic transition theory as developed and elaborated by Western population specialists. A comparison is then made of prerevolutionary and Soviet demographic developments. The conclusion is drawn that the Soviet Union has the *highest* birth rate of all industrial countries and one of the lowest death rates in the world and thus has realized "the most progressive type of population reproduction." The author indicates that the main reasons for the recent decline of the birthrate are the lowering of infant mortality, the mass migration from rural to urban areas, and the heavy involvement of females in industrial production. The bulk of the article reviews the birth, death, and natural growth rates in the various regions of the country.

The article on migration by Pokshishevskiy et al. provides an interesting, overall view of migratory movement in the USSR for the period 1959–62. Although the authors note the lack of migration data for the USSR in recent years, they do, however, present a general picture of migration by calculating a migration balance for each region by comparing actual population growth with the rate of natural increase. Kazakhstan, Central Asia, and the Northern Caucasus were found to be the most important regions of in-migration, whereas Siberia and the Far East experienced a negative migration balance. Other examples of interregional and intraregional migration are examined with a particular emphasis on the slowdown (and in some cases, reversal) of movement to Siberia and Kazakhstan. The cause of this problem is attributed to poor living conditions such as a low level of consumer services and housing.

The final article by Zayonchkovskaya and Perevedentsev is an extremely interesting study that looks more closely at the problem of poor living conditions in the eastern part of the Soviet Union. The study is concerned mainly with what the authors call individual migration rather than socially organized migration (the latter is movement of persons implemented by state institutions and social organizations). They note that individual migration in socialist countries differs from that in capitalist countries in that the former is really planned indirectly by the state when it establishes regional wage levels and allocates capital funds. In this study of Krasnoyarsk Kray, the authors indicate that since overall migration is tightly tied to manpower turnover, to understand migration it is necessary to discern the reasons why workers leave their positions. The analysis is based on a survey of 4,700 workers in the Krasnoyarsk area who left their positions of their own accord without a valid reason and migrated. The four factors found to be most important were dissatisfaction with (1) type of work, (2) wages, (3) living conditions, and (4) the desire to be with relatives. Another examination was conducted with regard to housing types that indicated that those individuals in their own housing were much less prone to move than workers in collective housing. The authors conclude that living conditions in this area are worse than many areas of the country, and this situation has much significance for future migration.

1. See for example N. N. Baranskiy, "Concerning the Economic Geographic Study of Cities," *Voprosy geografii* [Problems of Geography], no. 2 (1946), pp. 19–62, and R. M. Kabo, "Elements of the Geographical Study of the Population of the USSR," *Geografiya v shkole* [Geography in School], no. 3 (1941).

2. For example see Yu. G. Saushkin, "The Geographical Study of Rural Places in the Soviet Union," *Voprosy geografii*, no. 5 (1947), pp. 54–68; O. A. Konstantinov, "Changes in the Geography of Cities of the USSR after 30 Years," *Voprosy geografii*, no. 6 (1947), pp. 11–16; V. V. Pokshishevskiy, "International Migration as an Object of Geographic Study," in *Voprosy geografii* ("Collection of Articles for the Eighteenth International Geographical Congress") (Moscow, 1956); S. A. Kovalev, "Questions of Terminology in the Geographical Study of Rural Settlement," *Voprosy geografii*, no. 14 (1949), pp. 29–42; I. M. Mayergoyz, "Toward the Economic-Geographical Study of Cities," *Voprosy geografii*, no. 38 (1956), pp. 5–26.

3. The proceedings of the conference were published in six volumes, Geograficheskoye Obshchestvo SSSR, *Materials of the First Interdisciplinary Conference on Population Geography*, 6 vols. (Moscow, 1962). Volume 6 is an especially useful bibliography of research in population geography.

4. See V. V. Pokshishevskiy, *Population Geography in the USSR*, State of the Art Series (Itogi Nauki) (Moscow, 1966).

5. V. V. Pokshishevskiy, "Geography of Population and Populated Points," in *Soviet Geography: Accomplishments and Tasks*, Lawrence Ecker, trans., and Chauncy D. Harris, ed., American Geographical Society Occasional Publication, no. 1 (New York, 1962), p. 143.

6. A summary of this controversy can be found in V. V. Pokshishevskiy, "The Geography of Population and Its Tasks," *Izvestiya Akademii nauk SSSR*, seriya geografiya, [Proceedings of the Academy of Sciences of the USSR, Geography series] no. 4, (1962), pp. 3–11.

7. "Resolution of the First Interdisciplinary Conference on Population Geography," *Vestnik Moskovskogo universiteta,* seriya geografiya [Herald of Moscow University, Geography series], no. 3 (1962), p. 70.

8. V. V. Pokshishevskiy, "The Content and Basic Tasks of Population Geography," in *Population Geography in the USSR: Basic Problems* (Moscow-Leningrad, Nauka, 1964), p. 7.

9. For an elaboration, see V. V. Pokshishevskiy, "Geography of Population and Populated Points," p. 145.

10. The nineteen major reports of the conference were published in V. V. Pokshishevskiy, D. I. Valentey, and S. A. Kovalev, eds., *Scientific Problems of Population Geography,* (Moscow University, 1967). For English summaries of the conference see O. A. Konstantinov, "Some Results of the Second Interagency Conferences on Population Geography," *Soviet Geography: Review and Translation,* no. 8 (October 1967): 652–60.

11. See, for example, S. A. Kovalev and V. V. Pokshishevskiy, "The Geography of Population and the Geography of Services," in *Scientific Problems of Population Geography,* pp. 34–37; O. V. Leont'yev, "A Method for Calculating the Requirements for Mobile Public Services Under Various Geographical Conditions," *Vestnik Moskovskogo Universiteta,* seriya geografiya, no. 3, (1967), pp. 15–20.

V. V. POKSHISHEVSKIY

The Content and Main Tasks of

Population Geography

Throughout history population has been one of the important subjects of study for geographers. Countries and the peoples inhabiting them represented the object of study of geography at the earliest stages of its development. As economic geography became a separate unit in the system of geographical sciences, the main feature in the study of population became its [the population's] occupation, its economic activity. The treatment of population by geographers was always determined by the theoretical positions that science held at the moment. Soviet scientists from positions of Marxism-Leninism proceed in this interpretation from the following.

In the process of social reproduction, population as the bearer of labor occupies a very special place. All material and nonmaterial values needed for the life of society are created by labor; in this respect the inputs of labor are not analogous to other production components, such as raw material, fuel, auxilliary materials, and even the tools of production. The latter are passive participants in the productive process, whereas labor is its active core. "Labor is above all a process accomplished between man and nature, a process in which man by his own activity determines, regulates, and controls the exchange be-

Translated from *Geografiya naseleniya v SSSR* [Population Geography of the USSR] (Moscow-Leningrad, 1964), pp. 5–31.

tween himself and nature."[1] Therefore, the labor resources of the population differ in principle from all other productive forces. Lenin emphasized this special role ("The first productive force of all humanity is the worker, the toiler").[2]

Man for Soviet economic geographers is first of all the *subject of production*. At the same time, however, population forms the consumer pole of the process of social production—that pole on which, in the expression of Marx, production itself receives its completion.[3] In the course of personal consumption (in the broad sense, including satisfaction of cultural needs, needs for all types of service, housing, and so forth) the reproduction of the population itself is assured. "The economic geographer must not forget that the population is both the producer and consumer and that, in this manner, the economy is tied to the population, it may be said, from beginning to end."[4] Population, consequently, figures in the study of social reproduction in three capacities: (1) as the subject of production; (2) as the consumer of a definite part of the total social product (not only of material but also nonmaterial value); (3) as the most important link in the whole process of social reproduction, subject itself to reproduction.

The cycle of creation and consumption of values is brought about by the social division of labor, one of the forms of which is the territorial division of labor. Consumption of goods produced in one region often occurs in another; labor expenditures are often spent not where they were created. It is these spatial features of the processes of reproduction that are predominantly the subject of economic geography.

Because of the *all-permeating* role of the population in these processes, its labor activity and its interests must be examined in all fields of economic geography as for example, in the study of agricultural geography, the geography of fishing or forestry, industrial geography, and also in the composition of regional economic-geographical characteristics (especially in studying large populated points that are also centers of production and foci of consumption). In terms of the applicability to economic-geographical work, this author has had occasion to show, in his opinion, the special position of population characteristics in a separate article.[5] However, because of practical requirements, the differentiation of methods of scientific research have led to the fact that questions of study of the geography of population are becoming (within economic geography) a specialized field.

Below we shall return to this process of formation of population geography as a specific discipline; here we shall attempt to define the scientific content of this field: what has already been said provides an adequate base. We think that population geography as a discipline studying the composition and distribution of population and inhabited points[6] is useful only in a general sense. Lenin pointed out that concise definitions are always imperfect if only because they cannot "encompass the multifaceted connections of a phenomenon in its complete development"; "overly short definitions, even if convenient because they sum up the important (elements), are nevertheless insufficient especially if one has to derive the essential features of the phenomenon which must be defined."[7] Therefore, together with simplicity and brevity, a developed definition of population geography must be given.

We consider that this [population geography] *is a branch of economic geography, studying (dynamically, developmentally) the structure, distribution, and territorial organization of population examined in the process of social reproduction, establishing principles, especially spatial, and determining the changes in all these features of a population.* In this definition the following are emphasized: (1) the spatial aspects; (2) the examination of the population not for itself, but in its production-consumption relationships; (3) the necessity of a historical-genetic approach. It includes organically the geography of population within economic geography, understood in the spirit of the commonly accepted definition of it (economic geography) as a science studying the geographical distribution of production, and so forth. If one keeps in mind that social production in the broad sense of the word coincides with the concept of reproduction, then this definition of economic geography indisputably includes population in its subject matter. The most important categories of economic geography, for example, the territorial division of labor, economic-geographical situation, and many others, at closer examination prove to be also an instrument for the geographical study of population (the latter being engaged in certain branches of economic activity, and so forth).

In the USSR it is generally accepted that population geography is a branch of economic geography. This was emphasized in a special report on the discipline at the Second Congress of the USSR Geographical Society in 1955.[8] The resolution of the congress stated that the study of population geography is one of the main tasks of economic-geographical investigations. At the preceding congress (1947) questions of the geography of population were, in fact, also examined as a whole in the economic-geographical section. At present only three geographers can be named who, with certain reservations, do not support the idea of population geography being an organic part of economic geography.[9] We think that by narrowing and impoverishing the content of economic geography to the study of the distribution of industrial points or individual cultures alone, one can leave out the entire broad scope of questions of the geography of social interrelations of people and the territorial organization of settlement—without which population geography itself would amount to simple demographic information—presented in territorial cross section.

The enriching of economic geography with more socioeconomic content, bringing it to a higher stage that could arbitrarily be called the geography of society, overcoming the narrow departmentalization in economic geography: this is a valid process. It is rapidly being realized right now in the USSR, and there is no necessity to look at the system of geographic sciences for some special place outside of economic geography for population geography.

How can the tendency toward the separation of population geography from economic geography prove to be dangerous? N. N. Baranskiy expressed it well. Noting the growing tendencies to remove population geography from economic geography and to form a special independent geographic discipline bordering on ethnography, he warned that this could revive in population geography an anthropogeographical taint and stimulate the search for the "reasons for differences between countries . . . not in social conditions, but in the behavior of the people themselves, which are the prime causes producing differences between countries. And the behavior of the people, in turn, in the opinion of anthropogeographers, is determined by the natural environment surrounding them and does not change in time or changes exceedingly slowly. Such an opinion is close to racism."[10]

It must be noted that the few Soviet geographers who favored taking population geography out of economic geography, in their practical program, also placed its content into the usual economic-geographical framework. O. A. Konstantinov[11] noted this convincingly.

The negative experience of bourgeoise geography shows well what contrasting economic-geography and population geography could lead to. In their understanding of economic geography bourgeoise geographers usually combine such dehumanized theories of the distribution of industry, geography of markets, and the problems of spatial economics, whereas under the title of the geography of man a diffuse scope of questions deal with the relationship between abstract man to nature, his settlement forms, house types, some features of life, and so forth. Often the productive activity of man has not been fully taken into account, and often class relations are omitted, that is, in essence the idea of population as a productive collective, as the subject of production disappears. There is also no clear conception of population as the central link in the process of social reproduction.

Problems of the interrelationships between population geography and ethnography must be discussed separately; here we can point only to the fact that many characteristics of the population, which are not the subject of study by ethnographers but are completely in the sphere of geography of population, have many ethnographic characteristics (for example, features of production, habits of the population, some features of the diet, house types, and a number of the cultural features of the nationalities) that fully fit into the framework of population geography as well as economic geography; in the USSR other ethnographic characteristics define the national-territorial division of the territory and thereby also its economic-administrative organization, that is, they also lie in the sphere of interest of economic geography, as one of the important indicators for regionalization. There still remains, however, a certain range of questions that will not fit into the framework of geography and that is of interest only to ethnographers (the national features of folk art, folklore, and the like, as well as the study of the liquidation of some historical survivals in culture and life). From our point of view, these questions must be considered to be, for the most part, outside population geography and in the sphere of ethnography; however, for a final demarcation[12] of these sciences (especially as applied to the USSR) there remains only a rather limited sector.

Analogous problems of demarcation, of identifying areas of overlap, and establishing means of creative cooperation of scientific efforts exist with respect to population geography and demography. In this article it is not possible to discuss these problems in detail; in principle, however, the ways to interaction (and also demarcation), in our opinion, are sufficiently clear.

Demographic information is a necessary premise of any work in population geography. In the form in which it is published this (demographic) information is often insufficient for geographers studying problems of population (for example, [data for] territorial units) and demands refinement by conducting special field work or research. In some cases one can expect that under the influence of solving practical problems in population geography (and also labor economics) the collection and publication of statistical-demographic information will be increased to meet needs. This, for example, is the situation with regard to determining place-of-birth data (or place of residence at the time of the preceding census) that makes it impossible to study properly the process of migration and the redistribution of labor resources related to it. With respect to other problems it is difficult to expect full satisfaction from the usual demographic statistics to satisfy all inquiries in population geography. Even if there is improvement in the data concerning the place of work for people who live at a certain settled point, in order to study daily labor migrations, the actual number and composition of a population, the ties of a population with a major city, as well as other details of settlement, there will remain a sustained interest in special, even though sample, studies of suburban passenger traffic flows, the relationship between places of work and places of residence, the family composition of a "mixed population" where some of the family members work in agriculture and some commute to work in the city, and so forth. These special studies in population geography will be even more necessary if we want to look into the processes of formation of cadres for individual enterprises and industrial centers in connection with settlement.

Having its [own] important scientific-methodological problems, demographic statistics can, to a certain degree, serve the interest of population geography by creatively perceiving the problems formulated by the latter; but the fundamental scientific content of demography will always be limited to a combination of methods of collection and processing statistical information about population.[13] This determines also the pattern of interaction between the demographic statisticians, on the one hand, and the geographers studying the problems of population, on the other. It must be added that demographic statistics, of course, with its information on population, is called on to serve not only population geography but also many other branches of knowledge and spheres of planning.

We consider the interaction of population geography with labor economics very fruitful. The study of the utilization of labor resources is connected with the central place that population geography is called on to occupy in the study of the whole process of social reproduction. This makes the labor-resource line of development of population geography so promising. The creative interactions arising here have, in principle, the same character as the interaction between any branch of economic geography and the corresponding branch of economics. Without knowledge of technical-economic features of the latter, the principles of distribution inherent to it cannot be understood. However, because of the special place occupied by population geography in economic geography, the interactions also have a more synthetic character, in the final analysis encompassing all subdivisions of the national economy and the entire population. True, studies of a narrower, regional or sectoral type are possible, for example, on the geography of labor resources of some region or city, or within one sector such as agriculture.

The question concerning the interrelationship between population geography and historical geography (these interrelationships are fully realized within the limits of the system of geographical sciences themselves) is comparatively simple. Historical geography is called on to reconstruct a picture of the geography of past eras (selecting for this moments that are typical from the point of view of historical periodization or that are discontinuous); in the course of this reconstruction it tries to apply the whole arsenal (of tools), characteristic for geographical sciences, to disclose the relationships between phenomena, indicating their spatial complexes. Since "during a historical period of time natural conditions change comparatively slightly . . . the problems of historical geography in this area are very limited and, consequently . . . the content of historical geography . . . amounts mainly to changes in the historical past in the geography of population, the national economy, and the political economy," writes V. K. Yatsunskiy, the most prominent Soviet theoretician of historical geography.[14] Thus, there is a need for a special section of historical geography that specifically treats population in the process of its formation, migration, creation of a network of settled nodes, and subsequent transformation. It is valid, consequently, to speak of a historical geography of population.

One can also mention the existence of a sphere of interaction between population geography and medical geography. Previously we highlighted this question specifically, therefore, we will not dwell on it here but refer those who are interested to our work.[15]

Bourgeoise geographers often contrast population geography (classifying it primarily under the title geography of man, which is rejected by us) and settlement geography; sometimes they distinguish it as a special discipline as well as urban geography. Such contrasts, reflected sometimes in the sectional structure of international congresses, and in geographic bibliographies is, in our opinion, without foundation. Any settlement, including a city, is a definite form of the territorial organization of the population. In a broad sense settlement, as one of the main categories of population geography, encompasses all forms of localization of population, from the dispersed farmsteads to multimillion city concentrations. In life all these forms (of population settlement) are continually changing and are also related to other concepts of our discipline (general and differentiated density, migrational redistribution of population, and labor resources). In solving practical problems such as those of regional planning, for example, it is simply not expedient to deal simultaneously with a whole range of similar categories and to separate them according to seemingly independent subdisciplines. Proceeding from precisely these considerations the Moscow branch of the USSR Geographical Society, which formerly had a commission on the geography of population and cities, has recently deleted the words "and cities" from the name of the commission.

This does not mean, of course, that within the framework of population geography there cannot be a further specialization (the study of cities, villages settlements, and so forth). Experience has shown that such specialization guarantees the most efficiency. Moreover, the more specialized the investigations the more they intertwine with adjacent disciplines. Thus, the most valuable, practically constructive works of Soviet geographers on cities was done in close interaction with applied disciplines of the urban-development cycle: planning, economics, engineering. Such cooperation, in turn, often enriches population geography itself with new categories derived from an adjacent area, but geographically interpreted.

Having in mind what has been said above concerning the absence of sharp boundaries between the separate sections and directions within population geography, one can point out the following internal subdivisions: (1) general questions of population geography on the scale of the entire country or large parts of it, including theoretical principles concerning structure and distribution of population by territory and branches of activity in relation to the geography of production, ratio of urban and rural population, features of the migration processes, and so forth; (2) questions of urban geography; (3) questions of the geography of rural forms of settlement;[16] (4) questions of historical geography of population; (5) questions of ethnography;[17] (6) questions of cartography of population.[18] A very important connecting theme of research, which passes through at least the first three subdivisions, is the study of population from the point of view of labor resources; this question was discussed in detail in an earlier article.[19]

It is not difficult to see that each of the labeled directions or aspects are adjacent to a certain sphere of practical activity and can be used in planning, project or land management works, regionalization, and the like. The only exception is works in the historical geography of population, but they are organically necessary in order that the whole branch of economic geography will not have a descriptive nature but rather an explanatory analytical character that necessitates the introduction of genetic elements. As a whole, one may consider that population geography as a science has its own theoretical range of problems (establishment of general principles), regional cycles of research (where local manifestation of these principles becomes evident), and applied aspects (where, on the basis of recognized principles, practical problems are solved). It stands to reason that these three aspects do not exist in isolation but dialectically merge into one another.

The comparatively few prerevolutionary works specifically devoted to population geography[20] contained valuable factual material, but they were, as a rule, not satisfactory methodologically. The works of A. I. Voyeykov[21] alone are, to some degree, an exception; they stood far above the Western examples that influenced some Russian geographers. This influence was reflected in the strong spirit of anthropogeography, the imprint of which lies on the work of even such a great scientist as V. P. Semenov-Tyan-Shanskiy.

At the same time a large number of descriptive studies, without any special formulation of questions for population geography, were conducted not only by geographers but also by statisticians, economists, ethnographers, and physicians. These works accumulated vast information and, moreover, often treated their subject from progressive-democratic positions. Such works include the many *zemstvo* investigations and ethnographic and medicogeographical descriptions.

Thus, Soviet geography, in terms of questions of population study, has inherited from the prerevolutionary past rich descriptive material and traditions of progressive treatment, although they are characterized by an unsatisfactory methodology and utilized anthropogeographical approaches. In the course of twenty years, until a cadre of Marxist economic geographers took the leading position, the anthropogeographical tendencies continued to dominate the approach to questions of population geography; this was also due to the fact that among geographers working on these questions there were still many scientists trained in the prerevolutionary period. For example, from a theoretical standpoint the works by V. P. Semenov-Tyan-Shanskiy,[22] A. A. Kruber,[23] L. D. Sinitskiy,[24] I. Yamzin, and V. Voshchinin[25] were unsatisfactory. The appearance in the thirties of the first Marxist works, in particular, those reported at the First Geographical Congress in 1933, had still not created the necessary basic change. This can be explained primarily by the fact that the central theme of economic-geographical works at that time (and in part even later) were questions of the distribution of industry and the struggle with the bourgeoise legacy in geography. Under these conditions, "having killed the old anthropogeography, the new ideas did not create anything new in its place; the section on population, including the very substantial information in former geographical descriptions . . . disappeared without a trace, being lost between nature and economy. . . . Man had been forgotten," so N. N. Baranskiy[26] characterized the situation in the thirties and during the first half of the forties.

However, O. A. Konstantinov notes that even at the end of the twenties some problems of population geography had already been worked out from Marxist positions, namely the geography of cities. This seniority of urban geography was later of great significance in the formation of a Marxist population geography as a complete branch of economic geography.

N. N. Baranskiy, N. I. Lyalikov, R. M. Kabo, and Yu. G. Saushkin were significant in the renaissance in the USSR of population geography as a whole, based on new Marxist principles. They raised the interest in this subject at the Moscow State University and the Moscow State Pedagogical Institute, where under their direction many talented students worked (S. I. Bruk, V. L. Gerbov, F. V. D'yakonov, S. A. Kovalev, I. I. Tensina, [Chavchanidze], L. A. Ustinova, and others). Baranskiy theoretically substantiated a plan for the geographical study of cities;[27] this plan is still used by almost all Soviet geographers.[28] As early as 1941, Kabo made an appeal to include in the sphere of geographical study a wide range of questions in population geography,[29] and at the Second Geographic Congress (1947) he reported on a program for a special course on population geography worked out by him.[30] True, he had then not yet completely overcome the legacy of an anthropogeographical influence, and in his paper the correct Marxist premises were still combined with recommendations to continue the work in the spirit of V. P. Semenov-Tyan-Shanskiy and even follow some notions of Ratzel. His definitions of the limits of population geography were also diffuse;[31] later, however, R. M. Kabo himself successfully cleansed his earlier unclear ideas of the taint of anthropogeography.[32]

N. I. Lyalikov at the same time published a short course on the geography of population[33] and other studies.[34] Some of them, because of lingering traditions of leftism resulted in polemics in the course of which terminology became more precise (for example, the term capacity for defining the settlement of a territory).

In general, it must be pointed out that the status of Marxist geography at the end of the forties and beginning of the fifties led to many discussions. Expeditionary field work, widely developed at this time (especially by the geographers of the Moscow State University), became a living source (of information) and a method for putting everything in its place. The geographers, on the one hand, gave a concrete basis to theoretical ideas about the real character of those "connections between phenomena" that become apparent in the study of population, and, on the other, they showed how to use this study in practice for communist development. Generalization from the experience of field studies made it possible to create valuable programs that lent great purposefulness to further research.

Constructive investigations that were at the boundary between the study of urban geography and the solution of concrete problems of urban development achieved considerable development (primarily those of V. G. Davidovich). The study of rural settlement based on a very wide range of field studies (the volume of such studies, however, did not correspond at all to the diversity and multiplicity of objects subject to study), was undertaken by S. A. Kovalev and a number of other geographers; unfortunately, however, up to now it has not in practice been developed as well as the study of cities, although the groundwork has been laid.

Most attention was devoted to questions of typology of populated places (cities, villages, and those units occupying an intermediate place). As a result leading and practically significant criteria of typology (localization in populated places of specific national economic functions) were defined. The geographers studying settled points could offer to (government) organs in charge of population statistics, census materials well interpreted and processed. Mainly methodological questions of historical geography of population were solved, especially as the result of the excellent theoretically significant works of L. Ye. Iofa,[35] R. M. Kabo,[36] and V. K. Yatsunskiy.[37] The significance of historical-genetic elements for the typology of settled points in terms of practical utilization were also established. The study of population migration having initially also a predominantly theoretical and historical-geographic character was now being related to the practical planning of labor resources, especially most recently. At the Third Congress of the USSR Geographical Society, the significance of geographical characteristics of the population for economic regionalization was also demonstrated.[38]

The general upsurge of population geography was reflected in the appearance of a rather wide range of publications, for example, several collections (*Voprosy geografii* [Problems of geography]) devoted to this branch of economic geography, a number of monographs about individual cities or whole groups (of cities), many studies published in works and scientific journals of a number of schools of higher learning, and so forth.[39] With similar special publications (including also some studies on the population of foreign countries, not considered in this article) population geography has won for itself a secure place in textbooks and manuals on economic geography and in economic-geographical monographs of a geographical type,[40] including (monographs) on republics and large regions of the USSR.[41]

This brief historical survey cannot even touch, in a most general way, upon all aspects of the development of population geography (for example, questions of population mapping are omitted entirely); the purpose was only to give concrete content to the conception of the subject and problems of population geography discussed above, and also to show that, in the course of formation of this branch of economic geography, directions developed in a nonuniform way (for example, the early breakout of urban geography and the recent interest in geographical analysis of labor reserves).

The main ideas with which population geography deals are the socially productive structure of the population and its localization. A definite system of settlement is the external expression of the latter.[42]

The final conditionality of settlement in terms of territorial formations of social reproduction is the basic principle of population geography. "The law of correspondence of forms of settlement to the means of social production (and the superstructure) is inherent in all social formations."[43] Similar thoughts have been expressed earlier more than once, for example, by Yu. G. Saushkin, V. V. Varankin, and this author; it was also pointed out that *the geographic environment influences settlement through production,* that the size of individual populated points is usually related to the national economic functions performed; the density of population (that is, settlement) are functions of production (in particular, of the means of production), but also the structure of population (class, occupation, even age structure) reflects the means and character of social reproduction brought about at a given moment in the country.

From this initial thesis a number of derivative methodological aspects follow, in particular the necessity of a genetic approach for population geography (as well as for all of economic geography). It must be considered that the creation of material forms of settlement, which corresponds to specific features of territorial organization of production, as a rule goes on with a time lag. Therefore, an analysis of present localization of population and forms of settlement often demands that characteristics of previous eras be drawn upon. The inertia of earlier final forms of settlement, based on the preservation of material forms of populated places (buildings, roads, systems of water provision and the like) is an important concept that population geography constantly encounters in a concrete analysis of a region. Cases where the change of settlement forms outstrips the change in the character of the territorial organization of production are considerably rarer; such cases can occur in the course of a planned pioneering (type of) development where there is a conscious desire to utilize new forms of settlement for the acceleration of the territorial reconstruction of rural production.

Natural conditions, of course, influence the spatial distribution of population, but only as an intermediary, through production, creating conditions for its distribution and localization. Even in the sphere of rural settlement, together with the decisively significant foundations of agriculture there remain human factors: level of technology, the character of demands of society and the social organization of producers themselves, collective or state farm types of production considered spatially, size considerations of collective and state farms, the dimension and territorial structure of fields used by them, and so forth. One can show that the distribution of much urban production determining the forms of urban settlement are also under the influence of the natural environment (comparative wealth of raw materials in a region, fuel, and energy sources, and the like).

The direct influence of natural conditions on the forms of settlement can be traced only by the choice of areas themselves for communities of residential clusters. The mechanism of this influence is purely an engineering one: the technical possibility or impossibility of building up sections with specific gradients necessitates a certain engineering preparation of the territory. However, this mechanism too is historical: the engineering evaluation of an area depends, on the one hand, on the level of development of productive forces, on the technological capacity determining, for example, the possibilities for the amelioration of unfavorable factors; on the other hand, on the needs of society in the creation of populated points of given scales and type (which again are determined, in the final analysis, by the character of the geography of production). Areas where development is too difficult with a more weakly developed set of productive forces become sufficiently favorable with a higher level (of development) and a different method of production; and, conversely, the demands of social production and forms of settlement related to it can bring out in an evaluation of areas such criteria that earlier were nonessential. Thus, even direct engineering-geographical evaluations of the micro-geographical situation of communities and their areas must, without fail, *invisibly* include socioeconomic elements.

The principle of approaching population from a production point of view has determined the division of population into main functional groups (an approach that long ago had taken root in Soviet scientific literature). The first of the (functional) groups is related to national economic functions, the localization of which in some area or city cause also a corresponding localization of the population; these functions (industrial, agricultural, transport, and so forth) have a significance beyond local limits; they create the territorial division of labor. The second is the service group. If one digresses from local differences in the level of services (which would be legitimate only as a first approximation), then this group occupies a comparatively stable portion of the total population and by itself cannot determine its (the population's) distribution. The third group is the population not taking part in social labor; it too comprises a specific portion in the total population, depending mainly on its age composition and partly on the level and types of service. Consequently, only the first group is "demo-forming"; it actively determines the pattern of population distribution, and density of each of the populated points.[44]

The study of the relationships and structure of all three groups (in the country as a whole, within regions and individual populated points) is called the method of labor balance and connects population geography with the analysis of volume and distribution of production; that distribution which in the course of the territorial division of labor creates specialization of regions and individual centers.[45]

In the study of population geography the investigator consistently encounters the problem of the relationship between localization of places of residence and places of work. The simplest contrasting cases here are: (1) agricultural settlement when the application of labor, because of the character of agriculture itself, would be decentralized over a considerable area of fields, but the population itself concentrated in a place of residence, in the countryside, or small village; (2) settlement under urban conditions characterized by centralized places of work. On a broader plane it is possible to consider that the population, connected with one type of (economic) sector will, in its settlement, orientate itself with its (the economic sector's) areal distribution (agricultural or lumber economy, a number of extractive branches); branches of another type (mainly processing) concentrate production at individual points or cities best situated with respect to roads and having a number of historical premises (presence of funds, increased advantages in enterprise cooperation, and the like). Hence, in the total picture of population settlement two tendencies are combined [a tendency] toward concentration and toward decentralization, which create for each stage of the development of the country or region (in particular, considering also the development of means of transport) a specific settlement pattern.

The centripetal forces, forming urban clusters of production and population engaged in it, in turn create centrifugal forces at a specific level of agglomeration: overly large cities become complex for internal servicing and less economical: this includes also the concentration at one point of too many enterprises, which causes great difficulties.

The Program of the Communist Party of the Soviet Union points out that a rational distribution of industry will "remove the excessive congestion of population in large cities," that "small and average-sized, well-organized cities will be more and more developed which will make it possible to perfect and improve living conditions."[46] In the USSR group forms of urban settlement are actively being developed (city-satellites and other forms of urban agglomerations); this poses for population geography the task of solving a number of constructive problems. Agricultural settlement, on the contrary, is developing toward the substitution of small dispersed forms (for example, individual farmsteads) with greater centralization, which corresponds to the enlargement of the scale of socialistic production and the development of means of transport. On the whole, the premises for the prescribed Program of the Communist Party of the Soviet Union are being strengthened, and the essential differences between the city and countryside being overcome.

Among the principles established together by the economic-geographers and economists one may notice that the regions and centers with unidirectional development of heavy industry, especially those branches in which the use of female labor is limited, usually result in less than full utilization of the labor resources. Classic examples of this are the coal basins. Cases of predominant development of specifically female [oriented] branches of light industry are rarer. However, in a number of textile centers it was necessary to create special counterbalancing-industries in order to guarantee employment of male labor (Kamyshin, Narva, and others). Cases of increased turn-over of female cadres in the old textile centers, where the character of production does not guarantee a normal sex structure of the population, have often been described in geographical literature.

As a more general principle one can state that the more complete the production complex of any area, the more complete is the utilization of its labor resources. The disarrangement of some economic links or the existence of isolated, narrowly specialized enterprises makes it difficult to achieve high employment of labor reserves. Investigations in population geography have shown the dependence of the balance of labor resources on the territorial organization of the population, on the forms of settlement.

Principles have been established relative to the intensity of migrations and the degrees of permanence of the population migrating to rapidly developing regions and centers. There is an evident dependence of intensity and success of migration connections on the similarity of the industrial profile and natural conditions of the regions of out-migration and settlement; there is a higher rate of permanent settlement for migrants to cities from rural localities (this is partly due to the fact that they acquire as a result of migration, a profession that they formerly did not have); permanent settlement of migrants of all groups depends more on living conditions in the places they move to than on the difference in the nominal remuneration for labor.

Although a population has very varied contacts with the natural environment, it must be remembered that its interrelations with it exist only in the form of social relations between people,[47] therefore, *all phenomena that are the concern of population geography themselves lie within the scope of social principles.* This must be emphasized because bourgeois theories often bring biologism into the interpretation of questions of population geography, groundlessly introducing ecological approaches.

On the other hand, to recognize the social character of principles in population geography still more firmly connects it with that socioeconomic science—economic geography. By the way, we think that the existence of population geography itself—this most social wing of economic geography of the whole cycle of geographical sciences—vividly refutes the conception of a unified geography. V. A. Anuchin, the apologist for the latter, avoids completely the problems of population not by chance: if he only touched upon them he would have to enter the circle of production relations that he consistently considers to beyond the subject of geography, the "unified one" as well as economic geography.[48] It is evident that for a "unified geography" only one way is open for the study of population, the anthropogeographical; Anuchin, however, is a sufficiently experienced polemicist to understand how much this road would have exposed the defectiveness of the whole concept defended by him.

THE IMMEDIATE TASKS OF POPULATION GEOGRAPHY

Soviet population geography, the development of which is ever more closely connected with the practical needs of the economy, must intensify the working out of problems and aspects yielding conclusions important for the building of communism. The attention of population geography, as well as other socioeconomic sciences, must be first of all "directed to the discovery of ways of most effective utilization of material and labor resources in the economy, the best methods of planning and organizing industrial and agricultural production, to the development of principles of rational distribution of productive forces, and to the technological-economic problems of the building of communism."[49] Hereby, however, geographers studying population must not rush to supplant the work of planning and project organizations with their investigations; their work is to secure initial scientific bases for similar developments. By broadening and deepening the analysis of factual material that will yield new generalizations, fundamental theoretical investigations must also be created, or the practical aspects will be poorly grounded.

We think that the main urgent tasks of research work in the field of population geography are the following:

1. Creation of fundamental works on the theory of population geography. Despite successful investigations over a wide range of theoretical questions there are still no generalized works showing the complete system of principles at work here.

2. Creation of fundamental works (as yet absent) that would characterize the actual geography of population in the USSR as a whole as the result of the effect of specified principles: the distribution of population among regions and by branches of activity, the distribution in various parts of the country of specific types and forms of urban, as well as rural settlements, and so forth. All these phenomena must be illustrated dynamically, bringing out developmental tendencies also.

3. The development of geographic problems of the balance of labor resources, perfecting concepts and indices used or derived in the solution of these problems, and checking them with the largest possible factual material on the labor balance in the country, regions, separate areas, and populated places.[50]

4. Study of the geography of contemporary migration processes (especially in connection with problems of labor balance).

5. Development of methods of application of the results of investigations in population geography when drawing up projects for regional planning, planning of industrial centers and separate populated places, and also for land management, which foresees a reconstruction of the system of rural settlement, and similar problems. Although in this important practical field considerable experience has been accumulated, up to now it has not been sufficiently generalized.

6. Broad development of studies of the processes of reconstruction of a network of populated points of various types in the USSR.

Among these processes are the development of group forms of urban settlement, the development and transformation of small cities, the growth of settlements of an urban type and the processes of re-forming them into cities, the transformation of rural regional centers into semicity settlements, changes of a network of rural populated points and their material forms (in connection with enlargement and technological improvement of agricultural production and the reconstruction of its territorial organization), creation of new networks of populated points in pioneer regions under development. The course of investigations of such processes as a whole are already clear; there are successful examples of them, but the obtainment of mass results that could be introduced into the practice of actual planning and project work demands that these investigations be conducted everywhere on a broader scale and be aimed more toward the solution of constructive problems.

7. Intensification and close examination of works on population geography in foreign countries (geographical monographs as well as special investigations of individual problems of individual countries or whole groups of them).[51]

8. Active criticism of bourgeois theories of population geography, which demands, on the one hand, a systematic study of the works of bourgeois scientists and the disclosure of erroneous conceptions in them, and, on the other, confronting them with correct Marxist principles, also in the course of one's own analysis of problems in population geography in the USSR as well as in foreign countries.

In order to increase the international significance of Soviet concepts of population geography our geographers must decisively go on the ideological attack, and with all force contrast the scientific principles revealed by Soviet concepts with erroneous theories that the bourgeois scholars propagandize. What is necessary are daring and original works on the population geography of foreign countries (on such a level that they would compel the foreign geographers themselves to think, and would be convincing in terms of the completeness of factual material used) and an active fire of criticism against bourgeois theories of population geography, revealing the fallacy of Neo-Malthusianism in all its varities, racism, biologism, and so forth. Such criticism must be well linked with an analysis of shifts in the geography of population that have occurred in the USSR. Those shifts are sometimes very instructive from the point of view of countries recently freed from the yoke of imperialism and struggling for their national independence. Many geographers of these countries are trying to overcome the Western training they have received; the possibility of finding support in the methodological examples of Soviet investigations in this theoretically acute social sector—the geography of population—would, in principle, have great significance for them.

For Soviet geographers studying population it is pertinent to remember the words of the Program of the Communist Party of the Soviet Union that contain an obligation for them: "A point of honor for Soviet scientists is to secure for Soviet science the leading positions . . . and to occupy a leading position in world science."[52]

1. Karl Marx and Friedrick Engels, *Sochineniya* [Works], 23:188.

2. V. I. Lenin, *Pol'nyye Sochineniya* [Complete Works], 38:359.

3. It stands to reason that this is not isolated man, but population as a socially organized collective of producers. "Soviet geographers criticize bourgeois attempts to study population *by itself*, apart from its social organization and productive activity. The idea of a *geography of man* is considered unfortunate by Soviet geographers in that it is not oriented to the study of human collectives but opens the way to a metaphysical examination of the interrelations of individual man (as such) with the geographic environment while, in reality, from the earliest times this interaction serves the socially organized whole of the populace, who realize the production-labor activity, that is, not abstract *man*, but population." *Sovetskaya geografiya itogi i zadachi* [Soviet Geography: Accomplishments and Tasks] (Moscow, 1960), p. 235.

4. Compare also: "Production is directly consumption, consumption is indirectly production . . . . Intermediate between the two is movement." Marx and Engels, *Sochineniya*, 12:7.

5. V. V. Pokshishevskiy, "Content, Tasks, and the Place of Population Characteristics in Economic-Geographic Monographs, in *Voprosy geografii naseleniya USSR* [Problems of Population Geography of the SSSR] (Moscow, 1961).

6. Such a definition was given in Volume I of the *Kratkaya geograficheskaya entsiklopediya* [Short Geographic Encyclopedia,] 4:428–29.

7. Lenin, *Pol'nyyee Sochineniya*, 27:386.

8. V. V. Pokshishevskiy, "The Present State and Immediate Tasks of Population Geography," in *Materialy ko Vtoromu s'ezdu geograficheskogo obshchestva* [Material on the Second Congress of the Geographical Society] (Moscow, 1954).

9. I. M. Mayergoyz and B. S. Khorev express doubt that the first is only a part or branch of the second; A. M. Kolotiyevskiy also insisted on the independence of *demogeography*, as he prefers to call population geography.

10. N. N. Baranskiy, *Metodika prepodavaniya ekonomicheskoy geografii* [Method of Teaching Economic Geography] (Moscow, 1960).

11. O. A. Konstantinov, "History of the Formation of Urban Geography as a Special Branch of Geographical Knowledge in the USSR," *Materialy po geografii naseleniya* [Material on Population Geography], no. 1, (Leningrad, 1962).

12. This demarcation is necessary in order to secure the most successful cooperation between the creative efforts of geographers and ethnographers. Proceeding from the fact that a positive statement of constructive problems is most important for the development of science, we think that it is essential to define the range of questions that lie in the sphere of interest of both the geographers and ethnographers. In particular, one can point out that this range of questions is especially broad for some foreign geographers and ethnographers, particularly in the countries that have been backward and have freed, or are freeing, themselves from the yoke of colonialism.

13. We think that demography is, above all, a section of statistics. The opposite position—understanding it as *all encompassing* science of population, the subject of which are all phenomena, processes, and their regularities in the field of numbers, structure, distribution, and dynamics of population—was held by B. Ts. Urlanis, but was not supported by the majority of Soviet demographers.

14. V. K. Yatsunskiy, *Istoricheskaya geografiya* [Historical Geography] (Moscow, 1955), p. 10.

15. V. V. Pokshishevskiy, "The Interaction of Medical Geography and Population Geography," in *Problemy meditsinskoy geografii* [Problems of Medical Geography] (Leningrad, 1962).

16. This direction, in turn, encompasses (a) the study of rural (not urban) forms of settlement as a whole, (b) study of forms of agricultural settlement, and (c) study of forms of different rural nonagricultural settlement (French authors call it 'rurale non agricole') (lumbering, fishing, transport, resort, and other small settlements in rural areas).

17. The fourth and fifth directions are often practically connected.

18. This direction serves all preceding ones, and also has its own problems.

19. V. V. Pokshishevskiy, "Population Geography and Its Task," in *Izvestiya akademiya nauk SSSR,* seriya geografiya [Proceedings of the Academy of Sciences of the USSR, Geography series], no. 4 (1962).

20. We do not touch upon the earlier works of our geographers here; such an examination shows that the geographical study of population attracted great attention as early as the eighteenth century (the works by Tatishchev, Lomonosov, Zuyev, Rychkov, Krasheninnikov) and also the whole of the nineteenth century (works of Arsen'yev, P. P. Semenov-Tyan-Shansky, and others).

21. A. I. Voyeykov, "Distribution of the Population of the Earth in Dependence on Natural Conditions and the Activity of Man," in *Izvestiya Rossiyskogo geograficheskogo obshchestva,* [Proceedings of the Russian Geographical Society], nos. 2–3, (1906); and A. I. Voyeykov, "Density of Settlements in European Russia and Western Siberia," *Izvestiya Rossiyskogo geograficheskogo obshchestva,* nos. 1–2 (1909).

22. V. P. Semenov-Tyan-Shansky, *Antropogeografiya tsentral'noy promyshlennoy oblasti* [Anthropogeography of the Central Industrial Oblasts] (Leningrad, 1924), and *Rayon i strana* [Region and Country] (Moscow-Leningrad, 1927).

23. A. A. Kruber, *Obshcheye zemlevedeniye* [General Earth Science] (Moscow, 1922), part 3, book 3, *Antropogeografiya.*

24. L. D. Sinitskiy, *Ocherki zemlevedeniya (Antropogeografiya)* [Essays on Earth Science (Anthropogeography)] (Moscow-Petragrad, 1923).

25. I. Yamzin and V. Voshchinin, *Ucheniye o kolonizatsii i pereselenii* [The Study of Colonization and Migration] (Moscow-Leningrad, 1926).

26. N. N. Baranskiy, "The Study of Countries and Physical and Economic Geography," *Izvestiya Vsesoyuznogo geograficheskogo obshchestva* [Proceedings of the All Union Geographical Society], no. 1 (1946), p. 12.

27. N. N. Baranskiy, "Concerning the Economic-Geographic Study of Cities," in *Voprosy geografii* [Problems of Geography], no. 2 (1946).

28. It is especially important that N. N. Baranskiy connected this plan with the idea of the "economic-geographical position" worked out in 1939. O. A. Konstantinov soon afterwards applied this idea to cities; later many other geographers used this idea successfully in analyzing cities, series of cities and rural places; moreover, the idea itself became even more refined.

29. R. M. Kabo, "Elements of the Geographical Study of Population in the USSR," in *Geografiya v shkole* [Geography in School], no. 3 (1941).

30. R. M. Kabo, "Construction of a Program for a Course on the Population Geography of the USSR," in *Vtoroy vsesoyuznyy geograficheskiy s' yezd* [Second All-Union Geographical Congress] (Moscow-Leningrad, 1947).

31. For example, "geographical investigations of population do not fit into the framework of separate existing disciplines (economic geography, ethnography, history and others)," "in the regional section (or the course on geography of population) interrelations of people and nature are examined." A number of similar formulations of the Second Geographical Congress caused serious objections. R. M. Kabo at one time proposed the debatable idea of a special "socio-cultural geography, the subject of which must be nature and man in their mutual relations."

32. For this it is helpful to turn to his important practical work (the organization in 1945 of a commission on the geography of population in the Moscow branch of the Geographical Society, and the direction of many field studies resulted in a special collection).

33. N. I. Lyalikov, *Geografiya naseleniya SSSR* [Population Geography of the USSR] (Moscow, 1946).

34. N. I. Lyalikov, "Some Questions of Population Density in Geographical Literature," *Voprosy geografii*, no. 5, (1947), and "Sketches on the Population Geography of the USSR," *Geografiya v shkole*, nos. 3 and 5 (1948), nos. 2 and 3 (1949).

35. L. E. Iofa, *Goroda Urala* [Cities of the Urals] (Moscow, 1951).

36. R. M. Kabo, *Goroda Zapadnoy Sibiri* [Cities of Western Siberia] (Moscow, 1949).

37. V. K. Yatsunskiy, "Changes in the Distribution of Population of European Russia in 1724–1916," *Istoriya SSSR* [History of the USSR], no. 1 (1957).

38. V. V. Pokshishevskiy, "The Role of Population Geography in Questions of the Economic Regionalization of the USSR," in *Materialy k tret'yemu s'yezdu geograficheskogo obshchestva SSSR* [Materials of the Third Congress of the Geographical Society of the USSR] (Leningrad, 1959).

39. A very detailed bibliography of Soviet works on population geography, containing about 1200 items (only for 1955–61), has been published by S. S. Khorev. A special survey of works published in publications of higher schools of learning was done by S. A. Kovalev. Rather substantial for its time (but now, of course, out of date) is the examination of Soviet work on the geography of population that had appeared up to 1956–57, by L. Kosinskiy, covering 127 works. However, a perplexing disappointment is the range of Soviet works included on geography of population is the recently published special international bibliographic handbook by the American investigator W. Zelinsky (he gave only 32 items for the whole Soviet period).

40. In such monographs almost always "an important place is occupied by the chapter on geography of population containing information on its composition and dynamics, distribution and settlement, and populated places. To man, as the main productive force, much emphasis is rightly given in all other chapters. However, the quality of such characteristics is not always satisfactory from the point of view of the special demands on population geography."

41. The survey of elements of population geography in 17 large regional works on economic geography of the USSR was comparatively recently done by V. S. Varlamov. V. S. Varlamov, "Population Geography in New Monographs on Economic Regions and Union Republics," in *Voprosy geografii*, no. 45 (1959).

42. The notion of settlement is one of the main concepts of population geography. This concept is very capacious; it expresses simultaneously the process of settlement on a territory (in its complex conditionality by historical-economic premises), as well as the result of this process itself.

43. V. G. Davidovich, *Rasseleniye v promyshlennykh uzlakh* [Settlement in Industrial Centers] (Moscow, 1960), p. 17.

44. In the practice of long-range calculations of population for the needs of city building and planning this group has received the designation *city generating*; the name "demo-generating" used by us (which also was encountered in earlier planning literature) has a larger range of population of whole areas or regions and, as we think, can be included profitably in the scientific terminology of population geography.

45. These principles, formulated in the USSR by D. I. Sheynis and G. V. Sheleykhovskiy in connection with practical planning works on city building as early as the thirties, and since then generally accepted, bourgeois geographers have come to know only quite recently. Indicative is the work of the Swedish geographer G. Alexanderson who with the Americans, J. Alexander and H. Hoyt, has worked out a special conception of "basic and nonbasic branches" in cities. In 1956, Alexanderson wrote that the "classification of cities based on their economic structure and on definite statistical indices—is the child of recent years." It must be noted that with all the valuable concrete analysis performed by bourgeois geographers, in the sense of "basic and nonbasic branches," due to the ideological positions they hold they are not able to reveal social laws that determine the picture under analysis; they also are silent on many "troubles" of the capitalistic systems. Perfection of this concept is directed only toward excellence in the technique of statistical analysis, which some authors develop into a real game of numbers.

46. *Programma Kommunisticheskoy partii Sovetskogo Soyuza* [Program of the Communist Party of the Soviet Union] (Moscow, 1961), pp. 94–95.

47. "In order to produce, people enter into definite ties and relations, and only within the

limits of these social ties and relations does their relation to nature exist." Marx and Engels, *Sochineniya*, 6:441.

48.   The author is referring here to Anuchin's book *Teoreticheskiye problemy geografii* [Theoretical Problems of Geography].

49.   *Programma Kommunisticheskoy partii Sovetskogo Soyuza* [Program of the Communist Party of the Soviet Union](Moscow, 1961), p. 128.

50.   Experience accumulated in the course of territorial work on the labor balance in Siberia, the Baltic area, and other regions, related to questions of industrial distribution, has shown the significance of a more detailed local structure of these balances, which presumes a detailed study of the primary forms of settlement. The concept of settlement around which contemporary population geography is based has thus acquired a central significance also in the solution of balance problems. Moreover, on the character of settlement (concentration in cities or dispersion in rural areas, city size, and so forth) depends, to a large degree, on the composition of the population, the correlation between demo-forming and demo-servicing populations, the development of the service sector, and similar elements. A small city differs essentially from a large one in the very structure of its population, and both differ from a worker settlement and rural settlements (agricultural as well as nonagricultural). Here we are at the very sources of those socioeconomic phenomena that determine not only the design of the settlement but also many features of the economic fabric of an area, region, or country.

51.   *Tezisy dokladov Mezhudvedomstvennogo soveshchaniya po kapitalisticheskikh stran* [Theses of the Papers of the Inter-Disciplinary Conference on the Geography of Capitalist Countries](Moscow, 1961), pp. 136–43.

52.   *Programma Kommunisticheskoy partii Sovetskogo Soyuza* [Program of the Communist Party of the Soviet Union](Moscow, 1961), p. 129.

G. S. NEVEL'SHTEYN

Territorial Differences in the Natural

Growth of Population in the USSR

Changes in the distribution of the population on the territory of a country are due to mechanical (migratory) as well as natural changes in the population. Therefore, the study of *territorial characteristics* of demographic processes of birth and death rates and the natural increase of population is one of the tasks of population geography. This study must include mapped data on birth and death rates and natural growth of population by administrative-territorial units and economic regions, the clarification of factors explaining the differences of these processes, and the establishment of features manifesting the general law of population (corresponding to the method of production) in individual regions of the country.

The study of these questions is of special importance for the USSR, since the individual regions of our country are characterized by very great differences in their histories, economies, and natural resources. These differences lead to special features in the manifestation of the socialist law of population under the actual conditions of the various regions of the Union.

Before examining the territorial differences in the natural changes of the population of the USSR, we shall discuss briefly the principles

Translated from *Geografiya naseleniya SSSR* [Population Geography of the USSR] (Moscow, 1964), pp. 144–59.

of population growth in socialist countries. The essence of the socialist law of population has been discussed in a number of works by Soviet authors, who have developed the classics of Marxism-Leninism on this question. In the works by B. Ya. Smulevich,[1] A. D. Kuznetsov,[2] M. Ya. Sonin,[3] A. A. Dol'skaya,[4] D. I. Valentey,[5] and others, the socialist law of population is formulated in the following manner: "the law of population of socialism is expressed, first, in the full employment of the entire able-bodied population . . . at its rising cultural-skill level . . . , second, in a new type of reproduction of the population, in an expanded reproduction of the population based on a low mortality rate."[6] In addition, one of the basic features of this law is the "specificity of the character of settlement for socialism."[7]

The socialist law of population prevails at the present time in countries in which, before the victory of the socialist revolution, the capitalist law of population was in effect. The effect of this law led to considerable differences between the various countries and various regions of large countries, which were at different stages of capitalistic development with regard to the type of population reproduction. One must differentiate between (1) countries (regions) with a reduced reproduction of population (low birth rate, comparatively low death rate, and the development of depopulation processes); (2) countries (regions) with expanded reproduction of population based on "rapid replacement of generations" (indices of the birth and death rates are very high); for a number of reasons, after World War II the death rate in these countries fell, but the average life span remains short; (3) countries (regions) at the various stages of transition from the second group to the first.[8]

In prerevolutionary Russia the process of population reproduction was characterized by a rapid replacement of generations, typical for backward agrarian countries. During the period 1861–1913, the crude birth rate in 50 guberniyas of European Russia was 48.9 per 1,000, the death rate 34 per 1,000, and crude natural increase 14.9 per 1,000. The fluctuations of these indices were quite considerable in the various guberniyas. A number of investigations of this question by S. A. Novosel'skiy,[9] P. I. Kurkin,[10] A. G. Rashin,[11] and others show that there existed a definite relationship between the level of birth and death rates in the various guberniyas and the composition of their population with respect to sex and age, correlation of urban and rural population, the character of production relations in agriculture, the degree of industrial development, handicrafts and seasonal work, provision of medical help for the population, and other factors.

Thus the following relationship was formulated: the more industrialized the district, the longer marriage is delayed until the more mature years and the lower the indices of birth and natural population increase. The data for Yaroslavl' Guberniya showed that the development of seasonal work had an influence on the lowering of the birthrate. The materials for Voronezh Guberniya showed that the natural increase of population in each higher economic group increases in connection with the rise of economic strength of the economy.

Before the revolution in the European part of Russia, regions were distinguished by the most sharply expressed rapidity in the renewal of generations (the Ural and Lower Volga regions) and regions with an exceedingly low natural population growth (Baltic, Lake, and Lithuanian regions). In the Left-bank and Steppe Ukraine, in the Central agricultural region, and in Belorussia all demographic indices were above the average level. In most of the regions the natural population growth was higher in rural localities than in the cities; but in the Baltic area, in rural localities the natural increase of the population was 30 percent below that of the cities. This was connected with prevalence of individual farmstead agriculture and the development of a hired labor system in the countryside. In the Baltic area the period of the capitalist character of population reproduction remained longer than in most other regions of the country. Therefore, bourgeois Latvia was referred to as one of the most decrepit countries and resembled France in this respect.

In the Soviet Union the crude birthrate, compared to 1913, has dropped from 47 per 1,000 to 23.8 per 1,000 in 1961 (the average for 1957–61 was 24.9 per 1,000). The decrease in the birthrate was about 53 percent. At the same time the death rate per 1,000 inhabitants dropped from 30.2 in 1913 to 7.2 in 1961, that is, by 77 percent (the average for 1957–61 was 7.4). As a result, the natural population growth in our country increased for this period from 16.8 per 1,000 to 17.5 per 1,000 (average for 1957–61), that is, by 4.1 percent. At present the USSR has the highest birthrate of all industrial countries and one of the lowest death rates in the world. The infant mortality rate has dropped sharply: 133.0 deaths of children between age 0 to 4 per 1,000 born in the period 1896–97; in 1960–61, 9.9 died per 1,000. In 1913, 273 children less than one year old died [per 1,000 live births]; in 1961, 32.[12] The average life span, which was 32 years in 1896–97 in the 50 guberniyas of European Russia, reached 69[13] years in 1958–59 for the whole territory of the USSR. Thus, in the USSR as a whole the most progressive type of population reproduction has been realized.

The main reasons for the decline in the birth rate in the USSR was the lowering of infant mortality, the mass migration of rural population to the cities, and the wider involvement of women in industrial production. The percentage of urban population from 1926 to 1961 increased from 18 to 51 percent. The proportion of women, workers, and service personnel in industry was 28 percent in 1929; in 1961, 45 percent;[14] and for the whole national economy 27 percent and 48 percent respectively.[15] A sample investigation of women 15–49 years of age in the prewar period showed that there were 87.7 births per 100 women employed in production and 169.1 for unemployed women.[16]

With regard to the sex composition of the population, the lower birthrate was not connected with its change. So, from 1926 to 1939, the birthrate dropped from 44.0 to 31.3 per 1,000, but the proportion of women in the entire population remained almost unchanged (51.7 and 52.1 percent). Between 1939 and 1959 the birthrate dropped from 31.3 to 25 per 1,000, but the proportion of the basic child-bearing contingent (age groups 20–49 years) increased somewhat from 42.1 to 44 percent.[17]

Thus, the dynamics of birthrate and natural increase in the population are determined by the totality of a number of factors; moreover, at the various stages of development the significance of individual factors stands out. This contradicts the assertion by A. D. Kuznetsov that "the main factor determining the natural change of population during socialism . . . must be considered the degree of provision for material and cultural needs of the population."[18] In the opinion of Kuznetsov the higher the degree of provision for the needs, other conditions being equal, the wider the limits of natural growth in the total number of population and birthrate. By degree of provision for material and spiritual needs he means "the relation of consumption to the needs and demands."[19] Kuznetsov similarly explains the differences in the natural changes of population among the various economic regions.

Our country is interested in the rapid growth of the population. Therefore, the Soviet government conducts a policy of encouraging the growth of the birthrate on the basis of "conscious motherhood." This means that the birthrate increase is not encouraged in itself, but is encouraged under the condition of material security of upkeep and education of the children. Khrushchev said that "one must have in the family at least three children and rear them well. . . . If we were to add to the 200 million still 100 million more, even that would not be enough."[20]

Although there are general principles of population growth for the whole country there is a variety of concrete forms of their manifestation in the various regions of the Union. In our demographic, economic and geographical literature these differences between the various regions have not been analyzed, although suitable data have been published in recent years in oblast and republic statistical publications. One can note only individual comments on this question. Thus, in the work by M. Ya. Sonin it was pointed out that in 1957 the natural population growth in the Far East was 20–20.5 per 1,000 and in the Lithuanian SSR, 11.9; in Moscow Oblast, 13.1; in Leningrad Oblast 15.5;

and the Union average was 17.[21] But no explanations for these differences are given. In the work by I. Yu. Pisarev published in 1962,[22] there is a table with a demographic index by gubernias of European Russia that shows that the natural population growth is higher in the eastern regions and regions where communal landownership prevailed. For the Soviet Union the question concerning territorial differences in the natural population change has not been examined at all. Yet such a territorial (geographical) analysis makes it possible not only to show and explain the differences in demographic indices by regions of the country but to a certain degree makes it possible to check the validity of the statements about general factors influencing the dynamics of these indices in the country as a whole.

In order to study the territorial features of these processes in the USSR we utilized the data on the magnitude of birth and death rate and the natural population growth in the various republics, krays, and oblasts printed in local statistical yearbooks published by the agencies of the Central Statistical Administration from 1956 to 1962. Fifteen handbooks on Union republics and 58 handbooks on the autonomous republics, oblasts, and krays of the RSFSR and the city of Leningrad were utilized. There are no data on some northern regions (Yakut ASSR, Magadan and Kamchatka oblasts), for parts of the administrative-territorial units of Siberia and the Far East, the Center and Chernozem Center; in all [no data] for 14 krays and oblasts and one autonomous republic. Consequently, data used for all union republics and for the RSFSR account for four-fifths of all administrative-territorial units. For all the remaining territories, except the Yakut ASSR, Magadan and Kamchatka oblasts, indices for adjacent territories are assigned (for example, for Vladimir and Ivanovsk oblasts the data on the Yaroslavl' Oblast are used; for Vornezh Oblast, those from Lipetsk Oblast; for Kursk Oblast, those for Belgorod Oblast; for Altay Kray and Omsk Oblast, those for Kurgan Oblast, and so forth). In addition, the data on the composition of the population published in the 1959 census were utilized (in the summary edition and publications for the various republics). The analysis of these materials had a preliminary character.

For the various regions of the USSR the birthrate fluctuates from 45.6 per 1,000 in Pavlodar Oblast of the Kazakh SSR to 16.6 per 1,000 in the Latvian SSR. The highest birthrate (33–41 per 1,000) is in the republics of Central Asia, Kazakhstan, and the republics of Transcaucasia, excluding the Georgian SSR. Of these seven Union republics the Turkmen and Azerbaydzhan SSR have the highest birth coefficients. In the rest of the union republics the birthrate fluctuates from 16.6 to 29.3 per 1,000, the Latvian and Estonian SSRs have the lowest level and the Moldavian SSR has the highest level in this group of union republics.

Within the RSFSR, the birthrate index has a tendency to rise from the west and center to north and east. Thus, in the majority of krays and oblasts of the Urals, Siberia, and Far East it fluctuates between 26 and 31 percent, and in the majority of the oblasts of the Center, Northwest, Central-Chernozem region and the Northern Caucasus it is on the order of 20–22 percent. The oblasts of the Volga-Vyatka and Volga regions are between these two groups of oblasts and krays of the RSFSR. The national autonomous regions in the most diverse regions of the country (Dagestan, Kalmykia, Komi ASSR, Bashkiria, Chechen-Ingush ASSR, Chuvashia) are distinguished by their high birthrates in the RSFSR. The lowest birthrates in the RSFSR are in Pskov, Saratov, and Rostov oblasts and in Krasnodar Kray. In Kuybyshev and Volgograd oblasts, adjacent to Saratov Oblast, the birthrate is notably higher (24 per 1,000 as against 20 per 1,000).

The fluctuations of the mortality indices for the various regions are much smaller than the fluctuations of the birthrate index. Everywhere the index has dropped sharply as compared with the prerevolutionary period, and continues to drop. For the largest part of the territory of the Union this index varies between 7 and 9 per 1,000. In the RSFSR the lowest mortality indices are in the autonomous republics of the Northern Caucasus (Kabardino-Balkar, Northern Ossetian, and Chechen-Ingush republics) and in some northern regions (Sakhalin and Murmansk oblasts)—on the order of 6 per 1,000. Only in a very few oblasts is this index higher than 9 per 1,000, reaching 10 per 1,000 in Pskov Oblast. No definite territorial regularity can be detected in the change of the mortality coefficient. This coefficient is lowest in Tadzhikistan and highest in Estonia.

Since the fluctuations of the mortality index in the majority of regions is relatively small, the changes in the level of natural growth of population in the country repeat to a considerable degree the changes in the birthrate level. The index of natural population growth varies in the republics and oblasts from 5.8 per 1,000 in the Estonian SSR to 34.3 per 1,000 in the Turkmen SSR. In the RSFSR many of the central regions as well as the territory of the Northern Caucasus, excluding the national autonomous regions, have a lowered level of natural population growth. The highest level in the RSFSR is found in Siberia and the Far East, the northern territories of the European part of the USSR, and also the Urals. It must be noted that in prerevolutionary Russia, Siberia and the Far East also had a high rate of natural population growth.

Using approximate calculations based on the statistical yearbooks mentioned above, the natural population growth in the Union republics and large economic regions can be determined as shown in table 1. Examining the coefficient of natural population growth in the large

TABLE 1

NATURAL GROWTH OF POPULATION IN THE UNION REPUBLICS AND
LARGE ECONOMIC REGIONS (AVERAGE FOR 1957-1961)

REPUBLICS AND REGIONS	PROPORTION OF POPULATION OF THE USSR ACCORDING TO CENSUS OF 1959 (In Percent)	ANNUAL RATE OF NATURAL INCREASE (In Percent)	PROPORTION OF THE NATURAL INCREASE OF THE USSR (In Percent)	INDEX OF VITALITY OF THE POPULATION
RSFSR.	56.4	15.4	52.7	3.05
Northwest.	5.5	13.6	4.7	2.70
Center	10.7	12.0	8.2	2.51
Volga-Vyatka region. .	4.4	17.5	3.5	3.08
Chernozem Center . .	5.0	15.2	3.5	2.97
Volga region	6.0	16.5	5.5	3.06
Northern Caucasus. .	5.6	16.1	5.5	3.22
Urals.	8.4	19.1	10.1	3.39
Western Siberia. . .	5.4	20.0	5.5	3.33
Eastern Siberia. . .	3.3	20.1	3.5	3.57
Far East	2.1	22.3	2.7	4.28
Ukrainian SSR. . . .	20.0	13.2	14.5	2.86
Belorussian SSR. . .	3.8	17.5	3.3	3.53
Baltic	2.9	9.0	2.1	1.90
Lithuanian SSR . . .	1.3	12.9	1.2	2.59
Latvian SSR.	1.0	6.4	0.6	1.63
Estonian SSR	0.6	5.8	0.3	1.52

TABLE 1--Continued

REPUBLICS AND REGIONS	PROPORTION OF POPULATION OF THE USSR ACCORDING TO CENSUS OF 1959 (In Percent)	ANNUAL RATE OF NATURAL INCREASE (In Percent)	PROPORTION OF THE NATURAL INCREASE OF THE USSR (In Percent)	INDEX OF VITALITY OF THE POPULATION
Moldavian SSR.	1.4	22.6	2.0	4.37
Transcaucasia.	4.5	27.6	6.5	4.70
Georgian SSR	1.9	18.0	1.7	3.50
Azerbaydzhan SSR	1.8	35.3	3.3	5.86
Armenian SSR	0.8	32.6	1.5	4.64
Central Asia	6.5	33.8	11.5	6.04
Uzbek SSR.	3.9	31.5	5.7	6.08
Tadzhik SSR.	0.9	28.2	1.6	6.32
Kirgiz SSR	1.0	29.0	1.7	5.60
Turkenian SSR.	0.7	34.3	1.5	6.12
Kazakh SSR	4.5	29.7	7.4	5.07
Total for the USSR. .	100.0	17.5	100.0	3.46

economic regions the same spatial regularities mentioned above can be noticed: (1) a high coefficient in the republics of Central Asia, Transcaucasia, Kazakhstan, and Moldavia; (2) the average level of this index in the RSFSR and Belorussia; (3) a lower index in the Ukrainian SSR and a sharply lower one in the Baltic region; (4) a higher level of this index in the eastern and northern regions of the RSFSR than in its other parts. Within the individual large economic regions great fluctuations of natural growth can be observed, greatest where the composition of the region has autonomous national units. The Central and Central-Chernozem regions as well as Western Siberia and the Far East have the most uniform index.

With regard to the vitality of the population (the relation of the number born to the number that died), it is highest in the regions of a high birthrate and lowest in Estonia and Latvia with their very low birthrate.

What factors can explain the noted territorial differences? A comparison of the birthrate level and natural population growth with the population composition and the character of economic development of the individual republics and oblasts shows that this level depends on a large number of factors: age and sex structure of the population, the marriage coefficient, the relationship of urban and rural population, and its provision with agricultural lands, the degree of involvement of women in social production, historical conditions of population formation, and the economy of a given oblast of a republic. Of these factors the most important are the influence of the age structure of the population, the proportion of women in the structure of the entire population, and the proportion of women aged 16 and older who are married. In the Union republics these indices vary as shown in table 2.

A high birthrate and the high natural population growth are noted in the southern union republics with their earlier marriage age and traditions of large families. The high number of married people among the adult population is related to this. In these republics the male and female populations are more balanced than in the Baltic republics, the Ukrainian SSR and other areas with their lower population growth and lower birthrate. Between the level of birthrate and the portion of women in the age group 20–49 years there is a slight inverse relationship: in the republics with a birthrate above the Union average this age group's proportion is below the average Union index. However, this inverse relationship in republics with a birthrate below the Union average is less pronounced. Thus, in Lithuania and Estonia both these indices are below the Union average level. In these republics there is a large percentage of the population in the age groups above 49 years, especially women. Their proportion among the female population of Lithuania and Estonia is 26–27 percent as against 18.6 percent in the Union.

TABLE 2

RELATIONSHIP BETWEEN BIRTHRATE AND CERTAIN OTHER DEMOGRAPHIC INDICES

UNION REPUBLIC	BIRTHRATE (Per 1,000)	NUMBER OF WOMEN AGES 20-49 (In Percent)	TOTAL NUMBER OF WOMEN (In Percent)	MARRIAGE RATE PER 1,000 WOMEN, 16 YEARS AND OLDER	AVERAGE SIZE OF FAMILY
Azerbaydzhan.	42.1	40.4	51	58.3	4.5
Turkmenia	41.0	40.2	53	63.7	4.5
Armenia	39.5	41.6	51	59.7	4.8
Uzbekistan.	37.7	38.2	52	63.0	4.6
Kazakhstan.	37.0	41.5	53	58.1	4.1
Kirgizia.	35.3	39.4	52	61.6	4.2
Tadzhikistan.	33.5	46.0	51	64.9	4.7
Moldavia.	29.3	43.5	55	60.1	3.8
Georgia	24.6	44.1	55	55.7	4.0
Belorussia.	24.4	43.1	56	55.0	3.7
RSFSR	22.9	45.2	55	50.5	3.6
Lithuania	21.3	42.7	55	52.5	3.6
Ukraine	20.3	45.6	57	51.8	3.5
Estonia	16.9	42.6	58	48.8	3.1
Latvia.	16.6	43.2	57	50.0	3.2
Average for the USSR.	24.6	44.5	55	52.2	3.7

SOURCE: Zhenshchiny i deti v SSSR: Statisticheskiy sbornik [Women and Children in the USSR: Statistical Collection] (Moscow, 1961). Itogi vsesoyuznoy perepisi naseleniya 1959 goda. SSSR (svodnyy tom) [Results of the All-Union Population Census of 1959: USSR Composite Volume] (Moscow, 1962).

The relationship between the birthrate and natural increase and the proportion of urban population is very complex. Thus, in the city of Saratov the natural growth is 9 per 1,000 compared to 13 per 1,000 in Saratov Oblast; in the city of Gor'kiy, 12.6 per 1,000 compared to 15.2 per 1,000 in Gor'kiy Oblast; in Baku, 18.3 per 1,000 compared to 35.3 per 1,000 in Azerbaydzhan; in Yerevan, 24.3 per 1,000, compared to 39.6 per 1,000 in Armenia; and in Leningrad, 6.2 per 1,000, compared to 15.5 per 1,000 in the oblast. There are also other data, for example, in Riga the natural increase is 6.7 per 1,000 compared to 6.4 in Latvia. A number of oblasts in the RSFSR with a very high percentage of urban population have a high level of birthrate and natural growth. Thus, Murmansk Oblast (95 percent urban population), Kemerovo Oblast (79 percent), the Primorskiy and Khabarovsk krays (70–77 percent), and others have a birthrate and natural increase significantly higher than the Union average. In regions of new industrialization, high urbanization is not an impediment for a high level of natural population growth because of the great predominance of young age groups, a considerable population influx from other regions, and a higher wage level. In the regions of older settlement there is no direct relationship between the percentage of urban population and birthrate level and natural growth. An example of this can be seen in the Central Chernozem and Volga regions. Voronezh Oblast with 39 percent urban population has a natural growth of 16.5 per 1,000, and Kursk Oblast, with 24 percent urban inhabitants, only 13.8 per 1,000. Ul'yanovsk Oblast with 40 percent urban population, has almost the same index of natural increase (16 per 1,000) as Kuybyshev Oblast with 65 percent urban population. In conformity with this, the correlation coefficient between the two indices [birthrate and urban population] is not high (0.164).[23] We note that this coefficient calculated by K. Witthauer by the same method for 55 foreign countries was 0.43 (16).[24]

Between the density of population (total and rural) and birthrate there is no clearly expressed relationship. There is a high density of rural population in Moldavia, Azerbaydzhan, and Armenia, and for them the birthrate and the natural increase higher than the average for the country is characteristic. On the other hand, the Ukraine (especially the Right-Bank Ukraine) and the Central-Chernozem region, with a high density of rural population, have a birthrate and natural growth rate below the average level. Evidently this is related to a difference in specialization of agriculture. Hardly any of the rural population leave from the regions of labor-consuming agriculture. The dynamics of rural population, however, are of great significance for natural growth. For all regions with a decreasing rural population, a high proportion of lower indices of natural growth is characteristic: the oblasts of Pskov, Smolensk, Vologda, and Ryazan' serve as examples. In the oblasts of Kazakhstan and Uzbekistan, however, with a high percentage of rural population, the rural population grows and the index of natural growth is at a high level; the same is true for Armenia and Tadzhikistan, and other regions. It is characteristic that in Uzbekistan the birthrate in the regions of irrigation agriculture is higher than in regions of dry farming.

Within the limits of oblasts and republics, and within one natural-economic zone having common features of agricultural specialization, one can observe a certain relationship between the level of the birthrate and the natural population growth, on the one hand, and the agricultural lands and sown area provided for the rural population, on the other. Thus, in the Georgian SSR, where the natural increase is half that of the neighboring republics of Transcaucasia, there are 1.1 hectares of agricultural land per capita (compared to 1.6 in the Armenian and 2.0 in the Azerbaydzhan SSRs) and 0.33 hectares of sown area (compared to 0.6 in Armenia and Azerbaydzhan).[25]

The regions that before the revolution were industrially developed and had a high level of urbanization have, as a rule, a lower natural population growth. This historical inertia is now less significant than the conditions of present development of the socialist economy and socialist culture. However, the correlation of levels of natural population growth of Saratov Guberniya and the neighboring Samara Guberniya [Kuybyshev Oblast] observed before the revolution has remained the same. For the period from 1881 to 1913, the average yearly natural population growth in Saratov Guberniya was 14.5 per 1,000 and in Samara Guberniya, 17.0 per 1,000; at present, this index is also lower in Saratov Oblast than in Kuybyshev Oblast by approximately 25 percent.

The employment of women in the national economy and the distribution of employed women by social groups and branches of the economy have a great influence on the territorial differentiation of demographic indices. Thus, among the union republics, Estonia (45.1), Latvia (44.4), and Belorussia (48.6) have, as was pointed out, a lower population growth with their high percentage of employed women. In Georgia this index is higher that in Azerbaydzhan and Armenia (38.8 per 1,000, as against 34.5 per 1,000 and 31.7 per 1,000).[26]

It is characteristic that in the composition of the total number of women having an occupation the proportion of workers (nonagricultural) considerably exceeds the proportion of collective farm women in the republics with a lower natural population growth.[27] An inverse situation is typical for republics with a high natural population growth. Georgia, Estonia, and Latvia have the highest percentage of women with a higher, incomplete higher, and specialized secondary education (index of) 8–9; in Azerbaydzhan, Turkmenia, and Moldavia, this index is 5–6.

TABLE 3

	PROPORTION OF WORKERS (In Percent)	PROPORTION OF COLLECTIVE FARM WOMEN (In Percent)
Estonia.	46	25
Latvia	42	31
Armenia.	29	48
Turkmenia.	26	53
Azerbaydzhan . . .	23	57

The republics of Central Asia and the eastern regions of the RSFSR having the highest coefficient of natural population growth, have the highest money income per collective-farm household. Thus, in 1961 this index for Turkmenia was 1,771 rubles; for Uzbekistan, 1,544 rubles; for Tadzhikistan, 1,213 rubles; and for the Kirgizh SSR, 1,205 rubles; in the Far East it reached 2,324 rubles, and in Eastern and Western Siberia, about 1,300 rubles. The Union average was 830 rubles. The index is below average in the Ukraine, Belorussia, Georgia, Lithuania, and the Center of the RSFSR. These regions have a natural population growth below the average for the country. However, this relationship is not observed everywhere, in particular, because of the different agricultural commodities of the collective farms in the various regions of the Union. Similarly, one cannot establish a direct and constant relationship between the level of wages in the various regions and the level of natural population growth.

In connection with this it must be noted that the statistical data available do not confirm the expressed positions about the presence of a reverse relationship between incomes of the working people and birthrate (and, consequently, the natural population growth). Unconfirmed also is the position that "as a main factor determining the natural change in population during socialism . . . must be considered the degree of provision for the material and cultural needs of the working people."[28] This level is higher in the Center of the RSFSR and the Ukraine, but the natural population growth is below the national average here.

In order to explain the territorial differences in the natural change of population certain basic economic factors were examined although, of course, these factors are intertwined with "a whole set of intermediate sociopsychological and all kinds of other interlinear experiences and reactions."[29]

By character and type of population reproduction in the USSR, one can distinguish the following four groups of republics and oblasts: (1) regions where to a certain extent traces of "rapid shifts of generations" remain—the high birth and mortality rates are above the national average; to this group belong some oblasts of Kazakhstan, Siberia and the Far East, the Volga-Vyatka region, the Center and the Northwest; the number of these regions is rapidly diminishing in view of their greatly lowered mortality rate; (2) regions with the most progressive type of population reproduction—a high birthrate and low mortality rate; to this group belong the republics of Central Asia, Moldavia, Azerbaydzhan and Armenia, the autonomous republics of the North Caucasus, and the Northwest; (3) regions with an average index of natural population change; to this group belong many of the oblasts of the Central-Chernozem and Volga regions as well as Georgia, the Left Bank and Steppe part of the Ukrainian SSR; (4) regions of a lower birthrate, and average or above average mortality rate and, correspondingly, a low natural population growth; to this group belong the Baltic republics, Pskov, Yaroslavl', and Saratov oblasts, and the Right Bank of the Ukraine.

The measures undertaken during the present period of development of communism to thoroughly develop the economy and cultures of all oblasts and republics and to sharply raise the material and cultural level of the whole population have already led and are leading to a territorial expansion of the second type of population reproduction, to a considerable increase in the natural population growth in the fourth group of regions, to a further lowering of the mortality rate, and to a still more considerable increase in the average life span. The Program of the CPSU, adopted at the Twenty-second Congress, projects a wide program of measures for the further increase of the life span, for the combination of "happy motherhood with a more active creative participation of women in social labor and social activity, and in occupations in science and art."[30]

The data on the dynamics of natural change of population in the regions are of important practical significance. In particular, they are taken into account when planning a school and medical network, as well as in planning migration. In a number of socialist countries these data are published in statistical publications. K. Witthauer,[31] utilizing these data on the German Democratic Republic and the Polish People's Republic, noted that in the G.D.R. there are considerable differences between the indices of natural change of population in the old industrial districts and the northern districts that have recently started to develop industrially. The average population growth in the G.D.R. is 3.6 per 1,000 (1959), but in Rostock and Neubrandenburg it reaches almost 10, whereas in Halle and Magdeburg it drops to 2.8–

3.3. In Poland, according to 1960 data, the natural growth in the Lodz district was 11.5 per 1,000 and in the Koszalin district it was twice as high; in the country as a whole it was 14.9, and in 1961, 13.1 per 1,000.

The published data on the number of the population according to the census of 15 January 1959 and the census of 1 January 1962 in the republics and large economic regions of the RSFSR give an idea of the actual growth of the population in the last three years and make it possible to compare by regions the total population growth with the natural population growth for the same period. This comparison is shown in table 4.

This table shows that the actual population growth in the RSFSR (122.1 – 117.5 = 4.6 million) is about 80 percent of the natural growth

TABLE 4

CHANGE IN POPULATION AS A RESULT OF NATURAL
GROWTH AND MIGRATION (1959-1961)
(In Millions)

REPUBLICS AND REGIONS	ACTUAL POPULATION		POTENTIAL POPULATION FOR 1 JANUARY 1962 (1959 Natural Growth for 3 Years)	DIFFERENCES BETWEEN ACTUAL AND POTENTIAL POPULATION
	15 January 1959	1 January 1962		
Northwestern. . . .	11.5	11.9	12.0	-0.1
Center.	24.8	25.2	25.7	-0.5
Volga–Vyatka. . . .	8.3	8.3	8.7	-0.4
Central Chernozem .	8.6	8.9	9.0	-0.1
Volga	12.4	13.0	13.0	. . .
Northern Caucasus	11.8	12.7	12.4	0.3
Urals	18.6	19.5	19.7	-0.2
Western Siberia . .	10.2	10.7	10.8	-0.1
Eastern Siberia . .	7.0	7.4	7.4	. . .
Far Eastern	4.3	4.5	4.6	-0.1
RSFSR	117.5	112.1	123.3	-1.1
Ukrainian SSR . . .	41.9	43.5	43.5	. . .
Belorussia SSR	8.0	8.3	8.4	-0.1
Kazakh SSR.	9.3	10.9	10.1	0.8
Central Asia. . . .	13.7	15.2	15.0	0.2
Transcaucasia . . .	9.5	10.4	10.2	0.2
Baltic.	6.0	6.3	6.2	0.1
Moldavian SSR . . .	2.9	3.1	3.1	. . .
Total for USSR .	208.8	219.8	219.8	. . .

in the republic ($123.3 - 117.5 = 5.8$ million); the other 20 percent comprises the migration of population to other Union republics. In the Belorussian Republic migration constitutes about 25 percent of the natural population growth. In all other republics except Estonia, the actual population growth is higher than the natural growth. In Kazakhstan the natural population growth ($10.1 - 9.3 = 0.8$ million) covers only 50 percent of the total population increase in this republic ($10.9 - 9.3 = 1.6$ million), which is quite understandable and expedient because of the development of virgin lands and the huge industrial building within its borders. Less explainable is the influx of population to Central Asia with its high natural population growth.

In the RSFSR the only region where the total increase in population exceeds the natural growth is the Northern Caucasus. The migration from the Center is more than half of its natural growth and in the Volga-Vyatka region it equals the total natural increase. The migration during the three years under consideration was from the Chernozem-Center, Urals, Western Siberia, and Far East. Correspondingly, the population growth in these regions is lower than in the country as a whole. Yet during the period 1939–59, the population growth in the eastern regions was considerably above the average in the country. It is an important task to rationalize the interregional migration of labor resources.

1. B. Ya. Smulevich, *Kritika burzhuaznykh teoriy i politiki narodonaseleniya* [A Criticism of Bourgeois Population Theories and Policies](Moscow, 1959).

2. A. D. Kuznetsov, *Trudovyye resursy SSSR i ikh ispol'zovaniye* [Labor Resources of the USSR and Their Utilization](Moscow, 1960).

3. M. Ya. Sonin, *Vosproizvodstvo rabochey sily v SSSR i balans truda* [The Reproduction of Manpower in the USSR and the Labor Balance](Moscow, 1959).

4. A. A. Dol'skaya, *Sotsialisticheskiy zakon narodnaseleniya* [The Socialist Law of Population](Moscow, 1959).

5. D. I. Valentey, *Problem narodnaseleniya* [The Problem of Population] (Moscow, 1961).

6. B. Ya. Smulevich, *Kritika*, p. 41.

7. D. I Valentey, *Problema*, p. 115.

8. B. Ya. Smulevich, *Kritica*, p. 64.

9. S. A. Novosel'skiy, *Obzor glaveneyshikh dannykh po demografii i sanitarnoy statistike Rossii* [A Survey of the Most Important Data on Demographic and Sanitary Statistics of Russia](St. Peterburg, 1910).

10. P. I. Kurkin, *Statistika Izmeneniya naseleniya v Moskovskoy gubernii* [Statistics of Population Change in Moscow Guberniya] (Moscow, 1902).

11. A. G. Rashin, *Naseleniye Rossii za 100 let* [The Population of Russia through 100 Years] (Moscow, 1956).

12. *Narodnoye khozayaystvo SSSR v 1961 godu: Statisticheskiyy yezhegodnik* [The National Economy of the USSR in 1961: Statistical Yearbook](Moscow, 1962).

13. B. Ya. Smulevich, *Kritika*.

14. P. I. Kupkin, *Statistika*.

15. Ibid.

16. S. G. Strumilin, *Problemy eknomiki truda* [Problems of the Economics of Labor] (Moscow, 1957).

17. *Narodnoye khozyaystvo SSSR v 1960 godu: Statisticheskiy yezhegodnik* [The National Economy of the USSR in 1960: Statistical Yearbook] (Moscow, 1961).

18. A. D. Kuznetsov, *Trudovyye*, p. 38.

19. Ibid., p. 39.

20. N. S. Khrushchev, Speech at the Conference of the Komsomol and Youth Leaving for the Virgin Lands, *Pravda*, 8 January 1955.

21. M. Ya. Sonin, *Vosproivodstvo*, p. 27.

22. I. Yu Pisarev, *Narodonaseleniye SSSR* [The Population of the USSR] (Moscow, 1962).

23. The correlation coefficient (*r*) was calculated from the formula where x_o is the deviation

$$r = \frac{\sum x_o Y_o}{\sqrt{\sum x^2 Y_o^2}}$$

of the percentage of urban population in the various regions from the average level of this percentage in the Union, Y_o is the deviation of the birthrate coefficient (in percent) in the various regions from its average level in the Union. This coefficient shows the strength (closeness) of the relationship between the two indices. Usually the closeness of the correlation is considered satisfactory if the correlation coefficient is higher than 0.5.

24. K. Witthauer, "Bevolkerung und naturliche Bevolkerungsbewegus in der DDR" [Population and Natural Population Shifts in the German Democratic Republic] *Petermanns Geogr. Mitt.*, 4th qtr. (1961).

25. The provision of agricultural land for the rural population must be compared with the natural growth of the rural population. Due to the absence of the proper data, figures for the natural growth of the entire population are used, that to a great degree reflect the natural movement (change) of population in the Transcaucasian republics.

26. *Zhenshchiny i deti v SSSR: Statisticheskiy sbornik* [Women and Children in the USSR: Statistical Collection] (Moscow, 1961).

27. Ibid.

28. A. D. Kuznetsov, *Trudovyye*, p. 38.

29. S. G. Strumilin, *Problemy*, p. 204.

30. *Programma Komunisticheskoy partii Sovetskogo Soyuza* [Program of the Communist Party of Soviet Union] (Moscow, 1961), p. 97.

31. K. Witthauer, "Bevolkerung."

V. V. POKSHISHEVSKIY
V. V. VOROB'YEV
YE. N. GLADYSHEVA
V. I. PEREVEDENTSEV

On Basic Migration Patterns

The Marxist-Leninist concept of population of the working people as the main productive force require that data about number, distribution, and structure of the population (taken over time) be made basic to an understanding of the economic geography of a country. This is especially evident in the case of the USSR and other socialist countries where cities have no unemployment and rural areas do not suffer from overpopulation, two phenomena that often threaten the direct and principal relationship between the manpower contingents and their productive potential under the conditions of capitalism.

In terms of mobility, population occupies an intermediate position among all productive forces, between natural resources, which are territorially immobile, and machinery and manufactures, which in principle have absolute mobility (although they vary in transportability). At the same time, labor can be utilized in the production process only where it is localized, a fact that distinguishes it from all materialized labor. (This excludes commuting to work within limited areas.)[1]

Territorial redistributions of population and manpower are determined in the final analysis by shifts of production, which require a

Reprinted, with modifications, from *Soviet Geography: Review and Translation* 5, no. 10 (December 1964), by permission of the publisher.

substantial amount of time. At any given period, therefore, the existing geography of population constitutes one of the main factors in the current location of production. This is especially true in the case of a socialist economy, which in principle assures full employment in any region; the concept of a population surplus in this case has a relative meaning, namely that the manpower in a surplus region can be used with less effectiveness than in other regions. The phenomenon of depressed areas, which under capitalist conditions became areas of out-migration, is unknown in a socialist economy. (The present depressed areas of the United States, the French departments of the Massif Central that are losing population, and the depressed agrarian areas of northeastern Brazil had an analogue in prerevolutionary Russia in the provinces of central European Russia, which suffered from an acute agrarian depression. Lenin wrote about these areas: "The colonization problem in Russia is subordinated to the agrarian question in the center of the country."[2] This Leninist formulation reveals an important economic law of migration under the capitalist mode of production, namely the predominance of the role of the expellant factor in migrations, rather than the principle of equilibrium between push and pull factors favored by bourgeois scholars. Lenin's law also serves to explain related processes, such as urbanization in many underdeveloped countries, which is proceeding at a scale far from commensurate with the real economic potential of the cities of these countries.) The causal factors of the migration process in the USSR are not related to unemployment and overpopulation in the areas of origin (these phenomena now exist only on the pages of Soviet textbooks) but are related to the positive aims of resource development and to the realization that the use of the high level of socialist technology combined with labor will be more effective when applied to the rich resources of newly settled areas.

The geographic redistribution of population, reflecting shifts in the geography of a growing economy, has been one of the key features of the economic development of our country, both before the revolution and during the Soviet period.

Aside from migration, this redistribution has been distinguished by another important factor: territorial differences in the rate of natural increase of the population. (The importance of this factor was first stressed in general terms by Soviet specialists on the history of population geography.)[3] The relative importance of the migration factor and the differences in the rate of natural increase have varied at different stages in history. In the period of development of capitalism in Russia, a higher rate of natural increase was evident in the areas of colonization where the economy was being developed at that time, the steppe guberniyas of New Russia and Siberia; these areas were also the destinations of the principal migration flows. During this period the area of peasant out-migration expanded as the territorial extent of the agrarian depression increased.[4] Pokshishevskiy[5] indicates for the period 1863–1913, the following balances of the principal interregional migration flows, in millions of people; Siberia, +4.2; Far East. +0.7; Urals, primarily southern Urals, +0.8; Central Asia and Kazakhstan, excluding northern Kazakhstan that was included in Siberia, +1.3; steppe guberniyas of New Russia, +3.6; Northern Caucasus, +1.5; Chernozem and Nonchernozem Center, –4.5; Ukrainian wooded steppe, –3.2; Volga, –0.8; European North, –0.5. There was also a substantial out-migration from the Baltic region and the Polish territories that were then part of Russia, but it represented mainly emigration to other countries. (This study also contains migration data for the preceding period, starting in the seventeenth century, in the tabular appendices.)

During the Soviet period, the spread of socialist conditions throughout the country eliminated the social hardships that produced low rates of natural increase in some areas in the past. The gap between rates of natural increase in urban and rural areas was narrowed. Only in periods of intensification of the class struggle do we find a temporary decline of the growth rate in areas where that struggle was particularly intense. For example, whereas the total population of the USSR rose by 16 percent between 1926 and 1939, the increase was only 1.2 percent in Kazakhstan, despite in-migration. (In Kazakhstan, intensification of the class struggle coincided with the breakup of the previously dominant nomadic herding economy; this led to the loss of much livestock and a temporary decline in living standards. Some of the former nomads settled on neighboring lands of Uzbekistan.)

Interregional migrations assumed a larger scale in connection with the industrial construction of the five-year plans. Between 1926 and 1939, about 3 million people moved to Siberia and the Far East, and more then 1.7 million to Central Asia and Kazakhstan.[6] A big percentage of the new migrants settled in cities: the proportion of urban population rose from 18 to 32 percent, with the growth especially marked in large cities. The total population of the 25 largest cities (1939) rose from 8.77 million in 1926 to 18.12 million in 1939 chiefly as a result of interregional migrations, especially to Moscow, which grew by more than 2 million, Leningrad (by 1.5 million), Novosibirsk, Gor'kiy, Sverdlovsk, Volgograd, and Zaporozh'ye (increased three to fivefold).

The years of the Great Patriotic War (1941–45) [World War II] were marked, together with large absolute population losses (from military losses, extermination of civilians by the Fascist invaders, higher mortality, and a lower birthrate), by major territorial population shifts. Most important was the temporary eastward migration from territories occupied by the Hitlerite invaders and from the front-line zone (evacuation of civilians and of industries). Between June 1941 and 1 February 1942, 10.4 million people were evacuated.[7] This figure is based on a survey conducted by the railroads. According to a survey conducted by local authorities, almost 6 million found refuge by the spring of 1942 in oblasts of the RSFSR (mainly in the Volga, Urals, and Western and Eastern Siberia), and more than 1.5 million in Kazakhstan, Central Asia, and Transcaucasia. These data are apparently incomplete because some of the evacuated people settled in the new areas without passing through organized evacuation centers). The evacuation of 1941 was followed by the second evacuation wave in 1942. During the period of maximum eastern invasion, the population of nonoccupied territories did not fall below 130 million.[8] Voznesenskiy reports that the occupied territories held 45 percent of the Soviet population before the war;[9] we can therefore conclude that the total evacuation, from June 1941 to October 1942, was close to 20 million.

Although in principle it was temporary, this huge migration resulted in the partial settling of the evacuated population in the eastern areas, especially in the Urals, Western Siberia, and Kazakhstan. For example, in 1947, when the last count of displaced persons was made in Irkutsk oblast, 15 percent of the original arrivals were still there. Reevacuation extended over many years.

Territorial differences in the natural rate of increase of the population also play a role in the war years. The eastern areas were able to compensate, in part, for their human losses by natural population growth. A net population loss applied mainly to the European part of the country, where the occupied territories accounted for three-fourths of the total loss (including displaced persons).

In the postwar years, the natural population increase was consistently higher in the eastern and southern areas than in the western areas that had suffered especially from war destruction. At the time of the 1959 census, the prewar population level had not been reached in some formerly occupied western areas: Belorussian SSR, Lithuanian SSR, Pskov, Novgorod, Smolensk, and Kalinin oblasts of the RSFSR, and some oblasts of the Ukrainian SSR. (The recovery of the Moldavian SSR resulted from a combination of a high rate of natural increase and in-migration of Russians and Ukrainians in connection with industrialization; the proportion of Russians and Ukrainians in the population of the Moldavian SSR rose from 16 to 17 percent in 1941 to 24.8 percent in 1959. In the case of Estonia and Latvia, where the rate of natural increase was low, the recovery was related to an influx of Russians, Ukrainians, and Belorussians, who constituted 22.3 percent of the population of Estonia by 1959, and 30.9 percent in Latvia. This influx was related to the industrial rebirth of the Baltic region where ports assumed a key role in the geographic division of labor.) In the East and South, consistent population increases were insured by a high rate of natural increase and, in some areas, by continuing migration into the area.

The absence of questions in the 1959 census that would permit distinguishing between the migration factor and the natural rate of increase in overall population growth makes it difficult to judge the relative importance of these two factors in the redistribution of the population, both in the first postwar five-year-plan period and subsequently. The two factors can be distinguished only indirectly by comparing the actual population growth with the rate of natural increase, but this method at least yields an order of magnitude. (Current registration data of new arrivals is also inadequate because registration in rural areas began only in 1960, having been limited previously to cities; furthermore one would have to assume incompleteness of registration since registration does not take into account the length of stay in the previous place of residence. Available data on territorial differences in the rate of natural increase were recently generalized for major regions by G. S. Nevel'shteyn.[10] The author used more than eighty statistical handbooks published in republics, krays, and oblasts in the period 1956 to 1960. He notes that fluctuations in the rate of natural increase depend mainly on birthrate levels, "which vary to a much greater extent in individual regions than do death rates.")[11]

An overall picture of the territorial redistribution of population as a result of migration over the last four years is shown in the cartograms. (These have been compiled on the basis of the migration balance, the difference between the actual population growth and the natural rate of increase.) The redistribution of population by migration in the last four years apparently amounted to as much as one-seventh of the natural increase. The first cartogram (see figure 1) suggests the approximate scale of both out-migration and in-migration in various regions of the USSR. In 1959–62, for example, Kazakhstan had an in-migration of almost one million people (representing the main interregional migration flow). In-migration was also substantial into Central Asia and in the Northern Caucasus and was relatively noticeable in the Baltic republics and Transcaucasia. Contrary to the general impression, there was no net in-migration into Siberia and the Far East, where the migration balance is now negative. (This change in the direction of migration, which is hoped to be temporary, was demonstrated in the last few years in the work of V. I. Perevedentsev.)[12]

The main source of out-migration is still the central areas of the European part of the RSFSR (especially the Volga-Vyatka region, where out-migration is almost equal to the natural increase; out-migration is also high in several oblasts of the Central Region, although

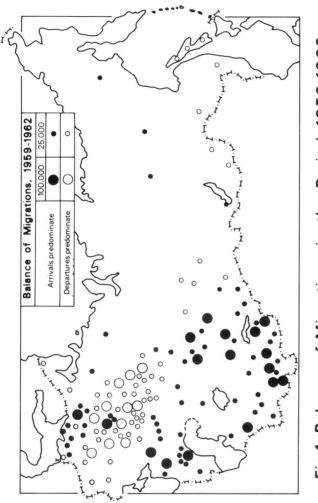

Fig.1. Balance of Migrations in the Period 1959-1962

for the region as a whole this is partially offset by the continuing attraction of population to the Moscow agglomeration). During this period, more than one million people from the central European RSFSR moved to other parts of the country. Another major source of out-migration is Belorussia, whose contribution to the replenishment of manpower reserves elsewhere is quite substantial compared with the modest size of the population of the republic itself. On the other hand, the contribution of the Ukraine is unexpectedly small. Although the Ukraine itself is well supplied with manpower, and even has a surplus in the western part, out-migration from the Ukraine was lower than that from Belorussia, whose population is five times smaller. (The low level of out-migration exchange from the Ukraine can be explained in part by the fact that in the migration exchange with Siberia, in the period 1956–60, arrivals in the Ukraine exceeded departures: for every 100 departures from the Ukraine for Siberia, 143 moved from Siberia to the Ukraine.)[13] The Urals, once a region of net in-migration, now also records an excess of departures over arrivals.

The second cartogram (see figure 2) depicts the relationship between the migration factor and natural increase in overall population growth. The sum of the differences between regional rates of natural increase and the national average is somewhat smaller than the volume of migrations, but it is a close approximation.

It is interesting to note that, except for the Belorussians, Russians take a far more active part in migrations than the indigenous peoples of national republics. This has been verified, for example, by a special study of the influence of the ethnic factor on the intensity of migrations, made by Perevedentsev at the Institute of Economics in Novosibirsk. The study investigated the participation of ethnic groups in the migrations to various parts of Siberia from areas with similar natural and climatic conditions. (For example, among migrants from the Ukraine who arrived in Novosibirsk in 1959, 65 percent were Russians and 31 percent were Ukrainians, although Ukrainians far outnumber Russians in the Ukraine itself; Russians also accounted for 65 percent of the migrants from Transcaucasia, 85 percent from Central Asia, 78 percent from Kazakhstan, and 82 percent from the Baltic republics. Only Belorussia supplied more Belorussians (54 percent) than Russians (36 percent). The proportion of indigenous peoples was also quite high among migrants from the Volga region, Russians accounting for two-thirds and indigenous groups for one-third of the migrants.)

Study of the geography of migrations should take into account the existence of several types of migrations (which are closely interrelated): (1) migration within rural areas (mainly from smaller rural places to larger ones, such as central settlements of collective and

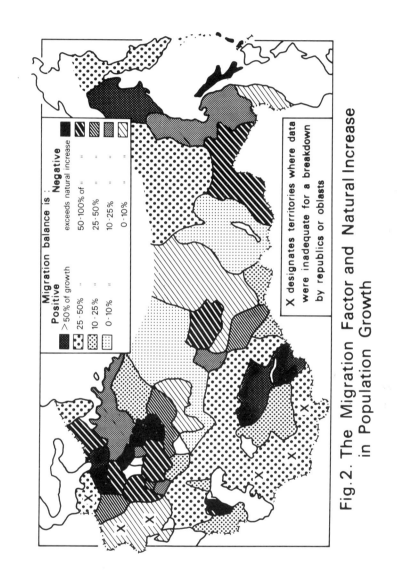

Migration balance is:

Positive		Negative
■	>50% of growth	■ exceeds natural increase
▨	25 - 50% "	▨ 50-100% of "
▧	10 - 25% "	▨ 25 - 50% "
▦	0 - 10% "	▨ 10 - 25% "
⬚		▨ 0 - 10% "

X designates territories where data
were inadequate for a breakdown
by republics or oblasts

Fig. 2. The Migration Factor and Natural Increase
in Population Growth

state farms) is caused by the consolidation of agricultural enterprises, which requires in turn a consolidation of settlements. A lack of data makes it impossible to analyze this process in quantitative terms; it can be inferred from the number of rural population centers and their size groups. (For example, in Irkutsk Oblast, the number of rural places declined from 5,000 to 3,200 in the period 1926–59, and their average size rose from 130 to 230 people.[14] The population of smaller rural places is usually limited to those [people] who are essential for work in associated livestock and crop sectors of the large farms. (2) Migration from rural areas to towns (of the same region) is the principal source of growth of urban population. This type of migration, related to the rapid expansion of industry and the need for manpower, occurs in all regions of USSR, although with varying degrees of intensity. The reverse flow, from town to countryside, is far smaller. (3) Migration between urban places (within a region), is usually from small urban places to more diversified, multifunctional cities and industrial centers. Small urban places often serve as an intermediate stopping point in the movement of rural population to urban centers (including urban centers of other regions) that are constantly experiencing an influx of population from the countryside. (This tendency, characteristic of many parts of the USSR, was illustrated by the example of Krasnoyarsk Kray in the study by Zh. A. Zayonchkovskaya and V. I. Perevedentsev.)[15]

All these types of migration are intraoblast or intraregional. They are being investigated less frequently than interregional migration although they also deserve attention in terms of volume and their role in changes in population geography. The type of migration that is most commonly being studied is (4) interregional migration. The traditional picture of migration from regions with surplus manpower to regions with a manpower deficit arose, as noted above, from the migration experience of the capitalist era, although even for that period such a generalization was too mechanistic. In a socialist economy, in addition to the *consciousness* factor, territorial differences in natural conditions and living levels play an important part. We do not mean here any single indicator, but the entire complex of living conditions such as consumer services (repair shops, public baths, and so forth), and availability of durable consumer goods and food. Interregional migration has two forms: (a) organized migration (government hiring for work in distant areas, planned rural settlement, assignments of young graduates, resettlement as a result of public appeals); (b) individual (unorganized) migration. The first form now accounts for 15 to 20 percent of the total, so that individual movements greatly predominate. Even these differ from the haphazard migrations customary under capitalism, however. Under Soviet conditions, individual migration is regulated indirectly by the state through the planned development of specific industries, the guided training of personnel, and other measures.

Many economic-geographic problems that are of practical importance and are related to all these migration types can be traced to regions that suffer from constant manpower shortage, namely Kazakhstan, Siberia, and the Far East.

Kazakhstan is the Soviet region with the most rapid population growth. Between the 1939 and 1959 censuses, the population of Kazakhstan rose by more than 50 percent (average growth in the USSR as a whole was 9.5 percent). From the 1959 census until 1 July 1963 (a period of four and one-half years), a further growth of 25 percent was recorded. An important factor in this growth is a high rate of natural increase, almost double the average rate for the USSR (in 1961, respectively 29.1 and 16.7 per 1,000). Natural increase alone accounted for 1.3 million people in the last four and one-half years; the rest (more than one million) resulted from in-migration.[16]

Even before the revolution, the vast, sparsely populated lands of Kazakhstan were among the principal areas of mass settlement of peasants, mainly from the Chernozem Center, the Volga, the Ukraine, and Belorussia. This migration flow has continued in the Soviet period, varying in intensity at different stages under the influence of changing factors. As the industrialization of Kazakhstan proceeded in the 1930s, demand was greatest for industrial workers and trained personnel at construction projects of the republic, which lacked its own industrial labor force. There was also some planned settlement of peasants, especially in northern Kazakhstan. Cities grew rapidly. From 1926 to 1939, urban population increased from 519,000 to 1,690,000. (In this period many large villages, regional capitals, and railroad settlements were transformed into urban places.) More than 321,000 people moved to the cities of Kazakhstan in 1939–40. During World War II, when Kazakhstan became one of the arsenals in the eastern part of the Soviet Union, many industrial plants with their labor forces were relocated there, the population of the republic's cities increased by 30 percent (by the beginning of 1945), despite mobilization into the armed forces and a low rate of natural increase.

Toward the end of the war, starting in 1944, reevacuation resulted in out-migration from Kazakhstan. A large in-migration in 1946–47 and a high rate of settlement of the arrivals (48 to 51.6 percent of them remained in Kazakhstan) were related to mass demobilization. Subsequently, the movement to cities in Kazakstan remained high until the opening up of the virgin lands (for example, in 1950 more than 100,000 people moved to the cities).

In 1959 there began a stage of unprecedented intensive migration processes related to the opening up of the virgin lands and an increased rate of industrial development. A chronic manpower-shortage region, Kazakhstan would have been unable, given its own human resources, to plow up 22 million hectares of virgin and idle lands and at the same time to develop its extractive industries and other branches of the economy and culture. In subsequent years, the cities of Kazakhstan recorded a drop in the net in-migration balance, not because of a decline in arrivals but because of an increase in departures as a lower percentage of arrivals settled permanently. (In 1955, 44.1 percent of the arrivals settled permanently; in 1963, 19.5 percent.) Table 1 shows the regions of origin of migration to Kazakhstan in 1960-62.

From the total number of arrivals and departures in Kazakhstan in the period 1960–62, almost half comprised movements within Kazakhstan. Of these, one-third represented city-to-city migration, mostly residents of smaller urban places moving to larger cities (especially to republic and kray capitals) and from slower-growing to faster-growing places. (In addition there was a migration flow from the cities of Kazakhstan to other large cities of the Soviet Union, especially the capitals of nearby oblasts: Omsk, Novosibirsk, Sverdlovsk, Chelyabinsk, and Tashkent.) Second in importance (25 percent of all movement within the republic) were migrations from the countryside to towns. The reverse flow, from urban to rural areas, accounts for 22 percent; migration between rural areas accounted for 18 percent.

Migration processes vary in different parts of Kazakhstan in terms of size, intensity, percentage of permanent settlement, and composition of the migrants. A distinctive role in intra-Kazkhstan migrations is played by the virgin lands, which in 1962 accounted for almost half of all migration in rural areas and one-third of all migration in urban areas. Here the population evidently has not yet *settled down.* Up to 90 percent of the migrations from countryside to towns took place within the kray.

The low rate of permanent settlement in Kazakhstan is due primarily to the lag in housing construction. The still lower rate of permanent settlement in urban areas results both from the fact that the cities contain a substantial grouping of skilled workers who cannot always find work in their field, not only in the countryside, but also in the cities, and from the fact that organized government hiring of industrial workers often brings rather unqualified people to Kazakhstan. Accustomed to a comfortable life in large cities, there migrants cannot accept the relatively lower standards of services in the villages and towns of Kazakhstan. The higher rate of permanent settlement of migrants originating in rural areas, on the other hand, is explained by the fact that they want to advance their technical training in urban institutions and therefore situate themselves in a permanent job and home. Migrants from rural areas of Belorussia, the Ukraine, and espe-

TABLE 1

REGIONS OF ORIGIN OF MIGRATION TO KAZAKHSTAN IN 1960–1962

	REGION'S SHARE OF ALL IN-MIGRATION	PERCENTAGE DISTRIBUTION OF ARRIVALS BY		PERCENTAGE OF PERMANENT SETTLERS*	
		Urban Places	Rural Areas	Urban Places	Rural Areas
RSFSR[a]	51.6	62	38	30	23
Ukrainian SSR[b] . .	28.3	50	50	38	42
Belorussian SSR . .	9.7	42	58	42	50
Moldavian SSR . . .	1.9	16	84	17	51
Others	8.5	..	..	..	..
Total	100.0	59	41	29	31

NOTE: *Percentage of arrivals who settled permanently.

[a]The principal regions of origin in the RSFSR were Siberian oblasts bordering on Kazakhstan (connected by the Turksib and Trans-Siberian railroads) and the Urals, as well as the Volga-Vyatka region. The areas with a net out-migration to Kazakhstan in 1952 were: Altay Kray, Sverdlovsk, Kemerovo, Perm', Novosibirsk, Chelyabinsk, Omsk, Irkutsk, Gor'kiy and Korov oblasts.

[b]The principal areas of origin were the coal oblasts of the Ukraine–Donetsk and Lugansk; compared with the relatively high proportion of migrants from Kemerovo Oblast in note [a]; this reflects the manpower needs of the Karaganda coal basin.

cially the neighboring RSFSR oblasts also find their customary surroundings in rural areas of Kazakhstan. As government-sponsored
rural settlers, they get a government loan for acquiring livestock and
for construction or purchase of a home. Migrants originating in rural
areas, in contrast to the more mobile city people, usually move with
their entire family and settle permanently. For example, in Kokchetav
Oblast, almost 90 percent of the government-sponsored rural settlers
who arrived in the period from 1956 to 1960 stayed permenently. The
results of government hiring of industrial workers in other parts of
the Soviet Union have been far less satisfactory.

Replenishments of the labor force from outside areas are expected
to continue in the future and will amount to half or more of the
planned manpower growth; this emphasizes the importance of studying the entire problem of migratory processes in Kazakhstan, including the composition of the migrants and factors influencing their settling in the new areas. Some of the measures that would help stabilize
the permanent settlement of new arrivals are: expansion of vocational
training, facilities in labor-shortage areas, more housing construction,
better consumer services, improved urban planning that would insure
employment for other family members, and expansion of child-care
institutions.

Siberia and the Far East are also short of population. It has been
estimated that the development of these regions alone will require an
influx of 5 to 15 million people in the foreseeable future.[17] Despite
such prospects, these regions have been characterized in the last
few years (1956–62) by: (1) a net westward trend of population movement, which constitutes a dangerous paradox in areas of labor shortage; (2) a very high out-migration rate from the countryside, mainly
to local cities, but also to cities of other regions; (3) a high overall population mobility, two to three times greater than in the Center; (4) a
substantial westward migration of city dwellers, and their replacement
by migrants from the countryside. In some areas resettlement has
resulted in a net in-migration (Irkutsk Oblast, Yakut ASSR), but in
others the migration balance has been consistently negative since
1956 (Chita and Tomsk oblasts, and in part, Altay Kray). In the Far
East, out-migration has been most pronounced in Sakhalin and Amur
oblasts.[18]

A peculiar feature in Siberia is the steplike pattern of migrations.
Siberia is the origin of migration flows to the Far East and the Far
North. Migrants arriving in Siberia from the West settle mainly along
the more developed southern railroad belt, where part of the population from northerly areas of Siberia has also been moving; this has
resulted in a concentration of population there and, at the same time,
an expansion of the southern railroad belt. The Far North and equivalent areas (Bodaybo, Aldan, and the like) have an especially high mobility of population, as many migrants remain for only three to five

years. In the last few years, increasing mechanization in the mining industry has insured a steady or even higher level of output with fewer workers, so that these areas also have begun to show a net out-migration.

The outflow of population from the Siberian countryside is far greater than the actual manpower surplus. Among those who are leaving are farm specialists, machine operators, and educated youths who have not yet had specialized training.

The principal factors in the present predominance of the westward movement and the outflow of population from the countryside are differences in living conditions, including those between the cities in the eastern regions themselves. The living-standard differential has been repeatedly pointed out in the economic literature.[19]

Very persuasive material has now been published in the handbooks of the Central Statistical Administration of the RSFSR showing that Siberia and the Far East are far below the RSFSR average in per capita housing and in the proportion of housing that is equipped with running water, sewers, or central heating,[20] that the eastern regions lag, especially behind the Central Region, in many important indicators.[21]

The westward movement from Siberia and the outflow from the countryside have had negative economic consequences. In particular, the constant replacement of city dwellers with rural migrants has inhibited the level of skilled labor in the cities; some industries, notably lumbering and construction, have suffered acute labor shortages; agriculture has lagged because of a shortage of manpower in general and of skilled labor in particular; great losses were incurred in the process of organized mass movements to Siberia, as well as in other types of organized and unorganized migrations.

A rationalization of migratory processes will require the provision of optimal relationships between the living standards of the following population groups: (1) city dwellers of various regions; (2) urban and rural population within regions; (3) various occupational and skilled-labor groups; (4) new settlers and old settlers. The basic principles of interregional and intraregional regulation of living standards should be: (1) establishment of a higher living standard for workers of given occupations and skills in labor-deficit as compared with labor-surplus regions; (2) reduction of the gap in living standards between urban and rural populations in the areas with a farm-labor shortage, compared with areas with a farm-labor surplus; (3) provision of living conditions for new settlers that would enable them to match the living comfort of old settlers within a short period. All this can be done only if natural and socioeconomic regional characteristics are taken into account and if the necessary living conditions are assured in the light of those characteristics.

The migratory processes of Sakhalin have been quite specific in character. The population of Sakhalin was 100,000 in 1939 (northern part only), and reached almost 650,000 in 1959 (including the southern part of Sakhalin and the Kurile islands). Since 130,000 people lived in the northern part of Sakhalin in 1959, the difference (520,000) was made up by new settlers and their children; the actual volume of migration is estimated at 400,000. Direct in-migration took place mainly in the period 1946–54; later, as the economy became organized and increasingly mechanized, the need for manpower declined, and the trend reversed itself. In 1955–57, out-migration was still compensated by natural increase, but since 1957, the population of Sakhalin has been declining. L. L. Rybakovskiy[22] speaks of a high turnover in the labor force, in addition to the negative migration balance: "The specific aspect of the Far East, that a large part of the population consists of migrants from other parts of the country, has assumed a classic character in Sakhalin."[23] A survey of eight enterprises in southern Sakhalin, conducted by Rybakovskiy in 1960, showed that 97 percent of the workers were not born in Sakhalin.[24]

In addition to the principal migration arenas discussed above, there are some secondary arenas that also were significantly affected by migration in the postwar period.

Settlement of the lumbering areas and mining centers of the northern part of European Russia was mainly the result of migration within the vast Northwest region. It was most intensive in the postwar boom areas of the Komi ASSR and Murmansk Oblast (where the Pechenga Rayon, returned to the USSR, was also settled), and in the Karelian ASSR, which had suffered heavily during the war. The population changes of these areas are shown in table 2. In the case of the two autonomous republics, the large volume of in-migration can also be judged from the ethnic composition of the population; in the Komi ASSR, Russians now make up 48.4 percent, Ukrainians and Belorussians 12.7 percent; in the Karelian ASSR, the share of Russians (who

TABLE 2

	POPULATION (In Thousands)			
	1939	1956	1959	1963
Komi ASSR.	319	670	806	902
Murmansk Oblast.	291	474	567	648
Karelian ASSR.	469	615	651	673

incidentally, made up half of the population even earlier) rose to 63.4 percent and there are also many Belorussians (11.0 percent) and Ukrainians (3.6 percent).

Another area in the Northwest that had to be settled was Kaliningrad Oblast. The mass-migration stage (mainly from the central and western oblasts of the RSFSR, and partly from the Belorussian SSR) was completed by 1954; by that time, more than 50,000 workers' and farmers' families had been settled; a certain number of demobilized servicemen were also settled in the oblast. Additional in-migration continued, especially into the cities (for example, the population of the city of Kaliningrad rose from 188,000 in 1956 to 204,000 in 1959, and 232,000 in 1962, that is, faster than would be possible simply on the basis of natural increase), but in the last few years the migration balance has been slightly negative. It can be assumed that out of the present oblast population of more than 600,000, as many as 400,000 are first-generation migrants.

Migrations that were local in character but quite sizable in volume occurred from mountain areas of Central Asia to the plain, mainly to newly irrigated lands (such as the Hungry Steppe and the Vakhsh valley). Generalized studies of these migratory processes are lacking, but a large amount of information about them can be found in regional monographs.

Another local development was the repatriation of Armenians from abroad, which started shortly before the war; in the postwar period about 100,000 persons came to the USSR and settled almost entirely in Armenia (especially in the late 1940s).[25]

A rather large, but unrepeated migration was the repatriation of Poles to Poland, mainly from the Western Ukraine, Belorussia and Lithuania; their number was about 1.5 million, and most were repatriated in 1945–46.[26]

Finally there is the continuing attraction of large urbanized agglomerations, for example, the redistributions within the Ukraine in favor of the Donetsk-Dnieper region. The attraction of the Moscow agglomeration must also be mentioned. There, the population flow was mainly to the cities of Moscow Oblast (except Moscow proper), since the rural population in the oblast declined and the growth of the city of Moscow was limited. The net in-migration into Moscow Oblast has been about 40,000 a year since 1950.

An examination of migration processes of the last few years suggests a certain decline in intensity; in any case, it is far from meeting the manpower needs of the economy. In many areas, production shifts create serious manpower shortages. Differences in natural increase, which have been playing a significant role in the postwar period in the territorial redistribution of population, are too slow in their effect to satisfy quickly enough the increasing labor needs resulting from the rapid concentration of industry in new projects.

The second factor was the lag of the *new* (especially the eastern) areas in consumer services, housing, and living comfort. This resulted in a migration stoppage (and a reverse trend from labor-shortage areas), even when there was an objective need for migration. Not until the last few years have the character and magnitude of this migration slowdown been analyzed (previously, under the continuing influence of the personality cult, these phenomena were hidden by a kind of veil of shame). Only on the basis of such an analysis can effective measures be taken to overcome the slowdown. A certain change is already noticeable; the housing construction plan for 1964–65, for example, provides for a higher construction rate in the eastern regions.[27]

Krushchev's statement at a meeting in Vladivostok still holds true as a means of insuring the required rate and direction of migration. He said; "You must attract new people not by higher wages compared with other cities, but by good living conditions."[28]

1.   V. V. Pokshishevskiy, "The Geography of Population and Its Tasks," *Isvestiya Akademii Nauk SSSR*, seriya geografiya [Proceedings of the Academy of Sciences of the USSR, Geography series], no. 4 (1962), p. 8.

2.   V. I. Lenin, *Sochineniya* [Works], 3d ed., 12:225.

3.   See A. G. Rashin, *Naseleniye Rossii za 100 let, 1811–1913* [The Population of Russia over 100 Years, 1811–1913] (Moscow, 1956); idem, "Shifts in the Territorial Distribution of Population of Russia in the Nineteenth and Early Twentieth Centuries," *Voprosy geografii* [Problems of Geography], no. 20, (1950); V. K. Yatsunskiy, "Changes in the Distribution of Population of European Russia, 1724–1916," *Istoriya SSSR* [History of the USSR], no. 1 (1957).

4.   See V. V. Pokshishevskiy, *Zaseleniye Sibiri* [The Settlement of Siberia] (1951), and "Essays on the Settlement of the Wooded Steppe and Steppe Regions of the Russian Plain," *Ekonomicheskaya geografiya SSSR* [Economic Geography of the USSR] no. 5 (Moscow, 1960).

5.   Ibid.

6.   *Trudovyye resursy SSSR* [Manpower Resources of the USSR] (Moscow, 1961), p. 209.

7.   *Istoriya Velikoy otechestvennoy voyny* [History of the Great Patriotic War], 2:548–49.

8.   N. Voznesenskiy, *Voyennaya ekonomika SSSR v period otechestvennoy vovny* [The War Economy of the USSR During the Patriotic War] (Moscow, 1948), p. 26.

9.   Ibid., p. 157.

10.   G. S. Nevel'shteyn, "Territorial Differences in the Natural Increase of Population of the USSR," *Materialy po geografii naseleniya* [Materials on Population Geography], vol. 1 (Leningrad, 1962).

11.   Ibid., p. 8.

12.   V. I. Perevedentsev, "Basic Aspects of Western Siberia's External Migration Links," *Materialy I soveshchaniya po geografii naseleniya* [Materials of the First Conference on Population Geography], no. 2 (1961); idem, "The Problem of Territorial Redistirbution of Manpower Resources," *Voporsy ekonomiki* [Problems of Economics], no. 5 (1962).

13.   Perevedentsev, "Basic Aspects of Western Siberia's External Migration Links," p. 12.

14. V. V. Vorob'yev, "Changes in the Distribution of Population of Eastern Siberia," in *Geografiya naseleniya Vostochnoy Sibiri* [Population Geography of Eastern Siberia] (Moscow, 1962), p. 15.

15. Zh. A. Zayonchkovskaya and V. I. Perevedentsev, "On the Question of Present-Day Migration Links of the Population of Krasnoyarsk Kray," In *Geografiya naseleniya Vostochnoy Sibiri.*

16. *Narodnoye khozyzystvo Kazakhskoy SSR v 1960 i 1961 gg.* [The Economy of the Kazakh Republic in 1960 and 1961] (Alma-Ata, 1963).

17. V. V. Pokshishevskiy, "Prospects of Population Migration in the USSR," in *Geografiya naseleniya Vostochnoy Sibiri,* p. 75.

18. L. L. Rybakovskiy, "On the Creation of Permanent Settlement in Sakhalin," in *Vorposy trudovykh resursov v rayonakh Sibri* [Questions of Manpower Resources in the Siberian Regions] (Novosibirsk, 1961).

19. See *Trudovyee resursy SSR*, pp. 220–22, and D. D. Moskvin, "The Effect of the Living Standard on the Location of Manpower Resources and Population Stability," in *Problemy razmeshcheniya proizvoditel'nykh sil v period razvernutogo stroitel'stva kommunizma* [Problems of Location of Productive Forces During the Full-scale Construction of Communism] (Moscow, 1960).

20. Central Statistical Administration of the RFSFR, *Zhilishchnyy fond v gorodakh i rabochikh poselkakh RSFSR* [Housing in the Cities and Workers' Settlements of the RSFSR] (Moscow, 1963).

21. *Bytovoye Obsluzhivaniye Naseleniya* [Consumer Services] (Moscow, 1963).

22. Rybakovskiy, "On the Creation of Permanent Settlement in Sakhalin."

23. Ibid., p. 157.

24. Ibid., p. 159.

25. *Atlas Armyanskoy SSR* [Atlas of the Armenian SSR] (Yerevan-Moscow, 1961), Map 59.

26. L. Kosinski, *Procesy Ludnosciowe na Ziemiach Odzyskanych v latsch 1945–60* [Population Processes in the Recovered Lands in the Years 1945–60] (Warsaw, 1963).

27. P. F. Lamako's speech at the Third Session of the Sixth Supreme Soviet of the USSR, *Pravda*, 17 December 1963.

28. *Pravda*, 8 October 1959. See also Khrushchev's comment concerning the importance of living conditions for encouraging migration, made at the Twenty-second Party Congress of the CPSU, *Materialy XXII s'yezda KPSS* [Materials of the Twenty-second Party Congress of the Communist Party of the Soviet Union] (Moscow, 1961), p. 80.

ZH. A. ZAYONCHKOVSKAYA
V. I. PEREVEDENTSEV

Basic Factors of the Migration and

Territorial Redistribution of Population

An analysis of the summary and primary materials on migration of population makes it possible to establish some regularities concerning migration. The main reasons for moving and, even more, the reasons for territorial redistribution of the population[1] cannot be directly determined from statistical materials on mechanical movement. A vast amount of additional data, including comparative indices of migration and indices of living conditions of the population, are needed for this.

Migration under present conditions includes socially organized and individual movements of the population, the latter involving the majority of movements. A socially organized migration is caused by common social needs, whereas individual migration results from the needs of the migrants themselves. The causes for socially organized and individual migration may not coincide. Inasmuch as socially organized transference of population has been discussed in the scientific literature, we limit ourselves to its general characteristics and examine in more detail the reasons for individual migration.

Socially organized migration of the population is caused by the desire of society to distribute the labor resources in correspondence with

Translated from *Sovremennaya migratsiya naseleniya Krasnoyarkogo kraya* [Contemporary Population Migration in Krasnoyarsk Kray] (Novosibirsk, 1964), pp. 56–75.

the distribution of manpower needs. Objective factors of industrial distribution cause such a redistribution of manpower where, in some places, a shortage arises, or in others, there is an excess. The arising discrepancies are removed by the territorial redistribution of labor resources, that is, as a result of socially organized and individual migration.

Socially organized migration must guarantee a socially expedient territorial redistribution of labor resources in those cases where individual migration does not provide for this. A system of organized economic measures and others implemented by state institutions and social organizations serve as the means of achieving a socially organized territorial redistribution of labor resources.

Individual migration under socialism differs radically from spontaneous migration under capitalism; the latter spontaneously follows the unplanned flow of capital. Under conditions of socialism, the distribution of the labor force is determined by a planned distribution of capital investment between branches of the national economy and regions, and by a planned distribution of means for the construction of housing. The systematic establishment of wage levels in the branches of the national economy and regions, the policy of distributing funds of public consumption, and the like, fulfill an important function in the correct distribution of manpower in the national economy of the country. Thus, an unorganized migration is organized by a system of planned measures outlined by the socialist government. Deficiencies in migration under these conditions must be taken as the result of deficiencies in planning and directing the national economy of the country.

There is a close connection between socially organized and individual migration. Individual migration can decrease as well as increase the need for socially organized migration. In Krasnoyarsk Kray individual migration results in a population decrease, consquently requiring a supplementary socially organized migration. To clarify the causal factors of migration, one must determine the factors that compel people to change their places of residence, and the conditions under which migration becomes possible.

It is known that there are various reasons for individual migrations: economic, moral-political, demographic, ethnographic, cultural, and natural-geographic. People move mainly because they want to improve their living conditions.

The *economic* factors of migration include such territorial distinctions as differences in the variety of working facilities, in the level of wages, nominal and real income, living conditions, the level of repairs and service, and so forth.

Demographic factors of migration include territorial differences in the age structure of the population, its distribution by social groups, family ties, and the like.

Ethnographic factors are territorial differences in language, customs, life, and the like.

Cultural factors of migration, in a broad sense, encompass territorial differences in the educational opportunities, possibilities of leisure time, and so forth. Clearly cultural factors are closely connected with the economic, ethnographic (they partially overlap), and other factors.

Natural-geographic reasons for migration appear in the form of differences in the influence of natural conditions of various places on a given populace, including their economic situation, and physiological and psychological condition.

Moral factors of migration are understood to mean the influence of a sense of duty and heightened awareness [on the part of individuals] on territorial transfers.

These factors of migration do not operate in isolation, but in complex interaction with each other. They entwine, heightening or reducing the effect of one on the other. For example, the migration to some unfamiliar regions of Siberia is stimulated by higher wages. Because of difficult climatic and living conditions, predominately young men come to these regions. An abnormal sex structure of the population is created. Not having the opportunity to raise a family, many are forced to leave soon after arriving. Thus, some economic factors attract people into these regions; economic, natural, and other factors create a specific composition of the population that in turn (together with the difficult life and adverse natural conditions) serves as one of the reasons for the rapid departure of new settlers. Factors of migration and territorial redistribution vary with different population groups. We shall examine only the causes of the more massive types of migration.

The majority of migrants are young, able-bodied people employed in the national economy. Therefore, the main role in population migration is, undoubtedly, played by labor migration, related to the change of place of employment in the economy of the country. Hence, the movement of manpower into enterprises is a necessary condition for labor migration. Migration of the labor force determines all indices of general migration. Table 1 shows the very close connection between the change in the intensity of population outflow from cities and the change in the intensity of workers leaving industrial enterprises. For example, let us take the administrative oblasts, regions, and autonomous republics of the Russian Federation and the sovnarkhozy [regional economic council planning areas] corresponding to them.[2]

Thus, the connection between the turnover of manpower and population migration proved to be quite close, in spite of the partial incomparability of the data, because: (1) the industrial workers of the

TABLE 1

RELATIVE INDICES OF TURNOVER OF MANPOWER BY DEPARTURE FROM
INDUSTRIAL ENTERPRISES AND THE INTENSITY OF POPULATION
MOVEMENT OUT OF THE CITIES OF THESE OBLASTS
(In Percent)

GROUP OF SOVNARKHOZY AND OBLASTS	INTENSITY OF MOVEMENT (Outflow)		RATIO OF COEFFICIENT OF POPULATION MOVEMENT TO THE COEFFICIENT OF TURNOVER OF WORKERS BY DEPARTURE
	By Workers From Industrial Enterprises	By Population From Cities	
1	100	100	25
2	140	146	26
3	186	165	22
4	244	200	20

sovnarkhozy make up less than one-half of the city workers, and the indices of departure from places of employment of workers of other enterprises and institutions of the sovnarkhozy and departments may differ considerably from the corresponding indices of sovnarkhozy industrial workers; (2) the industrial workers of the sovnarkhozy live not only in cities but also in rural places, and the mobility of industrial workers living in the village is, as a rule, much higher than for city dwellers; the less significant increase in the frequency of workers leaving the cities than of workers leaving the enterprises (see table 1) is due to the fact that there are many sovnarkhozy with a developed timber industry where workers live mainly in rural places. Thus, the turnover in manpower in the timber industry is very high.

The determining role of migration related to the movement of manpower in all migrations can also be corroborated by observing a similar change in the indices of manpower movement and population migration through time.

The intensity of manpower turnover in large industry of the USSR has decreased from the beginning of 1930 to 1960 by approximately 75 percent and the intensity of migration approximately three times. The lag in the reduction of migration intensity relative to intensity of manpower turnover in industrial enterprises is systematic and explained by the sharp increase in the number of moves related to educational attainment.

A connection between the turnover of manpower and migration of population is found also in the fact that the majority of those moving to the city can settle there only because a corresponding number of persons have left the city. A person can move to a given city only if he finds there the means for existence. It will be recalled that the statistics of migration are not the statistics of arrival and departure, but rather the statistics of registration and discharge.

As a rule, persons arriving in a city register when it becomes apparent that they can obtain work. This follows from the fact that the discharge [removal from registration rolls] of migrants who had registered in the city one to two months earlier is a rare exception. The majority of those who moved there cannot live in the city for two months without work.

As a rule, dependents of individuals move with those members of the family who are employed in the economy, or after them, but in connection with their moving. Consequently, migration to the city must usually be connected with obtaining work or entering an educational institution. The number of in-migrants to a city is determined by the number of work and educational places that the migrants may occupy. The number of places that can be occupied by the migrants are formed from newly created positions and those freed as the result of people who had formerly occupied them leaving the city (excluding those positions that are occupied by people originating in the given city's population as a result of natural growth and redistribution of labor reserves from the households). Therefore, all transfers of population to the city can be divided into those related to the increase in the number of work and educational positions and those related to the substitution of some workers and students by others.

The former group of migrants is a small minority of all those moving to cities. In 1960, for every 100 persons leaving the cities of the RSFSR, 127 moved in; and for the cities of the Krasnoyarsk region, this figure was only 119. In other words, for every 100 arrivals in the cities of the region, 16 represented the growth of urban population; the remaining 84 replaced those who had left, if one speaks of working population. Thus, the freeing of work and educational positions by persons leaving the city is a necessary condition for resettlement of the majority of migrants. Hence, the out-migration of population from cities is a key phenomenon in the whole complex of urban migration phenomena. The exodus of each person from a given populated point serves not only as an initial feature in the process of his own transfer but also as a condition for the movement of another person who will take the freed position. It makes no difference if this position is taken by a city dweller who has been living in the given city for a long time. The new migrant simply takes the place of the person who took the position of one leaving the city. However long the chain of transfers connected with the out-migration of one person may be, the essence of the phenomenon remains unchanged.

In order to clarify the conditions of population migration, it is necessary to understand the reasons why workers leave enterprises and institutions. Let us turn to the statistics of labor force migration from industrial enterprises of sovnarkhozy. As the data for 1958–60 indicate, about two-fifths of the workers discharged from industrial enterprises of the sovnarkhozy in the RSFSR, and somewhat more than one-fourth of the workers discharged from industrial enterprises of the Krasnoyarsk Sovnarkhoz leave the enterprises for reasons considered valid under the labor law (the armed services, study, illness, and so forth); the rest—75 percent of those discharged in the Krasnoyarsk Sovnarkhoz—leave of their own accord and in connection with the breaching of labor discipline. Terminations for these reasons belong, in the practice of accounting and in the economic literature, to the turnover of manpower in the narrow sense of the word.[3]

Between Krasnoyarsk Kray and the western regions of the country there are no great differences in the coefficients of worker movement *for valid reasons*, whereas the differences in the fluctuations of manpower are great. In most western regions of the country the manpower fluctuation in industries of the sovnarkhozy is approximately 50 percent lower, and in the Center approximately 67 percent lower, than in the Krasnoyarsk Kray. One cannot understand the causes of such migration differences in the region as compared with the western regions of the country without clarifying the conditions of the above-mentioned differences in the fluctuation of manpower.

More than half of all migration to the cities of the region, and the great majority of out-migration from these cities, is connected with manpower turnover. Turnover, in the narrow sense, does not include the urban in-migration of those people who increase the size of the urban population by replacing those who left for the armed services, those who left after finishing their education, or other types of socially organized redistribution of population, and the dependents of these migrants.

According to our calculations, out of the total migrants from the cities or the region only a small number leave for the armed service. Still fewer leave the cities in connection with the completion of school and assignment to places of work (graduates of the universities, technical schools, professional-technological schools). A certain number of the urban dwellers are assigned to rural places in connection with agricultural settlement and social work. A portion of the urban population move in connection with a job transfer. All types of socially organized, territorial population migration from the cities of the region constitute, in any case, no more than 15 percent of those migrating from them. Approximately six-sevenths of those leaving the cities of Krasnoyarsk Kray leave individually.

Thus, the 16 percent of the migrants making up the growth of urban population, plus approximately 13 percent of the migrants replacing those who left the city in a socially organized manner, are not directly connected with the fluctuation of manpower. Those leaving in a socially organized manner (called into the army, graduating from educational institutions, and so forth) are replaced mainly by people who do not have their own families (those entering educational institutions, demobilized soldiers). The number of their dependents is small. Thus, approximately 70 percent of the migrants into the cities of the region settle in them after leaving the cities of their previous residence of their own accord, that is, in connection with manpower turnover. Hence, about 75 percent of the volume of population migration in the cities of the Krasnoyarsk Kray is related to manpower turnover.[4]

Consequently, in order to understand the conditions of present day urban migration it is necessary to understand the reasons for the labor force turnover in enterprises and institutions directly connected with population migration. Migration of rural population is to a certain degree determined by the migration of urban population and cannot be understood without it.

In order to establish the reasons for the turnover and to work out measures to curtail it in the industrial enterprises of Krasnoyarsk, in 1960, a mass sample survey was conducted of persons leaving of their own accord. As a result, it became clear that the main reasons for the voluntary resignation of workers was the dissatisfaction with working conditions and wages, living conditions (primarily housing), and the desire to be with relatives (see table 2). The desire to move to relatives is, in most cases, conditioned by economic reasons.

As can be seen from the data in table 2, the primary reason for leaving the city is the desire to move to relatives; second, dissatisfaction with living conditions; third, wages; fourth, working conditions. It is significant that the majority of those leaving because of dissatisfaction with wages and working conditions find work in the same city, whereas the majority of those leaving because of dissatisfaction with housing conditions go to other places.

The desire to move to relatives is inconclusive. It is necessary to clarify the causes. There is no doubt that family ties have a definite significance. However, they alone do not explain the desire of many to leave the enterprises in order to move to relatives. In order to understand the relationship existing between these dependent phenomena, we turn to the composition of the migrants and those features of living conditions of migrants moving to relatives, and migrants leaving for other reasons.

The majority of those leaving the cities are the newcomers who have not taken root, and, among the in-migrants to the city, the great majority are young people. Hence it follows that going to relatives

TABLE 2

REASONS FOR LABOR FORCE TURNOVER AND MIGRATION OF POPULATION

MAIN REASON FOR TERMINATION (Voluntary)	PERCENTAGE OF TOTAL TERMINATIONS	PERSONS PER 100 TERMINATIONS		NUMBER OF PEOPLE WHO TERMINATED AND LEFT THE CITY AS PERCENTAGE OF TOTAL LEAVING THE CITY
		Left the City	Remained in the City	
Dissatisfaction with wages.	17.5	27	73	9.5
Dissatisfaction with kind of work.	13.9	23	77	6.4
Dissatisfaction with living conditions.	11.2	57	43	12.7
Distance of the place of work and poor transport.	4.2	23	77	1.9
Lack of kindergarten or nursery for children.	4.0	18	82	1.4
Desire to go to relatives . . .	17.5	96	4	34.0
Undesirable climate	0.6	86	14	1.1
State of health of the workers.	4.9	28	72	2.8
Illness of relatives.	4.3	73	27	6.3
Desire to continue education. .	5.2	37	63	3.8
Other reasons	11.3	55	45	12.4
Reason not given.	5.4	71	29	7.7
Number investigated	100.0	49	51	100.0

NOTE: These data refer to approximately 4,700 workers leaving industrial enterprises of their own accord in Krasnoyarsk in the second quarter of 1960.

means mainly a return to the family and to all the welfare that a family offers. Those who leave enterprises in connection with a desire to move to relatives have living conditions differing greatly in some respects from conditions for all those leaving enterprises. In particular, this is clearly seen from table 3 dealing with housing conditions.

The data in table 3 clearly reflect the actual differences in housing conditions of those leaving for various reasons, since the survey was conducted by workers of the personnel departments of enterprises who, as a rule, know very well what living space the leaving worker occupies.

Thus, more than one-third of those leaving industrial enterprises in connection with the desire to move to relatives had quarters rented from private owners, and one-fifth lived in dormitories. Their housing conditions were worse than that of any other group of migrants, except for those who left because of dissatisfaction with housing conditions.

Inasmuch as the majority of those leaving the cities were newcomers who had not taken root, these data on the living conditions indicate that the population newly arrived in Siberian cities finds itself in much worse housing conditions than old residents. Those living in their own houses are much less prone to move from the cities than those who live in collectivized housing; those living in dormitories leave more readily than those who have apartments or rooms; those renting from private owners leave most readily.

Thus, the main reasons for migration are economic. Material on the labor force turnover in various sovnarkhozy of many economic-geographic regions of the country show that the reasons for termination by personal desire are the same everywhere, and the significance of individual factors varies little in the majority of cases (see table 4).

Consequently, the reasons for the turnover are the same, but their effect in Krasnoyarsk Kray is considerably greater, causing a more intensive turnover of manpower. It is characteristic, however, that in the sovnarkhozy of Siberia a larger portion of workers leave because of dissatisfaction with wages and housing conditions than in all other sovnarkhozy (except for Arkhangel'sk [in the northern section of the European RSFSR]).

It will be recalled that the factors that compel people to change their places of work are closely interconnected. Thus, under good housing conditions a worker is satisfied with lower wages than under poor conditions, such as renting a room in a private house; dissatisfaction with working conditions depends to a considerable degree on amount of wages, and so forth. Nevertheless, the data in table 4 give an idea about the relative importance of various groups of factors in the turnover of manpower.

TABLE 3

DISTRIBUTION OF THOSE LEAVING ENTERPRISES BY CATEGORIES OF HOUSING
(In Percent)

REASONS FOR LEAVING	THE PROPORTION OF INDIVIDUALS OF THE GROUP HAVING HOUSING			
	Apartment and Room of the Enterprise or Economic Council	Places in Boarding Houses	Privately Owned Home	Rented from Private Person
Moving to relatives. . .	33.1	21.0	9.8	36.1
Insufficient living space.	23.5	12.4	2.9	61.2
All others	44.1	14.3	17.2	24.4

TABLE 4

DISTRIBUTION OF WORKERS HAVING LEFT INDUSTRIAL ENTERPRISES LISTED BY REASONS
(In Percent)

MAIN REASONS FOR LEAVING*	SOVNARKHOZY								
	Krasnoyarsk	Novosibirsk	Moscow Oblast	Lugansk	Sverdlovsk	Perm	Vologda	Arkhangel'sk	Komi
Dissatisfaction with wages.	18.6	18.8	13.0	13.3	11.9	10.5	8.1	27.5	7.8
Dissatisfaction with conditions and kind of work.	14.4	18.5	27.1	10.9	14.6	13.1	13.1	10.2	8.6
Dissatisfaction with housing. . . .	12.1	10.3	8.3	6.8	7.8	5.4	5.5	18.3	4.1
Distance of housing from place of work	4.4	8.5	4.7	2.6	3.4	8.0	3.4	0.9	2.2
Lack of kindergartens and nurseries	4.2	7.0	7.3	1.1	1.8	1.0	3.8	4.6	0.2
Desire to move to relatives	19.0	22.4	8.7	32.6	17.4	23.4	26.3	22.9	29.4
Dissatisfaction with climatic conditions	0.7	...	...	1.5	1.0	1.7	0.8	...	10.0
Other	26.6	14.5	30.9	31.2	42.1	36.9	39.0	15.6	37.7

SOURCES: I. Kaplan, "Questionnaire on Reason for Worker Turnover in Industrial Sovnarkhozy," in Trud i zarabotnaya plata, no. 4 (1961), p. 35. V. D. Patrushev, "Ways to Limit Cadre Turnover in Industrial Enterprises in the City of Krasnoyarsk," Nauchnyy otchet za 1961, god (Krasnoyarsk, 1961), p. 195.

NOTE: *These data are the result of an investigation by the Institute of Economics of the Siberian Section of the Academy of Sciences (Krasnoyarsk and Novosibirsk Sovnarkhozy) and the Scientific Research Institute of Labor of the State Committee on labor and wages (the other Sovnarkhozy).

The manpower turnover in enterprises and institutions is the basis of individual migration; motives of an economic order serve as motives for worker movement. Consequently, the regional differences in the indices of migration must be determined by the differences in the complex of living conditions.[5] This conclusion is fully verified in an example of differences in the settled (established) populations of three city settlements of the Krasnoyarsk region (city of Achinsk, the settlements of Nazarovo and Dzerzhinskiy). As indicated, after three years of settlement in Achinsk 61 percent of the newcomers aged 16 and over leave the city, from the settlement of Nazarovo, 49 percent, and from Dzerzhinskiy, 47 percent. These differences in the level of adjustment correspond to the differences in the living conditions of the population. Let us briefly characterize these differences.

Achinsk is one of the oldest cities of the province and a relatively large industrial center. Light industry, building-material industry, and the food industry are developed here, that is, branches with a relatively low level of wages. In recent years large aluminum-oxide and cement plants have been developed. Nazarovo is a young, rapidly developing worker village having a large coal pit and a thermal power plant, the largest in Siberia. Dzerzhinskiy is a narrowly specialized ore-mining settlement.

The average pay is highest in Dzerzhinsky, lowest in Achinsk, The builders of the Nazarovo GRES [State Regional Power Plant] and the administration of the Achinskaluminstroy [industrial combine], who made up a large portion of the newcomers to Nazarovo and Achinsk in 1956–60, have a lower average pay than the workers at Dzerzhinskiy by 15 and 30 percent respectively.

The differences in housing conditions are considerable. They are better in Dzerzhinskiy than in Nazarovo and Achinsk (see table 5). About 50 percent of the housing fund in all three settled points is the personal property of the citizens. This type of housing has a very low level of amenities. Differences in the quality of the communal housing are also great. In Dzerzhinskiy it is represented by well-equipped substantial buildings of the last decade; in Achinsk a considerable part of the communal sector has houses of prerevolutionary construction, the majority lacking modern conveniences.

In Dzerzhinskiy the plan for new housing construction is over-fulfilled year after year. This settlement has no workers in urgent need of living space. In Achinsk in recent years a large housing construction project was developed; however, it lags far behind the needs. In the Achinskaluminstroy combine, for example, in mid-1960 more than 30 percent of the workers urgently needed better housing conditions. The construction workers, among whom the turnover is highest, have especially poor housing in Achinsk and Nazarovo.

TABLE 5

HOUSING AND ITS EQUIPMENT ACCORDING TO THE
CENSUS OF THE HOUSING FUND FOR 1 JANUARY 1960
(In Percent)

INDEX	SETTLEMENTS	
	Nazarovo	Dzerzhinsky
Dimensions of living area per person. . .	109	115
Equipped with:		
Running water	175	124
Sewer system.	252	490
Central heating	139	141
Bath (shower)	177	430

NOTE: Data for Achinsk were taken as 100 percent.

In Dzerzhinskiy the need for children's preschool institutions are fully met. In Achinsk and Nazarovo there are not enough: in comparison with Achinsk in 1960, in Dzerzhinskiy there were 40 percent more places in kindergartens and nurseries per 1,000 inhabitants, and in Nazarovo, almost 20 percent less.

Communal dining facilities are well organized in Dzerzhinskiy. The quality of food is high and prices rather low. Therefore, not only single workers but whole familes come regularly to the dining halls; the sale of take-home meals is widely developed. In Achinsk and Nazarovo the organization of cummunal dining is worse in every respect than in Dzerzhinsky, and sometimes the organization is simply unsatisfactory.

In the stores of Dzerzhinskiy one can always find necessary goods. The stores are distributed uniformly in the village. One need not spend much time shopping for needed items. In the stores of Nazarovo (except the construction worker settlement of the GRES [State Regional Power Plant]) and in Achinsk there are often interruptions in the supply of even those goods that are in sufficient quantity in the wholesale (supply) bases.

In Dzerzhinskiy the urban transport operates on schedule; there are enough buses. In Nazarovo and Achinsk the workers experience great difficulty, especially those who live far from their place of work, because of irregular operation of the urban transport.

Dzerzhinskiy is behind Achinsk, an old and relatively large cultural center of the province, in opportunities for professional study and especially in recreational possibilities. However, much has been done in Dzerzhinskiy for the organization of cultural leisure. For 11,000 inhabitants there are three clubs, three libraries, an artificial lake with a bathhouse and boat station, equipped sport fields and a large pleasant park.

In the new construction areas, cultural service to the population is intolerably slow. Thus, the construction of the state regional power plant in Nazarovo was completed in 1956, but only toward the end of 1960 was a club built for the constructioner worker settlement. However, it is difficult for the construction workers to use the cultural establishments of Nazarovo, since their settlement is seven kilometers away from the old village and communal transport operates poorly, particularly in the evening.

The decisive advantage of Dzerzhinskiy in the majority of socio-economic living conditions compensates not only the less favorable natural conditions but also determines the superiority of the whole complex of living conditions.

From the sharp differences in the adjustment of migrants from city and village in the various cities, it follows that various categories of newcomers have different standards for living conditions and the various aspects of these conditions. The migrants from cities have much higher standards with regard to cultural living conditions than rural migrants.

Thus, a comparison of the adjustments of migrants in places of settlement with the living conditions existing in them fully corroborates the conclusion concerning the decisive significance of the complex of living conditions in the migration of people, a conclusion made on the basis of subjective opinions of workers voluntarily leaving their place of work.

The much higher mobility of newcomers compared to old residents confirms this conclusion. To a certain extent such a phenomenon may be caused by a greater inclination of newcomers to migrate. However, one clearly sees that the majority of newcomers, for some time after arrival, are placed in much less favorable living conditions than the old residents. The majority migrating can count on obtaining a place in dormitories. However, those who come independently often do not have such an opportunity.

The newcomers feel, much more than the old residents, the deficiencies in trade and dining facilities and painfully perceive the disparity between many living conditions in the new places and their earlier acquired habits. Not having their own household, they depend on state and cooperative trade and collective farm markets, which do not always satisfy their needs for such things as vegetables, melons, and fruits.

Among workers of resettlement organizations, the adjustment of agricultural migrants is currently determined primarily by material conditions of their life at the places of settlement, that is economic facilities (housing, cattle, garden plot) and wages, the composition of the migrants themselves (primarily social) and the degrees of correspondence between living conditions at the places of settlement and their needs.

Undoubtedly, the complex of noneconomic factors has an important influence on the differences in the migration indices for newcomers and old residents. First, one must evidently place emphasis on the newcomers' alienation from an accustomed social environment, from relatives, friends, and acquaintances. From the passport cards it can be established that one migrant is followed by others from the same place—relatives of those who came earlier or simply fellow countrymen. Consequently, connections of each city in Krasnoyarsk province with some cities and rural administrative regions of specific oblasts are intensive, whereas with others there are almost none. If communications with those living in places of origin are established upon arrival in the region, then they, undoubtedly, also communicate when leaving the region. Data show that the majority of migrants not settling permanently return to their place of origin.

The adjustment of migrants who have come as a production collective (organized industrial group) is exceptionally high. In this case friendship, family, and other ties are retained to a considerable degree, and the ties with the places they left become weaker than for those who migrate individually. Of course, stability of migrants having moved in a body is influenced by other factors, in particular, better material conditions as compared with those of other categories of population (retaining work of their speciality, attention given by the administration to maintaining cadres). However, the material conditions that, for example, the young specialists find are almost the same as those of the migrants who have come in a group. Nevertheless, the young specialists are much more mobile.

The main reasons for migrating individually are found in the desire to move to places with better living conditions; the population of Krasnoyarsk Kray has increased little in recent years in spite of the considerable socially organized redistribution of labor resources from other regions for the benefit of the province. From this one can draw the conclusion that living conditions of the population are worse in the province than in many other regions of the country and the living conditions in the cities are better than in the rural places.

Let us compare living conditions of urban and rural populations of the province with those of other regions of the country and with the urban and rural populations within the region.

The climate of Krasnoyarsk Kray demands higher outlays for housing, clothes and food, which to a certain degree is taken into account in the pay rate. The nominal wages of workers and employees of the province in 1959 were 15 percent higher than the average for the RSFSR. They exceeded those of the republic average in all branches of the economy in the province. However, the considerable advantage was characteristic only for industry (28 percent), which is due to the significant portion of industrial-production personnel employed in the Far North and the structural characteristics of industry.

In spite of the high nominal wages, an able-bodied member of a family of workers and service employees in the province receives an average real cash income from wages approximately the same as the average for the RSFSR. This is the result of a high proportion of unemployed able-bodied population in social production and higher retail prices in state and cooperative trade and catering establishments in the province as compared with the RSFSR. This conclusion is confirmed by the fact that the retail turnover in government and cooperative trade, including public catering, calculated on an average annual basis, per inhabitant, in 1959 was only 98 percent of the average level in the RSFSR. One must keep in mind that the proportion of collective farm population in Krasnoyarsk Kray is lower on the average than in the RSFSR. Income in rubles and expenditures of collective farmers are less than those of workers and service employees.

One of the most important indices of the material level of living of the population is housing conditions. According to the census at the beginning of 1960, housing of the urban population of the province was somewhat lower than in the republic as a whole and especially lower than in the Northern Caucasus and in the Northwest of the USSR. The supply of urban housing, with all types of communal convenience, is considerably lower than the average for the republic. Herein lies the reason for the higher labor expenditure in maintenance and the large number of able-bodied population engaged in it.

Commercial service of the population in the province is also somewhat worse than that for the republic on the average. The number of jobs in the trade enterprises per number of inhabitants in 1960 was 95 percent of the average for the RSFSR.

The development of public dining facilities lags in the province. If one takes the seating capacity in public dining establishments in the RSFSR in relation to the population as an index of 100, then in Krasnoyarsk Kray in 1960 this index equaled 80. Yet the need for public restaurants is undoubtedly higher in the province than the republic as a whole, inasmuch as in the cities of the province, and especially at the large construction sites, the proportion of single people not having a household is higher than the average for the republic.

A lag in medical service for the population is also observed. In 1959 there were 14 physicians and an average of 52 medical workers per 10,000 inhabitants in the province compared with 19 physicians and an average of 66 medical workers for the republic. There are still few sanitariums and rest homes in the province, as in the other eastern regions of the USSR.

Under present conditions one of the important indices of the level of living of the population is its educational level. If the portion of persons of each educational level in the whole population of the RSFSR is taken as 100 (using the census of 1959), then the indices for the Krasnoyarsk province will be, respectively: completed higher education, 65; incomplete higher education, 70; intermediate special education, 92; intermediate general education, 79; seven-year and incomplete intermediate education, 93.

In the 1959-60 school year there were 77 students in higher institutions and 80 in intermediate special institutions per 10,000 inhabitants of the province, compared to 96 and 94 per 10,000 respectively for the RSFSR as a whole. In schools for workers and rural youths, and adult schools (1959–60 school year) there were 107 persons studying per 10,000 inhabitants against 118 in the RSFSR.

It is generally accepted that one of the most important synthetic indices of the level of living of the population is the average life span. An investigation of population reproduction in Siberia conducted by the Institute of Economics, Siberian Section of the Academy of Sciences, has shown that the special coefficients of mortality for the population in the province in almost all age groups are considerably higher than the corresponding coefficients for the country as a whole. Because of this, the average life span in Krasnoyarsk province for the age group 0–3 years is lower than for the population of the country as a whole, comprising 66 years, compared with 69 for the USSR.

Thus, taking each living factor separately, and especially considering complex living conditions, Krasnoyarsk Kray at the present time lags behind the republic as a whole, except in nominal wages. The backwardness of Krasnoyarsk Kray in terms of the complex of living conditions is the only reason for a negative migration balance for the province in the exchange of population between cities and the very inconsequential total population influx to the Kray from other regions of the country. The results of population migration for the province and the correlation of living conditions between the province and the republic as a whole agree well with each other.

The lag of rural places behind cities in living conditions is generally accepted. In the Program of the Communist Party of the Soviet Union, the task of removing the remaining differences in living conditions between urban and rural populations is one of the most important tasks of the Party and the people. "The elimination of socioeconomic and cultural-living differences between the city and the countryside is one of the greatest results of the building of communism."[6]

Let us compare the living conditions of industrial workers and collective farmers, since the living conditions of other significant population groups are somewhere between these categories of workers. The average total income of the population in the USSR and any of its regions calculated per family and per inhabitant rose continuously and rapidly in the 1950s. The Party and the government took a number of important measures directed toward the relatively faster growth of income for the collective farmers as compared to other categories of population (the abolition of obligatory delivery of agricultural products from private plots, a sharp increase in the purchasing price of agricultural products during 1954–56, reorganization of the system of storing of agricultural products, and so forth). Therefore the total income per family, per worker, and per member of a family has grown faster for collective farmers than for industrial workers. From 1953 to 1960 the difference in income between these two categories has noticeably decreased, although the differences in the level of income of workers and collective farmers were still considerable. Therefore, in the middle of 1962 additional measures were taken to strengthen the material incentive of the collective farmer.

Corresponding to the great differences in the size of total income per family and per family member between workers and collective farmers of the province are the differences in the volume and structure of expenditures in these families. It is well known that the higher the level of living conditions the smaller the proportion of expenditures for food. In families of industrial workers, the food expenditures as a proportion of total expenditures in 1960 were 33% smaller than that of the collective farmers. Correspondingly, expenditures for the purchase of industrial nonfood items took a much greater share of the budget of workers than for collective farmers. Thus, a particularly large proportion of expenditures of workers were observed to go for cultural needs, furniture, and household goods.

Although the collective farmer spends a considerably larger portion of his income on food than the industrial worker, the structure of his food ration as a whole is worse than that of the industrial worker. The structure differs in the significant proportion of relatively cheap products of vegetable origin and the small amount of more valuable products of animal origin (with the exception of fresh milk and eggs). The collective farmer uses 50 percent more bread and potatoes than the worker and 33 percent less meat. Still greater is the difference in the consumption of nonfood industrial goods. Therefore, the desire of rural inhabitants to move to cities is quite natural.

Of course, the migration of the population from village to city is stimulated not only by economic factors. Of great significance are the cultural advantages of the city over the village. A considerable proportion of the youth, especially those with a higher level of education, have a desire for urban life with its relatively favorable opportunities for further cultural and professional growth, rather than the material advantages of the city. It is necessary to keep in mind, however, that the cultural lag of the countryside depends to a considerable degree on its lag in the material level of living.

Thus, the correlation between conditions of life in the urban and rural places of the province and the processes of labor flow from labor-deficient villages fully corroborates the assertion concerning the leading role of living conditions in the unorganized territorial redistribution of the population.

1. It is necessary to differentiate strictly between population migration as a process and territorial redistribution of the population as a result of this process. Mixing these two notions is, in our opinion, one of the most important reasons for erroneous assertions in questions of migration.

2. The names of the sovnarkhozy are taken according to their status in 1960.

3. In fact, turnover in the narrow sense of the word (unplanned movement of manpower) includes, to a certain extent, leaving an enterprise and institution for other reasons too (expiration of contract, transfer to another enterprise at the request of the worker, and so forth), so that the official data on labor-force turnover lowers somewhat the actual dimensions of labor-force turnover in industrial enterprises.

4. About 85 percent of the departures from the city and about 70 percent of the arrivals.

5. By living conditions we understand all objective phenomena having some significance for a person; the concept living conditions is much wider than the concept level of living. If the level of living is determined by the amount of material and spiritual goods used by a person, then, from the point of view of living conditions, many other indices are important—ethnographic, demographic, natural, and so forth.

6. *Materialy XXII S"yezda KPSS* [Materials of the Twenty-Second Congress of the Communist Party of the Soviet Union.

SECTION VIII

Urban Geography

Introductory Note

Urban geography, traditionally viewed in the Soviet Union as a branch of population geography, has emerged in recent years as an independent and active subfield. A growing interest in the topic can be traced to the major Party and governmental concern with the development of Moscow and other cities beginning in the early 1930s; further stimulus has been received from the immense scale of postwar urban reconstruction and the continuing rapid urbanization of the country.[1]

As is true of Soviet economic geography in general, urban geography demonstrates a close relationship to planning needs and reflects a Marxist-Leninist philosophical base. However, unlike most other subdivisions of Soviet economic geography, the ties to local and regional planning are more evident than are those to national planning. Also in approach "there is probably less discrepancy between the views of Soviet and Western urban geographers than in any other field of social geography."[2] Bibliographical citations, especially in recent years, indicate a keen awareness of Western, especially American, theoretical and methodological approaches.[3]

Although the range of research topics covered by Soviet urban geographers is not very different from that encountered in the American literature, a very considerable difference in emphasis has been evident.[4] In the past Soviet urban geography has produced proportionately more studies of individual cities and has tended to stress certain systematic topics less emphasized by Americans, yet has neglected other topics popular in the West.

Well into the postwar period, the majority of writings in Soviet urban geography consisted of comprehensive studies of individual or groups of Soviet cities. Although providing a great body of detailed information about Soviet cities, these sterotyped studies were too frequently unimaginative, encyclopedic renditions of Baranskiy's recommended outline (included in this section).

The period of emphasis on general studies of individual cities has given way to a period of greater emphasis on a number of systematic topics. Among subjects well covered in the Soviet literature have been urbanization patterns of the USSR, questions of urban definitions for census use, city classification, urban morphology, urban situation and intercity ties, suburbs and satellites, and problems of optimum city size.[5] Urban systems have been a popular subject of study but with analysis carried on in terms of intercity and interindustry ties rather than in the central place framework employed in Western studies.[6]

A continuing characteristic of Soviet urban geography has been the relatively slight attention given to large-scale study of intraurban functional and social structure. Contributing factors include a lack of sufficient statistical data by small areal units and restrictions on the use of large-scale urban maps. Another element may be the overriding concern with planning decisions at the regional level. Urban planning in the USSR has often been treated as derivative and dependent on the primary industrial location decisions made at a regional or national level.[7] Basic research, therefore, has tended to focus on the same levels. However some recent articles seem to indicate a growing interest in intracity structure and problems.[8]

Until the early sixties, Soviet urban geography could still be characterized by a lag in development of theory and a lack of application of advanced quantitative techniques.[9] This situation has changed dramatically in recent years as is demonstrated by the work of such geographers as Medvedkov, Gurevich, Saushkin, Blazhko, Matlin, and others.[10] As was true earlier in American geography, the work of the urban geographer appears to be the spearhead of a discipline-wide quantitative revolution.

Substantive concern of late has focused on such question as urban systems and settlement patterns, size, and spacing—topics that have also been at the center of attention in American urban geography for the last decade. Much of this Soviet work, if it is not solely abstract and theoretical, employs empirical data from foreign countries. Although this may simply reflect deficiencies in available Soviet data, it also demonstrates that in urban geography, at least, there is now interest in developing theory and methodology that may not be immediately related to the practical demands of Soviet planning.

The articles included in this section range from a traditional methodological statement to more recent mathematical, analytical approaches. The initial article is an abridgement of N. N. Baranskiy's "On the Economic Geographic Study of Cities," perhaps the most frequently cited methodological work in Soviet urban geography. In this paper Baranskiy sets down guidelines for urban research. He urges study of city distribution, size, function, and growth as well as such topics as urban definition, classification, and situation. The greatest part of the original paper is reserved for a detailed outline for the study of individual cities.

The second article in this section, "On the Patterns and Tendencies of Urban Settlement in the USSR," by Davidovich, is included for its informational content. The author discusses trends in urban growth, urban place distribution, the relationship between city functions and size, the urban hierarchy, and the development of urban clusters.

Blazhko, in the third article of this section, "Economic-Geographic Mathematical Modeling of Cities," illustrates the growing interest of Soviet geographers in territorial models. She presents a simple static city model that is based on the concepts of the interindustry and interregional balance of production and consumption. The model consists of six matrices: total city industrial output, city industrial consumption, city population consumption, outshipments, inshipments for industrial consumption, and inshipments for population consumption. The model is then applied to three urban places in the Dnieper region to establish the structure of basic and nonbasic activities, interindustry and interregional relationships, and the functional role in the urban hierarchy.

The fourth article of this section, "Construction of a Model of the Location of a Trade Network," by Matlin, is evidence of the awakening interest in consumption and services, an area that in the past received but slight attention on the part of Soviet economic geographers.[11] Matlin is concerned with the development of an optimum distribution of retail trade enterprises. He presents a linear-programming formulation of the problem with the objective of minimizing the shopping time of the population subject to limitations of capital and operating costs and the full satisfaction of consumer demand.

In the final article of this section, "The Population Density of a City and the Probability Density of a Random Variable," Gurevich contributes to a topic that has engaged the interest of a number of Western urban geographers. In this paper Gurevich attempts "to establish a systematic relationship between the concept of population density of a city and the general theoretical probability concept of the density of the probability of a two-dimensional random magnitude," and "to extend the analysis beyond single-center cities to various models of urban formations having other than a single center." Although the question has potential planning significance, immediate practical application is obviously less the concern than with most Soviet urban work, and so this paper illustrates the more abstract and theoretical approaches that have recently become acceptable.

1. Y. G. Saushkin, "A History of Soviet Economic Geography," *Soviet Geography: Review and Translation* 7, no. 8 (October 1966): 36–37, 45–47.

2. I. M. Matley, "The Soviet Approach to Geography." (Ph.D. diss., University of Michigan, 1961), p. 146.

3. For example Yu. V. Medvedkov, "The Regular Component in Settlement Patterns as Shown on a Map," *Soviet Geography: Review and Translation* 8, no. 3 (March 1967): 150–68.

4. For a detailed review of Soviet urban geography for the postwar period up to the early sixties see R. J. Fuchs, "Soviet Urban Geography: An Appraisal of Postwar Research," *Annals of the Association of American Geographers* 54, no. 2 (June 1964): 276–89.

5. For examples of the Soviet approach on these topics see O. A. Konstantinov, "Some Conclusions About the Geography of Cities and the Urban Population of the USSR Based on the Results of the 1959 Census," *Soviet Geography: Review and Translation* 9, no. 7 (September 1960): 59–75; V. G. Davidovich, "Urban Agglomerations in the USSR," *Soviet Geography: Review and Translation* 5, no. 9 (November 1964): 34–44; V. M. Kharitonov, "On the Definition of Conurbation Boundaries," *Soviet Geography: Review and Translation* 9, no. 10 (December 1968): 838–47; B. S. Khorev, "A Study of the Functional Structure of Urban Places of the USSR," *Soviet Geography: Review and Translation* 7, no. 1 (January 1966): 31–57; V. S. Varlamov, "On a Quantitative Assessment of the Economic-Geographic Situation of Cities," *Soviet Geography: Review and Translation* 7, no. 1 (January 1966): 52–29; F. M. Listengurt, "Prospects of Economic and Territorial Growth of Small and Medium Cities of the Central Economic Region," *Soviet Geography: Review and Translation* 6, no. 8 (October 1965): 51–59; The March, 1962 issue of *Soviet Geography: Review and Translation*, contains several articles on the topic of satellite cities.

6. See N. I. Blazhko, "On Methods of Studying the Place of a City in the System of Cities of the USSR," *Soviet Geography: Review and Translation* 3, no. 3 (March 1962): 69–74 and the paper by the same author included in this section.

7. See D. G. Khodzhayev, "The Planning of the Distribution of Production in Population Centers and Some Problems in Population Geography," *Soviet Geography: Review and Translation* 8, no. 8 (October 1967): 619–29.

8. V. I. Romashkin, "A Mathematical Model of the Density Distribution of Moscow's Population," *Soviet Geography: Review and Translation* 8, no. 9 (November 1967): 709–21. See also Matlin's article presented in this section of the reader.

9. Fuchs, "Soviet Urban Geography: An Appraisal of Postwar Research," p. 288.

10. Medvedkov, "The Regular Component in Settlement Patterns." See also by the same author, "Applications of Mathematics to Some Problems in Economic Geography," *Soviet Geography: Review and Translation* 5, no. 6 (June 1964): 36–53, and "Applications of Mathematics to Population Geography," *Soviet Geography: Review and Translation* 8, no. 9 (November 1967): 709–21; and B. L. Gurevich and Yu. G. Saushkin, "The Mathematical Method in Geography," *Soviet Geography: Review and Translation* 7, no. 4 (April 1966): 3–25. Relevant articles by Blazhko, Gurevich and Matlin appear in this section.

11. See S. A. Kovalev, "A Geography of Consumption and a Geography of Services," *Soviet Geography: Review and Translation* 7, no. 7 (September 1966): 65–73. In this article Kovalev proposes the establishment of new branches of economic geography to study these activities. See also B. B. Rodoman, "The Organized Anthrosphere," *Soviet Geography: Review and Translation* 9, no. 9 (November 1968), especially the discussion of a multi-level system of services, pp. 790–792, and O. V. Leont'yev, "A Method for Calculating the Requirements for Mobile Public Services under Various Geographical Conditions," *Soviet Geography: Review and Translation* 8, no. 9 (November 1967): 737–43.

N. N. BARANSKIY

On The Economic-Geographic

Study of Cities

Cities can be studied in various ways, from various points of view, with different approaches, and on differing scales. Cities can be an object of study for historians, geographers, statisticians, economists, and sociologists, as well as planners, designers, architects and builders, financiers, military specialists and others—all can be interested in cities, each in his own way. In order to realize fully one's own approach to the study of cities and the special problems in this study area, each specialist will do well to acquaint himself with other approaches even if only by an overview of the basic literature of other specialists on urban studies.

Geographers, particularly economic geographers, can study cities either in association with the description of a country or region or by distinguishing cities or a single city as the particular object of study. In its territorial scope, the geographic study of cities can be quite varied. Cities can be studied on a world scale, on a country scale, or on a regional scale. One can also subject cities of one type or one category to comparative study; finally, one can be engaged in geographic study of a single city.

Translated and abridged from N. N. Baranskiy, *Ekonomicheskaya geografiya. Ekonomicheskaya kartografiya* [Economic Geography. Economic Cartography] (Moscow, 1960), pp. 178–221.

The geographic approach to the study of cities suggests inquiry into questions on the distribution of cities in general (or within the limits of a given territory) and on the mechanisms of this distribution; questions on the distribution of cities of a definite category (by size, type, specialization); and questions on differences in the degree of development of cities in the various parts of the territory under study, on the differences in the types of cities, in the character of their ties, in their situations, and so forth. Questions concerning the comparative percentages of city population in various countries and regions[1] and questions about the vitality of cities, especially on the degree of concentration of city population in the capital of the country or in the main city of a region, can be considered also. The degree of the geographic orientation in works of this kind can vary greatly; in essence it is determined, not by the title, but by the treatment of the theme itself. If emphasis is put on the exposure and explanation of spatial differences (in the distribution of cities, their character, and so forth) and also on spatial combinations and relations with regard to geographic details, then they will be geographic works.

The major, general questions that arise in the study of cities and that are also of interest from the geographic point of view of urban study are: (1) the problem of the very definition of a city (which settlements must be considered cities); (2) the problem of the typology of cities.

The first—*the very definition of a city*—is far from being as simple as it may seem at first glance. Here one must, first of all, take into account the fact that for a settlement to be considered a city, different criteria are established in different countries, and even in the same country different definitions are adopted at various times. In every country one can observe how some settlements, which formerly had the status of villages, become cities, and how others, on the contrary, are demoted from the status of cities to villages.[2]

From the lack of agreement on this question it is quite evident that many have ignored the substance of the question and have been compelled to accept a purely formal criterion: population size,[3] or simply the presence of a given settlement in an official list of cities (that is, a city is what the proper authorities accept as such). Of course, we do not intend to go into this question and propose our own definition of a city (this could be the subject of a special investigation); we want only to note the narrowness of the prevalent idea of a city as primarily a trade-industrial center. In our opinion such a definition is narrow not only as a general definition but also as a definition for a city of the capitalist era. There are many cities in any capitalist country that no one would think of excluding from the rolls of cities, but that at the same time have no significance as centers of trade and industry—cities that grew fulfilling a different organizational function.

Such are the purely administrative-political cities (a clear example is Washington, D. C., and a number of state centers in the United States), cultural cities (for example the university cities in Germany), resort cities, military cities, and others.

The historical limitation of this point of view is emphasized by the fact that in countries with a strongly expressed monopolistic capitalistic system, like the United States, a different point of view on city definition is emerging, which no longer emphasizes trade and industry in general but financial activities and the processing industry;[4] the adherents of this point of view consider the extracting industry together with agriculture as the recovery of raw material directly from nature.

The development of monopolistic capitalism, accompanied by an increase in the role of finance capital, has led the United States to distinguish super cities, which are primarily commercial-financial centers, from the total mass of cities. Super cities of this type have a concentration of: (1) enterprises of wholesale trade with wholesale warehouses, elevators, and refrigerators; (2) transportation headquarters—rail, road, and water; (3) the largest financial establishments—banks, trusts, insurance companies; and finally (4) the largest newspapers with their editorial offices, bureaus, and printing plants.

The American trusts have created a special type of architecture. The American skyscraper is the heart of what is called *haute finance*; it is the specific architecture of financial capitalism just as castles were the specific architecture of the feudalism of the Middle Ages. The business center with its skyscrapers, having absorbed banks, trusts, insurance companies, and other financial institutions, is the citadel of the American super city, for which the United States has coined a special name, *metropolis*, not in the sense of a capital but precisely in the sense of a super city characterized primarily by financial functions.

The problem of the definition of the city is made more complex by the prevalent phenomenon that large cities usually spill over into the suburbs, merging into a completely *built-up area* that encompasses hundreds of square kilometers. Such agglomerations of inhabited places are also created in other ways, not only as a result of sparks scattered by the large city, but as a result of the fact that over a considerable area conditions are created favoring the formation of urban settlements that are independent of each other. Most often these are the areas where there are deposits of useful minerals, but they can occur by especially convenient sea bays in the deltas of large rivers and in other causal places, including resort regions.

In all these cases a large agglomeration develops completely, or almost completely, covered with urban structures. Such agglomerations are found in the Donets Basin, in Germany in the Ruhr basin, in large spaces in England, in the Northeast of the United States, and also in other countries.

The juridical design always adheres to tradition, operates in the old fashion, and therefore lags behind life. The internal structure of this kind of agglomeration can be quite diverse. It can consist of an accumulation of small new cities surrounding one old large city; it can be the merging of two or more cities placed close to each other, and tangent to the suburbs of more or less equivalent cities. Finally, it can consist of an agglomeration of many equivalent small cities, none of which can pretend to a leading position.[5] Between cities merged in this manner considerable areas remain that are either not built up at all or are built up less densely and have a population of a more or less agricultural character. It is here that a number of questions arise that are quite difficult to solve: where does one draw the boundary dividing the city from the noncity in an area filled with such an agglomeration; should one consider this whole agglomeration as one city or distinguish within it several cities, and in the latter case how many and which ones? Much depends on the solution of these questions: the total number of cities, the city population and the division of this population into groups according to size of population, the number of cities in each group, and so forth.

The question concerning *classification of cities* is to a certain degree related to the question of the definition of the city, because if one considers cities to be only economic centers, then, in the classification of cities according to their functions, one clearly does not have to distinguish between administrative-political cities, cultural cities, military cities, and so forth. The most common classification of cities applied by economists and statisticians, and of course very important also for economic geographers, is the classification according to the predominant functions fulfilled by each of the cities.

In referring cities to one type or another, the statisticians most often proceed from the structure of the gainfully employed population according to sectors of the economy.[6] For each type a definite qualification is established and then expressed in the form of a proportion of the corresponding categories of the gainfully-employed population. It stands to reason that all such definitions are quite arbitrary; it is also clear that in reality, according to the function they fulfill, the overwhelming majority of cities belong not to pure types but to mixed types. In addition, one must keep in mind that this classification of cities, according to function, can be further divided within each of the large sectors of economy. Thus, for example, industrial cities can be subdivided into cities where the mining industry predominates and cities where the processing industry predominates; moreover, both of these subdivisions can be further divided by individual industries.

The classification of cities according to their functions is the most prevalent procedure, if one does not consider, of course, the purely numerical classification by number of inhabitants. In our literature this kind of classification of cities according to functions (on the basis of the structure of the gainfully employed population)[8] was conducted by Kvitkin from the censuses of 1923 and 1926, and then this classification with a few corrections was applied to the cities of the Ukraine by Vologodtsev.[9] In the classification of the employed population, such branches of labor as agriculture, manufacturing, handicrafts, transport, trade, government service, and others were distinguished. By combining occupations according to their relative predominance, Vologodtsev obtained the following groups of city settlements: manufacturing, handicrafts, transport, administrative-trade-industrial, agricultural manufacturing, agricultural- handicraft-trade, agricultural-administrative, and agricultural-transport. The above-mentioned works were wholly statistical without the slightest attempt to put the classified cities on a map in order to find some regularities in their distribution.

Besides classification of functions, we can think of a number of other possible classifications, with other bases for division: for example, classification by the date of founding or origin, singularities in situation (here the reference to communication lines is of special interest), by administrative-political significance, by planning, and also by the character of architecture and of building materials. Anthropo-geographers are especially interested in the relation of the cities to physical-geographic elements, differentiating between cities with coasts, rivers, lakes, passes at the outlet of mountain valleys, at the edge of deserts, and so forth.[10] Cities are also categorized according to their height above sea level, distance from the sea, and so forth.

In working out the typology of cities, it is necessary to keep in mind that in various historical eras and also for different countries and peoples the main types of cities, undoubtedly, will be more or less different; in each concrete case one will find under the new, the traces of the old.

In the economic-geographical study of cities, the most important and interesting of those features that, in our opinion, are possible *principles for classification* of cities are: (1) The time of origin of the city; (2) The peculiarities of its economic-geographic position;[11] (3) The function fulfilled by the city (or more often the combinations of functions); (4) The size of cities (by population and area); and (5) The extent of the territory over which the city extends its influence.

In the economic-geographic study of any city, it is undoubtedly necessary to study the connections and interrelations between these features; especially interesting and important from the economic-geographic point of view is the determination of the influence of the economic-geographic situation of the city on its development, especially on its functions and its growth. This will make up an essential part of the economic-geographic characterization of each city.

It is simple to divide cities by types according to each of the five bases; there are many such divisions, and, undoubtedly, they are of some use. However, as we have already pointed out, such divisions are quite tentative and often even arbitrary[12] and, in addition, they cannot pretend to be universally applicable to all eras and all countries.

Of much greater interest, but at the same time more difficult, would be a combination classification taking into account all these features as a whole. In the combination of these most important features, in their interaction and interconnection, logical and genetic, lies precisely the substance of the characterization. Some of these relationships (3 and 4, 3 and 5) are clearer than others. Of special interest from the economic geographer's point of view are the relationships between the economic geographic situation of the city and all the functions it fulfills. A priori the number of all possible combinations seems to be exceedingly high, but by thoughtful and detailed examination of typical combinations it will prove to be not very high.

One can think of such a composite classification as one in which the time of origin of the city would be taken as the first basis of division and afterwards the other features would be taken into account, such as situation, functions, and size. This promotion to the foreground of the time the city's formation would also make it possible to account for the peculiarities of geographic situation and the character of the functions fulfilled by the city, inasmuch as both these elements had very different significance at different times. Thus, one could put things in their places and master and understand them with greater facility.

But of greatest significance, in our opinion, would be a classification by genetic principle. This thought is conveyed by numerous examples from the history of the development of a number of other sciences, highly diverse from each other, that show that genetic classifications[13] of that type had the most important and decisive significance whereas all others played only a secondary role.

As for the question of the geographic situation of cities it is necessary to point out that it has long attracted attention; in many cases one has had to note how, after destruction, cities rose again and again on the same or very nearly the same spot.[14] Hence the thought naturally arises of special advantages in the situation that must explain the rise and growth of cities on that very place. The question concerning the economic-geographic situation of a city, as that of a country or region, is one of the most important.[15] For a city the question of its situation is probably even more important than for that of a country or region. On this question there is much literature.[16]

The great majority of works in the field of geography of cities are monographs devoted to the characteristics (or description) of some one city. This kind of work is written by the tens and hundreds in the form of dissertations and journal articles.[17] For a monographic study to give a rather full picture of a city with all its specific individual features, large and ultralarge cities are most suitable. And this is not because of their individual importance, but because they are more individual than the small cities in which the typical features usually predominate over the individual features, and therefore their study must be more typological.[18]

Between works devoted to general questions of the geography of cities—typology, location depending on natural and transport conditions, economic-geographic situation, character of communications, and so forth—and works devoted to a description of a given city, there is a certain dialectic interaction according to the formula: "the whole gives a soul to the parts, the parts give life to the whole."[19] The works of the first type are of great methodological significance for the works of the second type; they enrich them with a number of common ideas, open new points of view, and new approaches to the study of the chosen object. Due to the works of the second type, there accumulates the vast, often very interesting, factual material necessary for wider scientific generalizations in a given field, which at the same time provides interest for the totality of methodological questions of the geographic study of cities.

Therefore, we think it quite expedient to concentrate our attention on the question of characterizing the city. There is nothing sacred about the scheme for the description of a city. Its primary significance lies in the fact that it defines a definite methodological arrangement. It can be useful as a reminder of all the important facets of the object that should not be omitted in the presentation or in the investigation itself. We repeat once more, there is no need to adhere strictly to a previously assigned scheme, especially as concerns the order of presentation.

Proceeding from a critical study of a number of monographs, essays, and articles devoted to the geographic character of cities and guided by previously presented general considerations, we could recommend the following scheme for an extensive description of a city:

1. Introduction;
2. Location and natural conditions of the city and its immediate environment;

3. Historical-geographic outline of the city;
4. Contemporary description of the city as a whole;
5. Microgeography of the city;
6. Immediate environs of the city;
7. A look into the future.[20]

[Sections dealing in detail with items 1, 2, and 5 of the proposed outline have been omitted from the translation, eds].

HISTORICAL-GEOGRAPHIC OUTLINE

If the second section characterizes natural possibilities, the third speaks of the utilization of these possibilities in the process of the historical development of the city. The essential question here is the establishment of a division by periods. Having studied the factual history of the city, it is necessary to bring out the turning points in its development, points when the city's function changed qualitatively, when the city changed its type, its face, or when, while retaining the type of development, it changed sharply in size and significance. Such turning points are caused most often by sudden changes in the economic-geographic situation of a city that in turn are most often related to changes in the transport situation but can also be caused by changes of another order, for example, changes in the international situation of the city. A sudden change in the development of the city may be caused by circumstances other than changes in the economic-geographic situation—the discovery near the city of new, previously unknown deposits of useful minerals or the invention of new technological methods making it possible to utilize natural resources, that could not be utilized by old techniques.[21] For capital cities, sudden changes in the fate of their countries are of great significance. Together with its country, a capital grows and rises, but, with it, it also withers and falls.

Having established a division into periods and having briefly mentioned the main stages of the historical development of the city, one must then give for each period the corresponding geographic description of the city at that period; into this delimitation enter the functions fulfilled by the city, its population size and the composition of its population, its external connections, its microgeography by major regions, the general appearance of the city, the character of its buildings, and so forth.

At the end of each period the motive forces that prepared the transition from that period to the next must be briefly pointed out; these will be brief historical insertions connecting the geography of one period with the geography of the following period.

The historic-geographic outline must not evolve into pure history; it is necessary to warn against this, because such errors have often been made in the past. In order to avoid this danger it is necessary in each period to focus upon the elements of geographic order—the economic-geographic situation (first of all, the relation to the transport network), the type of city, its external connections, the internal division of the city, and so forth—and not upon the course of historical development itself. Furthermore, in each period attention must be mainly directed towards the connection of phenomenon, on the functional dependence between the above-noted forces—situation, function, communications, microgeography—because these forces do not exist independently but are very closely connected, so that only one factor need change and all the others will be changed as well. Thus, in a historical-geographical outline one must not give so much history as such, but the geography of each of the historical periods.[22]

One should bring out in the past that which is necessary for the explanation of the present, remembering that a historical-geographic outline in a geographical work has no independent significance; similarly, one must not go too deeply into the centuries; there is no need to devote attention to archeological findings and other antiquities having no bearing on the present time.[23]

From this it does not follow, however, that in the historical-geographic chapter the final, contemporary period must be described in too great a detail. Because it can then happen that all remaining items of the presented outline will enter in separate paragraphs into this most recent, contemporary period. Furthermore in the remaining sections of the scheme there will be nothing more to write, and one will have to repeat himself. In order to avoid this repetition, it is suggested that the presentation of the last period be as brief as possible.

For cities that have gone through history, "by an abbreviated course," so to speak, in which changes in recent times occurred too fast for the new to absorb the old, the historical-geographical outline is especially important for the precise purpose of bringing out the elements of various eras in the contemporary city and to classify the various layers in their historical sequence.

CONTEMPORARY DESCRIPTION OF A CITY AS A WHOLE

This is, of course, the central, part of the whole work. The preceding sections devoted to nature and history have an introductory significance with respect to this section; they give the premises necessary for the understanding of the contemporary appearance of the city. The following sections, which give a description of the separate parts of the city and its surroundings, serve as essential additions to this central section; the commanding place remains for the chapter devoted to the characterization of the city as a whole. It is clear that the greatest share of attention must be devoted to this chapter; moreover, it should receive relatively greater emphasis the shorter the study. In a brief presentation this central section becomes the only one, and all the others—the preceding and following—are incorporated in it, losing their independent, special existence.

The central section can be composed of three parts.

1.   The *type and significance of the city*: the basic functions fulfilled by the city, their dimensions and relationships.

2.   The *population of the city*: its composition (national groups, occupational and other), its cultural level, and its life.

3.   The *outer appearance of the city*: organization of services, type of construction, type of buildings, interurban transport, the layout and its division into parts.[24]

With regard to the first point, it is most important to determine the type of the city in terms of its economic-geographic situation, the natural conditions of its surroundings, and its historical past. The determination of the type of the city is, of course, the central point in the whole description of the city, so special attention must be paid to this question. The main indicator for the determination of the type of the city is, at the present state of statistics, the occupational structure of its population. In the course of scientific research work, one must go to the origin, that is, the primary materials of the latest census, and then analyze, verify, and organize them in order to maximize results.[25] In cities of a simple structure with a clear predominance of some one function, the determination of the city-type is comparatively simple. It is quite different with cities of a mixed type.

To assign a city to a certain type is a large part of the description of the city, but it is far from being all. "Definition is done by referring to the closest genus and pointing out the specific singularity." To bring out this specific singularity, it is necessary to compare the given city with cities of the same type and hereby take into account all peculiarities of its situation, natural conditions, and history. Only in the combination of the typical with the individual does the description acquire the necessary completeness.

Then, when the main functions of the city have been established, a description of each function must be given in the order of its importance, so that the order of presentation for the various cities will vary according to the main formula of the city. It is especially important to emphasize in each function the specific features of the city and connect them with the general description. The analysis of the economic functions of a city—transport, trade and industrial—cannot, of course, be conducted apart from the external ties of the city and their character, content, direction, and intensity. External ties have an especially great significance for the geographic study of cities, since only an analysis of these ties of a city makes it possible to understand the city against the background of its surrounding geographical environment.

The interrelationships of the city with its surroundings develop quite differently depending on whether the region is industrial or agricultural. In an industrial region, the agricultural environment of the city turns into its appendage, serving it with vegetables, milk, and recreation. In an agricultural region, on the contrary, the service role falls to the city that sells the products of agriculture and processes them.

In order to bring out the external ties of the city it is necessary to study the traffic to and from the city, in goods as well as people. Just as the traffic of goods is an important consideration for the geography of trade and industry, the traffic of people is important for the geography of other functions of the city—administrative, cultural and others. The movement of goods is best studied from transportation statistics using accounts of trade and industrial organizations to supplement and make the transportation data more complete. In their totality these data make it possible to reveal the area over which extend the industrial connections of the city as a manufacturing center (in raw materials, semifinished products, energy, fuel, equipment) and its relations as a trade and distributing center.

It is more difficult to study the movement of people since passenger statistics are meager. In order to find the area from which the city attracts students, resort visitors, and tourists, it is necessary to turn to the proper institutions relating to each group. For cities in the process of vigorous growth, especially the young and rapidly developing ones, it is very important and interesting to find out from where the city attracts new population. Such data, although not exact, can be obtained from current statistics of the population. And for a longer period one can use primarily materials of the last population census (if place of birth is recorded there).

When examining the industry of a city from the point of view of economic geography, it is exceedingly important to distinguish the industry working for distant markets from industry working for the local market, especially that for the satisfaction of the needs of the city itself. Of major significance, of course, is the industry working for distant markets; this is what determines the appearance of the city. As for industry serving the population of the city, it is more or less the same everywhere (bakeries, butchershops, shoe manufacturing, tailors, and so forth) and is therefore of little interest.[26] The distinction is made easier for us since industries of different significance belong to different orders.

In analyzing the industry of a city, it is necessary to find out all the conditions of development for each of the industrial branches important to the city, namely: (1) sources of the main types of raw material, their quality, cost, and distance from the city; the conditions of transport and cost of procurement; the possibilities for development and improvement of the energy and fuel base; (2) equipment and technology of production; (3) cadres of manpower, their composition, quality, training, productivity, and conditions and level of pay; calculations per unit of production, the possibility of raising productivity of labor and lowering the cost of production; (4) possibilities of marketing arrangements made by taking into account prospective development; here it is necessary first of all, of course, to determine the market region of a given product, fixing the boundaries with the neighboring enterprises of the industrial branch under examination.

When analyzing all the factors of industrial development of each branch basic to a given city, it is necessary to keep in mind: (1) *peculiarities of local conditions*; (2) *productive ties* between branches of a given industrial node and its immediate surroundings; (3) *comparison of the given node with others* analogous to it or cooperating with it.

In the study of our Soviet cities, the analysis of the conditions of their industrial development must be conducted with special completeness and thoroughness and in all the detail necessary to take fully into account the current needs of the economy and the directives on the development of individual branches of the economy and individual regions. For the most part, industrial functions are closely connected with transportation and trade functions when an assessment is made of their significance in the life of the majority of cities.

Within transportation, one must distinguish its types (rail, auto, river and, for maritime cities, the sea); in addition, it is important to make a distinction within transportation according to the significance it has in fulfilling the functions of the life of the city (supply of raw material and fuel for industry, hauling away of finished products, supply of food for the population of the city, and transit). It is very important to take into account the connections between the various types of transportation, both functional and spatial (mutual location of railroad stations, wharfs, and so forth). Within trade one must distinguish wholesale and retail trade; in a number of cases warehouse operations can be of great significance to a city. The transportation functions and trade are of great significance for cities that are large transport centers, such as cities at the estuary of navigable rivers or at the intersection of a navigable river and a railroad, or at the hub of several railroads.

In describing a city one must not neglect the other functions of the city, primarily the administrative, organizational, and cultural. Here one must take into account the fact that the relative significance of this category of service population—officials, employees, professional people—continues to grow in a number of countries. As we already know, in a number of cities these other functions have distinctive and sometimes predominant roles; in addition these functions do not remain without influence on the appearance of the city as well as on its economic importance. Here one must take into account that the presence of these other functions increases the population of the city on a permanent basis (occupied in fulfilling these functions) as well as on a temporary basis (persons coming to the city for a time in connection with fulfilling these functions: study, medical treatment, legal services, or business affairs). This increase in the population of the city results in any case in the increased capacity of its consumer market. In addition, the fulfilment by the city of administrative, political and cultural functions can attract some branches of industry either because they are necessary for the city's functions (as, for example, a copying-machine industry in political and cultural centers), because these branches need the services of the scientific institutions (as, for example, a number of precise and complex branches of the chemical and machine-building industries), or, finally, because these branches find among this population the main market for their special product (as, for example, factories and stores of musical instruments in cities with conservatories and music schools).

The significance of these other functions—administrative, political and cultural—is greatest in capital cities in which the political and cultural life of the country is concentrated, and increasingly so as the government becomes more centralized.[27] Hence is clear that when describing capital cities and also other political-administrative centers special attention must be given to these functions.

In regard to population, one must first of all show not only the total number but also its dynamics for recent years, as well as the composition by class, nationality, and occupation. One must then point out the cultural level of the population and note the most characteristic features in the life of the city (considering the various nationalities and various social segments of the population); such elements as public entertainments, festivals, and street life can also be of interest here.[28]

In the section about population of cities in the USSR, it is important to note the shifts in the social composition of the population as compared to prerevolutionary times, the sharp rise in the percent of the service population and also its cultural level. In particular, it will be of interest to show with regard to new cities and cities that have grown rapidly during Soviet time, since there are many, where the new population of the city came from and what new contributions this population has made.[29]

For seaport cities of worldwide significance or cities standing on the boundary of different cultures, it is very important to emphasize the multinationality of the population and the diversity of cultures of the various nationalities. For many large cities (especially port cities) in capitalist countries, the concentration of each nationality in a definite part of the city is characteristic; there are whole blocks of individual nationalities (in New York, San Francisco, Chicago, and also in Montreal in Canada).

In analyzing the population of the city, one must give attention to the heterogeneity of the national composition of the population of the city itself and its surroundings. Such a situation usually arises in a zone, transitional from the area of one nationality to the area of another nationality. As a general rule, the dominant nationality in an area is more industrialized, and so predominates in the cities. The problem of culture is closely related to the question of population.

If there were an independent field called cultural geography, an economic-geographer need not occupy himself with this question but refer those that are interested in it to specialists. But since there is no cultural geography as yet, economic geography—even if only for a time—must take on this topic because the cultural appearance of a city offers a great deal towards understanding it.[30]

Passing on to the external appearance of cities, one must first give a general description of the city, using photographic illustrations. Of great value from the geographic point of view are wide panoramas of the city taken from some neighboring mountain or the sea, or still better from the air, which provide a vivid description of the arrangement of the city and which serve as a photo-commentary on its plan. For cities of the USSR, it is important to show new buildings and to give clear comparison of the new with the old. For cities of capitalist countries, it is important to show the contrast between rich and poor blocks. Then one must describe the main types of city services (paved streets, running water, sewers, lighting, and intercity transport).

City services by no means correspond to the wealth of a city and the value of production created in it. The mining towns of the mountain states of the United States serve as a striking example. The nonferrous ore is mined through the labor of their inhabitants to a value of tens and hundreds of millions of dollars, but these cities very often have a shabby appearance. Record-breaking in this respect is the city of Butte, Montana, an unattractive city of 40,000 inhabitants; in the middle 1920s it celebrated a strange jubilee: the first billion dollars had been made from the mining of nonferrous metals. The essence of the matter lies in the fact that the profit goes into the pockets of shareholders living in the large cities of the northeastern United States; but from the wages neither pavements, nor public gardens, nor hospitals can be created.

In describing the external appearance of a city, how it is built up, the type of buildings, and so forth, the geographer must, of course, give attention to its specific geographic character. For forested regions wooden houses are characteristic; for unforested regions, stone, brick, or clay houses; for mountainous localities, narrow streets with little houses clinging to steep slopes; for a hot and dry climate, houses having no windows toward the street with flat roofs and inner courtyards are typical; for a moist climate, on the contrary, houses with steep roofs are typical. All such details are of great value in a geographic description, since they establish the dependence on the natural environment and create in the reader a graphic notion about the city being described.

Finally, one must divide the city into its major sections, pointing out the functions that each one of them fulfills. One thereby prepares the transition to the next section devoted to the description of individual sections of the city.

THE IMMEDIATE ENVIRONS OF THE CITY

This section, like the preceding one, is of significance mostly for larger cities that exert a great and varied influence on their environment. The larger the city, the better developed is its transport, the wider and more diverse its connections, and the greater are its environs. Such a sprawl of large cities, which is due primarily to the improvement and development of convenient means of suburban communication, is a phenomenon prevalent in all countries where large cities have had time to grow. When one sees the haste with which the workers and employees try to break away from the city limits during the summer heat at the end of a workday or workweek to get to fresh air, then one understands vividly the conditions of modern city life. Heat and stuffiness from high stone walls, insufficient fresh air and greenery, especially the lack of playgrounds—all these conditions make summer life difficult in a big city.

In capitalistic countries there are a number of conditions compelling one to stay away from the central parts of the large cities at any time of year. First of all, there is the high price of living quarters and food. Moreover, beyond the city limits it is possible to have a household with a vegetable garden, fruit garden, and a cow. This is on one side of the ledger and on the other are the expenditures in money and time for daily commuting. Interurban trains, especially electric trains and in some places special ships, help to solve this problem. But the radical solution is the automobile that brings one from door to door.

The large-scale development of automobile transport in the United

States has greatly expanded the limits of the immediate surroundings of large cities.[31] The ties of a city with its nearest surroundings are of great importance. One must keep in mind here the totality of the various ties of the city, namely: (1) people who live permanently outside the city and work in the city;[32] (2) people who live in the city and work at some factory or other enterprise outside the city; (3) temporary (usually summer season) residence outside the city for rest (summer house, vacation home, sanatorium, resort); (4) supply of the city with vegetables, milk, berries, fruit, fowl; (5) supply of agriculture of the suburban zone with wastes from the city (field irrigation, garbage from city for pig feed); (6) daily trips to the city (to study) or occasional trips (to shop, go to the doctor, to litigate, to do business).

Due to the different kinds of ties, the limits of the sphere of attraction vary greatly. Therefore it is not simple to determine these limits.[33] The deciding factor here must probably be the element of density of communication and, above all, passenger communications. Scrupulous and hard scientific investigation is demanded to determine these limits, but at the same time the study is of great scientific and practical interest. Revealing the character, direction, and intensity of these communications can greatly add to the economic-geographic knowledge of the suburban zone and the city itself.

The description of the environs can be done in two ways: either in a systematic, logical way by indicating the types of communications and then describing the area encompassed for each type; or topographically, moving from place to place in sequence and noting for each place all types of communication this place has with the city.[34] The first method of description tells more about the communications themselves, but compels one to be repetitive; the second plays down the communications somewhat but makes it possible without repetition to describe sensibly the surroundings of the city. If the description is to serve also as a guidebook, then it is best to conduct it by routes, proceeding from a map scheme of the immediate communications of the city and arranging the environs along roads leading from the city.

A number of cities may have very important suburbs of some independent significance and interest (Versailles near Paris; Potsdam near Berlin, Petrodvorets, Kolpino, and others near Leningrad; Kuntsevo, Fili, and others near Moscow). It is quite clear that special attention must be given to their details if the total size of the work allows it.

A LOOK INTO THE FUTURE

This concluding section is of special significance, of course, for our cities in a socialist country with a planned economy, a country looking toward the future and a period of rapid growth. Showing the prospective development of the city has a direct, practical, and real significance for us in planning. With respect to cities of the USSR, a revelation of their prospective development is the logical conclusion to be drawn from the presentation that has preceded; such a revelation also serves as a conclusion in itself, for it summarizes the analysis.

The elaboration of the prospective development of a city must proceed along two lines of thought, logically connected but at the same time different. On the one hand, it should be a plan of development of functions fulfilled by the city, above all industrial; on the other it should be a plan of new construction: occupational (above all, industrial), residential and communal. It is quite clear that both these schemes are mutually dependent and serve each other as necessary supplements, because before anything is built anywhere, one must know what it is and why it is to be built; but on the other hand, one cannot accept a plan for the development of a city without giving thought to precisely where the proposed building is to be built and what further practical obligations these buildings will have with respect to intercity transport and other communal accommodations, and with respect to residential buildings. Thus, a plan of city development must encompass both at once.

The plan for development must proceed from a certain ideal of maximum development of all possibilities provided: *of natural resources* of the city itself as well as of its surroundings (energy, raw material, water, building areas, building materials); *of labor experience* of the population; *of requirements* of the region and of the country as a whole. In working out a plan of development for a city under present conditions, the rapid *perfection of technology* must be taken into account as well.

The structure of industry planned for the city must be selected in such a manner that its maximum comprehensiveness is assured (that is, maximum development of productive connections within the center, maximum utilization of raw material, and completeness of the entire productive cycle) together with a correspondingly low cost of production.[35]

From this it is clear that planning by industrial sector alone, which comes down to a scattering among industrial centers and then only in the regions where new construction has been planned for each sector, is far from sufficient to meet national demands, because in such planning each center receives a random selection of new buildings without assurance of comprehensiveness. It is also clear that to set up a plan for the development of a given city, having in mind only this one city without comparing its possibilities with those of a number of other cities, is also out of the question; in order to justify the plan for the development of a given industrial center, it is necessary to proceed continuously by comparing and "looking afar."

The next stage in setting up the plan for development, after the plan of maximum possibilities of development has been elaborated, will be the establishment of a definite sequence of the maximum planned construction.

In determining the dimensions of the planned growth of the city—by population and construction—it is methodologically important to distinguish: (1) the increase in new functions of the city in the interregional division of labor; (2) the increase in the function of serving the internal needs of the city. To the first category would belong new industrial and transport enterprises and also new cultural and administrative institutions that are designed for a circle of consumers outside the city, a large factory, new railroads, higher education institutions, oblast and administrative establishments (in the event that the city becomes an oblast-center). To the second category belongs the servicing of this new population with local industry, local transport, communal services, schools, and so forth. The increase in the first category must be calculated before the second category, which is a function of the first, can be known. The scheme for transforming development onto a map must be undertaken along with the development plan.

For each new building a definite area must be assigned that takes into account all conditions, namely: the conditions of the area itself (relief, soil, ground water, and so forth) and approach roads (for supply of raw material, fuel, the shipping out of production, and industrial ties with other allied industrial enterprises). And here again, a scheme must initially be established for planning the maximum utilization of all possibilities, and then this plan must be divided sequentially by order of construction.

For each plan one must also consider all the consequences that follow the industrial construction: the consequences for transport, for city service, for communal and residential construction, including the laying of new streets, new trolley lines, new water networks, and sewers.

The questions of rational planning and replanning can arise and often do arise, of course, not in connection with new construction, but as a need caused by the random growth of the city dating from prerevolutionary times.

Especially for larger cities, some problems can acquire an acute significance, since they are at the same time complex and difficult. This is especially true for cities situated not on an open flat plain that allows unhindered construction in various directions but situated in an area cut up by mountainous relief or water.

A scheme of rational planning or replanning of a city must simultaneously take into account a number of circumstances as, for example: the innercity regionalization and convenient connections between its parts; the space for the approach to the city of outer routes, that is, for railroad stations and sea, river or lake ports; and provision of all elements of urban services—sewers, running water, and lighting.

Due to the urgency with which such projects must be planned under Soviet conditions, we are accustomed to working in several stages: first generally and in its entirety, then more precisely; this is the so called "method of successive approximations." Of course, all cannot be anticipated, but the major and decisive elements must be foreseen or the entire undertaking will be useless.

The detailed projections for industrial development of cities, and schemes of planning and replanning, demand a very special technical knowledge that cannot be expected of economic geographers. Hence the need arises for a certain division of labor between the economic geographers, on the one hand, and engineers and architects on the other. The economic geographers can only assume responsibility for the general outline of development and replanning, leaving the elaboration of detailed working projects to the proper specialists—planners, engineers, and architects.

In drawing up the prospective development of a city, especially a large city, the interests of culture, in the widest sense of the word, must not be ignored. Attention must be given not only to a network of educational and public health institutions but also to children's playgrounds, parks of culture and rest, theaters, cinemas, and so forth.

This is the state of affairs with respect to Soviet cities. The more detailed and concretely elaborated the plans for development and new construction, the better it will be; for in our country there are people to realize these schemes. With respect to cities of capitalist countries it is a different matter. However, by studying its history, economics, and geography, some assumptions can be made concerning the future of a capitalist city. This can be especially interesting when a given city approaches a certain turning point in its development or runs into certain problems.

But these are only assumptions and have little to do with the planning assignments elaborated and implemented in our country.[36]

1. Of special interest are the cases where the percentage of urban population does not correspond to the degree of industrial development of a given country or region, as is the case in Australia, Canada, Argentina, and in our Far East or the Pacific Far West of the United States.

2. Since the Great October-Socialist Revolution up to 1,000 new cities and urban-type settlements have appeared, but simultaneously about two to three hundred provincial cities were demoted.

3. The qualification in different countries at different times has varied—2,000, 2,500, 5,000, 8,000.

4. See, for example, the article by Julie Mosches about Prague in the American journal *Geographical Review* 27, no. 3 (1937).

5.　The last example, not often found, is characteristic of mining regions.

6.　Here the following types of cities are usually distinguished: administrative-trade-industrial, trade, industrial-transport, administrative, city-villages. Also distinguished are: cultural, military, and resort cities.

7.　Whatever gradation is offered by size of population it would be difficult to prove their qualitative significance. Why, for example, must a large city have 100,000 population and a superlarge one, 500,000? Or why does a city come under the category of industrial with 20 or 30 percent industrial workers in the composition of the gainfully employed population?

8.　One has to use the indices of population because there are no other indices that more clearly express the economic significance of functions (as, for example, the index of pure production in a value-price expression).

9.　I. K. Vologodtsev, *Osobennosti razvitiya gorodov Ukrainy* [Characteristics of the Development of Ukrainian Cities] (Kharkov, 1930).

10.　The special features of the physical-geographical site are often taken into account in greater detail; thus, for example, there are estuary cities (including those at the head of the estuary) as, for example, Quebec, cities near falls, fords, islands, the remains of river terraces, and so forth.

11.　On this question see my article about economic-geographic situation published in this collection. The article by O. A. Konstantinov in *Geografiya v shkole* [Geography in School], no. 3 (1941), deals with the problem of the economic-geographic situation of cities. In it are elaborated many specific examples of how, and under what conditions, the economic-geographic situation of a city is changed and together with it the type and designation of the city itself.

12.　One can also think of a mixed classification proceeding from several bases of division (some combination of features).

13.　One type of clear and very enlightening example is the question of classification of shore types in the field of geography.

14.　Serving as examples of such cities are Istanbul (Constantinopole, Byzantium), Tunis (Carthage); in Russia, Rostov, Sevastopol', Odessa, Tbilisi, and Volgograd. Here the convenience of location may play a role (at a convenient sea bay, estuary of a navigable river, confluence of rivers, important deposits) and also the remains of material values and sometimes, one would think, simply tradition.

15.　A city can better create for itself a situation than can a country or region. In the article by I. M. Mayergoyz "The Geographic Situation of the City of Volgograd" (*Voprosy geografii* [Problems of Geography], no. 2) we read: "Especially significant is the situation for a city, where intense economic, and not only economic, activity is concentrated on a territory quite limited in dimensions, for a city that first of all grows not at the expense of its own territory, not "by itself," but grows out of the economic life of the district or country of which it is the focus. For the origin and development of a city at a given place the situation is, as a rule, more important than for the country or the region."

16.　Of older works we can mention Korl and Hassert, and of the new the works by Lavedan, *Geographie des villes* (Paris, 1936).

17.　Of recent major works devoted to the geographic characteristics of individual cities we point out the description of Quebec in the work by Raoul Blanchard, *L'Ouest du Canada francais* (1953).

18.　Of course, there are small cities with a clearly expressed individuality.

19.　In addition, there may be works devoted to some cluster of cities; this kind of cluster is most often found in large mining regions.

20.　One must, of course, distinguish between the scheme of presentation and the scheme of investigation. These are different things but are interconnected. The scheme of presentation predetermines to a considerable degree the content (but not the sequence) of the investigation itself.

21.　There can be reverse cases when the new technology undermines the branch of economy on which the development of the city was based. For example, Quebec, for a long time the largest center of wooden shipbuilding has lost significance since the 1870s when ships were built of metal.

22.　The best example known to us of such a historical-geographical outline of a city is the outline of Quebec in the work by Blanchard.

23.   In this case we have in mind the historical-geographic outline as one of the chapters in a historical-geographic monograph. There can be works specially devoted to the history of a particular city; for such works all periods of history are of equal value.

24.   A city is not only a point with a number of statistical indices but a definite area on the earth's surface and, moreover, an area developed in a definite way; the geographer must never forget this. Around large cities there must certainly arise satellites in the form of craft settlements and various other settlements—summer cottages, factories, railway station settlements. As they grow and the city too, they merge into one solid, built-up point so that there is no visible boundary between them. In essence such suburbs or outskirts are one and become parts of one large city. But the expansion of the official city boundary, as a rule, lags behind life, and many of these suburbs and outskirts formally continue their independent juridicial existence remaining outside city boundaries. With the juridical city lying within the official city limits there is formed another city, the greater city, having sometimes twice as large a population. For example, New York within the city limits has over 7 million inhabitants, and Greater New York within a 50-mile radius from Battery Park has about 14 million. In official London there are 4 million, and in Greater London about 9 million inhabitants. The same is true for Paris, Berlin, Leningrad, and Moscow. It is exceedingly difficult to determine in such cases the true limit of the greater city.

25.   The special value in the description of Quebec, mentioned above, lies precisely in the exploitation of the primary materials of the census.

26.   In the United States it is assumed that the number engaged in this kind of local industry is about 10 percent of the city's population; when industrial maps are drawn this 10 percent of the city's population is omitted from the number of industrial workers and the symbols for industry are plotted from the remainders.

27.   Goethe called attention to the difference in this respect between France and Germany when in one of his talks with Eckerman he pointed to the presence in Germany, in connection with its historical and political fragmentation, of almost twenty cultural-political centers, whereas France the whole clutural-political life of the people was concentrated in one center-Paris (I. P. Eckerman, *Razgovory s Gyote* [Talks with Goethe] [Moscow, 1934], p. 785).

28.   A strong development of street life is quite characteristic for cities in the torrid zone.

29.   For large cities with a more developed internal division and in general worthy of special attention to microgeography, it is of interest to follow this historical stratification in space also. In what sequence did the city grow? What were the reasons for creating new sections of the city? Who settled them? How, with the growth of the city did the functions of its old sections change, how were these old sections rebuilt and changed in appearance?

Especially rich examples of this process of stratification of new on the old can be found in the United States, the USSR, and also in the Chinese People's Republic.

30.   Robert and Helena Lynd in their book *Middletown* have given a remarkable picture, diverse and thorough, of the life of an average American city. In our literature good descriptions of the cultural appearance of cities in Czechoslovakia can be found in outline form by Il'ya Ehrenburg.

31.   The best example is Los Angeles, which, according to the competent witness of a German author in a monograph devoted to this city, together with its immediate surroundings has more automobiles than all of Germany and is the largest city in the world in the area it occupies. In order to understand fully these singularities of Los Angeles it is necessary to take into account at least three circumstances: (1) a large percentage of the inhabitants are well-to-do people who have moved there in order to live in a resort-like area; (2) Los Angeles is spaciously developed and widely dispersed, because a large number of the country houses are of a type of individual summer house with garden; (3) the city has its own fuel source (both in extraction of oil and its processing).

32.   According to data of the Moscow Oblast Project of 1939, one-fifth of those working in Moscow live in the suburbs and one-third of those living in the suburbs work in Moscow.

33.   For large cities these limits can be marked by proceeding from the diagram of suburban trains, taking the terminal stations of suburban communication as the extreme limit that cannot be crossed at all, and the limit one would use only in special cases.

34.   The character of communications of the city with its immediate surroundings, depending on the singularities of the city itself, as well as of its surroundings, can vary greatly. An interesting question arises here concerning the determination and study of these communication

types. Inasmuch as the communications of the surroundings with the city, and, consequently, the influence of the city on these surroundings is brought about by the rays of the transport network issue from the city, the limits of influence inevitably acquire the form of a star.

35. In order to show the prospective development of an industrial center, we have since the 1930s applied widely specialized schemes showing the nearest sources of raw material and fuel of the major existing industrial enterprise as well as of those projected, and also the industrial connections between them. However, under conditions of insufficient departmental coordination and haste in building, it has recently happened that instead of a planned radical rebuilding of the city as a whole, near the old city nucleus a piling-up of a number of villages has occured around new construction. Tula and Ulan-Ude can serve as examples.

36. However, it must be noted that in the United States since the beginning of the 1930s the years of the New Deal, various planning works, including replanning of cities were widely carried out. Replanning of cities, mainly capitals, was carried on also in England and France.

V. G. DAVIDOVICH

On the Patterns and Tendencies of

Urban Settlement in the USSR

How will the map of population settlement of the Soviet Union change in the future? A reply to that question requires, first of all, a study of the facts of the present. Many aspects of the future are already evident from the tendencies of development and distribution of urban and rural settlements since the Great October Socialist Revolution. An analysis of these tendencies and study of the existing patterns of settlement are essential for a solution of the problem of reconstruction of urban and rural settlement in the USSR. This is one of the most important tasks in the construction of a communist society. The present paper does not pretend to be a comprehensive study of such a complex problem as settlement patterns, but is essentially an attempt to analyze some of its aspects.[1]

The basic law of population settlement is the general historic law that settlement patterns correspond to the mode of production and to the superstructure, with the former playing the leading role. The influence of the natural environment (relief, rivers, seas,

Reprinted with modifications from *Soviet Geography: Review and Translation* 7 (January 1966):3–31, by permission of the publisher. The article originally appeared in *Voprosy geografii* [Problems of Geography], no. 66 (Cities of the World), pp. 6–33. Figures 3–12 have been omitted from the original article.

forests, minerals) on settlement patterns is indirect and varies with the mode of production. After the introduction of a new mode of production, the forms of settlement are also renewed, not immediately, but after more or less lengthy processes. The old forms of location of production and settlement continue to exist side by side with the new forms, which gradually transform and replace them.

The development of society affects the economic-geographic situation of every populated place or group of settlements, and consequently their economic functions, the development and location of industry, agriculture, transportation, economic-managerial and administrative-cultural institutions, and the pattern of settlement in relation to local natural conditions. All these conditions and processes of formation and development of forms of location of production and forms of settlement are interrelated (see figure 1).

In the USSR the development of a network of urban and rural places depends on the economic laws of socialism and its transformation into communism. Particular manifestations of those laws are the interrelated patterns and tendencies of settlement to which this article is devoted.

GROWTH OF URBAN AND REDUCTION OF RURAL SETTLEMENT

A continuous and rapid development of production through industrialization of the country and mechanization of agriculture causes a rapid growth of urban population and a decline of rural population (table 1 and figure 2). As a result, the share of urban population has grown rapidly: from 18 percent in 1926 to 52 percent in 1963–64, and in many oblasts even more. The magnitude of migration from rural areas to urban places between 1926 and 1939 and 1939 and 1959 (table 2) has been unprecedented in the history of mankind: in 32 years (1926–59) this migration amounted to more than 43 million people.

It follows from the figures of the long-range growth of production of industry and agriculture and increases in labor productivity envisaged in the Communist Party's Program, adopted at the Twenty-second Party Congress, that the urban population of the Soviet Union by 1980 will be about 190 million, or 68 percent of the total population.[2]

RELATIONSHIPS BETWEEN URBAN AND RURAL POPULATION

The following pattern of relationships exists between urban and rural population in the development of the network of populated places of the USSR: (1) The expansion of old and the creation of new industrial centers in agricultural areas; (2) the transformation

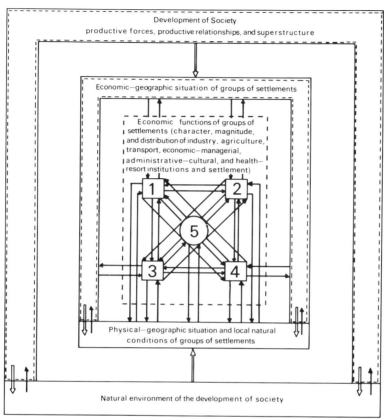

1—Industry, 2—Transport, 3—Agriculture, 4—Economic-managerial, administrative-cultural, and health resort institutions, 5—Populated places

Figure 1. Conditions and Processes of Formation of Settlement Patterns Origin and Development of Settlements Cities, Towns, and Villages

of many villages into urban places, usually in the following sequence —village, rural rayon [administrative region] seat, urban-type settlement, and later, in some cases, city; (3) the transformation of many villages into settlements of a mixed, transitional type combining agricultural and urban functions; (4) the concentration of rural population in the zones of large cities and the conversion of villages into satellites of these cities with intensive employment and cultural-service ties; (5) the concentration of rural population in the zones of

TABLE 1

YEAR	POPULATION IN MILLIONS			PERCENTAGES		
	Urban	Rural	Total	Urban	Rural	Total
1926.	26.3	120.7	147.0	18	82	100
1939 (prewar boundaries) . . .	56.1	114.5	170.6	33	67	100
1939 (postwar boundaries) . . .	60.4	130.3	190.7	32	68	100
1959.	100.0	108.8	208.8	48	52	100
1963 (1 January). .	115.1	108.8	223.1	52	48	100
1964 (1 January). .	118.6	107.7	226.3	52	48	100

SOURCE: Central Statistical Administration USSR, SSSR v tsifrakh v 1963 g. [USSR in Figures in 1963] (Moscow, 1964), pp. 9-10.

TABLE 2

PERIODS	NUMBER OF YEARS	RURAL–URBAN MIGRATION (In Millions)		PERCENTAGE OF MIGRANTS IN TOTAL URBAN POPULATION AT END OF PERIOD
		Total Migration	Annual Average	
1926–39	12	18.7	1.56	33.3
1939–59	20	24.6	1.23	24.3
1926–59	32	43.3	1.34	43.0

SOURCE: Data cited in the article by V. Starovskiy, "First Results of the Population Census," Kommunist [Communist], no. 7, p. 78.

new cities, and often the organization of new areas of rural settlement in previously undeveloped regions. These tendencies help eliminate substantial differences between town and countryside.

In the future the problem of planned reconstruction of the rural population will gradually be solved through an economic upsurge of collective and state farms, the development of intercollective and collective-state farm associations, joint construction of enterprises for

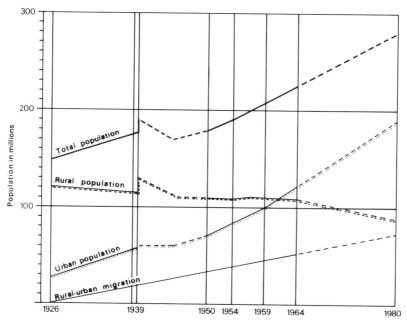

Figure 2. Population Growth of the USSR

the processing of farm products, cooperation in the construction of power stations and the development of a building industry, and an increase in the level of cultural, housing, and other services for the rural population. Most widespread in this process will be the transitional type of settlement, combining agricultural, industrial, and cultural-political functions, which will gradually develop into urban-type settlements, and sometimes cities. There will be an amalgamation of agricultural settlements and, at the same time, a reduction in the density of the network of such places. Although they will remain smaller than cities (because of the character of agricultural production), they will benefit from amenities of the urban type. Social-economic and cultural-service differences between city and countryside will be eliminated.

The basic trends of past and future reconstruction of the settlement pattern are the following interrelated processes: (1) a rapid growth of urban population; (2) a slight decrease in rural population as a result of heavy rural-urban migration; (3) a rapid growth of the relative share of urban population; (4) increase in size of urban places; (5) a growth of the number of urban places and increase in density of the network of such places; (6) increase in size of rural places; (7) reduction of the number of rural places and of the density of the network of rural places; (8) reduction of the number of rural places per urban place. These trends became evident in the period 1926 to 1959. The prospects are for even greater changes along these same lines.

THE TREND TOWARD A MORE EVEN DISTRIBUTION OF POPULATION

In accordance with the law of the planned (proportional) development of the national economy, it is essential: (1) to eliminate the historically formed discrepancies between the distribution of production and the distribution of natural resources; (2) to insure integrated development of each economic region; (3) to eliminate discrepancies in the levels of development of economic regions.

Shifts in the distribution of industry, agriculture, and transportation produce shifts in the distribution of population. The distribution of population of the USSR between western and eastern regions (Urals, including the Bashkir ASSR; Western Siberia, Eastern Siberia, Far East, Central Asia, and Kazakhstan) has changed as follows: the share of the eastern regions in the total population (see table 3) rose from 26 percent in 1926 to 31 percent in 1962, and in the urban population from 20 percent to 31 percent (see table 4).

To determine the probable net migration from the western regions to the eastern over the past period, we calculate first the approximate magnitude of natural increase in the two parts of the country. We will regard the rate of natural increase per 1,000 population as the same in the East and the West, except for the period 1939–59, when (because of the war) the rate of natural increase in the East is taken as twice as high as in the West (such a large magnitude will undoubtedly guarantee against exaggeration of the probable size of migration). The data are shown in table 5.

The size of migration thus obtained is not exaggerated. The migration flow of 5.1 millions in 1926–39 corresponds to statistical data: during these years 4.7 million people settled in the East (excluding the Urals).[3] Consequently, it is safe to assume that more than 15 million people migrated from the West to the East (including the Urals) in the period from 1926 to 1962.

In the long run, the distribution of population throughout the USSR is expected to become even more uniform in connection with a more rational location of industry, further leveling of the economic development of economic regions, and the development of the East and the North. The share of the eastern regions will substantially increase. Even if we assume no increase in the annual eastbound migration (in percent of the population of the western region), more than 11 million people may be expected to move eastward by 1980.

TABLE 3

POPULATION OF USSR AND THE WESTERN AND EASTERN SECTIONS

	TOTAL POPULATION* (In Millions)						TOTAL POPULATION* (In Percent)						TOTAL POPULATION GROWTH* 1926-62	
	1926	1939 Prewar Area	1939 Postwar Area	1959	1961	1962	1926	1939 Prewar Area	1939 Postwar Area	1959	1961	1962	(In Millions)	(In Percent)
Western.	111.5	123.9	144.0	145.7	149.9	151.5	75.8	72.7	75.5	69.8	69.3	68.9	19.9	34
Eastern.	35.5	46.7	46.6	63.1	66.3	68.2	24.2	27.3	24.5	30.2	30.7	31.1	32.7	62
Total.	147.0	170.6	190.7	208.8	216.2	219.7	100.0	100.0	100.0	100.0	100.0	100.0	52.6	100

SOURCES: Narodnoye khozyaystvo SSSR v 1961 godu [National Economy of the USSR in 1961] (Moscow, 1962) and other statistical year books.

NOTE: *Population growth does not include population gained in 1939 in the western parts of Ukraine and Belorussia and in the three Baltic republics.

TABLE 4

URBAN POPULATION OF USSR AND WESTERN AND EASTERN SECTIONS

| | URBAN POPULATION (In Millions) | | | | URBAN POPULATION (In Percent) | | | | | URBAN POPULATION GROWTH 1959-62 | |
	1926	1959	1961	1962	1926	1959	1961	1962	(In Millions)	(In Percent)
Western. .	21.0	69.2	74.2	76.5	79.7	70.7	68.5	68.4	7.3	66
Eastern. .	5.3	31.6	34.1	35.3	20.3	29.3	31.5	31.6	3.7	34
Total .	26.3	100.8	108.3	111.8	100.0	100.0	100.0	100.0	11.0	100

SOURCES: Narodnoye khozyaystvo SSSR v 1961 godu [National Economy of the USSR in 1961] (Moscow, 1962) and other statistical year books.

TABLE 5

POPULATION-MIGRATION DATA FOR THE WEST AND THE EAST, 1926-1962

PERIODS	NUMBER OF YEARS	WESTERN-EASTERN MIGRATION (In Millions)		MEAN POPULATION OF THE WEST IN GIVEN PERIOD (Including Migrants) (In Millions)	MEAN WEST-EAST MIGRATION OF WESTERN POPULATION (Including Migrants) (In Percent)
		Entire Period	Annual Average		
1926-39	12	5.1	0.425	120.2	0.35
1939-59	20	8.6	0.430	149.2	0.29
1959-62	3	1.7	0.567	149.5	0.38
1926-62	35	15.4	0.440	132.3	0.33

GROWTH AND DISTRIBUTION OF THE NETWORK OF URBAN PLACES

The development of production in existing industrial regions as the bases for the industrialization of the country will lead to a growth of existing cities and increasing density of the network of urban places in these regions.

The exploitation of the resources of undeveloped regions will lead to expansion of the areas of urban settlement. This expansion will take the form of bulges or wedges extending into undeveloped or little developed territories or cutting across them. In addition, new areas of urban settlement may arise far from presently settled territories.

There is a steady process of formation of many new cities and towns (see tables 6 and 7). More than 2,700 new cities and urban-type settlements were established between 1926 and 1963. The network of urban places of the USSR was transformed: by 1963 more than half consisted of new urban places, and in many parts of the East and North the urban network was created virtually from scratch. The process of formation of new cities and settlements is proceeding at a rapid pace (see table 8).

In connection with the development of new mineral sites, the creation of new steel and power centers, the irrigation of deserts and draining of swamps, and the construction of new railroads and highways, there has been a marked expansion of the network of cities and urban-type settlements; new urban centers arise even in undeveloped areas. The areas of urban life will further expand as individual prongs and islands merge into a single network enclosing some empty spaces, and as these spaces then fill in. In areas with separate, scattered urban networks of urban places, the process of concentration of these networks will continue. If we assume that the annual increase of the number of urban places will be even greater than in recent years (1959–63), then the total number of urban places can be expected to grow substantially.

THE RELATIONSHIP BETWEEN THE SIZE OF CITIES AND THE CHARACTER AND MAGNITUDE OF THEIR FUNCTIONS

The size (population) of each city and urban-type settlement depends on its economic specialization: the composition, character, and magnitude of its functions (branches of industry, transportation, economic-management and cultural-political functions). The majority of cities and many urban settlements fulfill several functions at the same time. For a quantitative analysis of this relationship, we will therefore use a typology of cities based on combinations of functions (see tables 9 and 10).

TABLE 6

YEARS	NUMBER OF URBAN PLACES		
	Cities	Urban-type Settlements	Total
1926 (December).	709	1,216	1,925
1939 (January)	923	1,450	2,373
1939 (Postwar area). . .	1,194	1,568	2,762
1959 (midyear)	1,679	2,940	4,619
1960 (1 January)	1,682	3,031	4,713
1962 (1 January)	1,723	3,220	4,943
1963 (1 April)	1,763	3,261	5,024

SOURCE: Narodnoye khozyaystvo SSSR v 1962 godu [The National Economy of the USSR in 1962] (Moscow, 1963).

TABLE 7

YEARS OF FORMATION OF NEW URBAN PLACES	NUMBER OF URBAN PLACES		
	Cities	Urban-type Settlements	Total
1926-39 (prewar area).	214	234	448
1939-63 (postwar area)	569	1,693	2,262
Total	783	1,927	2,710
Percentage of new places out of the total number of urban places in 1963	44.4	58.1	53.9

SOURCE: Narodnoye khozyaystvo SSSR v 1962 godu [The National Economy of the USSR in 1962] (Moscow, 1963).

TABLE 8

YEARS	URBAN POPULATION (Millions)	NUMBER OF URBAN PLACES	AVERAGE SIZE OF URBAN PLACE (Thousands)	NUMBER OF YEARS	TOTAL INCREASE OF NUMBER OF URBAN PLACES	AVERAGE ANNUAL INCREASE OF NUMBER OF URBAN PLACES
1926.	26.3	1,925	13.7			
1939 (prewar territory) . .	56.1	2,373	23.7	12	446	37
1939 (postwar territory). .	60.4	2,762	21.9		389	
1959 (January).	100.0	4,619	21.7	20	1,757	88
1963 (January).	115.1	5,012	23.0	4	393	98

TABLE 9

FUNCTIONAL TYPES OF CITIES AND THEIR SIZE CLASSES

	NUMBER OF CITIES BY SIZE CLASS (In Thousands)					TOTAL NUMBER OF CITIES OVER 50,000
	50-100	100-200	200-400	Over 400	
Capitals of union republics.	..	1	5	9		15
Other manufacturing cities:						
seaports.	10	3	8	3		24
river ports.	28	16	11	13		68
rail hubs.	30	26	7	5		68
rail branch (spur)	48	15	..	..		63
sea and river ports (without rail service)	4	..	..	..		4
extractive or extractive and manufacturing centers (with rail service).	37	15	10	1		63
resorts.	4	..	..	..		4
Total	161	76	41	31		309

TABLE 9--Continued

	TOTAL POPULATION OF EACH SIZE CLASS (In Thousands)				TOTAL POPULATION OF CITIES OF MORE THAN 50,000 (In Thousands)
	50-100	100-200	200-400	Over 400	
Capitals of union republics. . .	..	169.9	1,182.9	10,462.7	11,815.5
Other manufacturing cities:					
seaports	700.5	427.3	1,979.8	4,166.7	7,274.3
river ports.	2,069.5	2,424.9	2,970.5	8,130.8	15,595.7
rail hubs.	2,039.0	3,418.0	1,878.1	3,259.6	10,594.7
rail branch (spur)	3,382.8	2,131.9	..	..	5,514.7
sea and river ports (without rail service)	274.0	..	..	..	274.0
extractive or extractive and manufacturing centers (with rail service).	2,625.3	2,006.0	3,297.0	704.8	8,633.7
resorts.	299.8	..	..	..	299.8
Total	11,390.0	10,578.6	11,308.3	26,724.6	60,002.4

NOTE: Table includes all cities of USSR with population of more than 50,000 according to 1959 census.

TABLE 10

FUNCTIONAL TYPES OF CITIES AND THEIR POPULATION

FUNCTIONAL TYPES	POPULATION TOTAL (In Thousands)	NUMBER OF CITIES	AVERAGE SIZE (In Thousands)	PERCENTAGE OF	
				Population	Number of Cities
Capitals of union republics	11,815.5	15	787.7	19.7	4.8
Other manufacturing cities:					
seaports.	7,274.3	24	304.0	12.1	7.8
river ports	15,595.7	68	229.3	26.0	22.0
rail hubs	10,594.7	68	155.8	17.6	22.0
rail branch (spur).	5,514.7	63	87.5	9.2	20.4
sea and river ports (without rail service). . . .	274.0	4	68.5	0.5	1.3
extractive or extractive and manufacturing centers (with rail service)	8,633.7	63	137.0	14.4	20.4
resorts	299.8	4	75.0	0.5	1.3
Total.	60,002.4	309	194.2	100.0	100.0

An analysis of the data in these tables and bar graphs, relating to all 309 cities of the USSR with a population of more than 50,000 and a selective analysis of some smaller cities suggest the following relationships between city size and combinations of functions:

1. Usually, all capitals of Soviet Union republics not only perform an administrative-cultural role, but are also important manufacturing and transportation centers. Most of these capitals have a population of at least 400,000, so most of the total population of capital cities is concentrated in these very large cities.

2. Most very large cities with a population of at least 400,000 combine manufacturing with rail and water transport functions. These are usually river and seaports that perform cultural-political functions, or capitals of large autonomous republics, oblasts, or krays. Less common are manufacturing and rail centers without waterway connections (Khar'kov, Sverdlovsk, Chelyabinsk, Voronezh, L'vov). Only one oblast capital combines manufacturing and extractive industries (Donetsk). Among these large cities, none is situated on a single branch railroad or spur (except two republic capitals, Yerevan and Alma-Ata), and none is without railroad service.

3. Most large cities with populations of 200,000 to 400,000 combine manufacturing with rail and water transportation, but there are also quite a number that combine manufacturing and extractive industries (Krivoy Rog, Novokuznetsk, Nizhniy Tagil, Karaganda, Prokop'yevsk, Kemerovo, and others). Most of the cities in this category are oblast capitals. This class of cities also does not include places that are situated on single rail branches or spurs.

4. Most of the cities with a population of 100,000 to 200,000 combine manufacturing with rail services (without waterways) and are usually oblast capitals. Some of them are situated on single branch or spur lines, but all have railroad connections.

5. Most of the medium-size cities of 50,000 to 100,000 are situated on single branch or spur railroads. This category also includes cities without rail connections (Yakutsk, Petropavlovsk-Kamchatskiy, Magadan, Chistopol'). It includes several resorts (Sochi, Kislovodsk, Pyatigorsk, Yevpatoriya). However, most resorts (except cities combining resort functions with administrative-cultural and industrial-transport functions) are small or medium-size cities and urban settlements.

6. The number of centers of extractive industry increases markedly as one passes from large to medium-size and small cities and settlements. In the absence of significant manufacturing and administrative-cultural functions, an extractive industry usually gives rise to urban settlements or small and medium-size cities.

7. Manufacturing centers without major transport and administrative-cultural functions usually give rise to small and medium-size cities and settlements.

8. Administrative-cultural centers with a low level of industrial development are usually small cities.

THE SYSTEM OF COMMAND POINTS OF THE COUNTRY AND THE FATE OF LARGE CITIES

New, socialist forms of social life have given rise to a system of command points of the country concerned with economic management and cultural and political functions (capitals of union republics, of autonomous republics, oblasts, and krays). These are centers with a favorable economic-geographic situation and a diversity of industrial-transport and cultural functions. This produces a trend toward a rapid growth of population and a sharp increase in the number and relative share of large and, especially, very large cities (from 400,000 to 1,100,000 population). The distribution of such centers has become more uniform throughout the USSR (in 1926 there were none in the East; in 1939 the East contained one-fifth of the total, and in 1959 one-third). These big cities contain an especially large share of the urban population in Leningrad economic region and in Transcaucasia (Baku, Tbilisi, and Yerevan contain more than 42 percent of the total urban population of the three Transcaucasian republics). Large cities also play a dominant role in the Middle Volga region (Kuybyshev, Kazan') and in Western Siberia (Novosibirsk, Omsk).

At the same time there has been a tendency for the million-cities (Moscow and Leningrad) to slow their growth and to reduce their share in the total urban population of the USSR (from 13.7 percent in 1926 to 8.5 percent in 1959 (see table 11). This has been accompanied by a rapid growth of satellite cities around these centers.

In the long run, measures will evidently be taken to prevent excessive development of supercities and to limit the growth of existing large cities. Technical progress and the widespread introduction of new technology, integrated mechanization, and automation of production will reduce the labor requirements not only of new but also of modernized enterprises. A major role will be played by a rational location of industry, the integrated development of economic regions, specialization and subcontracting between widely separated enterprises, and restrictions on excessive growth of existing large plants and on the size of new plants.

The growth of very large cities is expected to slow. Little by little, part of the population of the million-cities will be induced to move

TABLE 11

SIZE CATEGORY OF URBAN PLACES	NUMBER OF URBAN PLACES		TOTAL URBAN POPULATION (Millions)		PERCENTAGE OF NUMBER OF PLACES		PERCENTAGE OF TOTAL URBAN POPULATION	
	1926	1959	1926	1959	1929	1959	1926	1959
Below 10,000.	1,446	2,528	5.2	12.5	75.10	63.20	19.8	13.4
10,000-400,000. . . .	473	1,444	15.7	55.1	24.60	36.07	59.7	59.0
400,000-1,100,000 . .	4	27	1.8	17.8	0.20	0.68	6.8	19.1
Moscow, Leningrad . .	2	2	3.6	7.9	0.10	0.05	13.7	8.5
Total.	1,925	4,001	26.3	93.3	100.00	100.00	100.0	100.0

NOTE: For comparative purposes, all figures apply only to the prewar territory of the USSR.

with the transfer of enterprises and institutions to other economic regions and to satellite cities. The population of the very big cities will also be reduced by modernization of existing large enterprises and by the increased movement of labor resources toward newly developed regions of the North and the East; that is, by a planned stimulation of migration from the million-cities to rapidly developing economic regions of the country (mainly in the East).

New enterprises needed to supply services to the large cities (construction materials, processed foods, and light industry) will be situated in satellite cities around these centers. These satellites are also expected to absorb some of the enterprises and institutions of the central cities. They will serve, too, as residential cities for people who are employed in the central city but prefer to live closer to nature. These people will benefit from an appropriate housing program in existing satellite cities and towns, situated in favorable natural sites and linked by convenient means of transportation with the central city. People who are employed in the satellite cities will also be encouraged to move away from the central city.

All these measures will help limit the growth of the million-cities and gradually reduce their population. This trend will also be promoted by the introduction of high-speed transportation and of new types of transport (for example, monorail systems) and improved provision of public services in the satellite cities.

Whereas the population of the old large cities will be limited and ultimately gradually reduced, the system of command points will be further developed, and economic-management, cultural, political, and research centers will be spread with increasing uniformity throughout the entire territory of the Soviet Union.

INCREASING COMPLEXITY OF FUNCTIONS AND
THE GROWTH OF URBAN PLACES

The needs of a planned integrated development of the economy and of a rise in culture will result in an increasing complexity of functions and a growth of urban places. Between 1926 and 1959 the average size of an urban place rose from 13,700 population to 23,300, that is by 70 percent; the average size of places in the below-10,000 class increased from 3,600 to 5,000; in the 10,000-400,000 category, from 33,200 to 38,200; and in the 400,000-1,100,000 class, from 450,000 to 659,000. The relative share of small urban places (below 10,000) in the total urban population dropped from 20 percent to 13 percent. Cities and settlements of optimal and permissible size, that is, from 10,000 to 400,000, are becoming more widespread; they have increased from 25 percent to 36 percent of the total number of urban places (table 11), and account for almost 60 percent and, in the East, for almost 70 percent of the urban population.

Increasing complexity of the functions of urban places will make almost all of them polyfunctional, that is, they will play the role of industrial, transport, cultural, and research centers of various sizes and combinations of functions. At the same time there will be a decline in the number of small cities and settlements, further reducing the percentage share of small places in the network of urban places. A dominant role is expected to be played by middle-size cities from 10,000 to 400,000 population, which will make up more than one-half of the total number for urban population.[4] In addition all regions (in all climatic zones of the country) will develop a network of specialized health and recreation centers: resorts, camp sites, children's camps, suburban hotels, and so forth.

DISTRIBUTION OF CLUSTERED FORMS OF URBAN SETTLEMENT

The exploitation of mineral resources, the construction of plants in the vicinity of large cities, and the creation of agglomerations of plants and mines have produced a clustered type of location of production in many industrial nodes. Dense industrial clusters sometimes extend over distances of tens or even hundreds of kilometers. This pattern of location of industrial enterprises and the need for workers to save commuting time have resulted in a trend toward increasing distribution of clustered forms of urban settlement, that is, territorial groupings of cities and settlements. Each such cluster consists of one or more centers of gravitation and their satellites related by commuting ties, either both workers' and cultural-shoppers' commuting or just cultural-shoppers' commuting, if it is sufficiently intensive. Territorial proximity or urban places connected by convenient passenger transport promote the development of commuting ties and thus combines these places into an urban cluster.

Almost 40 percent of all Soviet urban places now constitute territorial groupings that contain almost three-fourths of the total urban population of the country. Separate urban places are still quite numerous, accounting for more than 60 percent of all urban places. However, because of their small size they contain only a little more than one-fourth of the total urban population.[5]

Analysis shows a diversity of clustered forms of settlement depending on the magnitude and combination of functions, the character of the distribution of industrial plants, railroads and waterways, the maximum distance between plants, mines and other enterprises, and the physical site. There is a constant tendency for simple forms of settlement to become more complex: individual, isolated places become a system of two urban places, then linear groupings, and finally clusters of cities and settlements.

Clustered forms are unevenly distributed throughout the territory of the USSR; they predominate in coal basins (Donbas, Kuzbas, Karaganda), in the mining and industrial districts of the Urals, in the suburban zones of the country's largest industrial-transport and cultural-political centers (Moscow, Leningrad, and Khar'kov), and in many areas of pioneering development of the Far North and Far East.

The prevailing tendency in clustered forms of settlement is for satellite places to increase in numbers and in size and for entire clusters to become more complex with increases in size of the centers of gravitation. Large urban agglomerations have become widespread. The 150 largest groups of cities and settlements (with a population of more than 100,000 each) contained 64 percent of the total urban population of the USSR in 1959). About 42 percent of the total urban population lived in the thirty-seven largest groupings (with a population of more than 400,000 each), including ten groupings with a population of at least one million.[6] Large groups of cities and settlements arise as a result of a favorable economic-geographic situation of the central city and of the entire urbanized district.

Technical progress and a growth of commuting and intercity transport services will tend to expand the radius of settlement and the radius of daily cultural and shopping services. The zone of satellite places linked to the central city only by cultural and shopping ties will tend to expand. The radius of this zone may now exceed 70 kilometers and in some directions even 80 kilometers. Such a radius corresponds to a two-hour ride. Since the radius lengthens in most cases when the central city increases in size, the two-hour commuting criterion does not always apply. The development of suburban transportation plays a key role in the formation of groups of satellite places. There is a characteristic relationship between the present suburban settlement pattern around Moscow, the two-hour commuting zone, and the passenger flow on suburban railroad lines.[7]

In the long run, clustered forms of settlement will predominate in almost all industrial and transport nodes, in suburban zones of large and medium-size cities, and in resort and recreation zones. Isolated urban places will become a rare occurrence in networks of cities and settlements.

In accordance with the policy of achieving a more uniform distribution of productive forces, an effort will be made to limit the growth not only of the largest cities but also of excessively large urban agglomerations around Moscow, Leningrad, Donetsk, Gor'kiy, and Khar'kov. For this purpose, enterprises and institutions of the central city will be moved not only to satellite cities but also beyond the agglomeration. Construction of new industries in satellite cities will be limited to activities to meet the needs of consumers of the central city and its satellites. The growth of the very large agglomerations, especially those with a population of a million or more, will thus be slowed. Many new agglomerations of moderate size will arise throughout the country, especially in the eastern regions.

The role played by various conditions of settlement undergoes change under the influence of technical progress, the growth of economic capabilities, and the well-being of the people in the transition phase to communism. For example, as means of transportation develop and permissible radiuses of settlement increase, it becomes possible to combine the construction of housing and other services for a group of scattered plants to form single cities replacing small isolated settlements. There will be a continued rapid expansion of both the radius of settlement and that of daily cultural and shopping services in clusters of urban places. The boundaries of groups of populated places, united by common services, will thus keep moving outward.

Technical progress and economic prosperity will enhance the possibilities of transforming entire territories through swamp drainage, desert irrigation, climatic changes, transformation of land forms through large earth-moving operations, and so forth. This will induce us to review our attitude toward so-called unsuitable territories and help eliminate blank spots on the map of settlement.

CONCLUSION

A study of tendencies of development and distribution of urban places shows that such a development does not always proceed in a strict pattern and along harmonic lines. In the struggle between contradictions, negative tendencies may produce certain disproportions (for example, excessive growth of very large cities) that must be eliminated by making use of objective laws of settlement in the interest of society.

In the light of the magnificent program of construction of communism, there is a clear need for long-range planning of the development of the network of existing and new cities and urban settlements of the USSR in conjunction with rural settlement. This task must be solved as an integral part of the long-range plan of development of the national economy and culture over the next two decades. Planning of a network of cities of the entire country, its republics, and economic regions will help eliminate shortcomings and insure a correct course of urban development. Urban geography assumes great practical importance in this connection. The following areas of research should be emphasized: (1) a study of facts that characterize networks of urban places in the USSR as a whole and by regions—density of networks, functions and sizes of places, forms of settlement, history of formation—making use of census data and other statistical materials and of economic geographic research; (2) analysis of facts, explanation of cause-and-effect relationships, tendencies and patterns of settlement; (3) introduction into urban geography of the methods of mathematical and graph-analytical research, not only cartograms and cartodiagrams, but also mathematical programming, nomograms, chrono-diagrams, triangular diagrams of three-component structures, and so forth.

It is necessary to uncover many quantitative relationships and patterns, from the simplest to the more complex, and this will help solve problems in the constructive geography of settlement. The first attempts to study existing patterns have demonstrated the need for making practical use of the findings in economic planning and in the drafting of regional and urban plans. By further work on the theory of settlement, geographers should make their contribution to the building of a communist society.

1. The over-all problem has been discussed to varying degrees in the work of many economic geographers, economists, and statisticians such as N. N. Baranskiy, E. I. Bogorad, E. Davydov, P. I. Dubrovin, L. Ye. Yofa, R. M. Kabo, S. A. Kovalev, N. Ya. Koval'skaya, N. A. Kokovin, A. M. Kolotiyevskiy, O. A. Konstantinov, E. V. Knobel'skorf, G. M. Lappo, N. I. Lyalikov, I. M. Mayergoyz, A. A. Mints, O. R. Nazarevskiy, G. S. Nevel'shteyn, P. G. Pod'yachikh, V. V. Pokshishevskiy, M. I. Pomus, Yu. G. Saushkin, I. I. Spidchenko, S. G. Strumilim, Kh. Ya. Takhayev, S. I. Sharov and others.

2. Central Statistical Administration USSR, *SSSR v tsifrakh v 1963 g* [USSR in Figures in 1963] (Moscow, 1964), p. 11.

3. Scientific Research Institute of Labor, *Trudovyye resursy SSSR problemy raspredeleniya i ispol'zovaniya* [Labor Resources of the USSR: Problems of Distribution and Utilization] (Moscow, 1961), p. 209.

4. *Materialy I mezhduvedomstvennogo soveshchaniya po geografii naseleniya* [Proceedings of the First Interdepartmental Conference on Population] (Moscow-Leningrad, 1961), 1:55.

5. For more detail on this point see *Voprosy geografii* [Problems of Geography], no. 45 ("Satellite Cities") (1961), p. 15, which contains a table based on the census of 15 January 1959.

6. Ibid., pp. 16 and 29.

7. Ibid.

N. I. BLAZHKO

Economic-Geographic Mathematical Modeling of Cities

Mathematical model-building is one of the new research methods used in economic geography. A model of a given dynamic system makes it possible to study the structure of the system and its relationships and to carry out experimentation, which is especially effective in the compilation of long-term forecasts. The functional relationships between elements of the model reproduce the actual relationships of the real elements of the system. Thus, by carrying out theoretical and practical research with the model, one can also obtain information about the object that is being modeled. The functional interdependence between elements of the model is expressed in mathematical form, which makes it possible to interpret the quantitative aspects of phenomena and to obtain, on the basis of a quantitative analysis, new data of a qualitative character.

Ordinary economic-mathematical models cannot satisfy the needs of economic geographers. They require a somewhat different approach to the study of territorial economic systems, making it possible to tie in the study object (city or district) with a territory. In other words,

Reprinted with modifications from *Soviet Geography: Review and Translation* 6, no. 9 (November 1965):66–78, by the permission of the publisher. The article originally appeared in *Vestnik Moskovskogo Universiteta*, seriya geografiya, [Herald of Moscow University, Geography series], no. 4 (1964), pp. 18–27.

as V. S. Nemchinov and Yu. G. Saushkin worte,[1] what is needed in economic-geographic research are territorial models.

The construction of models in urban geography is a relatively recent development.[2] This method holds promise both for the study of individual cities and their component elements and for systems of cities, but it still requires further development.

It is worth emphasizing the unusual flexibility of the matrix form of a model, which makes it possible to make changes (expand, supplement, narrow) in accordance with the given problem. We can therefore have several models of a given city (or system of cities), for example, a model of the production structure, of the structure of consumption, of manpower structure, and so forth. In the present case, we will examine the model of a city from the point of view of its place in the territorial division of labor of the Soviet Union.

The basis for the construction of the city model, as for models of economic regions, is the interindustry and interregional balance of production and distribution described in the literature.[3] But in contrast to regional interindustry balances, which are usually broken down by industries, the city model requires an urban-planning approach: (1) the city-forming and city-serving industries must be distinguished; (2) analysis of inflow and outflow of goods must take account of the scope of the city's participation in the division of labor (on a regional, republican, national, and international scale); (3) the inflow of goods must be broken down by destination (for industry or for consumption) to help understand the character of the city's relationships compared with its functional structure.

The proposed model consists of two parts and several interdependent matrices. It is shown in tables 1 and 1a. The central part of the model is matrix-column X, which shows the city's industrial production. It is adjoined by the square matrix A of interindustry relationships within the city. This matrix differs from interindustry balances in the order of arrangement of the industries: the upper part of the matrix is occupied by city-forming industries, the lower part by city-servicing industries. It is preferable not to show these two categories in separate matrices because there may be productive relationships between them that also shed light on the production complex of the city. The horizontal rows of matrix A show supplier industries and the distribution of their output within the city. The vertical columns of matrix A show consumer industries, arranged in the same order as the suppliers. Matrix A adds up both along horizontal rows and vertical columns to show the volume of industrial consumption by industries and for the city as a whole.

The right part of the model (matrices B' and B'') shows the channels of distribution of the city's industrial production that remain after industrial needs within the city have been met.

TABLE 1

MODEL OF A CITY

NO.	INDUSTRY	X TOTAL OUTPUT	A — CONSUMPTION BY CITY'S INDUSTRIES	A — INDUSTRIES 1	2	3	n	B' CONSUMPTION BY CITY'S POPULATION	B'' TOTAL OUTFLOW	B'' REGIONS I	II	III	IV
1	Mining	X_1	$x_{1.}^{*}$	x_{11}	x_{12}	x_{13}	x_{1n}	y'_1	y''_1	$y''_{1\mathrm{I}}$	$y''_{1\mathrm{II}}$	$y''_{1\mathrm{III}}$	$y''_{1\mathrm{IV}}$
2	Metallurgy	X_2	$x_{2.}$	x_{21}	x_{22}	x_{23}	x_{2n}	y'_2	y''_2	$y''_{2\mathrm{I}}$	$y''_{2\mathrm{II}}$	$y''_{2\mathrm{III}}$	$y''_{2\mathrm{IV}}$
3	Machinery	X_3	$x_{3.}$	x_{31}	x_{32}	x_{33}	x_{3n}	y'_3	y''_3	$y''_{3\mathrm{I}}$	$y''_{3\mathrm{II}}$	$y''_{3\mathrm{III}}$	$y''_{3\mathrm{IV}}$
n	. . . . . .	X_n	$x_{n.}$	x_{n1}	x_{n2}	x_{n3}	x_{nn}	y'_n	y''_n	$y''_{n\mathrm{I}}$	$y''_{n\mathrm{II}}$	$y''_{n\mathrm{III}}$	$y''_{n\mathrm{IV}}$
	Total	X	x	$x_{.1}$	$x_{.2}$	$x_{.3}$	$x_{.n}$	y'	y''	y''_{I}	y''_{II}	y''_{III}	y''_{IV}

TABLE 1A

NUMBER OF REGION	C' INFLOW FOR CITY'S POPULATION	C'' INFLOW FOR CITY'S INDUSTRIES	INDUSTRIES			
			1	2	3	n
I	z'_I	z''_I	$z''_{I\,1}$	$z''_{I\,2}$	$z''_{I\,3}$	$z''_{I\,n}$
II	z'_{II}	z''_{II}	$z''_{II\,1}$	$z''_{II\,2}$	$z''_{II\,3}$	$z''_{II\,n}$
III	z'_{III}	z''_{III}	z''_{III1}	z''_{III2}	z''_{III3}	z''_{IIIn}
IV	z'_{IV}	z''_{IV}	$z''_{IV\,1}$	$z''_{IV\,2}$	$z''_{IV\,3}$	$z''_{IV\,n}$
Total	z'	z''	z''_1	z''_2	z''_3	z''_n

NOTE: X is matrix of total industrial output; A is matrix of city's industrial consumption; B' is matrix of city's population consumption; B'' is outflow matrix; C' is matrix of inflow for city's population; and C'' is matrix of inflow for city's industries.

*Following V.1. Romanovskiy's method we use (.) to fix the place of the index.

Matrix-column B' shows the consumption of the city's population that is satisfied by the city's own industries (this includes all industries that are even partly consumer-service in character). Matrix B'' is the outflow matrix. Its horizontal rows correspond to the city's industries. Matrix B'' has only five columns. The first shows the total outflow and the others the outflow by regions.

Like matrix A, matrix B'' is balanced by horizontal rows and vertical columns, making it possible: (1) to determine the share of each industry in the city's total outflow; (2) to show quantitatively the scope of outflow relationships; and (3) to determine the total outflow.

The second part of the model does not balance with the first. It consists of two matrices: C' and C'', characterizing the city's inflow. Matrix-column C' shows the city's inflow of goods for the supply of the population. Matrix C'' shows the inflow for the needs of industry. In both matrices, the horizontal rows correspond to the regions of origin and the columns to the city's industries. Matrix C'' adds up both by rows and by columns to show the total inflow for each industry, the total inflow from each region, and the total inflow for all industries of the city.[4]

Let us examine the model more closely: The total volume of the city's industrial production X is made up of the sum of output of individual industries X_i, where i is the number of the producing industry (row index), $i = 1, 2, 3, \ldots, n$. Each X_i, in turn, consists of: (1) the output of the ith industry directed into industrial consumption within the city—x_{ij}, where $j = 1, 2, 3, \ldots, n$, the number of the consuming industry (column index in matrix Q); (2) consumption of the city's population—y_i', and outflow to other regions—y_i'', the column indices, I, II, III, IV signifying, respectively, the city's own economic region, and other regions of the city's republic, other republics of the USSR, and export. Thus:

$$X_i = \left(\sum_{j=1}^{n} x_{ij} \right) + y_i' + y_i''.$$

The total volume of industrial output, as noted above, is made up of the sum of production of the individual industries:

$$X = \sum_{1} X_i = \sum_{i=1}^{n} \left(\sum_{j=i}^{n} x_{ij} + y_i' + y_i'' \right) = \sum_{i=1}^{n} \sum_{j=1}^{n} x_{ij} - \sum_{i=1}^{n} y_i' - \sum_{i=1}^{n} y_i''.$$

But industrial production and consumption within the city require also an appropriate inflow of goods. The total inflow consists of: (1) inflow for the needs of the population; (2) inflow for the needs of industry.

Total inflow Z would thus be equal to: $Z = z' + z''$,
where

$$z' + z_I' + z_{II}' + z_{III}' + z_{IV}' \tag{3}$$

and

$$z'' = z_I'' + z_{II}'' + z_{III}'' + z_{IV}'' = \sum_{i=1}^{n} z_i''. \tag{4}$$

[In order to be consistent with the other notation used in the equation and tables, the last expression should read,

$$\sum_{i=1}^{IV} z_i''$$

or alternately,

$$\sum_{i=1}^{n} \left(z''_{\text{I}_i} + z''_{\text{II}_i} + z''_{\text{III}_i} + z''_{\text{IV}_i} \right) \text{, Eds.]}$$

All parts of the model are intererelated. Matrices A, B', B'', which characterize the industrial complex of the city, its city-forming and city-servicing industries, the marketing of industrial output and the scope of the city's participation in the territorial division of labor in the form of outflow, are interrelated by formula (2).

Matrices C' and C'', which characterize the inflow of goods, show the scope of the city's participation in the territorial division of labor in terms of inflow. Matrix C' is related to matrix A because the inflow for the needs of industry insures the required volume of industrial output. Matrix C' is related to matrix B', since Z' and Y'' [together] constitute the consumption of the city's population. All data in the model are expressed in value terms. For comparative purposes, weight indices can be used for the inflow and outflow matrices.

The proposed model can be used for the study of specific cities. In that case the letter symbols are replaced by numerical indicators relating to the specific characteristics of the given city.

The model considered here is a static model. It shows the city at a given stage of development, in an "instantaneous photograph." But it is useful for research despite its static character. The model makes it possible to establish in strictly quantitative terms: (1) from matrices A, B', B'', the structure of city-forming and city-servicing industries and their relationships: (2) from matrices B'', C', C'', the structure and scale of the city's external relationships. The model permits the solution of a number of economic problems: for example, determining the volume of industrial output required for a given inflow and outflow of goods, establishing the volume and structure of output of city-servicing industries needed to satisfy a given level of demand, and so forth. In the case of a system of cities, the static model makes it possible to identify functional types of urban places, and to establish the position of each type in the hierarchy of the system and of each city in the territorial division of labor.

Let us now examine some geographic aspects of the proposed model when applied to the specific cases of the cities of Zaporozh'ye and Krivoy Rog and the urban settlement of Rakhmanovka in the Dnieper economic region. Although these three places vary in size and level of development, they are in many respects similar. All three are relatively young urban places. Their principal city-forming activity is heavy industry, which constitutes the core of the production complex of both Zaporozh'ye and Krivoy Rog (Rakhmanovka is part of the Krivoy Rog complex).

The similarity of these places makes it possible to compare their models and, what is especially important at the present level of economic-geographic research, allows one through the use of quantitative characteristics to establish the position of each in the hierarchy of urban places of the Dnieper region.

The models of Krivoy Rog and Zaporozh'ye are also interesting because the two places are not run-of-the-mill cities of the Dnieper region. They are centers of local territorial production complexes and the cores of local urban agglomerations. Krivoy Rog was formed on the basis of the Krovoy Rog iron-ore basin, and Zaporozh'ye at the Dnieper hydro station and at the intersection of major transport routes. Both cities are industrial centers of national rank, around which gravitate smaller places supplying mostly industrial raw materials, semifinished goods, or food products. Among these satellite places is Rakhmanovka. Places like Rakhmanovka are found in all industrial and mining districts of the Soviet Union; in the Ukraine, they are especially common in the Donets Basin. Rakhmanovka is a small place that arose in the Krivoy Rog industrial district as a mining center. Its functional structure is extremely simple. Rakhmanovka exists entirely on the basis of its relationships in the territorial division of labor (see tables 2 and 2a).

The model yields the following conclusions:[5] matrices X and A show that the industry of Rakhmanovka is highly specialized. In view of the lack of integration of Rakhmanovka industries, there are no relationships among them. The town has no service industries. Therefore the needs of the population must be satisfied entirely through inflow (matrix C'). Matrix B'' shows that the entire industrial output is shipped out of the town. The inflow and outflow matrices show, moreover, that Rakhmanovka participates mainly in the territorial division of labor within its own economic region (predominance of exchanges with region I). All this tends to suggest that Rakhmanovka and other places like it are of the lowest rank in the hierarchy of the system of urban places of the Dnieper region.

Another picture is revealed by the model of Krivoy Rog (tables 3 and 3a). This is a city with diversified industry, clearly reflecting the industrial complex that includes Krivoy Rog and its satellite towns, among them Rakhmanovka. The basis of the complex is iron-ore mining, the primary city-forming factor that led to the origin and development of Krivoy Rog. The iron-mining industry, in turn, formed the basis for ore-sintering, iron-and-steel, steel-fabricating, and machine-building industries. Development of the iron and steel industry led to creation of a coke-chemical industry, and the new construction led to establishment of a construction materials industry. All these industries together account for more than 90 percent of the total value of industrial output of the city.

These characteristics are reflected clearly in the model, which not only shows the presence of the production complex described above, but uncovers the quantitative relationships among its principal components (matrix A). About 40 percent of the total industrial output is consumed by industries within the city, including 15 percent of the mined ore, 100 percent of the sintered ore, 90 percent of the machinery, 60 percent of the coke chemicals, and so forth. At the same time, the model shows the inadequate development of machine building, and the limited character of food-processing and especially light industry, which have almost no links with the basic industries of the city. Light and food industries in Krivoy Rog are primarily service industries, but meet only 25 percent of the city's needs (see matrices B' and C').

Matrices B'', C', C'' show the broad development of external ties, especially in terms of outflow (about 50 percent of the city's total output). In contrast to Rakhmanovka, Krivoy Rog operates mainly on a republic and national level in exchange (more than 80 percent of the city's outflow leaves the Dnieper economic region). All this suggests that Krivoy Rog is of a high rank among cities and occupies one of the highest levels in the heirarchy of urban places of the Dnieper region.

A still more important center in terms of population and industrial output is Zaporozh'ye. Like Krivoy Rog, Zaporozh'ye has a diversified industrial structure and extensive internal and external ties. These characteristics, inherent in all highly developed cities, are even more pronounced in the case of Zaporozh'ye, as can be seen from the model (tables 4 and 4a).

Matrix A in Zaporozh'ye has almost no gaps, suggesting the high degree of integration of the city's industries. In contrast to Krivoy Rog, food and light industries, in addition to the city's basic industries, also play a role in the ties among the city's industries. The share of industrial output that supplies other city industries is lower in Zaporozh'ye than in Krivoy Rog because the industry of Zaporozh'ye uses raw materials shipped in from the outside, but the value of output is higher, suggesting a higher level of industrial development. Service industries are also more highly developed in Zaporozh'ye, satisfying 43 percent of the city's needs (see matrix B'). The external-relations matrices (B', C', C'') show the broad range of the city's participation in the territorial division of labor. About 70 percent of its output is shipped out, of which 90 percent goes to other economic regions of the USSR and abroad. Zaporozh'ye also receives goods from other economic regions of the USSR and from abroad.

To sum up the results of analysis of the three models of Dnieper cities, we can make the following conclusions: (1) the cities vary in economic significance, as clearly emerges from a comparison of the basic indicators of the models (see table 5); (2) they differ markedly

TABLE 2

RAKHMANOVKA MODEL
(In Arbitrary Units)

NUMBER	INDUSTRY	X		A		B'	B"	REGIONS			
		X	x	INDUSTRIES		y'	y"				
				1	2			I	II	III	IV
1	Iron Mining. . .	240	...	...	...	...	240	80	60	...	100
2	Building-materials mining	150	...	...	...	...	150	150	...	...	...
	Total.	390	...	...	...	...	390	230	60	...	100

TABLE 2A

NUMBER OF REGION OF ORIGIN	C'	C"	INDUSTRIES	
	z'	z"	1	2
I	70	8	8	..
II	120	29	19	10
III	50	10	10	..
IV	..	..	..	..
Total	240	47	37	10

TABLE 3

KRIVOY ROG MODEL
(In Arbitrary Units)

No.	INDUSTRIES	X	x	A (INDUSTRIES)								B' (y')	B'' (y'')	REGIONS			
		X		1	2	3	4	5	6	7	8			I	II	III	IV
1	Iron Mining . .	13,600	2,150	...	2,135	7,820	.15	...	...	...	...	...	11,450	120	360	10,260	700
2	Ore-sintering . .	7,850	7,850c	...	...	...	...	...	30	...	...	...	...	...	...	...	...
3	Heavy metallurgy .	26,400	720	...	...	70	...	570	80	...	...	...	25,680	2,200	10,000	9,180	4,300
4	Chemicals, coke .	13,000	8,600	...	5,000	3,500	30	40	20	10	...	10	4,390	800	2,300	1,290	...
5	Metal-working and machine—building. .	14,000	13,000	13,000	...	...	...	...	...	...	...	30	970	500	...	470	...
6	Construction materials . .	6,000	3,000	1,000	430	1,500	...	...	...	70	...	2,000	1,000	750	70	180	...
7	Light industry . .	800	...	...	...	...	...	...	...	...	...	800	...	...	...	...	...
8	Food industry . .	7,500	100	...	...	...	30	...	...	...	70	5,900	1,500	1,000	500	...	...
	Total.	89,150	35,420	14,000	7,565	12,890	75	610	130	80	70	8,740	44,990	5,370	13,240	21,340	5,000

TABLE 3A

NUMBER OF REGION OF ORIGIN	Z'	C' Z''	INDUSTRIES C''							
			1	2	3	4	5	6	7	8
I	8,600	2,800	...	850	...	20	...	80	150	1,700
II	11,200	31,000	1,600	1,400	7,000	12,000	4,000	2,000	2,200	800
III	5,700	4,200	1,300	...	...	30	...	1,100	1,500	270
IV	...	...	...	...	...	...	...	...	...	...
Total.	25,500	38,000	2,900	2,250	7,000	12,050	4,000	3,180	3,850	2,770

TABLE 4

ZAPOROZH'YE MODEL
(In Arbitrary Units)

No.	INDUSTRIES	X	X	A INDUSTRIES							B'		B'' REGIONS			
		X	x	1	2	3	4	5	6	7	y'	y''	I	II	III	IV
1	Heavy metallurgy	138,000	32,000	13,300	...	...	17,000	1,700	...	...	...	106,000	10,000	34,000	50,000	12,000
2	Light metallurgy	20,000	6,300	700	600	800	4,200	...	...	...	...	13,700	800	5,000	6,700	1,200
3	Coke and chemicals	15,000	9,500	8,500	600	330	30	40	...	...	500	5,000	500	1,500	1,200	1,800
4	Metal-working and machine-building	33,400	1,200	200	100	...	900	...	...	...	600	31,600	2,600	8,200	20,000	800
5	Construction materials	8,300	3,000	1,500	200	300	60	840	40	60	2,500	2,800	1,450	800	500	50
6	Light industry	11,600	1,600	...	...	...	100	...	1,500	...	5,100	4,900	3,000	1,200	700	...
7	Food industry	28,000	2,000	...	...	400	...	...	...	1,600	14,300	11,700	2,500	3,800	5,000	400
	Total	254,300	55,600	24,200	1,500	1,830	22,290	2,580	1,540	1,660	23,000	175,700	20,850	54,500	84,100	16,250

TABLE 4A

| NUMBER OF REGION OF ORIGIN | C' | | C" INDUSTRIES | | | | | | |
	Z'	Z"	1	2	3	4	5	6	7
I	8,100	44,300	31,500	100	...	150	750	500	11,300
II	13,400	17,000	3,700	100	8,300	1,000	50	1,500	2,350
III	8,500	18,300	10,800	1,700	...	1,200	150	3,600	850
IV	...	4,200	...	4,200	...	...	...	...	...
Total.	30,000	83,800	46,000	6,100	8,300	2,350	950	5,600	14,500

TABLE 5

PLACE NAME	1959 POPULATION (Thousands)	X	x	y'	y"	z'	z"	Z
Rakhmanovka. . .	4.8	390	. . .	. . .	390	240	47	287
Krivoy Rog . . .	189.0	89,150	35,420	8,740	44,990	25,500	38,000	63,500
Zaporozh'ye. . .	433.2	254,300	55,600	23,000	175,700	30,000	83,800	113,800

in functional structure and relationships, so that they must be assigned to different types of urban places. These differences are evident in the structure of the city-forming branches of industry, the volume and scale of external relations, and in the volume and character of internal interindustry relationships. The latter can serve as an additonal criterion for a functional classification of cities.

Despite the importance of static models in geography, such models do not exhaust all the potentialities inherent in model-building and in the search for ways of optimal development of cities and systems of cities. Such an objective, which is extremely important from an economic point of view, can be achieved only through the use of dynamic models. But the construction of such models still requires additional study.

1. V. S. Nemchinov, "A Model of an Economic Region," in *Primeneniye matematiki v ekonomicheskikh issledovaniyakh* [Application of Mathematics in Economic Research], no. 2 (Moscow, 1961); and Yu. G. Saushkin, "The Construction of Economic-Geographic Models of Regions for a Rational Location of Industry," in *Trudy konferentsii po voprosam razmescheniya promyshlennosti i razvitiya gorodov* [Proceedings of a Conference on Problems of Location of Industry and Urban Development] (Vilnius, 1962).

2. We know of only one much study, a model of the city of Oslo: R. Frisch, "A Survey of Types of Economic Forecasting and Programming and a Brief Description of the Oslo Channel Model," Memorandum from the Institute of Economics (Oslo, 1961).

3. V. S. Dadad'yan and V. V. Kossov, *Balans ekonomicheskogo rayona kak sredstvo planovykh raschetov* [The Balance of an Economic Region as a Means of Making Plan Calculations] (Moscow, 1962); B. Leont'yev, *Issledovaniye struktury amerikanskoy ekonomiki* [A Study of the Structure of the American Economy] (Moscow, 1958); V. S. Nemchinov, "A Model of an Economic Region"; idem, "Theoretical Problems of the Interindustry and Interregional Balance of Production and Distribution," in *Trudy nauchnogo soveshchaniya o primenenii matematicheskikh metodov v ekonomicheskikh issledovaniyakh i planirovanii* [Proceedings of a Scientific Conference in the Application of Mathematical Methods in Economic Research and Planning], no. 3 (Moscow, 1962); A. G. Aganbegyan and V. D. Belkin, eds., *Primeneniye matematiki i elektronnoy tekhniki v planirovanii* [The Application of Mathematics and Computer Technology in Planning] (Moscow, 1961); Yu. G. Saushkin, "On the Construction of Economic Models of Regional and Local Territorial Complexes," *Vestnik Moskovskogo Universiteta*, seriya geografiya [Herald of Moscow University, Geography series], no. 6 (1960); idem, "The Construction of Economic-Geographic Models of Regions for a Rational Location of Industry."

4. Such a construction of the city model is most convenient for economic-geographic research, the model was built by analogy with the models of economic regions, but this turned out to be inadequate for the analysis of actual cities. N. I. Blazhko, "Model Building of Cities," in *Razvitiye novykh issledovaniy prirodnykh resursov* [Development of New Research on Natural Resources] (Odessa University, 1963).

5. It should be noted that this and other city models show only the principal patterns of existing relationships; complete models required for the solution of complex economic problems would require the collaboration of economists, mathematicians, and economic geographers with the help of the statistical authorities; however, these schematic models are quite adequate for preliminary economic-geographic research.

I. S. MATLIN

Construction of a Model of the

Location of a Trade Network

At present one of the new applications of economic geography, the geography of services, is rapidly developing. The task of this branch of geography is not only to describe the existing location of services using quantitative indices but also to solve problems of the optimum location of enterprises in this sphere, which includes the disciplines of regional planning and economics.

In view of the importance and the complexity of the solution to such a problem, it has become necessary to use mathematical methods and calculating techniques. The use of the mathematical method for prognostication in the sphere of economic services becomes especially urgent in the light of the Directives of the Twenty-third Congress of the Communist Party of the Soviet Union for a new five-year plan, which is based upon the expectation of considerable expansion in the volume of services for the general public and the transformation of "domestic services into a highly mechanized branch."

In considering the methods that can be applied to the solution of the

Translated from *Vestnik Moskovskogo Universiteta*, seriya geografiya [Herald of Moscow University, Geography series], no. 6 (1966), pp. 38–46.

problems arising in the area of economic services, one must first of all point to the methods of mathematical statistics and the theory of mass services, which even in name corresponds to the activity to be realized and emphasizes the mass servicing of the population. However, because the theory of mass services does not provide a methodology for the analyses of the location of entering flows of claims for many types of service, especially of the multistage type, a better result can be obtained in many cases by using the method of statistical tests,[1] especially when one takes into account the fact that simulation models for the processes of mass services have been well developed.

In our opinion, to produce the optimum distribution of enterprises in the service sphere, models of linear and nonlinear programming can be used, but to derive parameters for these models various statistical methods must be extensively employed. The purpose of this work is to provide methods for planning the location of enterprises in one of the largest-scale branches of the service sphere, marketing.

The location model of trade network enterprises is a linear programming model; consequently, the first question arising in construction of the model is the question of the criterion of the optimum. Two such criteria can be offered: (1) minimization of expenditures for the construction and operation of the trade enterprises; and (2) minimization of the total time spent by the population in purchasing goods.

It is evident that a significant constraint in utilizing both criteria is the necessity to give full satisfaction to the demands of the population. In utilizing the first of the criteria, one must be careful to ensure that, in addition to satisfying the volume of demand, a definite level of quality is maintained in the service. Since, as in the other branches of the service sphere, an index to the quality of marketing is the amount of time spent by the purchaser in making a purchase the problem posed by utilizing the first of the criteria will have the following (verbal) formulation: minimize the expenditures in the trade network subject to the constraints of the volume of demand and the time used in the purchase of commodities.

The greatest difficulty in the formulation of this problem is the determination of the time factor. In order to formulate constraints of this type one must be able to assign some limit of time (moreover one differentiated by the type of good) in the course of which the demand for this group of goods must be satisfied. No valid method can as yet be offered to derive such a limit. This and some other circumstances make it impossible to utilize the first of the two criteria described above.

In this work the second formulation is used, that is, minimization of the total time spent by the population to purchase goods under a limitation of capital and operating expenditures in the trade network and subject to full satisfaction of the demand of the population.

We introduce the following notations:

X_{ij} — the amount (in rubles) of a commodity bought by an inhabitant of locality j at place i; $(i = 1 \ldots i^*)$, $(j = 1 \ldots j^*)$;

K_{iq} — the specific investments in a unit of turnover of merchandise q at place i;

r_{iq} — conditionally constant operating costs per unit of turnover of merchandise q at place i;

C_{iq} — variable operating costs per unit of turnover of merchandise q at place i;

K — the limit of investment;

C — the limit of operating costs;

t_{ij} — time spent to move between place i and place j (and back);

t_i — average time spent within the trade enterprise at place i;

L_{iq} — the number of purchases of commodity q at place i per unit of time (frequency of purchases);

T_{iq} — demand for commodity q at place i.

Before we turn to the mathematical formulation of the problem, we shall describe briefly the method of forming the parameters T_{iq}, r_{iq}, C_{iq}, K_{iq}, L_{iq}, t_i.

A study of population demand must serve as the basis for planning the location of separate enterprises in the trade network. The formation of demand by the population for particular groups of commodities depends on a number of factors: average per capita income, social composition of the population at a given place, and such demographic parameters as family situation and age-sex composition.

At the present, the most suitable statistical materials for the study of demand of the population are the data from budget studies. As experience shows, for the forecast of the volume of consumption of a certain commodity it is sufficient in most cases to study the relationship between the demand for this commodity and the income of the population. If sufficiently large groups of homogeneous (in terms of consumer properties) commodities are selected then one can disregard other factors such as change in style, and so forth in studying demand, since these factors have a greater influence on the redistribution of the demand within separate groups than on the redistribution of demand from group to group.

As investigations have shown, the dependence between consumption of a certain commodity and the income of the consumers can be adequately approximated by a linear function

$$f_q(x) = \mathbf{a}_q x + \beta_q, \tag{1}$$

[in this equation x = income] where $f_q(x)$ is the quantity of the product q used in the course of one year by each consumer.

If $g(x)$ is the function of population distribution by income, and M the upper limit of incomes, then the volume of personal consumption of commodity q at place j equals

$$T_{jq} = \int_O^M f(x)g(x)dx. \tag{2}$$

In this notation the population in place j equals

$$D_j = \int_O^M g(x)dx \tag{3}$$

and the total sum of incomes equals

$$I_j = \int_O^M xg(x)\,dx. \tag{4}$$

Then the consumption of commodity q in place j will equal

$$T_{jq} = \int_O^M f_q(x)g(x)dx = \int_O^M (\mathbf{a}_q x + \beta_q)g(x)dx =$$

$$\mathbf{a}_q \int_O^M xg(x)dx + \beta_q \int_O^M g(x)dx. \tag{5}$$

From equations (3) and (4) it follows that

$$T_{jq} = \mathbf{a}_q I_j + \beta_q D_j. \tag{6}$$

Thus, in order to make a forecast of the change in demand for a certain commodity, it is sufficient to have the data from budget studies, the data on the number of people in all places j, and the data on the total incomes at these places.[2]

In view of the limitations of the computer, it is impossible at present to solve the problem for the optimum location of a trade network by using the data on the consumption of each commodity; moreover, such a formulation of the problem is not expedient.

In connection with this, questions arise concerning the groupings of individual commodities, which is to be expected since this study is concerned with the location of the trade network. The grouping of commodities must conform to the specialization of the stores or their departments in such a manner that a definite combination of commodity groups would clearly correspond to any of the possible types of stores. In order to avoid the introduction of additional designations, q will designate the groups of commodities $(1 \ldots q^*)$.

Each type of store e $(1 \ldots l^*)$ can be characterized by a definite collection of commodity groups $(1 \ldots q^*)$ and by a definite number of working places. In addition, we introduce the following notation:

T_{ieq} —   the turnover of merchandise of all stores of type e at place i of commodity group q;

T_{in_eq}—   the turnover of the n_e store at place i for the commodity group q; where $n_e = (1 \ldots n_e^*)$.

By knowing the value of investments in one store of type e at place i and the value of the conditionally constant costs in this store, by using the method of averaging the coefficients accepted in problems of such type, and by comparing the investments and operating costs by commodity groups, one can find the values, of interest to us, K_{iq} and r_{iq}.

The amount of commodities of group q bought at place i by the population in place j was designated above by X_{ijq}. The sum of these magnitudes for every j gives the turnover in merchandise of the group q at place i.

$$T_{iq} = \sum_j X_{ijq} . \tag{7}$$

For simplification let us assume that the variable operating costs for the turnover of merchandise of a specific group are composed of transport costs.

Let C_{siq} be the net cost (or transport tariff) for the transfer of 1 unit of commodities q from place s to place i, then the total cost of transfer of commodities q from all places to place i is equal to

$$C_{iq} = \sum_s \sum_j C_{siq} X_{ijq}. \tag{8}$$

All investments at the populated place i (for all commodities) will comprise

$$K_i = \sum_q \sum_j K_{iq} X_{ijq}. \tag{9}$$

Investments in all populated places comprise

$$K' = \sum_i \sum_q \sum_j K_{iq} X_{ijq}. \tag{10}$$

Thus, one can formulate the first of the constraints of the problem

$$\sum_i \sum_q \sum_j K_{iq} X_{ijq} \leq K, \tag{11}$$

where K is the limit of investments.

The conditionally constant operating costs for turnover of all commodities at all places will equal

$$R' = \sum_i \sum_q \sum_j r_{iq} X_{ijq}. \tag{12}$$

If the total transfer costs of commodities q from all places s to place i equals $\sum_s \sum_j C_{siq} X_{ijq}$ from (8), then all variable operating costs comprise

$$R'' = \sum_i \sum_q \sum_s \sum_j C_{siq} X_{ijq}, \tag{13}$$

and all operating costs will equal

$$C' = R' + R'' = \sum_i \sum_q \sum_j r_{iq} X_{ijq} + \\ \sum_i \sum_q \sum_s \sum_j C_{siq} X_{ijq}. \tag{14}$$

Now it is possible to formulate the second constraint of the problem,

$$\sum_i \sum_q \sum_j r_{iq} X_{ijq} + \sum_i \sum_q \sum_s \sum_j C_{siq} X_{ijq} \leq C, \tag{15}$$

where C is the limit of operating costs.

The following group of limits must be determined by the requirement of satisfying the demand of the population in each of the populated places with each of the groups of commodities q.

$$\sum_i X_{ijq} = T_{jq}. \tag{16}$$

The number of such constraints is $q^{*}\cdot j^{*}$.

We proceed now to the formulation of the linear form. Using the notation introduced earlier one can note that the expenditures of time by the population at place j equal

$$\sum_{i} (t_i + t_{ij}) L_{iq}, \tag{17}$$

and the total expenditures of time by the whole population equal

$$\sum_{i} \sum_{j} (t_i + t_{ij}) L_{iq}. \tag{18}$$

Most complex is the question of determining the value L_{iq}, the frequency of purchases of commodities q at place i. One must first determine the frequency of visits to the individual store of type e: L_{n_e}.

The value L_{n_e} is clearly not equal to the sum of frequencies of purchases of all commodities, and at the same time it is not less than the frequency of purchases of the commodity most often bought.

The simples, but not a very exact formula for calculating L_{n_e} is

$$L_{n_e} = \frac{T_{n_e}}{\lambda_{n_e}} \tag{19}$$

where T_{n_e} is the turnover n_e of the store, and λn_e is the average norm of commodities bought there.

The indices L_{n_e} can be obtained by a sample examination of stores (by counting the number of purchasers).

Now, from the value L_{n_e} obtained by a sample, we proceed to the value L_{iq} that is of particular interest to us.

We formulate a statistical function showing the dependence of the frequency of visits on the turnover of merchandise. Since the problem of location of a trade network is a problem of linear programming, we shall have to approximate this function as a linear one,

$$L_{n_e} = a_e \sum_{q} T_{in_eq} + b_e \tag{20}$$

where a_e and b_e are parameters of the functions and

$$\sum_{q} T_{in_eq}$$

is the turnover of merchandise of the n_e store of all groups of commodities.

Summing the values

$$\sum_q T_{ineq}$$

for all stores of type e and taking into account that

$$\sum_{n_e}\sum_a T_{ineq} = \sum_q T_{ieq}; \; \sum_{n_e} b_e = n_e^* \, b_e$$

we can write

$$\sum_{n_e} L_{n_e} = a_e \sum_q T_{ieq} + n_e^* \, b_e. \tag{21}$$

The parameters a_e and b_e refer to a specific type of store e. The significance of these parameters for place i depends on the proportion of the various types of stores at this place and also on their turnover. We find the weighted average value of parameters a_e and b_e for place i:

$$a_i = \frac{\sum\limits_e a_e \sum\limits_q T_{ieq}}{\sum\limits_q \sum\limits_e T_{ieq}}, \tag{22}$$

$$b_i = \frac{\sum\limits_e b_e \sum\limits_q T_{ieq}}{\sum\limits_q \sum\limits_e T_{ieq}}, \tag{23}$$

Designating

$$\sum_{n_e} L_{n_e}$$

by L_e and substituting in formula (21) for the values a_e and b_e their average values from (22) and (23), one can write

$$L_e = a_i \sum_q T_{ieq} + b_i. \tag{24}$$

Hence the frequency of visits of all trading enterprises of the place is

$$L_i = \sum_e L_e = \sum_e \left(a_i \sum_q T_{ieq} + b_i \right) =$$

$$a_i \sum_e \sum_q T_{ieq} + e^* b_i. \tag{25}$$

The value

$$\sum_e T_{ieq}$$

is the turnover of the group of commodities by trading enterprises of place i,

$$\sum_e T_{ieq} = \sum_j X_{ijq},$$

and the total value

$$\sum_q \sum_e T_{ieq} = \sum_q \sum_j X_{ijq}.$$

Hence

$$L_i = a_i \sum_q \sum_j X_{ijq} + e*b_i. \tag{26}$$

Weighing the value L_i of turnover of the individual groups of commodities and designating the relation

$$\frac{\sum_q L_q}{L_q},$$

where L_q is the frequency of purchases of the group of commodities q, by δ_{iq} for all those commodities that are sold at place i, we get

$$L_{iq} = \left(a_i \sum_q \sum_j X_{ijq} + e*b_i \right) \delta_{iq}. \tag{27}$$

Now, after determining the value L_{iq} one can write the problem of linear programming for the determination of the optimum scheme of location of trading enterprises in the following manner

$$\sum_e \sum_q \sum_j K_{iq} X_{ijq} \leqslant K$$

$$\sum_i \sum_q \sum_j r_{iq} X_{ijq} + \sum_i \sum_q \sum_s \sum_j C_{siq} * X_{ijq} \leqslant C \tag{28}$$

$$\sum_i X_{ijq} = T_{jq}$$

$$X_{ijq} \geqslant o$$

$$\delta_{iq}(t_i + t_{ij}) \left\{ a_i \sum_q \sum_j X_{ijq} + l_i * b_i \right\} \to \min.$$

The method for finding all parameters entering into the problem, except the parameter of average time spent in a trading enterprise at a place, was given above. These parameters can be determined by special calculation (for which data are needed on the average time spent by salesmen for each purchase at a given organization, data on the number of purchases, and so forth) or they may be obtained by sample observations.

We have previously stated the most common form of the problem of location of a trade network. The problem in this form is applicable to regions where a network of trade services is only beginning. In a number of cases where there is already a network of trading enterprises, the location problem for new enterprises can be solved more simply.

We introduce the following notation:

G_e — the number of stores of e-type, which must be built in the region;

K_e — the average investments per store of type e;

C_e — the average operating costs per store of type e.

Using the notation introduced, we formulate the constraints

$$\Sigma\ K_e I_e \leqslant K \quad \Sigma\ C_e J_e \leqslant C \quad I_e \geqslant 0$$

and the linear form

$$\Sigma\ t_e I_e \to \max.$$

Here t_e is the time that is saved by the population in building one store of type e. It is evident that in this problem t_e is a function of I_e. Approximating this dependence by a linear function $t_e = \alpha_e I_e + \beta_e$, the functional then takes on the form

$$\Sigma\ (\alpha_e I_e^2 + \beta I_e) \to \max.$$

The problem changes into a problem of quadratic programming and can be solved, for example, by the Hindrep-D'Ezopo method.

To find the coefficients α_e and β_e one can use the cartometric method.

Let there be i^* possible points of distribution of a store of type e and in this store there are q_e^x groups of commodities.

Assuming that the trading enterprise has been built at place i, and, considering that each consumer acts to minimize his own time spent for travel, one can find for each of the groups of commodities, those places whose populace will buy these commodities at place i.

Let the district encompassing the sphere of influence of the planned store for the group of commodities q equal P_{iq}; then the whole sphere of influence of the planned store is equal to

$$P_i = P_{i1} \cup P_{i2} \cup P_{i3} \ldots \cup P_{iq}*.$$

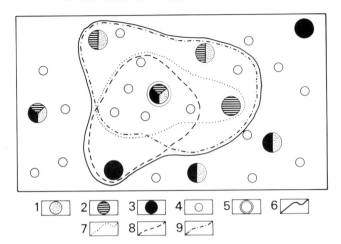

A   Map-diagram   for   Calculation
of   the   Value t_e

1-q=1;   2-q=2;   3-q=3;   4-others;   5-populated point in
which a store is planned;   6-P_i;   7-P_1 ;   8-P_2 ;   9-P_3

Knowing the number of people and the sum of their incomes at places encompassed on the map by the border of the district P_i, it is not difficult from formula (6) to calculate the future turnover of merchandise of this store at place i and its frequency of being visited.

By using these data, one can calculate the total economy in time for travel. Knowing how the frequency of visits to the store will change, and from which stores the loss of purchasers will occur, by having a model of these types of stores one can calculate the economy in time for the population through the process of providing a service. Having also obtained, as a result of modeling, the value of average time spent within the projected store, one can easily find the value t_{ie}, which represents the economy in time for the population as a result of building at place i a store of type e (see figure 1).

Carrying out the investigation and arranging the values $t_e = \sum_i t_{ie}$ in order of progression, one can calculate the parameters of the equation $t_e = a_e I_e + \beta_e$. After solving the problem one must place the stores at places with max t_{ie}.

The method described is a *synthesis of methods of optimum programming with the cartometric method*; the work with the map is rather labor-consuming. There is, however, the possibility of introducing data from the map (in the form of a matrix of distances t_{ij}, a matrix describing the existing system of trade services, and so forth) into the computer and thereby automating the cartometric work.

The solution of problems similar to the one above does not yield impressive sums in economy of material and financial resources; however, it does make it possible to economize on a no less precious resource, the free time of the population. Since this makes it possible to use the free time that is gained for rest and cultural growth of the population, it will help to improve living conditions.

1.   The idea of the method of statistical tests (the Monte Carlo method) consists of "losing" certain physical processes on the computer. See N. P. Buslenko, *Matematicheskoye modelinovaniye proisvodstvennykh protsessov tsifrovykh mashmakh* [Mathematical Modeling of Productive Processes by Computers] (Moscow, 1964).

2.   The results obtained by Valdsoo at the Institute of the Academy of Science of the Estonian SSR have been employed here and in other places as shown by the asterisk.

B. L. GUREVICH

The Population Density of a City

And the Probability Density of a

Random Variable

The diversity of cities in various countries appears to defy analysis: some have a single center, some no center, and some several centers. However, the methods of mathematical geography make it possible to uncover regularities in the population distribution in cities and thus to obtain a more ordered picture of them. Such an ordering is of great practical importance both for geographical city studies and for urban planning.

Until now geographers have judged the presence or absence of a single center in cities by descriptive means, and to a large extent intuitively. Through the use of the methods of mathematical geography, the intuitive approach is increasingly supplanted by precise concepts and supporting computations. This, in turn, will undoubtedly lead to a somewhat different geographical study of cities and will enrich mathematical models and diagrams by a large number of examples drawn from real life.

A previous article on this subject[1] dealt with a single-center (simple and generalized) city, the typology of single-center cities, and the

Reprinted with modifications from *Soviet Geography: Review and Translation* 8 (November 1967):722–30, by permission of the publisher. The article originally appeared in *Vestnik Moskovskogo Universiteta*, seriya geografiya [Herald of Moscow University, Geography series], no. 1 (1967), pp. 15–21.

structure of the density of population in the proposed city types. In this connection, the use of theoretical probability concepts required the introduction of special terms, for example, Poisson and Gaussian distributions of population density, and even the distinction of parallel and radial forms of these distributions. But the use of theoretical probability concepts in this article was episodic rather than systematic. And since the model of a single-center city omits from the analysis of the structure of population density a rather large number of other cities, the present paper attempts: (1) to establish a systematic relationship between the concept of population density of a city and the general theoretical probability concept of the density of the probability of a (two-dimensional) random variable;[2] and (2) to extend the analysis beyond single-center cities of various models to urban formations having other than a single center.

It should be noted that "The Mathematical Method in Geography"[3] stressed the need for using theoretical probability structures in the study of various important problems of mathematical geography. Hereafter we will be using the symbols adopted previously,[4] namely, δ for the population density of a city at any point M within the city. Thus, $\delta = \delta$ (M). We will use the letter Ω to designate the segment of the earth's surface on which the city is situated. No matter what the size of the city, we will assume that segment to be planar. Therefore the coordinates of any point M of the city may be of any type as long as they describe a continuous field (for example, polar coordinates r and ϕ). It should be noted, finally, that our analysis will not extend beyond the limits of segment Ω. Hereafter we will not assume that the city situated on the plane segment Ω is of the single-center type.

First we will establish a single-valued relationship between the density δ of population and the density of the probability f of a certain fully defined two-dimensional random variable.

Let us arbitrarily select within segment Ω a certain city area ω (see figure 1). Within area ω we will then take an elementary (infinitely small) area $d\omega$ and designate by $\delta = \delta$ (M) the population density at the center M of that elementary area. Since δ is expressed in population per unit area, $d\omega$ is the population within the area $d\omega$.

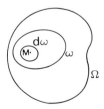

Figure 1

By summing up the expressions of the type $\delta d\omega$ over the entire area ω, we can obtain the population $N\omega$ of the area ω in the form of a double integral distributed over that area:

$$N\omega = \iint\limits_{\omega} \delta\, d\,\omega = \iint\limits_{\omega} \delta\,(M)\,d \tag{1}$$

In particular, if the area ω coincides with segment Ω, which encompasses by the entire city, we obtain the population of the entire city, namely:

$$N = N_{\Omega} = \iint\limits_{\Omega} \delta\,(M)\,d\Omega. \tag{2}$$

We assume, of course, that population density $\delta(M)$, which we can call the point density of population, is a fully defined continuous function of point M belonging to segment Ω.

If we now ask, what is the probability of any resident, taken at random, living within city area ω, and if we designate that probability as $P\{A\in\omega\}$, the natural answer would be:

$$P\{A\in\omega\} = \frac{N_{\omega}}{N}. \tag{3}$$

Here A designates the random point of segment Ω in which the randomly selected city resident lives. The symbol $A\in\omega$ designates the falling of A within area ω. We thus assume that the probability depends neither on the randomly selected resident nor, therefore, on the point A in which he lives, but rather entirely upon the area ω and is thus expressed by the ratio of the population N of that area to the population N of the entire city.

The probability $P\{A\in\omega\}$ can be written briefly in the form $P(\omega)$ or even P_{ω}. On the basis of equations (3) and (1), note that

$$P\omega = \frac{\iint\limits_{\omega} \delta\,(M)d\omega}{N} = \iint\limits_{\omega} \frac{\delta(M)}{N}\,d\omega = \iint\limits_{\omega} \frac{\delta}{N}\,d\omega, \tag{4}$$

so that, on the basis of (2),

$$P_{\omega} = \frac{\iint\limits_{\omega} \delta\,d\,\omega}{\iint\limits_{\Omega} \delta\,d\,\Omega}. \tag{5}$$

By assuming

$$f(M) = \frac{\delta(M)}{N},$$ (6)

we can replace (4) by

$$P\omega = \iint\limits_{\omega} f(M)\,d\omega = \iint\limits_{\omega} f\,d\omega.$$ (7)

It will be easily seen that when $\omega = \Omega$

$$P\Omega = \iint\limits_{\Omega} f(M)\,d\Omega = 1,$$ (8)

because on the basis of (2)

$$P\Omega = \iint\limits_{\Omega} f(M)\,d\Omega = \frac{1}{N}\iint\limits_{\Omega} \delta(M)\,d\Omega = \frac{N}{N} = 1.$$

The relationship (8) signifies the trivial fact that the probability of a randomly selected resident of the given city living in that city is equal to 1 since we are dealing here with a reliable event.

Thus the function $f(M)$, defined by equation (6) satisfies three conditions: (7), (8), and the requirement that $f(M) \geqslant 0$ for any point $M \in \Omega$. In probability theory this kind of function is called a probability density function; in this case it refers to the density of the probability of a randomly selected resident of our city living in area ω. In other words, it is the density of the probability of a two-dimensional random variable, namely, the probability of point A of plane segment Ω falling within plane area ω.

Formula (6) is of considerable interest. In the first place, it simply allows us to compare the population density $\delta(M)$ of the city and the probability density $f(M)$, assuming $M \in \Omega$. All we need to know is the distribution of the population density $\delta(M)$ in the given city and its segment (say, its map outline) Ω. The quantity N, needed to proceed from $\delta(M)$ to $f(M)$, is fully defined by formula (2), that is, by the function $\delta(M)$ and segment Ω. Consequently the population density of a city occupying segment Ω quite naturally gives rise to a definite probability density function $f(M)$.

Secondly, formula (6) enables us to proceed from the function $f(M) \geqslant 0$, which satisfies (8) and is defined for the given segment Ω (that is, from the probability density function), to the population density $\delta(M)$ of the city occupying that segment.

However, here the transition from $f(M)$ to $\delta(M)$ is not simple, since on the right side of (6) we have the ratio of $\delta(M)$ to N, and it is clear that by changing both $\delta(M)$ and N proportionally, that is, by taking $\lambda\delta(M)$ and λN (where $\lambda = $ const > 0), we would obtain the same $f(M)$. Therefore, if we want to make the transition from $f(M)$ to $\delta(M)$ single-valued, we have to give, say, the value δ at any one point A of the city. For such a point $A \in \Omega$, we will then have on the basis of (6)

$$f(A) = \frac{\delta A}{N},\qquad(9)$$

And, of course, here the population N of the city does not depend on A. In that case, (9) makes it possible to obtain that population:

$$N = \frac{\delta(A),}{f(A)}\qquad(10)$$

whence the probability density function $f(M)$ now has a strictly single-valued relationship to the population density within the segment Ω. In fact, on the basis of (6) and (10),

$$\delta(M) = N \cdot f(M) = \frac{\delta(A)}{f(A)}\, f(M).\qquad(11)$$

The foregoing enables us to use various densities of the probability of continuous distributions, which are well known and have been studied (and even tabulated) in probability theory in order to construct densities of population distribution that may be desirable from certain points of view—perhaps in the planning or replanning of cities. The structure of the population density of a particular city may, in turn, be regarded as a source suggesting new models or formulas of probability density in probability theory. The expression (6) thus links two completely separate areas of knowledge: urban geography and urban planning, on the one hand, and probability theory, on the other. In other words, (6) relates certain general theoretical probability structures to population-density structures (perhaps even to structures of the density of distribution of geographical formations in general).

This relationship is expressed systematically by (6) and not episodically and is therefore mutually beneficial both for geographical city studies and for probability theory.

Let us now examine in more detail some aspects of the relationship between $f(M)$ and $\delta(M)$ in the light of formulas (6) and (11).

Following the concept developed previously,[6] we will note that the transition from $\delta(M)$ to $f(M)$ can be effected by assigning a rate T of change of the population density of the city. Introducing the polar coordinates (r,ϕ) (with a geographically meaningful choice of pole 0 and the polar axis) of point $M(r,\phi)$ of segment Ω, we obtain population density $\delta(M) = \delta(r,\phi)$. If we then call the logarithmic derivative, with sign reversed, of the population density with respect to radius vector r the rate $T = T(r,\phi)$ of change of the city's population density, we can write:

$$T = \frac{1}{\delta}\frac{\partial \delta}{\partial r}. \tag{12}$$

The magnitude T characterizes (in fractions of density δ) the rate of change of the density along a ray $\phi = $ constant. We will assume $T = T(r, \phi)$ to be a continuous function of point (r, ϕ) and we will not require T to be nonnegative.

If the study city is such that we can select a pole 0 so that for any point (r, ϕ) of the city the rate of change will satisfy the condition $T(r,\phi) \geqslant 0$, then such a city, in accordance with the definition given by Gurevich and Saushkin[7] will be a single-center city.

Otherwise we will regard the city as a non-single-center city. The typology of non-single-center cities can be constructed similarly to the typology of single-center cities proposed by Gurevich and Saushkin. But in the case of non-single-center cities we confront the specific problem of the choice of pole 0. We will not deal with this problem here since it requires separate analysis (the point is that any change in the pole of a non-single-center city may change the type of city if we assume a typology of non-single-center cities similar to the typology of single-center cities proposed by Gurevich and Saushkin). For the sake of the present discussion, we will therefore assume that, as before, a geographically and economically meaningful point is selected as pole 0.

If $T = T(r)$, that is, the rate of change is a function only of the radius vector r, in other words, the rate will be constant for any given r at any angle ϕ within the limits $0 \leqslant \phi < 2\pi$, then the city can naturally be called a city of the radial or concentric type (in the sense of the distribution of population density). The density isolines, in this case, will coincide with circumferences of radius r whose centers are at the pole. In any direction at a constant angle ϕ, the rate of change of density within a city of the radial type would be the same, and therefore the distribution of population density will be identical and indistinguishable along concentric circumferences having a general center at the pole.

Let us examine the simplest case of a radial-type city, namely $T = T(r) = $ constant $= -1$. This is a non-single-center city (even if, on the basis of natural intuition, we cannot regard it as a multi-center city). In this city the population density increases from the pole toward the boundary in any direction and moreover at a uniform rate at any point of the city (r, ϕ). In this case the population density can be expressed by the exponential expression

$$\delta = \delta_o e^r. \tag{13}$$

The symbol δ_o designates the population density at the pole 0 of the city. And, of course, we must assume that $\delta_o \neq 0$. The distribution of population density (13) is shown in figure 2. At the pole, the popula-

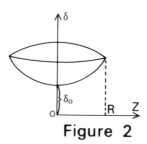

Figure 2

tion density reaches an absolute minimum. If we assume that the city extends from pole 0 in all directions for an identical distance $R =$ constant, we find, in accordance with formulas (2) and (13), that the population is:

$$N = \int_0^{2\pi} d\phi \int_0^R \delta r \, dr \, d\phi = 2\pi\delta_o \int_o^R re^r dr,$$

and, after integrating by parts

$$N = 2\pi\delta_o (1 + Re^r - e^r). \tag{14}$$

In view of (6), the probability density function $f(M) = f(r, \phi)$ takes the form

$$f(r, \phi) = f(r) = \frac{e^r}{2\pi(1 + Re^r - e^r)} \tag{15}$$

provided that the radius vector r of point (r, ϕ) does not exceed the radial distanct R at the border.

The city of Split in Yugoslavia, where the old low-density center dating back to Roman times is surrounded by groups of high-rise buildings, is a non-single-center city in the sense just described. We can construct a model of Split as a radial-type city in which the population density increases toward the boundary in any direction from the pole, which would coincide with the center of symmetry of the old low-density town. However, the exponential expression (13) is not likely to apply in this case. The population density would most probably follow the equation

$$\delta = \mathbf{a}r + \delta_0, \text{ where } \mathbf{a} = \text{const} > 0, \tag{16}$$

which expresses a linear relationship between the population density and the distance r of point $M(r, \phi)$ from the center. Non-single-center cities in the precise sense of (13) and (16) are intuitively single-center cities.

Of special interest is the following example of a non-single-center city occupying segment Ω (figure 3), whose outline, that is, city boundary, is expressed in polar coordinates by the equation.

$$r = \cos 2\phi. \tag{17}$$

If we assume a population density

$$\delta = \delta(r, \phi) = r(\cos 2\phi - r), \tag{18}$$

we obtain a distribution of population density in the form shown in figure 4. The city is thus of the non-single-center type (in the strict

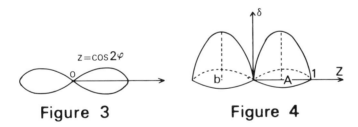

Figure 3   Figure 4

sense introduced earlier). But it is obvious that this non-single-center city is intuitively two-centered. Its centers lie at the points where function (18) reaches its maximum, that is, at the points $A(\phi = 0, r = 1/2)$ and $B(\phi = \pi, r = 1/2)$.

Finally, let us take the case of a radial-type city that, though a non-single-center city in the strict sense, can be assigned neither to the intuitively single-center nor the intuitively multi-center categories.

Let the population density of this city be

$$\delta = \delta_0 + \delta_0 \sin r, \tag{19}$$

where δ_o is the density at pole 0. It is easily seen that the rate of change of density is

$$T = \frac{-1}{\delta} \cdot \frac{\partial \delta}{\partial \gamma} = -\frac{\cos r}{1 + \sin r}. \qquad (20)$$

We will assume that the radius vector r changes within the limits of $r = 0$ (at the pole) and $r = 4\pi$ for any direction of angle $\phi = $ const. Formula (20) shows that this city is of the non-single-center category (in the strict sense established earlier). And it follows from (19) that the population density in this city reaches a maximum at two points in each directions ϕ, $0 \leqslant \varphi \leqslant 2\pi$, namely when $r = \pi/2$ (≈ 1.57 km) and $r = 5\pi/2 (\approx 7.85$ km). Along each direction $\phi = $ const. the distribution of population density is shown in figure 5. By rotating the curve

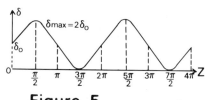

Figure 5

of figure 5 around a vertical axis over 360 degrees, we would obtain a graphic representation of the distribution of population density (19).

The concept of a non-single-center city is thus rather broad, including cities that are intuitively single-centered (formulas [13]and [16]) and cities that are intuitively multi-centered (formula [18])and even cities that cannot be assigned either to the intuitively single-centered or the intuitively multi-centered category (for example, the city defined by formula [19]).

It goes without saying that in each of the cases considered here we can compute the population N and, by means of formula (6) thus find the corresponding probability density function $f(r, \phi)$.

1. B. L. Gurevich and Yu G. Saushkin, "The Mathematical Method in Geography," *Vestnik Moskovskogo Universiteta*, seriya geografiya [Herald of Moscow University, Geography series], no. 1 (1966).

2. See for example, B. V. Gnedenko, *Kurs teorii veroyatnostey* [Course in the Theory of Probability] (Moscow, 1965).

3. Gurevich and Saushkin, "The Mathematical Method in Geography," pp. 17–18.

4. Gurevich and Saushkin, "The Mathematical Method in Geography," and L. I. Vasilev-skiy, "Mathematical Methods in Economic Geography," in *Kratkaya geograficheskaya entsiklo-pediya* [Short Geographic Encyclopedia], vol. 5, supplement.

5. See Gnedenko, *Kurs teorii veroyatnostey*.

6. See Gurevich and Saushkin, "The Mathematical Method in Geography."

7. Ibid.

8. Ibid.

SECTION IX

Historical Geography

Introductory Note

It can be easily argued that in the course of development of Soviet geography the most paradoxical subfield or branch is that of historical geography. The merits of studying the geographies of the past or the development of the cultural-economic environment from a genetic point of view have been vehemently argued by a few Soviet geographers, and deferential support has been professed by most of their colleagues. In actual fact, however, historical geography in the Soviet Union has experienced a history similar to that in the United States—an erratic record of development marked by occasional studies of high quality.

Despite the venerable roots of historical geography as a field of study in the prerevolutionary period, there was almost a complete lack of interest in the subject in the period following 1917. This is somewhat understandable given the overwhelming concern with practical problems concerning the building of the fledgling socialist nation and the need to disown the bourgeois past. By the 1940s, however, the cause of historical geography was championed by Yatsunskiy beginning with an article included in this section. Even at that time Yatsunskiy admits to a lack of study in the area of historical geography despite great expressed interest on the part of geographers.[1] Similarly, an early article by a leading Soviet economic geographer, Saushkin, emphasizes the great need for the study of changes through time, a historical approach.[2] Some twenty years later the same author characterizes historical geography as a *boundary discipline* that "tends to be more inside [the confines] of geography, but assumes a wide range of forms."[3] Still other contrasts may be cited. For example, from 1950 to 1960 there have been three special issues of *Voprosy geografii* [Problems of Geography] dedicated to historical geography,[4] but not one chapter on this subject was included to the comprehensive report on the state of the arts published in *Soviet Geography: Accomplishments and Tasks.*[5] Moreover, there is not one chair of historical geography in the geography departments of Soviet universities, although a section on the History of Geographical Knowledge and Historical Geography has been established in the Geographical Society. Iofa, one of the more recent proponents of historical geography, also notes the importance of the field for the contemporary and future needs of geography as a discipline but indicates that "no serious, broad research in this field [historical geography] is now under way."[6] The content or definition of the subfield has also been subject to a number of interpretations. The definitions given or implied have often included the history of geography and geographical thought, and even the history of exploration.

The range of topics covered in historical-geographical studies in the Soviet period is extensive even if the number of published works is comparatively small.[7] Population and migration, man's effect on the environment and past agricultural settings have been the dominant research themes. Moreover, historical geography, like population geography, is assuming an increasingly interdisciplinary character.

The three studies in this chapter include two philosophical methodological articles, one representing the earlier exhortations by Yatsunskiy and one by Iofa reflecting a recent position, and an interesting substantive article concerned with grain yields in the prerevolutionary period.

The lead article by Yatsunskiy is now a classic and played an important role in focusing attention on problems in historical geography. He traces the course of development of historical geography from Clüver in the Netherlands (seventeenth century) to the present, discussing the variations in content and definition of the field. The most important historical-geographical literature is reviewed in some detail allowing the author to trace the subject matter as it evolved, including exploration and discoveries, toponymy, the role of the natural environment and, most recently, the study of production and economic relations.

Yatsunskiy also offers a definition of the field: "the study and description of the geographic aspect of the historical process." He adds that the subject matter must include (1) the natural landscape, (2) population, (3) production and economic relationships, and (4) political phenomena and political relations. The author pleads for more attention to historical geography and cites a number of proposals for its development. The original appended statement of support for Yatsunskiy's views made by the editors of *Istorik-Marksist* [Historian-Marxist] is left intact.

The article by Iofa is also an argument for the cause of historical geography. The major point made is that geography is as much a chronological science as an areal or spatial one, and that historical geography provides a much needed genetic approach to the study of phenomena. The author sets himself three main tasks in the paper, (1) to differentiate history from geography, (2) to assess the importance of the study of geography of the past for contemporary geographic problems, and (3) justify a reorganization of the field of geography to take account of the past. Basically, Iofa's argument rests on the premise that without an historical-geographical approach, many of the common geographic concepts (such as the region) cannot be fully understood, nor can the general direction of development be properly detected.

The last study by Nesmeyanova is an excellent example of the application of newer, more sophisticated methods to a problem in historical geography. The objective of the study is to analyze the yield capacities of grains in an area of European Russia in a prerevolutionary period. The analysis is, in fact, a reexamination of data used in an earlier study, although the current examination utilizes smaller study units (uyezds instead of guberniyas).

1. V. K. Yatsunskiy, "The Subject Matter and Problems of Historical Geography," *Istorik-Marksist* [Historian-Marxist], no. 5 (1941), p. 29.

2. Yu. G. Saushkin, "The Cultural Landscape," *Voprosy geografi* [Problems of Geography], no. 1 (1946).

3. Yu. G. Saushkin, "A History of Soviet Economic Geography," in *Economic Geography of the USSR, History and Present Development*, ed. N. N. Baranskiy et al. (Moscow, 1965), pp. 54–172.

4. *Voprosy geografii*, no. 20 (1950); no. 31 (1953); no. 50 (1960).

5. Lawrence Ecker, trans., and Chauncy D. Harris, ed., *Soviet Geography: Accomplishments and Tasks*, American Geographical Society Occasional Publication, no. 1 (New York, 1962).

6. L. Ye. Iofa, "On the Significance of Historical Geography," *Geografiya i khozyaystvo* [Geography and Economics], no. 11 (1961), p. 101.

7. For a summary of past emphases in Soviet historical geography and an excellent bibliography of selected items, see R. A. French, "Historical Geography in the USSR," *Soviet Geography: Review and Translation* 9, no. 7 (September 1968):551–61.

V. K. YATSUNSKIY

The Subject Matter and

Problems of Historical Geography

Historical geography as a branch of knowledge has existed for several centuries. The founder of historical geography among the German geographers and historians (in the scientific literature of other countries it may be said that this question was rarely posed) was Clüver, the former professor at Leyden University in the Netherlands in the first quarter of the seventeenth century.[1]

The Belgian Professor Van der Linden, in an introductory speech at the opening of the first international congress on historical geography in 1930, expressed a different point of view: he pointed to Ortelius, the famous Flemish geographer of the second half of the sixteenth century, the author of the first historical atlas, as the "forerunner of historical geography." An analagous opinion was expressed in 1935 by Professor Almagia, the competent Italian specialist on the history of geographical science, characterizing Ortelius as one of the founders of historical geography. Quite recently, in 1938, the American Barnes, in his book, *A History of Historical Writing*, noted that the English historian and geographer of the twelfth century, Giraldus Cambrensis, also "was studying historical geography."

In this article I cannot undertake a special investigation regarding the genesis of historical geography.[2] But, in any case, the cited statements by historians and geographers allow one to confirm that his-

Translated and abridged from *Istorik-Marksist* [Historian-Marxist], no. 5 (1941), pp. 3–29.

torical geography in Western Europe has existed for more than three centuries, even if its beginning is related to Clüver. In our country the history of its development is shorter, corresponding to the younger age of Russian historical science. Nevertheless, there are the beginnings of historical geography in Tatishchev's works, and our prerevolutionary historians usually date the beginning of the development of our historical geography as a special discipline from the works of Nadezhdin.[3] Thus, in our country too, historical geography is not an especially young science.

In the course of several centuries many works on historical geography have accumulated. At international historical congresses usually a special section on historical geography is organized. Such a section, as a rule, is organized also at international geographical congresses. In 1930, in Belgium a special international congress on historical geography convened at which scholars were present from Belgium, France, Germany, England, Italy, Spain, Holland, and Poland. According to press accounts[4] at the congress 55 papers were read in seven sections, and the congress resulted in very lively discussions.

In recent years the interest in historical geography has noticeably increased in foreign countries. The number of works has grown. In many schools of higher learning a course in historical geography is given. Thus, historical geography is an old scientific discipline with a large literature, and, moreover, a discipline in which interest is growing.

If we turn to historical-geographical literature, however, we find a great diversity of opinions concerning the meaning of the idea historical geography. This diversity was less clearly expressed in the discussion on the subject of historical geography that was organized in 1932 in London by the historical and geographical societies.[5] It must also be noted that specific works on historical geography often contradict the definitions of the subject given by the authors themselves.[6] Sometimes, in order to get out of the existing difficult position, the authors give two definitions: one more general, the other more narrow and conforming to their presentation. S. M. Seredonin in his course on the historical geography of Russia is an example. One must note that the content of his book is even narrower than his narrow definition.

Consequently historical geography even before World War I had a reputation as a science with a very indefinite content. S. K. Kuznetsov began his course in Russian historical geography at the Moscow Archeological Institute in 1907–8 with the words: "I shall hardly be in error if I say that the content of the science that I am to present, Russian historical geography, is exceedingly indefinite, the very idea concerning it is hazy."[7]

Similar references are heard in our time; for example in 1932, Professor Gilbert wrote in an article: "What is Historical Geography?: The term 'historical geography' does not have a completely definite meaning for the historian and the geographer. Works designated by this term are very diverse and differ significantly between their character and aims."[8] Quite recently a well-known medievalist, Marc Bloch, in his review of collective works by English scientists under the editorship of Darby, *Historical Geography of England Before A. D. 1800,* wrote: "Our dictionary is as yet so imperfect that to call the book 'historical geography' means to risk not giving beforehand an exact idea of its content."[9] In our Soviet literature an attempt was even made to deny the very expediency of existence of historical geography.[10]

One hardly needs to prove that the described lack of clarity in the understanding of the subject of historical geography hinders successful work in this field. On the other hand, however, a simple addition of another definition to these formerly expressed can hardly set things right. Therefore it seems to me, it will be more expedient to take a more complicated path. Putting aside for the moment the difinitions of the subject of historical geography proposed by various authors, let us try to make clear in fact what content the authors included in their historical-geographical works.

In the systematization of the factual content of historical-geographical works I shall, in my presentation, arrange brief characteristics of individual trends chronologically as much as possible and attempt to relate these trends to the development of historical and geographical science.[11] Such a survey will help me to substantiate better my own views on the subject and problems of historical geography and will be of some interest since there is no proper resume either in our own or foreign literature. It is natural that in view of the abundance of accumulated literature much will have to be left untouched and much can be touched only lightly.

The most elementary problem faced by the historian-geographer is the location on the map of geographic names of the past. He tries to determine the places where the ancient peoples lived, the site of ancient cities, places of battles and other points connected with historical events. (Where, in fact, was it located?) Thus one can define, paraphrasing the well-known expression by Ranke, the problem that historically was faced before all others, and in the formulation and first attempts to solve it lies the genesis of historical geography as a science.

In the last quarter of the sixteenth century Ortelius, working with maps of his atlas, the first in the world, saw his main task to be helping his contemporaries to read the ancient authors.[12] On the cover of his atlas in the form of an emblem he put the words: "Historiae oculus geographia." In cases where Ortelius was at variance regarding place names used by the ancient authors he often gave corresponding names on the map itself. In order to interpret the ancient geographic names and to establish their links with contemporary names Ortelius composed a historical-geographical dictionary under the title *Thesaurus Geographicus.*[13]

A large number of subsequent investigators of the seventeenth, eighteenth, and nineteenth centuries continued the work begun by Ortelius. In the seventeenth century, Clüver, who studied the geography of ancient Italy, Sicily, and Germany, and Valois, who worked on the geography of ancient Gallia, were distinguished in this field. Experts in historical geography of the ancient world valued their works highly as late as the nineteenth century.

In the eighteenth century there were completed the authoritative works by d'Anville whom Niebuhr called "the great d'Anville, one of the greatest geniuses I know."[14] In the nineteenth century Heinrich Kiepert was widely known for his atlas of ancient Greece, an atlas of the ancient world[15] and a number of maps with ancient Roman inscriptions published by the Prussian Academy of Science,[16] as well as excellent school wall maps on ancient history in wide circulation before World War I. He also wrote *A Textbook on Ancient Geography* devoted mainly to the study of ancient geographic nomenclature.

In addition to the general works mentioned above many specific studies were written ascertaining the site of a given geographical point of the past or the place where a given historical event had occurred. In our country Tatishchev posed and tried to answer questions of this kind. In his first book, *History of Russia* he worked on the problem concerning "the name, origin, and place" of the various peoples that inhabited our country in the past. When, at the end of the eighteenth century, Musin-Pushkin completed a special study on the question "concerning the site of the ancient Russian Tmutorokan princedom"[17] he was not the first to investigate this question, which had been studied by Tatishchev, Prokopovich, Bayer, Shcherbatov, and Boltin before him.

In the nineteenth century questions of this kind were studied by a number of investigators such as Lerberg,[18] Brun,[19] and particularly Barsov who compiled a *Geographic Dictionary of the Russian Lands of the 9th to 14th Centuries* and *Essays on Russian Historical Geography.* The first of these works is similar in structure to the *Thesaurus Geographicus* by Ortelius. In the second, the author examines the geographic names found in the first chronicles, determines the site of the corresponding points, investigates the settling of tribes, the boundaries of lands and princedoms, establishing the geographical scope of the chronicler, but provides no maps.

Clearing up questions of ancient topography is the object of study of present Soviet historians.[20]

From the determination of the sites of places, remarkable with respect to history, it was natural to proceed to the routes of historical voyages and campaigns of famous warriors. Geographic maps of sea routes, which served as a guide to the seafarers, have existed since ancient times and usually included the shores of the countries adjacent to the sea routes. This was especially developed the fourteenth century in Italy (the so-called Portolans). Then, the sea routes began to be marked by a line on the maps. In the geographic atlas outlined on parchment paper by Agnese in 1546[21] is a map of Magellan's voyage and the route that the Spanish flotilla took to Peru. For historical maps this method was used for the first time by Ortelius in his atlas by tracing the route of voyages of the biblical patriarch Abraham. Clüver, in his *Italia antiqua*, investigated "by what route Hannibal crossed the Alps."[22] This method was widely used in the historical atlas by the French geographer DuVal, who depicted the sea routes of Odysseus and Aeneas, the retreat route of the Ten Thousand Greeks, based on the story by Xenophon, and the campaign routes of Alexander of Macedonia.[23]

From the location on the map of places historically interesting, the natural step was to study the political borders of the governments of the past and the changes they suffered in the course of the historical process. Ortelius already indicated interest in this when in his "Parergon" he marked the borders of governments and sometimes delimited political divisions within the various countries. In France in the seventeenth century it was Nicolas Sanson who called attention to internal political divisions; he is considered by the French historians as the initiator of historical geography in France.[24] Sanson, however, as well as Ortelius, considered the political borders of the past as static (boundaries) without trying to trace their dynamics.

The first attempt to study the dynamics was made also in the seventeenth century in France by DuVal, a nephew and student of Sanson. DuVal constructed three maps of the growth of the territory of the Roman empire: *Imperii Romani Infantia,*[25] *Imperii Romani Adolescentia*[26] and *Imperii Romani Inventus.*[27] Later the study of the evolution of political boundaries became the most popular problem of historical geography. In France this question has and is receiving special attention not only in the scientific literature but also in textbooks. From the time of the July Monarchy, textbooks on historical geography were widely used in which the history of unification and territorial growth of France and the changes in its administrative divisions are given. In a scientific respect much had been done in this direction in France at the end of the nineteenth century by Longnon in his painstaking investigations.[28] In 1881, the English scientist Freeman published a similar course on the historical geography of Europe in two volumes—an atlas and a text—describing all the basic changes in the political and, in part, the religious geography of Europe from ancient times to the nineteenth century. Freeman's book was very well known: three editions were printed in England, and it was translated into French and Russian. The Russian edition appeared under the editorship of I. V. Luchinskiy in 1892. Of a similar character is the classical work on colonial countries outside of Europe by the German geographer Supan, *Die territoriale Entwicklung der europaischen Kolonien mit einem Kolonialgeschichtlichen Atlas von 12 Karten und 40 Kärtchen im Text.* The author examines in sequence, from the geographic point of view, the history of the division of the world between the European countries up to 1900 and contains a number of maps of colonies in connection with the most important moments of this history.[29] In contrast to the book by Freeman, written in the form of a handbook, Supan's is a monographic historical investigation. Lenin valued this work highly and he used it in his work, *Imperialism as the Highest Stage of Capitalism.* In our country in 1793 a *Historical map of the Russian Empire* was published where the territorial growth was described from Peter I to Catherine II, inclusive.

In the nineteenth and twentieth centuries a number of investigators worked on specific questions of history of our external and internal boundaries. The work by Nevolin "Concerning the Five Administrative Regions and Villages of Novgorod in the Sixteenth Century" should be noted. More often, however, such questions were studied not in special works but rather in the works of those studying the local history of a certain part of our country or investigating the organization of local governments. Thus, for example, M. K. Lyubavskiy in his work *District Division and Local Government of the Lithuanian-Russian Government to the Time of the First Lithuanian Statute* devoted a whole section of the book to the political geography of the Lithuanian-Russian government in the fifteenth and sixteenth centuries; Yu. V. Got'ye, as an appendix to his investigation on the Trans-Moscovian region in the nineteenth century, drew up a map of the Trans-Moscovian region of the middle seventeenth century from record and census books and enumerated the camps and volosts. In another investigation, *History of the Oblast Administration in Russia from Peter I to Catherine II,* Got'ye devoted a special chapter to the oblast division in the period 1725–75.

Old historical maps are an exceedingly valuable source for the determination of the places to which the geographic names of the past belong and for the study of the boundaries of former states and provinces. It is natural that the publication and study of these maps has become one of the tasks of historical geography. At the end of the sixteenth century Marc Velser, a member of a famous Augsburg merchant family and a learned humanist, found in the library of the humanist Peutinger an ancient Roman map known later in science under the name "Tabuia Peutingeriana." Velser sent the map to Ortelius in

Antwerp for study and publication. Ortelius did not finish this work and the *Tabulla Pentingeriana* appeared after his death.[30] Since then a huge literature has accumulated concerning it. In the USSR academician Ya. A. Manandyan worked with this map when studying the trade routes of ancient Armenia.[31] The publication and study of ancient maps in the nineteenth century was widespread. The French-man Jomard[32] and the Portugese Santarem[33] did much in this respect in the middle of the century, and at the end, the famous Swe-dish explorer of polar countries and historian of cartography Norde-skield[34] was important. At the present time a great deal is being done in this field abroad; for example, in Italy,[35] Czechoslava-kia,[36] and Yugoslavia[37] *Monuments cartographica* of these coun-tries were published. Especially splendid in design and exceptional in completeness of material is the multivolume edition *Momumenta cartographica Africae et Aegypti*[38] by Usuf Kamal in Egypt.

In our country the work by V. A. Kordt, *Materials on the History of Russian Cartography* is receiving deserved recognition and has ap-peared in three editions consecutively, 1899, 1906, and 1910. He also published *Materials on the History of Ukrainian cartography* in 1931. In the same group of historical-geographical works must be counted the publication of *The Book of the Great Chart*[39] and maps by Re-mezov.

The study of geographic memorials of the past as a historical source had induced the investigators to study the historical develop-ment of geographical views. On the one hand, this had to be the direc-tion of scientific thought and the expansion of historical science and development of geography, but on the other much time was needed for all these influences to yield concrete results. The humanists looked upon ancient culture as a whole, not knowing how to differen-tiate between periods in its development. In the geographical thought of antiquity, they did not see any periods of development.

The content of historical works of the sixteenth and seventeenth centuries amounted to an exposition of only political events of the past. The situation changed in the eighteenth century. In the Age of Enlightenment in France the bourgeoisie put before the historians broader tasks. As the Spanish historian Altamira remarked, the eighteenth century put forward the principle that "history is not a history of rulers, but a history of peoples."[40] The history of culture was born. Freret[41] at this time initiated the study of the history of geographic views of antiquity. In the nineteenth century the subject of study became the development of geography in the Middle Ages. The Polish historian Lelevel who, according to Marx, "did much more for the clarification of the enslavement of his fatherland than a whole crowd of writers whose whole baggage amounts simply to curses at Russia,"[42] as an emigrant wrote a major work *La geo-*

graphie du moyen age, which is still significant. In our country I. D. Belyayev published a study entitled *On Geographical Information in Ancient Russia* in 1862. Much more has been done in this field since then.

It is hardly necessary to prove that the history of geography is not the same as all historical geography, although, of course, there are many points of contact between these branches of knowledge; and, in particular, the geographical works of the past, such as old maps, often can serve as historical sources. However, historical geography and history of geography are often confused, and moreover, specialists confuse them; thus, for example, S. M. Seredonin in his course on *Historical Geography* characterizes the above-mentioned article by I. D. Belyayev as a work on the historical geography of our country.

The complex of problems relating to content that has faced historical geography from the very beginning include: the location on a map of important historical sites, determination of routes of military campaigns, investigation of the history of political boundaries, and the related study of ancient maps as one of the historical-geographic sources. The content of this complex fully corresponds to the demands that the so-called political history presents to historical geography.

The next problem that is usually also assigned to the field of historical geography is the question about the population and its distribution in a given country in the past. This concept was familiar to the scientists of the sixteenth and seventeenth centuries. They attempted to determine the living area of some peoples mentioned by ancient writers. In the nineteenth century under the influence of a national upsurge in Germany and a national renaissance of the Czechs, Slovaks, and Slovenes and also as the result of the growth of historical knowledge and the development of scientific linguistics, the work in this sector of historical geography grew. A new source was introduced, toponymic information.

The inclination to interpret the meaning of geographic names existed in antiquity. During the Renaissance and later, historians often tried to explain geographical names, but the absence of linguistic training led to the most arbitrary conclusions. In the first half of the nineteenth century, with the growth of scientific linguistics, toponymics found firm support for the historian's research. In the second half of the century in the countries of Western Europe, comprehensive work was organized for the collection of geographical names. This work is still in progress. In England there is a special scholarly organization, the English Place-Names Society, that publishes systematized lists of geographic names by duchies. Similar publications exist in Germany, France, and some other countries. In Germany a special journal on toponymics is published: *Zeitschrift fur Ortsnamenforschung;* in Belgium, *Bulletin de la commission de toponymic et diatectologie.* Toponymics is, of course, not historical geography, but its data are widely used in historical geography.

In 1821 one of the founders of scientific linguistics, Wilhelm Humboldt,[44] published a work *Prüfung der Untersuchung über die Urbewohner Hispaniens vermittelst der Vaskischen-Sprache* (Analysis of the Investigation on the Spanish Original Inhabitants by Means of the Basque Language), in which he tried by means of the Basque language to analyze the geographical nomenclature of Spain in order to determine the national composition of the original population of Spain. The savants of Slavic renaissance turned their attention to this historical source early: Kollar[45] and Shafarik[46] used it for investigation. Since then, in Western Europe very much has been done in this direction. A number of methodological difficulties[47] were explained; methods were worked out for the use of the data of toponymics; scientific trends were formed which accumulated a considerable literature; there exist toponymic atlases among which should be mentioned the substantial *Atlas nazw geograficznych Slowianszczyzny Zachodniej* by Kozerovsky,[48] the untiring investigator of Slavic toponymics of Eastern Germany.

In our country more than a hundred years ago N. I. Nadezhdin pointed to the significance of toponymic data for historical geography. In his article: "An Experiment in the Historical Geography of the Russian World" Nadezhdin wrote: "The first page of history must be a geographic map not only as an aid in order to know what happened where, but as a rich archive of documents and sources."[49] He points out further that, for a historian, the important thing is not the meaning of the name but the definition, to which language it belongs, in order to determine what people settled at a given place in the past. On the basis of the analysis of names of rivers in Eastern Europe, he drew a sketch of Slavic and Finnish tribes distributed there in the past. In this article Nadezhdin mentions Humboldt's work on the ancient population of Spain. The article had a considerable influence on Russian geography from the point of view of methodology as well as thematics. The problem of population has for a long time occupied a central place. For example, in the well-known book by Barsov[50] a great deal of attention is placed on the settlement of Eastern Slavs according to data from the chronicles. The course on historical geography by S. I. Seredonin is devoted exclusively to the shift and distribution of peoples in European Russia from the times of Herodotus to the Mongolian conquest. From the problem of the ethnographic composition of the population our historians proceeded to the study of the history of colonization of the territory of Eastern Europe and northern Asia by Russian and Ukrainian peoples. S. M. Solov'yev emphasized the significance of colonization in the history of Russia. He studied colonization from nationalistic positions. This nationalistic bias was characteristic for many later historians of the prerevolutionary period. In our historiography many special works are devoted to the colonization of individual parts of the country; great emphasis is also placed on it in general courses. Thus, for example, V. O. Klyuchevskiy emphasizes colonization as "the main fact" of Russian history.[51] M. K. Lyubavskiy designed his course on historical geography of Russia as a history of colonization.[52]

The next problem often connected with historical geography is the investigation of the influence of natural conditions of the course of the historical process in a given country. Even the writers of antiquity discussed the influence of nature on man and on the course of history. Strabo connects the successes of Roman conquests with the geographical position and environment of Italy.[53] The influence of nature occupies an important place in the historical-sociological theory of one of the greatest Arab historians, Ibn Khaldun.[54] During the Renaissance the French historian Bodin[55] dwelled on this theme. In the eighteenth century Montesquieu and a number of other thinkers of the Age of Enlightenment ascribed great significance to the influence of nature on human society.

Thus, the question of the role of the geographic factor in history is a very old one. However, until the nineteenth century this question was usually put in a general form and was usually resolved in terms of accepting one of the natural conditions as a decisive influence, that is, the influence of the climate in a given country on the psychology of man and, through it, on society and the entire historical process.

In the nineteenth century under the influence of the famous German geographer Ritter who, in the expression of the Spanish historian Altamira, "confirmed the study of geographic phenomena as an element of social history."[56] The problem was more clearly formulated. Natural conditions began to be studied as external circumstances in which the historical process develops. A student of Ritter, the historian Kurzins, in 1851–52 wrote a monograph on Peloponnesus where, with the skill of an artist, he thoroughly characterized the geography of the area and its influence on the history of this country in antiquity. However, as Fueter[57] points out, Ritter influenced only a few historians.

In the 1880s when modern geography was already formed as a branch of natural science, the German geographer Ratzel attempted to develop a new branch of geography—anthropogeography—which was to study the influence of the geographic environment on the social life of humanity.[58]

In France, somewhat after Ratzel, a similar system of ideas was brought forth by Vidal de la Blache.[59] His ideas were further developed by his students.[60] Anthropogeography, or the geography of man as the French and English call it, has received considerable attention in Western Europe and America since then. Suffice it to point out that its development resulted not only in greater attention paid to the influence of nature on the course of the historical process by historians and geographers, but also in a number of attempts to trace this influence in actual examples in individual countries. Such attempts were undertaken frequently in the United States,[61] a young country with exceptionally rich natural resources. Among these attempts, the works of Turner[62] are distinguished; on a base of factual material he evolved a unique conception of North American history in which he cited as the main facts in the North American historical process, the colonizing movement to the West and the development of natural resources.

In our own scientific literature the question of the influence of the geographic environment is also an old problem that was familiar to the historians of the eighteenth century such as Boltin. In the nineteenth century S. N. Solov'yev, who had attended Ritter's lectures in Berlin,[63] began his *History of Russia* with an essay on natural conditions; he returns to their role later when studying the era of Peter I. A student of Solov'yev, V. O. Klyuchevskiy also began his course with an essay on the nature of the East European plain. It is of interest to note that both these introductory essays, by Solov'yev and Klyuchevskiy, contrast with the ideas of A. P. Shchapov who, in 1864, expressed his opposition to such an approach. In his article "Ethnographic Organization of the Russian Population" Shchapov wrote, "In our multivolume Russian histories there are a few words written in the first chapters on Russian tribes and peoples, or they are merely enumerated, and, similarly, a few words are devoted to Russian geography or the geographic influence on history—as if the tribes and peoples suddenly disappear without a trace from the face of the Russian land having had no influence on the Russian people or Russian history, and as if geography did not accompany history at every step, in each area. Where then are the land and the people? Could they have collapsed somewhere and only the government remained?"

Shchapov also tried to trace the influence of the geographic environment on Russian history in an article entitled "Historical-Geographical Distribution of the Russian Population" where he examined the dependence of population distribution on natural conditions in Russia. He considers the geographic environment to be a decisive factor of historical development that determines not only the economic life of man but also his psychology. Shchapov, however, ignores the production relationships and the state of the productive force. In his opinion the geographic environment acts directly on the economy and on man's nature itself. As a result, Shchapov moves toward idealism in the understanding of history and belongs to the group of materialists who could not bridge the materialistic view on nature in order to materialistically explain social phenomena.[64]

In our time I. I. Polosin[65] has made an interesting attempt to trace systematically the influence of natural conditions on the historical process; he sees the task of historical geography precisely in the working out of this problem.

Closely related to the problem of the role of natural conditions in the historical process is the question concerning the study of the state of these very conditions in the past, a question concerned with the reconstruction of what the German geographer Wimmer[66] fittingly called "die historische Naturlandschaft." The question is, what was the nature of a given country in the past, how much has it changed, particularly during that time in which the history of humanity developed: this question has always interested scientists-naturalists. From the point of view of a specialist in physical geography the task of historical geography consists first of all in answering this question, and all the other problems follow. The historian of Russian geography L. S. Berg,[67] in his work devoted a paragraph to the historical geography of our country beginning with this problem. The specialists-naturalists have worked hard to explain the evolution of the vegetative cover,[68] hydrography, and similar questions in our country as well as abroad. Natural-scientific material as well as historical sources were used.

To a lesser degree these questions were the subject of study of historians who usually relied only on historical sources, but who sometimes used the works of the naturalists. As examples one can point in the West to Desjardins, who produced a detailed and careful reconstruction of the physical geography of Gaul;[69] in our country Zamyslovskiy[70] attempted to do the same for Muscovy of the sixteenth century based on Gerberstein's data. Similarly, Bartold studied the changes in the direction of the course of the Amu-Dar'ya River in the historical past.[71]

In summing up what has been said about the study of the changes of the geographic environment in the historical past and its influence on the historical process one must confess that considerably more has been done on the former problem than on the latter.

Until recently questions of the geography of production and geography of economic relations attracted little attention from specialists in historical geography. The increase in attention to the question of economic history by historians and economists began in the last third of the nineteenth century. The work of Marx had a considerable influence on Western European science with regard to problems of socioeconomic history. Even the bourgeois scientists do not deny this; acknowledgement by an historian like Dopsh[72] is characteristic. But the development of economic geography lagged considerably, and continues to lag even now, behind the development of economic history.

The founders of bourgeois anthropogeography, in building their system, paid little attention to the problems of economic geography. The economists-geographers, instead of studying the geography of the productive force and production relations (a problem of little interest in the West even now), continued to describe the state of the national economy by sectors as they had done long before the nineteenth century. It is understandable that with the absence of an economic geography of the present there could be no economic geography of the past.

When the investigator Desjardins, in the work cited above, tried to describe the economic geography of Gaul during Roman occupation he found a sectoral characterization of the economy that was approximately similar to the economical-geographical characterists of France. The authors of historical-economic works were aware of the necessity to study the history of the economy by regions. Several works devoted to the economic past of certain localities appeared, but until recently, there were no works that gave the historical-economic geography of any country.

During the last fifteen to twenty years economic geography in the West has achieved considerable success. From sectoral descriptions of the economy the center of gravity has now shifted to regional characterizations. Economic history has accumulated considerable material in this field. As a result, in works on historical geography attempts have been made to create an economic geography of the past. East, in his *Historical Geography of Europe* published in 1935, had this as his aim.

In the Soviet literature interesting works on the economic geography of our past are associated with the late P. G. Lyubomirov. The most interesting work is his attempt to devise an economic regionalization of Russia of the seventeenth and eighteenth centuries.[73] Unfortunately, this attempt has the character of a sketch.

If one proceeds from individual problems studied by historical geography to composite works having the goal of giving a historical-geographical characterization of some country or territory, then one finds a great variety. The first work of this kind was *Italia Illustrata* by Biondo, an important Italian historian of the fifteenth century;[74] *Italia Illustrata* is a regional description of Italy. For each of the described regions of Italy Biondo gave information on its location, river names, population characteristics for past periods, the most important historical events that had occurred on the territory; he then enumerated the cities, former as well as contemporary, and something of their history,[75] but there were no maps in Biondo's work. As can be seen from this brief characterization in *Italia Illustrata,* the facts of local history were mixed up with elements of historical geography contemporary to the author. Therefore, *Italia Illustrata* can be considered a germ of historical geography, and it made a deep impression on the author's contemporaries. As Fueter points out, in Germany attempts had been made to imitate Biondo with a "Germania Illustrata" which ended in failure.

In 1586, a well-known English historian, Camden, published his *Britania.*[76] It utilizes a broader [more] improved scheme than Biondo's work. Camden had previously used a historical periodization. In the middle of the eighteenth century an Alsace scientist, Schoepflin,[84] utilized a scheme analogous to Camden's in a two-volume work, *Alsatia Illustrata.*

The confusion of facts on local history with some elements of historical geography became a characteristic feature of general, composite works in historical geography. Inclusion of facts on local history took, and is taking, place in historical-geographical works abroad as well as in our country. Moreover, in the prerevolutionary literature, the majority of works devoted to local history were referred to as historical-geographical works. This, of course, contributed to the fact that historical geography has the reputation of a scientific discipline with undefined content.

At the end of the nineteenth century, after contemporary scientific geography had evolved, there appeared attempts to create a geography of the past as a system of knowledge analogous to the geography of the present. In 1876 Desjardins placed before historical geography the task "of studying a defined past period of a country on the same principles, methods, and plan as that for a contemporary study."[78]

Analogous views were expressed two years before Desjardins by our scientist L. N. Maykov,[79] although he did not attempt to realize them. Zamyslovskiy did this in his book, *Gerberstein and his Historical-Geographical Information on Russia,* which appeared in 1884.

In German science since the end of the nineteenth century,[80] under the influence of Ratzel and his followers, the content of historical geography was divided into three main sections: (1) Historische Naturlandschaft, (2) Historische Kulturlandschaft, and (3) Historisch-Politische Landschaft.[81] The first term refers to the history of changes of the natural geographical landscape, about which I spoke above, and the third to the historical-political geography most familiar to historians. The second term means the study of how settled areas appeared in the past, their roads, fields and gardens, and how they were distributed. The main task of this section is to focus on the influence of natural conditions in the past on the distribution of the economy of the country under study and also on disposition and psychology of its population. However, in fact, in the German works on historical geography, in the chapters devoted to "Historische Kulturlandschaft" these problems are usually not solved to any extent. An example is the work by Kretschner, *Historische Geographie von Mitteleurope,* in which for a number of dates general essays are given on the state of agriculture, forests, mining, and communication in middle Europe, almost without any indication of the territorial distribution of these economic phenomena and why this distribution has a given form.

At the International Congress of Historians in Brussels in 1923[82] and the International Congress on Historical Geography in 1930, Pergameni, the president of the Belgian Geographical Society, expressed the thought that historical goegraphy is the "geography of man transferred to the past."[83] During a discussion organized in London by the historical and geographical societies on the question concerning content and tasks of historical geography,[84] Gilbert pointed out that "the main task of historical geography is the reconstruction of the regional geography of the past." Of the new works written in the spirit of this direction of greatest interest is the collective work under the editorship of Darby, *Historical Geography of England Before 1880 A. D.,* published in 1935. The authors examine the changes in the natural landscape of England, the composition and distribution of its population, and the economic geography of the country beginning with prehistorical times to the eighteenth century inclusive. They utilize written sources as well as archeological data. For the history of settlement toponymic data are used. In the study of the changes in the geographic landscape under the influence of man, great attention is devoted to the history of draining swamps in the seventeenth and eighteenth centuries. In studying the economic geography of the past they treat the geography of production as well as the geography of trade. The question of boundaries is included. The methodology of the book is common to English historical works. Class struggle remains untouched by the authors. Yet the book is positively the best achievement of contemporary historical geography abroad.

A few words must still be said about historical cartography. Beginning with Ortelius and almost to the end of the nineteenth century, the task of the compiler of a historical map amounted to establishing the site of historically notable places, fixing political boundaries and tracing their changes, and establishing the disposition and routes of military movements. Such is the content of maps of the known historical atlases of the nineteenth century by Spruner,[85] Droysen,[86] Schrader.[87] In the twentieth century special atlases appeared on colonial countries: Joppen[88] on India; Walker[89] on South Africa; Hermann[90] on China. They do not differ essentially in character from those cited above. Basically, the content of the most popular new school historical atlases by Putzger[91] and Shepherd[92] is the same, although they have some special features, particularly the atlas by Shepherd that has maps of trade routes, a plan of an English manor, historical-ethnographical maps, a map on taxation in prerevolutionary France, and an economic map of England in the period of the industrial revolution.

The development of scientific historical cartography proceeded in the twentieth century in two directions: Richter[93] in Austria and Fabricius[94] in the Rhine region of Germany working on the documentary material with meticulous accuracy try to give exceedingly detailed maps of the administrative and church divisions of settled places of the past. During the world-war years in Holland there were published in separate sheets a *Historical Atlas of the Netherlands*[95] compiled by the same method and edited by Beekman. In the postwar years the same type of atlas was produced by a Polish scientist, *Atlas historczny Polski.*[96]

The German *Geschichtlicher Atlas der Rheinprovinz* published in 1926 under the editorship of Aubin, and the *Atlas of Historical Geography of the United States* under the editorship of C. O. Paullin, published in 1932, represent a different direction in cartography. In these atlases, in addition to historical-political maps, are many maps on economic history and cultural history.[97] These two atlases, however, do not have the degree of detail as those cited above.

In our country little attention was given to historical cartography. True, in some scientific monographs there are excellent historical maps such as in the works of Yu. V. Got'ye,[98] M. M. Bogoslovskiy,[99] M. K. Lyubavskiy.[100] We do not, however, have a historical atlas. Of the few elementary school atlases, the best is the atlas by Zamyslovskiy (last edition 1887). The author of this article attempted from 1923 to 1925 to introduce into our historical cartography historical-economic maps by publishing an atlas on the history of the national economy of Russia in the eighteenth through twentieth centuries.[101]

This brief survey of the state of historical geography shows that the reputation of economic geography as a science with undefined content is to a degree deserved for a number of reasons. Historical geography is developing slowly; up to now few historical geographers have worked in this field. Finally, the conception of economic geography for a long time was distinguished by its imprecise definition due to the prolonged so-called sectoral direction prevailing in this science. Even now economic geography cannot be called a fully developed science.

On the other hand, however, a survey of the state of historical geography makes it possible to establish a definite trend in its development. The development of historical geography is closely connected with the development of historical science and the development of geographical science. When historical science amounted mainly to the "actions of kings and military leaders," to the actions of "conquerors" and "subduers of countries,"[102] and, in geography, the most developed parts were mathematical geography and cartography, then, naturally, the content of historical geography amounted to locating on a map historically notable places and to studying boundaries of countries and their approach routes. The investigation, however, of the influence of nature on the historical process could not go beyond general discussions, since in the geographical sciences the investigation of nature itself was little developed, and in the historical sciences the study of economic science was absent.

The development of economic history, on the one hand, the formation of physical geography as a natural scientific discipline, and the simultaneous development of economic geography, on the other, could not but cause an expansion of the content of historical geography, the introduction of new problems, and the appearance of attempts to construct historical geography as a system of knowledge similar to that of contemporary geography.

What must be the content of Marxist historical geography? "The history of the development of society," it says in the *History of the All-Russian Communist Party: Short Course* is first of all the history of development of production, the history of methods of production, replacing each other in the course of human relations."[103] Historical science "must first of all be engaged in the history of producers of material goods, the history of the toiling masses, the history of peoples."[104]

The main task of historical geography must be the study and description of the geographic aspect of the historical process. Historical geography, being an auxilliary discipline of historical science and not pretending to discover basic rules for the course of history, must, on the basis of a periodization accepted in historical science, give a number of characteristics of economic and political geography of a given country or territory for the corresponding periods of time. The main elements of the above-mentioned characteristics and descriptions must be: (1) the natural landscape in a given era, or a historical physical geography; (2) population from the point of view of its nationality, distribution, and movement on the territory, or historical geography of population; (3) geography of production and economic relations, or historical economic geography; (4) a geography of external and internal political boundaries including the most important historical events, that is, historical political geography. All these elements must be studied not in isolation, but in an interconnected interdependent manner.

The geographic environment, according to indications by the classics of Marxism, "is one of the constant and necessary conditions in the development of society," and it has an influence on human society.

However, "its influence is not a determining influence."[105] Fully taking into account this decisive indication of Marxist theory, historical geography in the study of phenomena of the past must nevertheless take into consideration and investigate the role and influence of the geographical environment on the history of society.

The range of natural resources exploited by man gradually expands in the course of history. During the feudal period coal was rarely mined whereas during the period of capitalism it acquired a great economic significance. Oil was being extracted in considerable quantity only in the second half of the nineteenth century and phosphorus iron ores became useful minerals only after the invention of the Thomas process. On the other hand, a natural phenomenon, harmful at one level of economic development can become useful with the appearance of new technology. The waterfalls before the development of hydroelectric stations were only an impediment to navigation whereas now they are a source of white coal. Thus, the role of the same geographic environment at various stages of historical development can vary. Consequently, the role of the geographic environment must be taken into account in historical geography at each historical stage.

As pointed out in chapter 4 of the *History of the All-Russian Communist Party: Short Course,* during the brief period of time in which the history of man developed there were no radical changes in the geographical environment. Therefore, at first glance it may seem superfluous to include in the number of tasks of historical geography the reconstruction of the natural landscape of the past. However, this is only at first glance. First, as the same *Short Course* points out, changes in the geographical environment did occur in the course of the historical period of the life of human society. These changes in some cases could be of some significance, and there is no basis to ignore them. Such changes include alterations of shore lines (for example, the formation of the Zuider Zee in the Netherlands in the twelfth-thirteenth centuries),[106] changes in the direction of rivers, the filling of estuaries with sand, and the like. An example of a change in the course of a river is that of the Yangtze. In the course of Chinese history it changed its bed many times and its estuary changed for 700 kilometers.

Man's activity has a very strong effect in changing geographic environment and this must be considered. This effect of man on nature is greatest on the soil-vegetation cover. The soils of Western Europe today are quite different from the soils of Europe in the Middle ages. Many swamps have been drained. The destruction of forests is a well known fact, and there is no need to dwell on it. One may mention canals such as the Suez and Panama.

The geography of production and economic relations are the problems that contemporary economic geography is engaged in. Historical geography must investigate these questions with respect to the past. This is the most complex and difficult task, but, at the same time, the most rewarding since these questions tie together and transform all the elements of historical geography from a collection of single facts, necessary only for the understanding of political history, into a special branch of historical science.

As pointed out above, the influence of the geographic environment on society can be studied only through the investigation of the influence of this environment on the economy of society. The geography of population is closely connected with the geography of the economy. The connection between the political boundaries of a territory and its economy is also clear. In studying the economic geography of the past, one has to conduct an investigation by branches [sectors] of economy as well as by regions.

Sometimes the opinion is expressed that present economic regions are the creation of capitalism (if it concerns capitalist countries), and in the precapitalistic period economic regions are absent. Of course, before capitalism, regional differences were smaller but they were present, even in the distant past. A number of historical works prove this.

In order to show more vividly the content of historical geography let us turn to a specific example: the content of historical geography in our country. It stands to reason that in a journal article it is not possible to give a full exposition. I shall limit myself to a brief characterization of the general scheme of its construction and two specific examples.

Periodization of the historical geography of the USSR must be the same as the periodization of our historical process. For each period a picture of the natural landscape must be given as well as the distribution of population and nationalities, the geography of productive power, productive relations and economic relationships, indication of the borders of governmental bodies on the territory under study and, in cases when this is possible and important, the internal administrative division of these governmental formations.

All these elements are examined in their interrelations, investigations are conducted as a whole on the territory of our country (the general picture of distribution of population, agriculture, industry, political and administrative borders) as well as for the most important regions. Naturally, this division into regions cannot be uniform for the whole history of our country, for at different times of the historical process it varies. It would be incorrect to be limited only to a comparison of historical-geographic characterizations for a number of years. It must be shown how they succeed one another; consequently, the characterizations must be dynamic.

Proceeding from this scheme let us pose two specific questions on the historical geography of our fatherland: on the general historical-geographical characterization of our country in the eighteenth century and the historical geography of the Central-Chernozem region in the sixteenth through eighteenth centuries.

For the eighteenth century a dynamic historical-geographical characterization of our country must be given separately for the following territories that, in terms of economics, are economically somewhat related and were at that time politically weakly connected:[107] (1) Eastern Europe and Siberia, (2) Caucasus, (3) Kazakhstan and Central Asia.

The historical-geographical characterization of Eastern Europe and Siberia, in a general survey, must include the changes in political boundaries in Eastern Europe caused by the Russian conquest of the shores of the Baltic and Black seas together with the nationalities on these territories and the majority of the Ukrainian peoples. The center of attraction must be concentrated on the territorial changes themselves and not on military and diplomatic facts that directly caused these changes. The process of colonization of the South and the East of the country in this period as well as the founding of new cities must be subjected to careful survey. The nationality composition of the colonists must be examined. Serious attention must be devoted to the development of new natural resources not involved in earlier economic development: the vast areas of chernozem soils and mineral wealth of the Urals, Siberia and, in part, central Karelia. The scientific expeditions of the eighteenth century must be mentioned here. Colonization and development of new natural resources are closely related to the change of governmental borders but they are due, of course, not only to these changes, and this must be correctly accounted for in the study. The geography of industries in the eighteenth century must be given and explained. Further, attention must be devoted to the differences between the consuming and producing zones. In the study of economic connections, the construction of canals must be considered.

In the area of the geography of industrial relations the main attention must be turned to the geography of serfdom and the geography of corvee and quit-rent. The latter must be connected with the formation of consumption and production belts. The history of administrative divisions in the eighteenth century also finds a place in the historical geography of our country. Hereby the interest of historical geographers lies not in the study of local administrative organizations, but in the changes of actual administrative borders.

To illustrate the content of historical geography of a region, we briefly dwell on problems of the Central-Chernozem region from the sixteenth to the nineteenth centuries. In the sixteenth and seventeenth centuries the main questions concerning this territory will be the land-scape of the "Wild Field" [original natural landscape] of the sixteenth century, the direction of construction of fortified lines in this "Wild Field," the character and direction of colonization, the state of agri-culture, and the building of cities and the composition of their popu-lation.

For the eighteenth century the historian must note the end of colo-nization and the beginning of the transformation of this region into an agrarian center of the country. The geography of agriculture, the ge-ography of the social composition of the population, the geography of corvee and quit-rent, the distribution of the developing manufactures, and the formation of administrative divisions are the main questions of historical geography of the Central-Chernozem region for this period.

For the first half of the nineteenth century, when the region had be-come the agrarian center and granary of the country, the main atten-tion of the investigator must be turned to the geography of agricul-ture and serfdom, to the distribution of wool mills and sugar refineries, to the geography of fairs, and to the geography of the economic con-nections of this region with the neighboring territories and especially with the Central-Industrial region. An important task will be the in-vestigation by districts of the dynamics of virgin-land cultivation and of the population growth of the region during this period.

After the reform of 1861 the main factor influencing the life of the region was the so-called impoverishment of the Center. This phenom-enon left a heavy imprint on its economic and population geography. The roots of this impoverishment lay initially in the conditions of the liquidation of serfdom. Naturally, the geography of these conditions in the central chernozem provinces must be examined. This situation must be compared with the distribution of survivals of serfdom in the region after 1861. Changes in the sown area, the growth of which stops, must be studied as well as the geography of other agricultural phenomena. The fertility of soils of the region begins to decline, and this process too, must be studied. Of course, the geography of rail-roads and industry must receive attention. Finally, migration and sea-sonal work must be studied.

Two examples of historical-geographic study of our country have been discussed, but there are many examples from more remote per-iods. I need only point to an unpublished work by S. B. Veselovskiy that describes the geography of feudal divisions in northeastern Europe in the fourteenth and fifteenth centuries. Veselovskiy compiled an exceedingly interesting map that was the only one of its kind in our science.

Historical documents are the primary sources for historical geogra-phy. Specifically these documents are ancient maps and ancient geographical descriptions of a given country or place, such as the works of ancient geographers for the historical geography of the ancient world, the geography of Georgia by Prince Vakhushti for the historical geography of Georgia, and so forth.

In addition, archeological data are sources for historical geography and are especially necessary for the reconstruction of the economic goegraphy of the remote past. For the study of the succession of nationalities on a territory in the remote past toponymic data are a valuable source. For the reconstruction of the natural landscape, natural-historical data must also be utilized.

The character of the sources also determines the method of investigation in historical geography, which involves primarily the historical method (criticism and analysis of historical documents and archeological data).

In studying the historical statistical sources, the statistical method must be applied as this is usually done in historical-economic investigations. Using toponymic data the historical geographer, if he has no special linguistic training, must utilize the results of analyses of these data made by linguists. In processing the natural-historical data for the reconstruction of the natural landscape of the past it is sometimes necessary to use the methods of these branches of natural science.

The historians usually consider historical geography an auxilliary science. This is almost the only point where the opinions of the majority of historians agree with respect to historical geography. Bernheim in his well-known *Lehrbuch der Historischen Methode* considers historical geography an auxilliary science together with paleography, heraldry, and numismatics. Almost all bibliographic handbooks on history relegate it to the auxilliary sciences, as for example, the *International Bibliography of Historical Sciences* published by the International Committee on Historical Sciences; the German handbook *Quellenkunde der Deutschen Geschichte von Dahlmann-Waltz,* the Czech *Bibliografie (Eske' Historie Zirbe),* the Polish *Bibliografja Historji Polskiej,* and others. Almost the only exception in this respect is the Swedish *Svensk Historisk Bibliografi 1875-1920,* by Setterwall, which places the works on historical geography in the section on local history. Geographers usually are not inclined to consider historical geography only a servant of history but rather give it a more independent position.[108]

The traditional opinion of historians at the present time is doubtless obsolete. Historical geography is developing into a separate branch of historical science. This can easily be seen from what has been said above on the evolution of its content. Indeed, such disciplines as paleography are usually called auxilliary sciences because the results of their investigations are of little independent interest, but are needed as additional means for historical investigation in the true sense of the word. Paleography interests us as a means of reading ancient manuscripts, but not as history of writing. The historian needs the study of diplomacy, not for itself, but for critique of documents.

As long as the content of historical geography did not go beyond the locating of historically notable places and the determination of boundaries, the results of its work had mainly an auxilliary significance. For the science of political history it was important to know the location of places where the described events occurred and the location of the boundaries that had arisen as the result of wars. Courses in historical geography in those times were, in essence, handbooks and in the sixteenth to eighteenth centuries took the form of dictionaries with alphabetically arranged material. Even later, in the form of a systematic survey, these courses did not resemble the usual historical courses, but rather were guides. It is sufficient to read one chapter in such books as *Historical Geography of Europe* by Freeman in order to be convinced of this. Historical geography as defined above by me is no longer a collection of information of a handbook character but a definite system of knowledge representing an independent interest.

Does historical geography belong to the field of historical or geographical sciences? Basically it processes historical sources by the historical method. It is evident that historical geography is a historical science. Geographers such as Oberhummer[109] have long ago accepted this. This does not mean, however, that the work in this field is the monopoly of historians. Geographers too can work fruitfully and have worked in the field of historical geography. One can draw an analogy with another branch of historical science, economic history. After the instructions cited above in the *History of the All-Russian Communist Party: Short Course* one can hardly maintain that the science of economic history, or the history of the national economy as we usually call it, is not an organic part of historical science and that work on its problems is not one of the urgent tasks of historians. But, on the other hand, in addition to historians, economists also work successfully in this field.

Geographers themselves point to the fact that the work of geographers on problems of historical geography does not exclude the latter from historical sciences; for example, Oberhummer says that "the geographer, as soon as he leaves the field of geographical investigation, and begins to work on history stops being a natural scientist and becomes a historian."[110] Supan who wrote an outstanding work on the division of the world was a specialist in physical geography, but this does not exclude his study from historical works.

As mentioned above, the reconstruction of the natural landscape of the past demands, in addition to the utilization of historical documents, the use of natural-historical material. Here the investigator must use the methods of natural science. Therefore, this work can be done more successfully by specialists in physical geography than by historians. The study of the influence of the landscape of the country on the economic and political geography of the past is the province of historical geography.

One cannot close one's eyes to the difficulties of the development of historical geography. These difficulties lie primarily in the field of economic geography of the past. Objectively, the greatest difficulty is in the insufficient historical-economic material by regions. The subjective diffiiculties involve the fact that, on the one hand, historical geography has attracted few workers and, on the other, there is a need for a broad preparation on the part of the investigator in order to work in this field. A specialist in historical geography must not only know history, but must have training in geography. These subjective difficulties are, of course, surmountable.

The development of historical geography as described above can be of great benefit for historical science as a whole. Historical geography provides a spatial localization of the historical process and thereby it (1) contributes to the realization and deepening of our ideas concerning many facets of the historical process, and (2) makes it possible to discern and explain a number of local features in its development. In this manner many incorrect generalizations can be avoided. This is especially important for the history of national economy where, as Lenin emphasized, regional study is necessary. The development of historical geography will also make it possible to investigate, from methodologically correct positions, the role of the geographic environment in the actual historical development of individual countries.

Historical geography can play an important role in the formation of economic geography as a scientific discipline. Presently, economic geography as a science is still in the process of growing. Historical geography must help economic geography in determining actual economic-geographical regions. This is important in itself but can also contribute to the establishment of a principle in the formation of economic-geographical regions.

In spite of an animated interest in historical geography in the USSR we are doing very little work, in fact, much less than in bourgeois countries. Bourgeois historical geography must be anti-Marxist. Historical-geographical themes must be included in the plans of the scientific research institutions of history and geography. Of particular importance is the creation of an historical atlas of the USSR. Additionally, it is necessary to organize a data collection system and to study our geographical names as is done in Germany, England, and France. Finally, historical geography must receive the right to be recognized in our institutions of higher learning.

FROM THE EDITORS [OF *Istorik-Marksist*]

It is necessary to note the importance of the questions raised by Comrade Yatsunskiy. The situation with regard to historical geography in our scientific research institutions and higher educational institutions is quite unsatisfactory.

The editors of this journal feel that the proper secondary school and university committees should take the necessary measures to change the existing situation with regard to historical geography in the higher educational institutions. Our historical, scientific-research organizations must finally introduce into their plans historical-geographic problems.

The editors request that historical institutes and scientific workers speak out on the questions touched upon in this article as well as about the practical measures necessary for the development of scientific research and the teaching of historical-geographic information in the system of historical education.

1. See *Handbuc'. der Alten Erdbeschreibung von d'Anville zum Gerbauch seines Atlases Antiguus in 12 Landkarten;* and Wimmer, *Historische Landschaftskunde* (Innsbruck, 1885).

2. The opinion of Barnes is, in any case, incorrect: Giraldus Cambrensis wrote geographical works, but did no historical-geographical studies. Elements of historical geography in works of general geography are found in such studies as Biondo's *Italia Illustrata* in the fifteenth century (more about him appears below); Ortelius in the sixteenth century was the first to separate historical geography from general geography. Lack of space does not allow elaboration of this theme.

3. For example, S. M. Serdonin and S. K. Kuznetsov.

4. See *Journal des savants*, August–October 1930; *Annales de Géographie*, 15 January 1931.

5. Materials from the discussion are published in *Geography*, Vol. 17 (1932) Part I, p. 95.

6. This can be clearly seen in the work by Kretschmer, *Historische Geographie von Mitteleuropa.*

7. S. Kuznetsov, *Russkaya istoricheskaya geografiya* [Russian Historical Geography] (Moscow, 1910).

8. In *Scottish Geographical Magazine* 3 (1932).

9. Published in *Annales d' histoire économique et sociale* 44 (March 1937).

10. See Saar, *Istochniki i metody istoricheskogo issledovaniya* [Sources and Methods of Historical Inquiry] (Baku, 1930).

11. In spite of the fact that historical geography has existed for more than three centuries and a huge amount of material has accumulated, in the scientific literature there is not a single attempt to study the history of its development in connection with the development of historical or geographical science. The author of this article attempts to fill this gap in a monograph being prepared for publication, *Istoricheskaya geografiya, istoriya yeye razvitiya kak nauchnoy distsipliny, predmet i metod* [Historical Geography, the History of its Development as a Scientific Discipline, Its Subject Matter and Method].

12. Ortelius, a Flemish geographer of the second half of the sixteenth century (1527–98), became known in Europe by the publication of a fundamental geographical atlas entitled *Theatrum orbis terrarum* published in 1570. This atlas had 21 Latin editions and several

editions in French, German, Spanish, Flemish, Italian, and English. Along with Mercator, Ortellus is considered the outstanding representative of the Flemish cartographic school. In addition to his geographical atlas Ortelius prepared the first historical atlas, *Parergon theatri orbis terrasum.* There is a considerable literature on Ortelius as a geographer (primarily in the book by Bagrow, Abrahami Ortelli *catologus geographorum* [Gotha, 1928]); on the other hand, however, the historical-geographical works of Ortelius, of such great significance at one time, were not scientifically analyzed in the literature of the nineteenth and twentieth centuries.

13. Published in 1578 under the title of *Synonimia geographica.* In the second edition the title was changed to *Thesaurus Geographicus.*

14. Neibuhr, *Vorträge über alte Länder-und Völkerkunde*; d'Anville was an honorary member of the Russian Academy of Science.

15. *Formae Orbis Antigui.*

16. *Corpus Inscriptionum Latinarum.*

17. Published in 1794.

18. Lerberg, *Issledovaniya, sluzhashchiye k ob'yasneniya drevney russkoy istorii*, [Research Which Serves in the Explanation of Ancient History], 1819.

19. Brun, *Chernomor'ye Sbornik issledovaniy po geografü yuzknoy Rossu*, 2 vols.

20. See for example, Kudryoshov, "Historical-Geographical Data on the Polovian Land from Chronicles of the Campaign of Igor' Severskiy Against the Polovtsy in 1185," in *Izvestiya Gosudarstvennogo geograficheskogo obshchestva*, vol. 69, part 1.

21. A copy of this atlas is in the manuscript section of the Saltykov-Shchedrin Public Library in Leningrad.

22. See Clüver, *Italia antigua*, p. 363.

23. Du Val, *Cartes géographiques dressées pour bien entendre les historiens, pour connoistre les entendres des anciennes Monarchies et pour lire avec fruit les Vies, les Voyages, les Gueeres et les Conquestes des grands Capitaines* (Paris, 1660).

24. For example, Jullian in the introduction to the book by Mirot, *Géographie historique de la France* (Paris, 1930).

25. In the atlas entitled *Diverses cartes et tables pour la géographie ancienne, pour la chronologie et pour les itineraires et voyages modernes* (Paris, 1665).

26. Ibid.

27. Ibid.

28. Longnon is the author of *Atlas historique de la France depuis César jusqu'à nos jours* (updated to 1380); *La formation de l'unite française, Geographie de la Gaule au VI siècle; Les noms de lieux de la France*, as well as other works.

29. After the publication of Supan's work several other large studies appeared that attempted to trace the history of political boundaries in colonies and dependent countries. Among the most important of these are two works; a three volume by Hertslet, *The Map of Africa by Treaty* (London, 1909), in which the author examines the history of Africa on the basis of treaties between European powers, illustrating cartographically the boundaries established by these treaties. The second is a recently published study by Ireland Gordon, *Boundaries, Possessions and Conflicts in South America* (1938).

30. The map was printed by Moretus, the owner of the famous Plantin publishing house under the title *Tabula Itineraria ex illustri Peutingerorum bibliotheca quae Augustae Vindelicorum est beneficio Marci Velseri septemviri Augustiani in lucem edita.*

31. See his study, *O torgovle i gorodakh Armenii V–XV, Vekov* [The Trade and Cities of Armenia, 5th to 15th Centuries].

32. Jomard, *Les monuments de la géographie ou recueil d'anciennes cartes européennes et orientales publices en fac-similé de la grandeur des originaux* (Paris, 1842–62).

33. Santarem, *Atlas composé de mappemondes et de portulans et d'autres monuments géographiques depuis le VI siecle de nôtre ere jusqua'a XVII-me* (Paris, 1842–53).

34. Nordenskild, *Atlas to the Early History of Cartography* (Stockholm, 1889); *Periplus, An Essay on the Early History of Charts and Sailing Directions* (Stockholm, 1897).

35. *Almagia Monumenta Italiae Cartographica* (1930).

36. *Monumenta cartographica Bohemiae.*

37. Sindik, *Stare karte jugoslavenskikh zemal'a* (Belgrad).

38. This is not sold, but rather sent to the largest libraries in the world. In the USSR it is found in the Saltykov-Shchedrin Public Library in Leningrad.

39. Published the first time by Novikov in 1773 and issued several times since.

40. Altamira, *La Ensenanza de la historia*, p. 131.

41. Freret, *Observations générales sur la géographie ancienne.* The work was preserved in the papers of the Academy of Inscriptions in Paris and published in 1850 in *Memoires de l'Institut national de France.*

42. Marks and Engels, *Sochineniya* [Works], vol. 11, part 1, p. 508.

43. Almost simultaneously the publisher mentioned above published, *Essai sur l'histoire de la cosmographie et de la geographie pendant le moyen age*, 3 vols. (Paris, 1849-52.).

44. The older brother of Alexander Humboldt who, with Ritter, is considered the founder of modern geography.

45. Kollar, *Rozprawy o qmienach, počatkach i starožitnostech narodu Slawskiego a geho Kmenu* (1830).

46. Safarik, *Slovanske starožitnosti* (1836, 1837).

47. For their characterization see the work of D. Yegorov, *Kolonizatsiya Meklenburga v XIII veka*, [Colonization of Mecklenburgh in the Thirteenth Century], vol. 1, chap. 9.

48. Poznan, 1934-37.

49. Published in *Biblioteke dlya chteniya*, vol. 22 (1837).

50. Barsov, *Ocherki russkoy istoricheskoy geografii. Geografiya nachal'noy letopisi* [Essays in Russian Historical Geography. The Geography of the First Chronicles], 1st ed. (1874); 2d ed. (1885).

51. V. Klyuchevskiy, *Kurs russkoy istorii* [Course in Russian History] (1914), 1:23.

52. M. Lyubavskiy, *Istoricheskaya geografiya Rossii v svyazi s kolonizatsiyey* [Historical Geography of Russia in Connection with Colonization]. (A course read at Moscow University in 1908-9, Lithographed edition.)

53. Strabon [Strabo], *Geografiya* [Geography], pp. 286-87 (translated by Mishchenko).

54. See Belyayev, "The Historical-Sociological Theory of Ibn-Khaldun," *Istorik-Marksist*, nos. 4-5 (1940).

55. Bodin, *Six livres de la République* (1576).

56. Altamira, *La ensenanza de la historia*, p. 166.

57. Fueter, *Geschichte der neueren Historiographie* (1911), p. 497.

58. Ratzel's main work includes *Anthropogeographie* (Stuttgart, 1882), and *Politische Geographie oder die Geographie der Staaten, des Verkehrs und des Krieges* (München, 1903).

59. The main works of Vidal de la Blache include *Principes de Géographie humaine* (Paris, 1918) and *Tableau de la Géographie de la France.*

60. Concerning the school of Vidal de la Blache there is an article in the Soviet literature, I. A. Vitver in *Uchenyye zapiski Moskovskogo universiteta* [Teaching Notes of Moscow University], no. 35.

61. The work of the popularizer of Ratzel's ideas must be noted here. See E. Semple, *American History and Its Geographical Conditions*, 2d ed. (1933); and *The Geography of the Mediteranean Region, Its Relation to Ancient History* (1932).

62. The main works by Turner include *Rise of the New West, 1819-1829* (New York, 1906); and *The Frontier in American History.*

63. See *Zapiski S. M. Solov'yeva* [Notes of S. M. Solov'yev], p. 65.

64. A Sidorov "The Petit-Bourgeois Theory of the Russian Historical Process (A. P. Shchapov)," in *Russkaya istoricheskaya literatura v klassovom osveshchenii* [Russian Historical Literature in Class Terms].

65. In a course on historical geography of the USSR given by I. I. Polosin in 1939 at the Moscow Historical-Archival Institute. The course has not been published. I express my gratitude to I. I. Polosin for making it possible for me to see a copy of the lectures.

66. Wimmer, *Historische Landschaftskunde*. After Wimmer, this term became established in the German literature.

67. L. Berg, *Ocherk istorii russkoy geograficheskoy nauki* [Essay in the History of Russian Geography] (Leningrad, 1929).

68. In our country they studied the famous struggle between forest and steppe in the direct sense, and not figuratively as the historians did.

69. Desjardins, *Geographie historique et administrative de la Gaule Romaine*.

70. Zamyslovskiy, *Gerbershteyn i yego istoriko-geograficheskiye izvestiya o Rossii* [Gerberstein and His Historical-Geographical Work on Russia].

71. See the work of V. V. Bartol'd, "On the Amu-Dar'ya Falling into the Caspian Sea," in *Zapiski Vostochnogo otdela Russkogo arkheorlogicheskogo obshchestva*, Vol. XIV, Vypusk 1, 1902; "Essays on the Aral Sea and the Lower Reaches of the Amu-Dar'ya from Ancient Times to the 17th Century," in *Izvestiya Turkestanskogo otdela Russkogo geograficheskogo obshchestva* [Proceedings of the Turkestan Section of the Russian Geographical Society] vol. 4, 1902; and *K istorii orosheniya Turkestan* [On the History of the Irrigation of Turkestan] (St. Petersburg, 1914). Recently, this question, which was often studied earlier, was again investigated by a specialist in physical geography, A. S. Kes'. See A. S. Kes' "The Riverbed of the Uzboy and Its Genesis," in *Trudy Instituta geografii Akademii nauk SSSR* [Works of the Institute of Geography of the Academy of Sciences of the USSR], no. 30, 1939.

72. See the collection, *Historie et historiens depuis cinquante ans*, 1:13, published by the French journal *Revue historique*.

73. Under "Russia" in *Entsiklopedicheskiy slovar' Granat*, vol. 36, part 3.

74. Biondo was born in 1392 as his most recent biographer, Hogara, points out. Usually his year of birth is incorrectly given as 1388. Biondo died in 1463. According to Fueter's evaluation, in which the Soviet investigator Vaynshteyn concurs, "Biondo did more for the study of the Middle Ages and ancient Rome than all his contemporary humanists together" (*Geschichte der neueren Historiographie* [1909], p. 109).

75. Fueter, for some reason, considers that *Itallia illustrata* was written in dictionary form, which, in fact, it is not. Evidently following Fueter, Vaynshteyn repeats this opinion. One also cannot agree with Vaynshteyn that Biondo was some kind of scribe in the papal curia. The offices of notarius of the papal chamber and "apostolic secretary" that Biondo held were not the offices of a scribe, but it is true that Pope Nicholas V treated Biondo badly. See Masius, *Flavio Biondo, sein Leben und seine Werke,* and also Foygt, *Vozrozhdeniye klassicheskoy drevnosti* [The Renaissance of Classical Antiquity].

76. Britannia was written not only under the influence of Biondo's work, but also under the influence of Ortelius, with whom Camden became acquainted in 1577 when Ortelius was in England. See the work "Camden" in the *Dictionary of National Biography*, ed. Leslie Stephen, vol. 8; and Denuce, *Oud nederlandsche Kaartmakers in betrekking met Plantyn*, 2:41.

77. He was an honorary member of the Russian Academy of Sciences.

78. Desjardins, *Géographie historique et administrative de la Gaule Romaine*.

79. L. Maykov, *Zametki po geografii drevney Rusi* [Notes on the Geography of Ancient Russia] (St. Petersburg, 1874).

80. If I am not mistaken, Wimmer was the first to propose such a division in his book *Historische Landschaftskunde*.

81. This concept of historical geography found a sympathetic response even in the Polish scientific literature in the postwar years. See Arnold, "Geografja historyczna, jej zadania i metody", *Przeglad Historyczny* 8 (1929).

82. See *Compte-rendu du V congres international des sciences historiques* (Brussels, 1923).

83. *La géographie humaine transporteé dans le passé.*

84. See page 697 above.

85. Spruner, *Handatlas für die Geschichte des Mittelalters und der neueren Zeit.* Engels used Spruner's atlas when he wrote *O razlozhenü feodalizma i razvitü burzhuouzü.* See Marx and F. Engels, *Sochineniya,* vol. 16, part 1, p. 433.

86. Droysen, *Allgemeiner historischer Handatlas* (1886).

87. Schrader, *Atlas de la géographie historique* (Paris, 1896).

88. Joppen, *Historical Atlas of India,* First edition, 1907, later edition, 1934.

89. Walker, *Historical Atlas of South Africa* (1922).

90. Hermann, *Historical and Commercial Atlas of China* (1935).

91. Putzger, *Historischer Schulatlas,* in very many editions.

92. Shepherd, *Historical Atlas,* in several editions.

93. Richter, *Historischer Atlas der Osterreichischen Alpenländer* (1906).

94. Fabricius, *Geschichtlicher Atlas der Rheinprovinz,* published in sheets since 1895.

95. Beekman, *Geschiedkundige Atlas van Nederland.*

96. For the material on Polish historical geography, the author expresses his appreciation to the historical faculty of L'vov State University, and especially to the director of the section of auxiliary sciences, Professor T. T. Model'skiy. The author also expresses gratitude to comrade Birzhishka of the library of Kaunas University.

97. Judging from the published program, a work of the same type is the *Atlas historico de la America hispano-portuguesa* by J. Dantin Correceda and Loriente Cancio, which began to appear in Spain before the Civil War. Evidently, only one number of the second volume appeared.

98. See Yu. Got'ye, *Zamoskovnyy kray v XVII veke* [The Land Beyond Moscow in the Seventeenth Century], and *Istoriya oblastnogo upravleniya v Rossii ot Petra I do Yekateriny II* [The History of the Oblast Administration in Russia from Peter I to Catherine II].

99. M. Bogoslovskiy, *Zemskoye samoupravleniye na russkom severe v XVII veke* [Zemstvo Self-Government in the Russian North in the Seventeenth Century].

100. M. Lyubovskiy, *Oblastnoye deleniye i mestnoye upravleniye Litovsko-russkogo gosudarstva* [Provincial Division and Local Administration of the Lithuanian-Russian State].

101. They were published by the educational section of the Chief Political Education Committee [Glavpolitprosveta] under the title, *Naglyadnyye posobiya po istorii narodnogo khozyaystva Rossii* [Visual Teaching Aids on the History of the National Economy of Russia].

102. *Istoriya VKP (Bol'shevik); Kratkiy Kurs* [History of the All-Russian Communist Party [Bolsheviks]: Short Course].

103. Ibid.

104. Ibid.

105. Ibid., p. 113.

106. See Demangeon, *Belgique-Pays Bas-Luxembourg,* p. 24.

107. The allegiance of some of the Kazakh khans to the Russian government established a purely nominal dependence on Russia.

108. See Kretschmer, *Historische Geographie von Mitteleuropa;* Oberhummer, "Die Aufgaben der historischen Geographie," a paper published in *Verhandlungen des neunten deutschen Geographentages in Wien.*

109. Oberhummer, *Die Aufgabe der historischen Geographie.*

110. Ibid.

L. YE. IOFA

On the Significance of

Historical Geography

V. N. Tatishchev, who founded scientific geography in Russia, divided geography in terms of properties into mathematical, physical, and political, which is, in present-day terms, social geography. In defining the latter, Tatishchev stressed that its object should also be studied chronologically "because its circumstances change with time."[1] In his *Leksikon*, he says even more clearly that this branch of geography "falls into ancient, medieval, and modern or present-day divisions."[2] Later, M. V. Lomonsov asserted that, in addition to history, there was need for "comparing ancient geography with the present" to understand "that visible material things on earth and the entire world" have undergone "great . . . changes."[3]

Tatishchev's ideas were based on familiarity with Russian and West European science and on his extensive practical experience in directing the economic development of the huge outlying territories of the Russia of his time (the Urals, the lower Volga), where he dealt with applied geography concerning the construction of cities and plants, the building of roads, and the introduction of agricul-

Reprinted, with modifications, from the translation by Theodore Shabad appearing in *Soviet Geography: Review and Translation* 4, no. 1 (January 1963), pp. 3–12, by permission of the publisher. This article originally appeared in *Geografiya i khozyaystvo* [Geography and Economics], no. 11 (1961).

ture. In generalizing his own ideas as well as those of Varenius, the Western historians, and the erudite school, Tatishchev formulated theoretical principles of geography that were far advanced for his time. (It is strange that some writers have recently asserted that Tatishchev merely mechanically applied Varenius's views to Russia[4] and that no one has mentioned Tatischev's chronological approach to geography. The erudite school and the works of its representatives in the field of historical geography are discussed in V. K. Yatsunskiy's very valuable book,[5] which is not being sufficiently used by geographers).

In this paper we would like to stress especially the value of the views developed by Tatishchev and Lomonosov regarding geography not only as an areal but a chronological science. They have left, however, no theoretical justification for the chronological aspect of geography. It is possible that their views were intuitive in character and derived from an inadequate differentiation of history and geography, from a feeling that these two sciences had a common origin, from the similarity of their approach to reality on the basis of generalization of experience, and from a desire to place the experience of the past at the service of the present and the future. In any case the study of geography in its chronological aspect was not further developed.

True, in the middle nineteenth century outstanding geographers such as Humboldt and Ritter spoke about the need for introducing historical elements into geography. "Geography also cannot manage without the element of history," wrote Ritter[6] in a paper especially devoted to this subject (his "cannot manage" is characteristic). But these were platonic wishes without a formulated methodology, and they gave a narrow interpretation to the importance of historicism in geography. Geographers did almost no work in that direction, and little was done by historians for their own purposes. At the end of the nineteenth century and the beginning of the twentieth, under the influence of Kantian and Neo-Kantian philosophy, and its separation of time and space, the view became prevalent among geographers that geography dealt only with the present day in contrast to history, which was concerned with the past; history, so to speak, monopolized time, and geography space. (These concepts seem to have penetrated everyday speech and are being applied to any object such as "history of an illness," "geography of fashions"; such use of words does not, of course, reflect the content of the respective sciences.) Geography began to be regarded as the last page of history, that is, geography and history were being differentiated not in terms of their objects of study, but in terms of time. Geography was proclaimed a descriptive (idiographic) science that was not concerned with general laws. I would say that its relationship to the past was treated only for educational purposes. The following examples show how geography and history were treated in those days. Elisée Reclus said the following in the epigraph of his last six-volume work *L'Homme et la Terre* (1905–8): "Geography is nothing but history in space just as history is geography in time." Ye. Chizhov, the Russian theoretical geographer, said at the end of the nineteenth century that "history is the geography of the past and geography is the history of the present."[7] Chizhov's article was highly praised by L. S. Berg, who cited Chizhov's quotation without comment in his article "Geography and Its Position Among Other Sciences."[8] V. E. Den said the following in his textbook on economic geography: "Economic geography supplements economic history; the latter studies economic life of the past, and the former that of the present."[9] These views were apparently also shared by such an authoritative geographer as D. N. Anuchin; in his article "On the Teaching of Geography and Related Questions," he said that geography dealt only with the "present moment" and that notions about the past of the objects studied by geography must "for a more intelligent understanding of the present,"[10] be supplied by geology and history (as if they all had the same object of study!). Many other such examples could be cited. It might be noted that these views of geographers corresponded to those of historians who did not consider the history of the present day to be their concern. But these views were apparently based not on theoretical considerations, but rather practical ones such as, "the materials are too recent" and so forth (although historians did have a strong foundation in the study of past eras). It is interesting that such considerations did not worry the geographers, who made the present day the study object of the entire discipline (which the historians felt unable to do despite, I repeat, the more solid basis at their disposal).

As Soviet geographers adopted the Marxist dialectic philosophy, they increasingly realized the need for a genetic approach to phenomena. However, most geographers retained the view that geography of the present day was all there was to the subject (unfortunately even N. N. Kolosovskiy paid lip service to this view in his interesting and important article "The Scientific Problems of Geography").[11] Genetic elements, in the geographers view (which was not always clear on this point), were expected to help explain individual geographic study phenomena and were to be taken from the work of historians. There was no interest in organizing the study of these genetic elements, which are so necessary for contemporary geography. The attitude of leading geographers toward historical geography (meaning the geography of all periods except the present) is convincingly illustrated by two examples. Volume one of the *Kratkaya geograficheskaya entsiklopediya* [Short Geographical Encyclopedia] in the article "Geography" states with regard to physical geography "a new branch of physical geography, paleogeography, has now become established," and with regard to economic geography there is no mention of any need to study past periods, let alone of any separate discipline (this discrepancy between physical and economic geography is of course strange, but such is the "logic" of the encyclopedia). Just as indicative is the following fact. The symposium *Soviet Geography: Accomplishments and Tasks*, which claims "to give a comprehensive survey of the present state of Soviet geography," contains articles on rather narrow subjects, but no article on historical geography!

All this is the result of the idea that has become deeply rooted, perhaps not so much in the conscious as in the subconscious (which is more serious), that geography is opposed to history as the present is opposed to the past. But from the point of view of dialectical materialism there can be no classification of sciences that distinguishes between disciplines concerned only with the present and those concerned only with the past. This contradicts the idea of development. Disciplines are classified in terms of their objects of study, which are investigated from the point of their origin, their development and their present state.

It is evidently important to dwell on three questions:

1. How does history differ from geography in terms of the time of study or the object of study, and would geography become history in departing from the present (not in the everyday but in the scientific sense)? Is it correct to say that accounts of travels, geographic regional descriptions, maps, and so forth, all the things that constitute geography for contemporaries, would become historical, and not geographic, documents for posterity?

2. What actual importance has study of the geography of the past for the geography of the present? Is it just a question of studying the genesis of individual present-day objects (cities, types of production, and so forth), or is there a broader significance for geographic theory and geographic thought? (It is understood that geography of the past must be studied in the interest of present and future generations just as history is, no matter how far back it may go.)

3. How should geography be reorganized to be able to make use of the geographic experience of past millennia, to take account of its wisdom and of its positive and negative sides?

We will try to answer these questions.

History and geography are old disciplines; they are twins, having originated from a common root and possessing a common approach to reality. They originated as a result of the natural and, possibly, earliest scientific need of human society to think about the experience of life since such experience is the earliest and most natural school of mankind.

The subject of history in general is clear: a statement and generalization of the experience of mutual relationships of people in social life, in political organizations, and in the field of production (the social aspects of production). Such a concept of the subject of history emerges, it seems to me, quite clearly from concrete historical studies (sometimes more important than formal definitions).

In addition to relationships among people, there are others that are important for people and require study — they are the relationship to the surrounding world, to nature, and to the objects produced by previous generations; the latter become for each succeeding generation just as important a means of labor as the objects of nature. (Marx wrote: "Examples of such means of labor [like the earth] that have already undergone the process of labor are industrial buildings, canals, roads, etc." He continues: "But the circumstance that flax and the spindle are the products of past labor is just as irrelevant to the actual process as the circumstance that bread is the product of past labor of the farmer, the miller and the baker is irrelevant to the act of eating.")[12] This aspect of man's experience, requiring knowledge, in certain combinations, of the social and physical aspects of life, has been taken over by geography. It is thus another aspect of history, moving parallel with the first and in close contact and mutual penetration. For a long time that contact was being maintained; it was felt by the ancient scholars, who were often both historians and geographers. In Russia both Tatishchev and Lomonosov were such people. (It has become customary lately to define the subject of geography in rather complex terms that are the result of the long development of the discipline and constitute, in essence, hypotheses and theories. It seems to me that the subject should be formulated very primitively and should not preordain the direction to be taken by research. [Chemists define their subject as the "study of matter and its transformations."])

As we have seen, false philosophical concepts later diverted geographers from the chronological path. The old feeling of unity with history took the wrong form; it was no longer the idea of traveling together through time (geography existed at all times) but the idea that geography took over from history in its present-day stage (the last page of history). The result was that no one studied geographical experience, an aspect of man's experience that has tremendous importance, and no one made use of the huge accumulations of geographic data of the past.

We are hopeful that the reduction of geography to the present and the related chronological approach are a passing phase; although there have been achievements, it was a period of lag compared with the general progress of science. The absence of a chronological aspect in geography at the present time is the principal brake retarding its development as a science able to meet present-day methodological and practical needs.

It seems to me that no one any longer disputes the importance of geography of the past for an understanding of the origin and development of many present-day phenomena. Any person thinking in dialectical terms understands how important such an approach is to understanding the essence of phenomena and forecasting future developments. The importance of the genetic approach is sufficiently clear to many.[13] We will therefore concentrate on the importance that study of the geography of the past has for the formulation of fundamental geographic concepts and laws, which continues to be one of the weaker aspects of the discipline (precisely because past geographical experience is being ignored). The point is that historical geography shows how geographical relationships arise and what the *mechanism* of their movement is in general. It makes it possible to trace that movement from cause to effect and even to effect of effect, something that cannot be done within the framework of present-day geography alone. Uncovering the logic of movement in such clear and definite forms (when the study object may be examined not only from the point of view of the cause but from the point of view of the effect) is, of course, also important for developing a strictly scientific method (especially since experimentation is impossible in this field).

I would like to point out that the few general geographic concepts that we have, such as geographical situation, geographical division of labor between regions, the regions themselves, are not speculative concepts but the result of generalization of the few scattered and accidental facts that we know from the geography of the past. (Some geographers do not realize this in the same way as the Molière character who does not realize that he is speaking prose.) When we talk about the importance of the maritime situation of a country or a city, we base our judgment to a large extent on generalization of specific examples of development of maritime countries and cities. As to how profound and correct our understanding is of such concepts based on individual examples rather than on study of entire periods, that is another question.

There is no doubt that our concept of geographical situation could be greatly deepened on the basis of data taken from geography of the past. We have not been working, in particular, on the social aspect of this problem, and it does exist.[14] The advantages of geographical situation are often regarded as a geometrical problem of determining the shortest distance and as something absolute. However, these advantages contain an element of relativity and are born in struggle, since many points may be advantageous for various parts of a country (when we talk of domestic links) or for countries (in case of foreign links). Account must, of course, also be taken in all cases of characteristics of the site, but this usually offers several possibilities and, consequently, a struggle is possible. In the seventeenth and nineteenth centuries, for example, Siberian merchants and the merchants of European Russia (and the administrations connected with these groups) had different views on the advantages of various points for organizing commercial fairs; these differences were settled by compromising on Irbit (such compromises often give rise to national capitals, such as Washington, Brasilia and, to some extent, perhaps even Moscow), but usually one side takes the upper hand and forces its selection on the other (of course, there are limits to this since the disadvantages of the selected place may be so great that relations will be broken, thus hurting the dominant side). At any rate it is important to remember that a certain political strength and a certain degree of independence are needed to turn the advantages of geographical situation to one's own benefit (for example, the city of Tver' [now Kalinin] occupied a similar and advantageous geographical situation both in the fourteenth and in the eighteenth century; however, in the fourteenth century it derived the benefit for itself and in the eighteenth for St. Petersburg). Our inadequate understanding and formulation of the concept *geographical situation* are also demonstrated by cases in which opposite conclusions are derived from similar situations. For example, when it suits us, we tend to explain the development of a region in terms of the intersection of routes linking powerful neighbors, but sometimes we use the same argument to explain the underdevelopment of a region on the ground that the more powerful neighbors tend to attract the regional economy to themselves. The explanations in both cases may be basically correct, but they are not carried through to the point where the differences between two apparently identical cases are made clear.

All that has been said about geographical situation is based on the much broader problem of mutual relationships among regions or among places and between regions and the country (state) as a whole. These complex questions can obviously not be understood without systematic study of such relationships in the past. The valuable idea of geographical division of labor needs further elaboration, and many aspects of interregional relationships have not been studied at all despite the fact that contact with geographical material of any period demonstrates the practical and theoretical importance of these aspects and the need for deeper investigation.

The point is that in every region there are forces interested in the development of a given aspect of the regional economy usually based on the use of local resources (natural and manpower resources), geographical situation, the historical accumulations of labor, traditions, and so forth. This often leads to a clash of interests between one region and other regions or between one region and the country as a whole. (In the geographical division of labor there is often an element of compulsion in which one place compels another to produce what the first needs for exchange purposes. There are regions that live on the resources of other regions. These are evidently two aspects of the same problem.) At the same time local initiative also contains many progressive, vital aspects without which a state cannot expect to achieve full-fledged development. Excessive pressure on local interests in the name of narrowly understood national interests often turns like a boomerang against those very national interests. In short, things must be weighed against one another. Under the Soviet system all this should be based on scientific principles that have not yet been sufficiently worked out.

Related to this is another problem — the shifting of economic centers in the country, the rise or fall (relative or absolute) of individual regions and places. Especially interesting and important for study are periods of sharp and relatively rapid change in mutual relationships between regions, in their relative strength and significance; these would be periods of "geographical revolutions" (for example, rapid and profound changes in interregional relationships resulting from the rise of St. Petersburg or the construction of railroads and the emancipation of the serfs). It seems to me that these complicated and important problems can hardly be investigated in sufficient depth simply on the basis of familiarity with the thin veneer of present-day life. Generalizations should be based on study of these phenomena through the ages, taking account, of course, of differences in the social and economic structure of society.

Because of the absence of geographic work on past periods, we are not making any progress on such an important geographic problem as the interaction between the productive activity of man and the natural environment (the problem remaining in its philosophic "cradle"). Available geographic generalizations on this problem are based on incidental facts and on frequently inadequate interpretations of quotations from the classic works of Marxism-Leninism. It seems to us that I. M. Zabelin oversimplifies when he writes that "the process of social development is accelerated by *unfavorable* and *diversified* natural conditions and slowed by *favorable* and *uniform* conditions."[15] This looks like a paradox or a word-play ("unfavorable conditions" are favorable for development, and what makes them unfavorable is not clear. Marx said that tropical nature was unfavorable for the development of capitalism, but that is something else). (Incidentally, Lenin objected to the view that primitive man supposedly received all he needed as a "free gift of nature.")[16]

As a matter of fact, can anyone maintain that the natural conditions of the United States with an abundant supply of coal and metals were not favorable for the development of capitalist production? Can anyone deny that the availability of iron and lumber made it possible in the feudal period of the eighteenth century to transform the Urals into one of the world's leading centers of metallurgy with what were for that time a substantial population and large plants with an advanced technology? On the other hand, it is interesting to see what R. Lapp, the American scholar, writes about present-day America: "A number of factors tend to explain why the United States does not have a more vigorous nuclear-electric power program. First, we live in a nation which has been blessed with great reserves of natural resources, especially in such fossil fuels as coal, natural gas and petroleum. Our beds of coal are so rich that they hold fuel for centuries. . . . Therefore, we have had less motivation for the development of domestic atomic power."[17] With regard to the Urals, too, it can be said that when metallurgy shifted to the use of coal, the presence of lumber in the Urals became a hindering factor.

The point is apparently that each stage of productive development has its distinctive "combination" of favorable resources and natural conditions, and it is within the framework of that developmental stage that favorable conditions have a favorable effect, that is, tend to speed development (all other conditions being equal). But as soon as development passes into another stage, technological progress and the growth of productive forces may be such as to transform the previously favorable conditions into a hindrance. Furthermore, some elements of the new combination of needed resources may be missing. In this connection, it must be stressed that the effect of nature on economic development must be investigated from the point of view of the sum total of all natural elements since each may affect development differently and pull it in another direction (in present-day practice, resources are often evaluated on a selective basis and from a technological point of view for a given type of production, with other elements of the natural environment neglected).

These are the first conclusions that suggest themselves, but, I repeat, the question is complex and requires concrete, and not speculative, research and thorough investigation of continuous layers, and not of selected examples, of past experience. This can be done only if the study of the geography of past periods is seriously approached.

There is no need to show the practical and theoretical importance of chronological study of the experience of man's influence on nature, that is, of questions such as the settlement and economic development of areas, the construction of roads and canals, the raising of different types of livestock and crops, the rise and development of individual industries, the process of formation of all economic elements of a region in their various combinations and oppositions to each other. All this would not only yield theoretical generalizations but provide direct, practical *counsel,* especially since in dialectical development there are often similar situations (at different levels, of course).

Problems that require a historical-geographical solution are closely related to those synthesizing studies that should be regarded as the highest form of geographical research (at the present time, such synthesizing work is rather elementary and lacking sufficient foundation). Without a historical-geographical approach, we cannot work out a geographic concept of area (region) or understand its specific characteristics (and that is the basis for synthesis). Without such an approach we cannot understand the actual significance of individual elements of the geographic complex, the degree of close connection or antagonism between them, the degree of stability or mobility of each and, finally, the general direction of development. And all these things we must know if we are to work out a system of concepts covering all the diversified elements of the geographic complex and their interrelationships from a single point of view. Only such an approach can justify combined investigation. Such generalized concepts must also be used in regionalization; it is quite inadequate to limit oneself, as is sometimes done, to combinations of current data that often reflect temporary factors. The system of geographic sciences has become unwieldy, and this makes it unsatisfactory from the point of view of organization. It is not a working unit (and that is the point of dividing a single science into branches.) The central objectives of geography are obscured by problems that are only of particular interest (for geography). A pulling together of the geographic sciences and a clear delimitation of objectives assigning the leading role to synthesizing geography would be useful for the further development of the discipline.

What has been said above is, it seems to me, sufficient to demonstrate the vast range of the geography of past periods and its importance for present-day geography and for the future of the discipline. Nevertheless, no serious broad research in this field is now underway, and the past, according to the false prejudices rooted in our mind, contines to be placed under the jurisdiction of history. This gives rise to a strange situation: research on major geographic problems turns out to be not within the competence of geography. Furthermore, is it not clear that work on the geography of past periods requires knowledge of geographic methodology, study of geographic documents of the past (maps, travel accounts, geographic descriptions), and knowledge of the natural environment to a far greater extent than historians need?

We have seen that leading ideologists of Soviet geography do not regard it necessary to include study of the geographic past within geography (although an exception is made arbitrarily, without logic, in the case of paleogeography) and that this view is expressed in articles in official publications (encyclopedias and symposia such as *Soviet Geography: Accomplishments and Tasks*) and in the organization of research in leading institutes. However, some gains are being made in the geographic community. The Geographical Society of the USSR has published several collections of articles on historical geography, thanks to the efforts of V. K. Yatsunskiy; some geography faculties in universities are beginning to offer, under the guise of historical geography, a somewhat geographized economic history.

If we are to bring geography back to its proper course, we must, in the first place, abandon the idea that present-day geography is all there is to geography. Disciplines are classified in terms of their objects of study, and such objects must be studied over time. Tatishchev was right in saying that geography is divided into ancient, medieval, and modern. Geographic education and research must be reorganized. Universities should offer geography specializations by historical periods closely related to work in archives, reading of old documents, and so forth. A specialist on ancient geography is needed just as much by society and science as a specialist on ancient history. Research institutes, and especially the Institute of Geography of the Soviet Academy of Sciences should make provision for broad research on the goegraphy of the past periods. Social geography should increase its ties with history; this also applies to present-day geography. Of course, the main effort should be concentrated on present-day geography, and most of the attention should be devoted to it. But such attention to this phase (and, one might say, to present-day culture) requires the proposed reorganization.

As for the question of how much attention should be given to some periods and how much to others, we would make the following comment: It is false to assume that more recent periods are of greater importance than earlier periods. That would be the case if development were evolutionary. But in actual, dialectical development, with its law of "negation of negation," problems and geographic situations of earlier periods may be more interesting and important than those of later periods. (It is no accident, for example, that literature, movies, and theater have recently displayed great interest in the periods of Ivan the Terrible and Peter the Great.) We must therefore develop a historical geography that could cover all periods of social development, starting from remote antiquity.

1. V. N. Tatishchev, *Izbrannyye trudy po geografii* [Selected Works in Geography] (Moscow, 1950), p. 211.

2. V. N. Tatishchev, *Leksikon Rossiyskoy* [Gazetteer of the Russian], part 2 (St. Petersburg, 1793), pp. 38–40.

3. M. V. Lomonosov, *Izbrannyye filosofskiye proizvedeniya* [Selected Philosophic Works] (Moscow, 1950), p. 396.

4. For example I. M. Zabelin and V. A. Anuchin.

5. V. K. Yatsunskiy, *Istoricheskaya geografiya* [Historical Geography] (Moscow, 1955).

6. K. Ritter, "On the Historical Element in the Earth Sciences," in *Magazin zemlevedeniya i puteshestviy* [Compendium of Earth Sciences and Travel] Part 2 (Moscow, 1853), p. 482.

7. Ye. Chizhov, "A Classification of Sciences," *Severnyy Vestnik* [Northern Herald] no. 12 (St. Petersburg, 1896).

8. L. S. Berg, "Geography and Its Position Among Other Sciences," *Voprosy stranovedeniya* [Problems of Regional Geography] (Moscow-Leningrad, 1925), p. 13.

9. V. E. Den, *Kurs ekonomicheskoy geografii* [A Course in Economic Geography] (Moscow-Leningrad, 1925).

10. D. N. Anuchin, *Izbrannyye geograficheskiye raboty* [Selected Geographic Works] (Moscow, 1949), p. 104.

11. N. N. Kolosovskiy, "The Scientific Problems of Geography," *Veprosy geografii* [Problems of Geography], no. 37 (1955), p. 137.

12. K. Marx and F. Engels, *Sochineniya* [Works] (Moscow, 1960), 23: 191-94.

13. See, for example, the work of N. N. Baranskiy, I. A. Vitver, Yu. G. Saushkin, and also V. A. Anuchin's *Teoreticheskiye problemy geografii* [Theoretical Problems of Geography] (Moscow, 1960), pp. 168–69.

14. L. Ye. Iofa, *Goroda Urala* [Cities of the Urals] (Moscow, 1951), p. 11.

15. I. M. Zabelin, *Teoriya fizicheskoy geografii* [Theory of Physical Geography] (Moscow, 1959), p. 256 (Zabelin's italics).

16. Lenin, *Sochineniya* [Works], 4th Russian edition, 5:951.

17. R. Lapp, *Atomy i tyudi* [Atoms and People], Russian edition (Moscow, 1959), p. 196.

G. YA. NESMEYANOVA

Geographical Variations in the Dynamics of Grain Yields in the Nonchernozem Center In the Period of The Development of Capitalism

The increase in intensity of agricultural production, which is observed everywhere, does not occur uniformly in the various parts of a territory. The most favorable conditions for intensification arise at times in one, and at times in other regions. Intensification of agricultural production shows up initially in increased yields. Therefore, it is of interest to trace how the geography of [agricultural] yields change with time and what the rates of increase in yield are in the various natural regions.

The dynamics of grain crop yield in provinces of the European part of Russia have been studied by V. M. Obukhov and published in a series of works.[1] Our goal is to analyze the yield of grain crops, including winter rye and oats, by the same method and for the same years as did Obukhov, but by smaller territorial units—districts [uyezds] instead of provinces [guberniyas]. This makes it possible to discern whether the statistical material on yields in pre-revolutionary Russia are suitable for the study of the historical geography of yields.

We used the statistical data for the districts of Moscow, Tver', Yaroslavl', Smolensk, Kostroma, Vladimir, Ryazan', Kaluga, and Tula

Translated from *Vestnik Moskovskogo Universiteta,* seriya geografiya [Herald of Moscow University, Geography series], no. 2(1967), pp. 113–19.

provinces, and a few districts of Orel, Chernigov, and Tambov provinces, that is, the territory corresponding approximately to the present Central Economic Region. Uniform data on yields of the most important crops were published in the works of the Central Statistical Committee. We processed the information for the years from 1883 to 1915.

Since it was of interest to trace the changes in the level of yields that occurred as a result of the evolution of agricultural production, it was necessary to exclude the influence of meteorological conditions and other circumstances under the influence of which the yield may fluctuate considerably. In order to do this we performed a fitting procedure on the thirty-three-year dynamic series by means of least squares, which is the same method used by Obukhov. In calculating the theoretical yield levels we used the equation for a straight line: $y_t = a + bt$, where y_t is the theoretical level of yields. The parameter a determines the initial level, parameter b, the yearly absolute increase in yield. Parameters a and b are found by using the equations,

$$a + b\Sigma t = \Sigma y$$
$$a\Sigma t + b\Sigma t^2 = \Sigma ty,$$

where y is the actual yield, t is the ordinal number of the year.

The straight line reflecting the stable increase (or decrease) in the yield capacity was chosen by Obukhov to describe the dynamic series given the following considerations. First, the thirty-three-year period is a comparatively short series, therefore it is quite admissible to make use of a straight-line (linear) fitting procedure. Secondly, Obukhov, who processed the same material by provinces, compared yield norms calculated on a straight line with the yield norms calculated with the use of a parabola of the second power. It proved that there are some differences only in Smolensk, Tver, and Yaroslavl' provinces, whereas in the other cases both these series are almost identical.

The theoretical levels of grain yields in 1883 and 1915 obtained by the calculations as well as the rates of growth (or loss) of yield capacity during this period are shown in table 1. Since the data on grain yields as a whole were not published by the Central Statistical Committee, the table indicates the average productivity calculated on the basis of actual harvests of individual crops. In order that the changes in the ratios of crops would not affect the size of harvest, the yield was weighted on the sown areas of 1916.

At the end of the nineteenth and the beginning of the twentieth centuries there were considerable shifts in the agriculture of Russia. Expansion of fodder areas with the utilization of legumes, cultivated crops, better agricultural implements, increase in the number of livestock, and, in relation to these, the fertilization of fields [were] all reflected in yield levels. Obukhov, who studied the relationship between yield and the evolution of agriculture through time, found that, on the average for European Russia, the coefficient of correlation between them for the grains as a whole is +0.612, for rye +0.526, and for oats +0.527. Among the central provinces, a relatively higher correlation coefficient was observed in Smolensk Province. In the rest of the provinces the increase in yield was below the average for Russia. A more detailed processing of the statistical data by district makes it possible to establish the differences in the rates of growth in yields capacity in the various types of localities.

In the 1880s the fallow system of farming predominated on the entire territory described. In terms of the composition of crops, only the spring planting differed somewhat in the various regions. In addition to the main crops, oats, barley, millet, and buckwheat were sown in various proportions. Other crops (potatoes and vegetables) occupied small areas, 2 to 4 percent of the sown areas.[2] They were grown, like hemp, on the same land every year, near the farmstead. Exceptions occured only in the area around Moscow and in Rostov District where potatoes and vegetables occupied 12 to 17 percent of the sown area. Perennial and annual grasses were almost absent.

During this period low grain yields were obtained over most of the territory: 5 to 8 centners per hectare. It is interesting to note that in the densely populated industrial districts of Moscow and Yaroslavl' provinces the yields on the peat-podsol soils were somewhat higher than in the forest-steppe zone.[3] This is evidently the result of widespread application of manure fertilizers. The relationship of pasture and arable land was such that 2 to 3 times as many cattle could be raised per hectare of arable land. In the forest regions all manure was used as fertilizer. The peasants in the south of Ryazan' and Tula guberniyas only began to use fertilizers on chernozem soils in the 1870s.[4]

The lowest yield of grains in 1883 is noted in the driftless regions of Meshchera and Poles'ye, 3.5 to 4.5 centners per hectare. In the course of the entire period under consideration these regions with sandy, infertile soils are noteworthy for their yields.

By 1915, a noticeable intensification of agricultural production had occurred. An analysis of the structure of sown areas allows one to conclude, that even in the prerevolutionary years, the farming systems everywhere no longer were characterized by features typical for the fallow three-field system. Potatoes were grown in the fields and annual and perennial grasses appeared in the sown areas. Clover, in combination with timothy grass, became most extensive. At first it was sown on sections close to the farmstead, on purchased or rented land, or in the corners of fields. Not until 1892 in Volokolamsk District was clover introduced into the four-field crop rotation. By 1915 the largest areas under clover were in the flax areas and the districts surrounding Moscow where they reached 15 to 25 percent of the sown area.[5]

The area under cultivated crops increased over the larger part of the territory from 5 to 10 percent. In the districts of Rostov and those surrounding Moscow it rose to 30 percent. Vegetable and potato cultivation near other large cities developed, as well as in the belt near the Oka River (Spasskiy and Muromak districts). A considerable por-

TABLE 1

NAME OF DISTRICT	GRAIN YIELD (Centners Per Hectare)		INCREASE OR LOSS IN YIELD IN 1915 (As Percentage of 1883)
	1883	1915	
Moscow Province			
1 Bogoroditsk. . .	8.4	8.5	1
2 Bronnitsky . . .	7.4	7.8	5
3 Vereya	7.0	8.6	23
4 Volokolamsk. . .	7.2	9.3	30
5 Dmitrov.	8.5	9.9	17
6 Zvenigorod . . .	7.6	9.1	20
7 Klin	8.8	9.2	4
8 Kolomna.	5.9	8.8	49
9 Mozhaysk	7.4	8.0	8
10 Moscow	8.7	10.7	23
11 Podol'sk	7.3	8.2	13
12 Ruza	6.8	8.1	20
13 Serpukhov. . . .			
Vladimir Province			
14 Aleksandrov. . .	8.5	9.0	6
15 Vladimir	7.2	8.9	23
16 Vyazniki	5.3	8.2	57
17 Gorokhovets. . .	5.9	8.5	42
18 Kovrov	6.3	7.1	13
19 Melenki.	4.0	6.3	58
20 Murom.	7.0	9.6	37
21 Pereyaslav . . .	8.6	9.8	14
22 Pokrov	6.6	7.4	12
23 Sudogda.	4.5	6.5	47
24 Suzdal'.	6.8	8.9	30
25 Shuya.	7.4	9.2	24
26 Yur'yev.	8.6	10.0	16

TABLE 1--Continued

NAME OF DISTRICT	GRAIN YIELD (Centners Per Hectare)		INCREASE OR LOSS IN YIELD IN 1915 (As Percentage of 1883)
	1883	1915	
Kaluga Province			
27 Borovsk.	5.9	6.9	16
28 Zhizdra.	4.7	6.3	34
29 Kaluga	6.6	7.8	19
30 Kozel'sk	5.8	7.8	36
31 Likhvin.	6.5	7.7	17
32 Maloyaroslav . . .	6.0	7.3	22
33 Medyn'	6.4	7.0	8
34 Meshchevsk	6.3	7.4	17
35 Mosal'sk	5.6	6.7	20
36 Peremyshl'	6.8	7.8	14
37 Tarussa.	6.2	7.4	21
Tula Province			
38 Aleksin.	6.4	8.2	27
39 Bogorditsk	8.3	9.9	19
40 Belev.	6.0	8.4	40
41 Venev.	7.1	9.4	32
42 Yepifan'	8.3	10.8	30
43 Yefremov	8.6	10.2	18
44 Kashira.	6.0	8.4	41
45 Krapivna	7.5	9.6	28
46 Novosil'	7.3	9.8	34
47 Odoyev	6.6	8.1	23
48 Tula	7.3	9.6	31
49 Chern'	7.8	9.4	20
Tver' Province			
50 Bezhetsk	8.6	9.3	8
51 Ves'yegonsk. . . .	7.7	9.3	21
52 Vyshniy Volodchek.	7.5	7.5	..
53 Zubtsov.	7.9	9.0	14
54 Kolyazin	9.0	8.2	- 9
55 Kashin	9.7	9.0	- 7
56 Korchev.	8.7	8.6	- 2
57 Novotorzhek. . . .	7.2	7.4	3
58 Ostashkov.	5.8	6.8	17
59 Rzhev.	6.4	9.2	44
60 Staritsa	7.1	8.8	24
61 Tver'.	7.7	10.3	34

TABLE 1--Continued

NAME OF DISTRICT	GRAIN YIELD (Centners Per Hectare)		INCREASE OR LOSS IN YIELD in 1915 (As Percentage of 1883)
	1883	1915	
Smolensk Province			
62 Belyy.	9.1	11.6	27
63 Vyaz'ma.	8.0	10.5	31
64 Gzhatsk.	7.8	7.1	- 9
65 Dorogobuzh . . .	7.6	10.3	35
66 Dukhovshchina. .	7.7	10.2	38
67 Yel'nya.	6.5	9.0	39
68 Krasnoye	9.2	7.7	-19
69 Poretsk.	6.7	9.6	42
70 Roslavl'	6.0	8.3	38
71 Smolensk	7.5	10.8	44
72 Sychevka	8.9	9.9	11
73 Yukhnov.	6.0	7.7	30
Yaroslavl' Province			
74 Danilov.	9.1	9.8	8
75 Lyubim	8.7	9.0	3
76 Mologa	11.4	10.5	- 8
77 Myshkino	11.7	10.9	- 7
78 Pyushekhonsk . .	8.9	9.7	9
79 Romanovo-Borisov	8.8	9.0	3
80 Rostov	9.1	10.0	9
81 Rybinsk.	10.1	10.4	3
82 Uglich	8.4	8.9	6
83 Yaroslavl' . . .	9.0	8.6	- 4
Kostroma Province			
84 Buy.	7.6	8.8	15
85 Varnavino. . . .	6.2	8.0	30
86 Vetluga.	6.1	7.9	30
87 Galich	6.9	7.5	10
88 Kineshma	6.9	8.5	24
89 Kologriv	6.7	8.1	22
90 Kostroma	9.2	8.9	- 3
91 Makar'yev. . . .	5.6	7.5	33
92 Nerekhta	7.8	8.7	12
93 Soligalich . . .	7.2	8.8	23
94 Chukhloma. . . .	6.3	6.9	10
95 Yur'yevets . . .	7.0	7.6	8

TABLE 1--<u>Continued</u>

NAME OF DISTRICT	GRAIN YIELD (Centners Per Hectare)		INCREASE OR LOSS IN YIELD IN 1915
	1883	1915	
Ryazan' Province			
96 Dan'kov	7.7	9.8	28
97 Yegor'yevsk . .	7.1	8.2	16
98 Zaraysk	7.2	9.0	26
99 Kasimov	5.9	7.3	25
100 Mikhaylov . . .	7.2	8.8	23
101 Pronsk.	6.9	8.3	20
102 Ranenburg . . .	8.6	11.3	31
103 Ryazhsk	8.4	10.6	26
104 Ryazan'	8.5	9.6	12
105 Sapozhok. . . .	8.1	10.5	30
106 Skopin.	7.2	9.0	24
107 Spassk.	8.1	10.6	31
Tambov Province			
108 Yelat'ma. . . .	7.1	9.4	33
109 Temnikov. . . .	4.7	7.0	49
110 Shatsk.	5.7	9.2	60
Orel Province			
111 Bryansk	4.5	5.6	25
112 Karachev. . . .	5.1	7.0	37
113 Sevsk	4.9	8.4	71
114 Trubchevsk. . .	3.6	5.5	54
Chernigov Province			
115 Mglin	3.9	5.9	51
116 Novozybkov. . .	3.3	5.3	61
117 Suvazh.	3.6	6.5	79
118 Starodub. . . .	3.8	5.9	56

tion of the cultivated crops was raised outside the field rotation system.

By 1915 the areas of comparatively high productivity (9 to 11 centners per hectare) expanded for all districts situated in the forest-steppe zone. In the forest zone the highest yields were obtained: (1) on grey forest soils of the subzone of broadleaf forests and the Vladimir Opel'ya; (2) in regions where loams predominate with a close underlayer of carbonate rocks (Bezhets and Rzhevskiy and others); and (3) in those districts where the three-field system had been discontinued longest (near Moscow, a large part of Yaroslavl' Province) and also on the territory of Smolensk Province with rapidly developing flax industry. In these parts of the forest zone the same absolute level of yield was reached as in the forest steppe, but with more intensive means of farming.

The lowest yields (5.5 to 6.5 centners per hectare) occurred on the light peat-podsol soils of the Poles'ye and Meshchera and also in the regions of hillmoraine relief of the Valday glaciation.

On the territory under consideration the average yield of grains for 1883–1915 rose from 5.8 to 7.9 centners per hectare, that is, by 36 percent. The minimum grain yield rose from 3.3 to 5.3 centners per hectare. The highest limit, however, 11.6 centners per hectare, remained the same. The average yield of winter rye increased from 6.3 to 8 centners per hectare, or by 27 percent; oats, from 5.9 to 8.4 centners per hectare (43 percent).

The rate of growth of yields was very uneven in the various parts of the territory. A considerable increase in grain yields, 1.5 to 2.5 centners per hectare, could be observed in the forest-steppe and broadleaf forest areas, although here the three-field system had been retained to a larger extent than in the zone of peat-podsol soils. By 1915, the nongrain crops occupied only 10 to 15 percent of the sown area. Very few grasses were sown (2 to 5 percent). Evidently, the growth of yields in the forest-steppe regions was related mainly to improved land treatment with use of better equipment and the increase in the amount of fertilizer on the fields. According to the data in the zemstvo statistical investigations, in the districts where leached and podsolized chernozems predominated, the peasants fertilized 20 to 30 percent of the fallow land annually by the beginning of the World War I. This became possible because the density of livestock increased. For example, in Ryazan' Province during this period, cattle increased by 34 percent and the area of grains decreased by 10 percent.

The number of improved agricultural implements increased very rapidly. Thus, in Yepifan District of Tula Province, the number of plows increased from 155 to 6,018[6] in twelve years (1899–1911).

On the larger part of the territory of the forest zone the growth of grain yields was slower and in some comparatively small areas it [productivity increase] was absent altogether. The farther northward from the zone of grey forest soils to the zone of peat-podsol soils, and the farther into the zone of podsol soils, the less evident is the growth in yields.

The relatively greater growth of yields in the western part of the forest zone, in the region where at that time flax growing developed, merits attention. During this period, in contrast to the central parts of the zone, a noticeable increase in the sown area can be observed. The development of flax growing caused the introduction of crop rotation with clover which, in turn, helped to develop livestock-raising and, consequently, better fertilization of the fields. As a result, the rate of growth of the productivity of grains in the majority of districts of Smolensk Province reached 30 to 40 percent and, in the level of yields, this province does not yield to the forest-steppe zone.

1. V. M. Obukhov, "The Evolution of the Yield of Rye in the Period 1883–1915 in European Russia," in *Vestnik statistiki* [Herald of Statistics], book 15; idem, "The Change in Yields of Grain Crops in European Russia in the Period 1883–1915," in *Vliyaniye neurozhayev na narodnoye khozyaystvo Rossi* [The Influence of Crop Failures on the National Economy of Russia] (Moscow, 1927); idem, *Urozhaynost' i meterologis-cheskiye faktory* [Yield Capacity and Meteorological Factors] (Moscow, 1949).

2. *Statistika Rossiyskoy imperii* [Statistics of the Russian Empire], vol. 22 (St. Petersburg, 1888). Most important data on land statistics from the 1887 survey.

3. This article makes use of the scheme of natural regions published in the book *Fiziko-geograficheskoye rayonirovaniye Nechernozemnogo Tsentra* [Physical-Geographic Regionalization of the Nonchernozem Center] (Moscow).

4. *Sbornik statisticheskikh svedeniy po Tambovskoy gubernii* [Collection of Statistical Information on Tambov Guberniya], vol. 6 (Shatsk Uyezd, Tambov, 1884).

5. *Predvaritel'nyye itogi Vserossiyskoy sel'skokhozyaystvennoy perepisi 1916 g.* [Preliminary Summary of the All-Russian Agricultural Census of 1916], 1st ed. (Petrograd, 1916).

6. Z. M. Svavitskiy and N. A. Svavitskiy, comps., *Zemskiye podvornyye perepisi 1880–1913 gg.—Pouyezdnyye itogo* [Zemstvo Household Censuses of 1880–1913—Results by Uyezds] (Moscow, 1926).

A Selected Glossary of Terms

Chernozem. From the Russian *chernaya zemlya* [black earth], fertile prarie soil of the Russian Steppe. The Central Chernozem Region is so named because of its chernozem soil and central location in the European section of the USSR.

COMECON. Council for Mutual Economic Assistance. (Sometimes the term CEMA is used.) Founded in 1949 to attempt economic integration among the European socialist countries, it includes today the following countries: USSR, Bulgaria, Romania, Hungary, German Democratic Republic, the People's Republic of Mongolia, Poland and Czechoslovakia (Albania, formerly a member, withdrew in 1965).

Gosplan. Gosudarstvennyi planovyi Komitet Soveta Ministrov [State Planning Commission of the USSR Council of Ministers]. Established in 1921, this has been the full name of this committee since 1957.

Kolkhoz. (Kollektivnoye khozyaystvo) collective farm—an agricultural producers' cooperative jointly operated by its membership. Each member of the kolkhoz shares in its output in proportion to the quantity and quality of labor days (*trudodni*) worked. Some part of the annual output of the kolkhoz is sold or delivered to the state procurement organization. Additional allotments are made for capital and reserve stocks as well as for special funds, seeds, fodder, insurance, and so on. The remainder is distributed among the kolkhoz members, partly in money and partly in kind.

Kray.   Territorial administrative unit, similar to an oblast, but usually containing an autonomous oblast within its boundaries. There are seven krays in the USSR.

Oblast.   An administrative-territorial unit, usually based on the principle of economic integration corresponding to a province. The oblast areas and borders change frequently following economic development. Autonomous oblast is an administrative territorial unit, often forming part of a kray. Autonomous oblasts represent territorial autonomy for people who are not sufficiently numerous for the creation of an autonomous republic. Internal administration is similar to that of an ordinary oblast. There are nine autonomous oblasts.

Okrug.   Name of an administrative territorial unit forming part of an oblast or an autonomous republic. There are no okrugs in the USSR at present except national okrugs. There are now nine of these, all for the peoples of Siberia and the far north. They are populated primarily by minority groups.

Rayon (raion).   Administrative territorial unit forming part of an oblast or kray, and itself subdivided into village soviets. In the Union republics that are not divided into oblasts, the rayons are directly subordinated to the republican governments. Larger towns are also subdivided for administrative purposes into a number of rayons. All activities within the rayon are controlled by the rayon party committee.

Sovkhoz.   *(Sovetskoye khozyaystvo)* state farm—a state-owned and state-operated agricultural enterprise. In contrast to the "kolkhozes" the workers and employees of the sovkhoz are wage earners, but as in the kolkhozes these wage earners are permitted to cultivate their own private plot.

Sovnarkhoz.   Economic council—territorial units of economic administration set up in 1957. They were subordinated directly to the councils of ministers of the union republics, and were responsible for the administration of industry and building (except that of purely local relevance). The jurisdiction of most of the councils extended over one oblast, kray, or autonomous republic, but some, especially outside the RSFSR, covered a group of oblasts. The smaller union republics had only one economic council each. The establishment of the councils was aimed at overcoming the problems of centralization inherent in administering industry by branches through ministries.

Uyezd.   A former territorial administrative unit, larger than the present rayon. They were abolished in the late 1920s, but continued to exist in the Baltic states and Bessarabia until the 1940s.

Zemstvo.   Institutions first introduced by Ivan IV as a means of collecting taxes through the transference of certain financial responsibilities from appointed to elected officials. Alexander II revived the zemstvos in 1864 as organs of rural government. District zemstvos were elected on a restricted franchise, and they elected the provincial zemstvos.

The activities of the zemstvos were supervised by provincial governors, but they had authority in economic and educational matters, public health, etc. The liberal and radical intelligentsia found the zemstvos to be the most approprate field for their practical work, and the constitutional movement prior to the revolution of 1905 chiefly expressed itself through the zemstvos. They were prominent in the war effort and in demands for a responsible government during the First World War. The sphere of zemstvo authority was widened after the February Revolution of 1917, but they were abolished after the Bolshevik seizure of power and replaced by the soviets.

Index